# HIV/AIDS

# HIV/AIDS

# OXIDATIVE STRESS AND DIETARY ANTIOXIDANTS

*Edited by*


VICTOR R. PREEDY

*King's College London, London, United Kingdom*

RONALD ROSS WATSON

*University of Arizona, Mel and Enid Zuckerman College of Public Health, and School of Medicine, Arizona Health Sciences Center, Tucson, AZ, USA*

Academic Press is an imprint of Elsevier
125 London Wall, London EC2Y 5AS, United Kingdom
525 B Street, Suite 1800, San Diego, CA 92101-4495, United States
50 Hampshire Street, 5th Floor, Cambridge, MA 02139, United States
The Boulevard, Langford Lane, Kidlington, Oxford OX5 1GB, United Kingdom



**Library of Congress Cataloging-in-Publication Data**
A catalog record for this book is available from the Library of Congress

**British Library Cataloguing-in-Publication Data**
A catalogue record for this book is available from the British Library

ISBN: 978-0-12-809853-0

For information on all Academic Press publications visit our website at https://www.elsevier.com/books-and-journals



*Publisher:* Sara Tenney
*Acquisition Editor:* Linda Versteeg-Buschman
*Editorial Project Manager:* Fenton Coulthurst
*Production Project Manager:* Kiruthika Govindaraju
*Cover Designer:* Limber Matthew

Typeset by TNQ Books and Journals

# Contents

## 11. Genistein and HIV Infection

JIA GUO AND YUNTAO WU



## 12. Opportunistic Infections in HIV Individuals and Enhanced Immunity by Glutathione

ALEXIS ALEJANDRE, LESLIE GONZALEZ, PARVEEN HUSSAIN, JUDY LY, ANAND MUTHIAH, TOMMY SAING, ANDDRE VALDIVIA AND VISHWANATH VENKETARAMAN



## 13. *Plectranthus barbatus*; Antioxidant, and Other Inhibitory Responses Against HIV/AIDS

PETRINA KAPEWANGOLO AND DEBRA MEYER



## 14. Methyl Gallate as an Antioxidant and Anti-HIV Agent

TZI B. NG, JACK H. WONG, CHIT TAM, FANG LIU, CHI F. CHEUNG, CHARLENE C. W. NG, RYAN TSE, TAK F. TSE AND HELEN CHAN

# List of Contributors

**Ajibola I. Abioye** Rhode Island Hospital, Providence, RI, United States; Brown University, Providence, RI, United States

**Jamal Ahmad** Aligarh Muslim University, Aligarh, India

**Alexis Alejandre** Western University of Health Sciences, Pomona, CA, United States

**Deepika Anand** University of Delhi, New Delhi, India

**Olatunbosun G. Arinola** University of Ibadan, Ibadan, Nigeria

**Onyemaechi O. Azu** University of KwaZulu Natal, Durban, South Africa; University of Namibia, Windhoek, Namibia

**Maria D. Borges-Santos** Sao Paulo State University, Botucatu, Brazil

**Roberto C. Burini** Sao Paulo State University, Botucatu, Brazil

**Charlene C. W. Ng** King's College London, London, United Kingdom

**Andrea Calcagno** Department of Medical Sciences, University of Torino, Torino, Italy

**Daniele Canale** University of São Paulo School of Medicine, São Paulo, Brazil

**Helen Chan** Vita Green Health Products (HK) Ltd, Tai Po, Hong Kong, China; Hong Kong Institute of Medical Research, Central, Hong Kong, China; Genning Partners Company Limited, Causeway Bay, Hong Kong, China

**Chi F. Cheung** The Chinese University of Hong Kong, Shatin, Hong Kong, China

**Joanna J. Chmielinska** The George Washington University, Washington, DC, United States

**Zephy Doddigarla** Mayo Institute of Medical Sciences, Barabanki, India

**Justin Dong** Western University of Health Sciences, Pomona, CA, United States

**Heike Englert** Muenster University of Applied Sciences, Muenster, Germany

**Muhammad Faisal** Aligarh Muslim University, Aligarh, India

**Wafaie W. Fawzi** Harvard T.H. Chan School of Public Health, Boston, MA, United States

**Pedro H. França Gois** University of São Paulo School of Medicine, São Paulo, Brazil

**Lizette Gil-del Valle** Pedro Kourí Institute of Tropical Medicine (IPK), La Habana, Cuba

**Leslie Gonzalez** Western University of Health Sciences, Pomona, CA, United States

**Rosario Gravier-Hernández** Pedro Kourí Institute of Tropical Medicine (IPK), La Habana, Cuba

**Jia Guo** George Mason University, Fairfax, VA, United States; Merck Sharp & Dohme Corp., Boston, MA, United States

**Jack H. Wong** The Chinese University of Hong Kong, Shatin, Hong Kong, China

**Parveen Hussain** Western University of Health Sciences, Pomona, CA, United States

**Petrina Kapewangolo** University of Namibia, Windhoek, Namibia

**Theodoros Kelesidis** David Geffen School of Medicine at UCLA, Los Angeles, CA, United States

**Jay H. Kramer** The George Washington University, Washington, DC, United States

**Rapalli Krishna Chaitanya** Central University of Punjab, Bathinda, India

**Shashank Kumar** Central University of Punjab, Bathinda, India

**Lingidi J. Lakshmi** Mayo Institute of Medical Sciences, Barabanki, India

**Fang Liu** Nankai University, Tianjin, China

**Jean-Pierre Louboutin** Thomas Jefferson University, Philadelphia, PA, United States; University of the West Indies, Kingston, Jamaica

**Judy Ly** Western University of Health Sciences, Pomona, CA, United States

**Ivan T. Mak** The George Washington University, Washington, DC, United States

**Debra Meyer** University of Johannesburg, Auckland Park, South Africa

**Yong Ming-Yu** Shriners Burns Hospital – Massachusetts General Hospital, Harvard Medical School, Boston, MA, United States

**Fernando Moreto** Sao Paulo State University, Botucatu, Brazil

**Ilaria Motta** Department of Medical Sciences, University of Torino, Torino, Italy

**Anand Muthiah** Western University of Health Sciences, Pomona, CA, United States

**Edwin C.S. Naidu** University of KwaZulu Natal, Durban, South Africa

**Tzi B. Ng** The Chinese University of Hong Kong, Shatin, Hong Kong, China

**Germaine S. Nkengfack Nembongwe** University of Dschang, Dschang, Cameroon

**Abdulfatah A. Onifade** University of Ibadan, Ibadan, Nigeria

**Vasiliki D. Papakonstantinou** Department of Science Nutrition-Dietetics, Athens, Greece

**Vinood B. Patel** University of Westminster, London, United Kingdom

**Victor R. Preedy** King's College London, London, United Kingdom

**Seema Puri** University of Delhi, New Delhi, India

**Sheu K. Rahamon** Edo University, Iyamho, Edo State, Nigeria

**Rajkumar Rajendram** King's College London, London, United Kingdom; King Abdulaziz Medical City, Ministry of National Guard Health Affairs, Riyadh, Saudi Arabia

**Tommy Saing** Western University of Health Sciences, Pomona, CA, United States

**Antonio C. Seguro** University of São Paulo School of Medicine, São Paulo, Brazil

**Chris F. Spurney** Children's National Medical Center, Washington, DC, United States

**David S. Strayer** Thomas Jefferson University, Philadelphia, PA, United States

**Chit Tam** The Chinese University of Hong Kong, Shatin, Hong Kong, China

**Ryan Tse** Vita Green Health Products (HK) Ltd, Tai Po, Hong Kong, China; Hong Kong Institute of Medical Research, Central, Hong Kong, China; Genning Partners Company Limited, Causeway Bay, Hong Kong, China

**Tak F. Tse** Vita Green Health Products (HK) Ltd, Tai Po, Hong Kong, China; Hong Kong Institute of Medical Research, Central, Hong Kong, China; Genning Partners Company Limited, Causeway Bay, Hong Kong, China

**Anddre Valdivia** Western University of Health Sciences, Pomona, CA, United States

**Lizette Gil-del Valle** Pedro Kourí Institute of Tropical Medicine (IPK), La Habana, Cuba

**Vishwanath Venketaraman** Western University of Health Sciences, Pomona, CA, United States

**William B. Weglicki** The George Washington University, Washington, DC, United States

**Yuntao Wu** George Mason University, Fairfax, VA, United States

**Miya Yoshida** Western University of Health Sciences, Pomona, CA, United States

# Preface

The World Health Organization has estimated that globally there are approximately 36 million people living with HIV and about half of these are on antiviral therapy. Oxidative stress can arise in HIV/AIDS during the progression of the disease and/or as a result of therapy. There is thus a need to understand the basic steps and pathways. This offers a focal point for further study. In some conditions oxidative stress can be ameliorated with pharmacological or natural agents.

While many immunologists, physicians, and clinical workers understand the processes in HIV/AIDS, some are less conversant with the science of nutrition and dietetics. On the other hand, some nutritionists and dietitians are less conversant with the detailed science of HIV/AIDS. Thus, immunologists, physicians and clinical workers, food scientists, and nutritionists are separated by divergent skills and professional disciplines that need to be bridged to advance medical sciences.

The book **HIV/AIDS: Oxidative Stress and Dietary Antioxidants** aims to cover in a single volume the science of oxidative stress in HIV/AIDS and antioxidants. However, the processes within oxidative stress are not described in isolation but in concert with other pathways and cell components such as apoptosis, the mitochondria, membranes, cell signaling, genomic instability, transcription, receptor-mediated responses, and so on. This approach recognizes that diseases are often multifactorial and oxidative stress is a single component of this. However, it is important to point out that the usage of any component or regimen requires scientifically vigorous trials and investigations. Treatments and pathways seen in modeling systems or in vitro need to be verified in vivo. Adverse effects also need to be investigated.

The book **HIV/AIDS: Oxidative Stress and Dietary Antioxidants** imparts information in two main sections.

1. **Oxidative Stress and HIV/AIDS**
2. **Antioxidants and HIV/AIDS**

The section **Oxidative Stress and HIV/AIDS** covers the basic processes of oxidative stress from molecular biology to whole organs. The first chapter describes antioxidant status in HIV infection in different clinical conditions. There follows chapters covering TB–HIV coinfection, dysfunctional high-density lipoprotein, aging, breast milk, and a chapter on oxidative stress in HIV in relation to metals.

The second section **Antioxidants and HIV/AIDS** begins with a chapter on HIV and gender differences in diet. This is followed by chapters on India and Africa. Thereafter, there are chapters on gene delivery of antioxidant enzymes, genistein, glutathione, *Plectranthus barbatus*, methyl gallate, taurine, magnesium, selenium, and vitamins D and E. Finally, a chapter on practical techniques for assessing antioxidant capacity and oxidative stress is followed by a chapter on recommended resources.

**HIV/AIDS: Oxidative Stress and Dietary Antioxidants** is designed for nutritionists, dietitians, food scientists, immunologist, physicians and clinical workers, health-care workers, and research scientists. Contributions are from leading national and international experts including those from world renowned institutions.

*Ronald Ross Watson*
*Victor R. Preedy*
The Editors

SECTION 1

# OXIDATIVE STRESS AND HIV/AIDS

CHAPTER

# 1

# Antioxidant Status in Human Immunodeficiency Virus Infection in Different Clinical Conditions

*Lizette Gil-del Valle, Rosario Gravier-Hernández*

**Pedro Kourí Institute of Tropical Medicine (IPK), La Habana, Cuba**

OUTLINE



## Abstract

Depletion of different antioxidant concentrations persisting during infection by human immunodeficiency virus (HIV) is associated with chronic inflammation, micronutrients deficiency, mitochondrial impairment, and oxidative stress. HIV entry to cell and replication occur in a highly oxidized condition characterized by antioxidant depletion. CD4+ T lymphocytes can be activated via a cascade of internal oxidative pathways as well.

Essential mechanisms of reactive oxygen species generation and antioxidants pathways, oxidative damage, and cellular function and how these responses change could mediate pathophysiological situations involved in HIV disease conducing to diverse comorbidities are discussed. Accrual clinical reports analyses for a better understanding of various interrelated factors are reviewed. These aspects are critical factors influencing HIV outcomes with or without antiretroviral treatment. These events should be managed during therapy, and it should be the focus of intense ongoing investigation. Currently, the most practical advice is to evaluate the antioxidant status early and to manage traditional risk factors of diverse conditions.

**Keywords:** Antioxidant status; Ascorbic acid; Electron transport chain; HIV; Oxidative stress; Reactive oxygen species.

## List of Abbreviations

**AA** ascorbic acid
**AIDS** acquired immunodeficiency syndrome
**ARE** antioxidant response elements



HIV/AIDS
http://dx.doi.org/10.1016/B978-0-12-809853-0.00001-8

**ART** antiretroviral drugs
**CO** carbonyl
**eNOS** uncoupled endothelial nitric oxide synthase
**ETC** electron transport chain
**GPx** glutathione peroxidase
**GSH** glutathione
**$H_2O_2$** hydrogen peroxide
**HAART** high active antiretroviral therapy
**HIV** human immunodeficiency virus
**HO-1** heme oxygenase-1
**HOCl** hypochlorous acid
**IFN** interferon
**isoP** isoprostanes
**LTCD4** lymphocytes T CD4+
**MDA** malondialdehyde
**NF-κB** nuclear factor κB
**NO** nitric oxide
**NOX** NADPH oxidases
**Nrf 2** nuclear factor (erythroid-derived 2)
$O_2^{\cdot -}$ superoxide anions
**·OH** hydroxyl radicals
**$ONOO^-$** peroxynitrite
**OS** oxidative stress
**PO** peroxides
**ROS** reactive oxygen species
**Se** selenium
**SOD** superoxide dismutase
**TDR** thioredoxin reductase
**TOC** tocopherol
**TRX** thioredoxin
**VL** viral load
**XO** xanthine oxidase

## INTRODUCTION

Cellular redox status is a normal physiological variable that may elicit cellular response such as transcriptional activation, proliferation, or apoptosis.[1] Reactive oxygen species (ROS) generation occurs as by-product of molecular oxygen ($O_2$) metabolism or by specialized enzymes in diverse organelles and as immunological effector mechanism also for neutralizing the ever-changing virulence of microorganisms. The ability of immune system to sterilize a site of infection by rapid production of ROS (superoxide anions ($O_2^{\cdot -}$), hydrogen peroxide ($H_2O_2$), hydroxyl radicals (·OH), hypochlorous acid (HOCl), peroxynitrite ($ONOO^-$), etc.) can keep organisms alive. The subjects, who could mount a robust immune response with vigorous yet coordinated ROS production, would be selected for survival.[2,3]

Sustained stimulation condition influences on the capacity of antioxidants to counteract oxidants and leads to a condition called oxidative stress (OS), which in a physiological setting can be defined as an excessive bioavailability of ROS, the net result of an imbalance between production and destruction of ROS (with the latter being influenced by antioxidant defenses).[4] The results of oxidative injury caused by ROS become themselves the sources of OS; the damaging of the membranes and protein structure can further promote ROS propagation, leading to enhanced oxidative impairment. This status could be also better defined as a perturbation of redox signaling.[3,5,6]

There are a number of recent studies encouraging the need to apply additional systemic biomarker as redox indexes to both recognized and characterized pathogenic role of redox status on the development of different pathologies.[3]

Persistence of parasitic viruses requires infecting host cells, pirating host resources, outmaneuvering host immune components, and replicating.[7] Human immunodeficiency virus (HIV) infection primarily induces chronic activation of innate immune responses, fulfilling the double function of limiting viral replication in the initial stages of infection and enhancing the generation of efficient adaptive immune responses.[8] In particular, production of interferon (IFN)-α and IFN-β by plasmacytoid dendritic cells may be triggered through Toll-like receptor engagement and have both immune-stimulating and antiviral activity.[9] During the HIV infection, ROS generation ($O_2^{\cdot -}$, $H_2O_2$, $ONOO^-$, nitric oxide (NO), and others) has been recognized as chronic, overlapping the antioxidant system and related to oxidative molecular damage, viral replication, micronutrient deficiency, and inflammatory chronic response.[10] Exposure to oxidants challenges cellular systems, and their responses may create conditions including changes in glutathione (GSH), thioredoxin (TRX), superoxide dismutase (SOD), catalase (CAT), ascorbic acid (AA), glutathione peroxidase (GPx), tocopherol (TOC), and selenium (Se), which are favorable for the replication of HIV through nuclear factor κB (NF-κB) mechanism.[2,11] Recent evidence demonstrates that some of these mechanisms may have negative regulatory effects on T cell function that subsequently results in its functional impairment, despite the persistence of an activated T cell phenotype. This imbalanced response has two major consequences: (1) progressive depletion of T cell subsets due to deregulated production of proapoptotic cytokines and (2) progressive loss of T cell function due to immune suppressive mechanisms associated with innate immunity[9] and have been assessed in both pediatric and adult patients, naïve or treated with antiretroviral drugs (ART).[12–17]

As a result of mechanism connections, complex immune dysfunction occurs and predisposes infected individuals to pathogenic, degenerative, and opportunistic infections during HIV evolution contributing to cardiovascular disease, cancer, kidney and liver disease, osteoporosis, and neuronal-cognitive impairment. Up to now these factors resound on life expectancy of HIV-infected individuals in comparison to HIV-negative counterparts.[18,19]

How viral and human interrelation conduces to polypathies and diverse clinical conditions in HIV infection is a revisited topic and has not yet been completely defined. This review is intended to analyze original investigations and review articles focused on the effect of oxidative metabolism in antioxidant status during HIV infection related to diverse clinical condition defined as asymptomatic, acquired immunodeficiency syndrome (AIDS) in naïve and treated subjects. Also ROS, antioxidant, and redox physiological aspects are considered. This work presents data from 70 reviews and original research about HIV, antioxidant status, and OS, in naïve and antiretroviral-treated patients. In an attempt to identify the relevant literature, a comprehensive search was performed using PubMed and Google Scholar. The following search terms were included in multiple combinations: OS, HIV and antioxidant status, OS, HIV, and clinical conditions. Further PubMed search was performed by selecting the "See all related articles" function, thus providing an additional extensive list of publications. Further search was performed by manual scanning of reference lists. Search was conducted from December 2015 to April 2016.

## REACTIVE OXYGEN SPECIES GENERATION, BIOLOGICAL FUNCTIONS, AND ANTIOXIDANT STATUS RELATED TO PHYSIOLOGICAL CONDITION

ROS are highly reactive molecules that originate from diverse process on cellular diverse organelles as mitochondrial electron transport chain (ETC). Almost all cells and tissues continuously convert a small proportion of $O_2$ into $O_2^{\cdot -}$ by the univalent reduction of $O_2$ in the ETC. ROS are produced by other pathways as well, including the respiratory burst taking place in activated phagocytes, ionizing radiation's damage on cell membranes components, and as by-products of several cellular enzymes including NADPH oxidases (NOX), xanthine oxidase (XO), and uncoupled endothelial nitric oxide synthase. Endoplasmic reticulum also produces ROS.[5,20]

ROS are related to various physiological processes, to maintain a state of homeostasis; living organisms are striving to keep those highly reactive molecules under tight control through an intricate system of antioxidants. Low-molecular-weight antioxidants, named the nonenzymatic defenses, and the antioxidant enzymes constitute the total antioxidant repertory that defends the organism from OS. Organisms continuously must confront and control the presence of both prooxidants and antioxidants. Balance of both is regulated and extremely important for maintaining vital cellular and biochemical functions.[6] This balance often referred to as the redox potential is specific for each organelle and biological site, and any interference of the balance in any direction might be deleterious for cells and organisms. Cells present potent gene machinery that is triggered under prooxidant conditions and regulates the expression of antioxidant proteins and enzymes. A variety of enzymatic and nonenzymatic antioxidants present in human cell become insufficient to avoid cellular ROS interaction and might lead to oxidative damage. Changing the balance toward an increase in the reducing power, or the antioxidant, might also cause damage and can be defined as reductive stress.[2,20]

Organisms respond to oxidative stimulus by organizing a stress response to prevent further injury. An increase in the intracellular levels of antioxidant mediators and at the same time the removal of already damaged components are both part of this response. ROS levels are controlled by endogenous antioxidants such as SOD, GPx1, GSH, and CAT. The tripeptide glutathione (γ-L-glutamyl-L-cysteinylglycine) is the key low-molecular thiol antioxidant involved in the redox mechanism of defense in diverse cells. A decrease in GSH levels has been connected to physiological processes such as aging and disorders.[1,21] Although GSH is the primary molecule involved in detoxification of ROS in the body, antioxidant enzymes such as GPx1 are also known to play a role in this process. During detoxification of peroxides (PO), the enzyme GPx1 converts GSH to glutathione disulphide (GSSG). A high ratio for GSH/GSSG is important for the protection of the cell from oxidative damage.[20,22] Disruption of this ratio causes activation of redox sensitive transcription factors, as well, such as NF-κB, AP-1, nuclear factor of activated T cells, and hypoxia-inducible factor 1, which are involved in the inflammatory response. Activation of transcription factors via ROS is achieved by signal transduction cascades that transmit the information from outside to the inside of cell.[6,23]

Interaction of ROS with other tissue components produces a variety of other reactive products; following activation of iNOS, NO can bind superoxide anion to form the highly reactive $ONOO^-$.[7] The latter may attack lipids, proteins, and DNA, to enhance oxidant-related injury. Several kinds of molecules could contribute to the antioxidant capacity of plasma from exogenous or endogenous origin.[6] An association between vitamins A, E, C, vegetables and fruits intake, minerals, and total antioxidants status suggests a role of antioxidant–prooxidant balance on T cells stabilization and functions, which resound in immune response. The mechanisms underlying the present observations are not clear, but multiple pharmacological effects of antioxidant vitamins and flavonoids have been reported, including vascular protection, antiinflammatory, antitumor, and antihypertension activities.[20]

ROS generation represents one of the first lines of defense mounted against invading pathogens and constitutes essential protective mechanisms that living organisms use for their survival. ROS are deeply involved in both arms of the immunological defense system, the innate and the acquired responses.[23] Also ROS contribute to transduction cascades within cells and thereby upregulating the cellular response. In regulatory mode, it also contributes to decrease the cell activation threshold. When ROS are present at high concentrations they could produce oxidative reactions but at low levels could modulate transcription factor activation.[3,6]

Some physiological roles in immune defense, antibacterial action, vascular tone, and signal transduction are demonstrated.[2] Transient over production of ROS in viral infections is generally related to prooxidant effect of inflammatory cytokines and/or polymorphonuclear leukocyte activation exerting also intracellular signal. Reductive cellular environment creates the electrochemical gradient necessary for electron transfer in oxidation–reduction reactions occurring in biological systems.[1]

Every biological molecule had risk of damage by ROS. The extension of oxidative process depends on nature and severity of the stimulus and varies greatly among different tissues and species.[24] Such damaged cell molecules can impair cell functions or can lead to cell death ultimately resulting in disease. A growing number of molecules, such as many kinases, phosphatases, and transcription factors, in a wide range of signal transduction pathways, are thought to be modulated by intracellular redox status. Moreover, a few transcription factors, such as the small GTP-binding protein Rac, are known to activate ROS-generating enzymes (e.g., NOX) and produce ROS as a modulator of downstream molecules.[6]

ROS regulation of transcription factors can occur by direct modification of critical amino acid residues, primarily through the formation of disulphide bonds, at DNA-binding domains or via indirect phosphorylation/dephosphorylation as a result of changes in redox-modulated signaling pathways.[22]

Activation of AP-1 by in vivo adenovirus administration is redox-modulated and involves the participation of redox factor-1 (Ref-1). Ref-1 is a unique molecule that has two distinct enzymatic functions, a DNA repair enzyme and a redox regulatory transcription factor.[2]

Furthermore, it is believed that phosphorylation of IκB, inhibitory subunit of NF-κB, is a key step in NF-κB redox activation. ROS-mediated phosphorylation of IκB, leading to its ubiquitination and degradation, allows the NF-κB complex to be translocated to the nucleus and acts as a transcriptional activator.[7]

At the cellular level, OS generated by ROS and ROS-modified molecules can influence also a wide range of cellular functions. Direct consequence of OS is the damage to various intracellular constituents. For example, when lipid peroxidation occurs, changes in cellular membrane permeability and even membrane leakage can be manifested.[4] Oxidative damage to both nuclear and mitochondrial DNA has detrimental effects, leading to uncontrolled cell proliferation or accelerated cell death.[5] Several cellular processes including proliferation, differentiation, and apoptosis also can be regulated by oxidants.[23]

Apoptosis, also known as programmed cell death, plays an important role in all stages of an organism development. In the intrinsic apoptotic pathway, it has been shown that proteins of the mitochondrial permeability transition pore complex, which controls mitochondrial membrane potential, are the direct targets of ROS.[2,20] These proteins include the adenine nucleotide translocator in the inner membrane, the voltage-dependent anion channel in the outer membrane, and cyclophilin D at the matrix. Prooxidants capable of induction of mitochondrial permeability potential include not only chemicals, such as t-butyl peroxide and diamide, but also lipid peroxidation products such as 4-hydroxynonenal.[7] Moreover, it has been increasingly recognized that oxidative damage to organelles, such as lysosomes and endoplasmic reticulum, stimulates cross-talk between these organelles and mitochondria and induction of apoptosis via intrinsic signaling pathway.[2]

ROS are also tumorigenic by virtue of their ability to increase cell proliferation, migration, and survival and by inducing DNA damage, all contributing elements to tumor initiation, promotion, and metastasis.[1,7,25]

As molecular and cellular defects accumulate during life span of organisms, resulting in redox balance perturbation and ROS endogenous generation, it will influence further the regulation of a number of physiological functions (e.g., metabolism and stress tolerance).[24,26]

## ANTIOXIDANT STATUS AND OXIDATIVE STRESS IN HUMAN IMMUNODEFICIENCY VIRUS NATURAL INFECTION

Several studies have demonstrated that humans infected with HIV are under chronic OS characterized by chronic persistent infections, increased T-cell activation, increased levels of many inflammatory markers, reduced T-cell proliferation, and perturbations of micronutrients and antioxidant system (Table 1.1).

In Table 1.1 are compiled studies that show oxidative stressed HIV-infected populations and having significantly lower antioxidant concentrations than non-HIV individuals.[12,14–16,27–34] In the literature disturbs in the metabolism of GSH, TRX, AA, TOC, and Se (seric and tissue antioxidant) diminished concentrations are reported. Increased PO, isoprostanes (isoP), malondialdehyde (MDA), and carbonyl (CO) concentrations and altered GPx, CAT, and SOD activity have also been reported. In addition, altered levels were found in both pediatrics and adult HIV patients.[12,14–16,27–34]

There are some reports to appoint oxidative pathways convoluted in different aspect of HIV infection. The role of beta-chemokine receptors (CCR2b, CCR3, and CCR5) and the alpha-chemokine receptors (CXCR1, CXCR2, and CXCR4) as entry coreceptors for HIV-1 has been elucidated, and it is also demonstrated that oxidative tone favors the structural mediates interaction through disulphide bounds with HIV sites.[35,36]

Cells of myeloid lineage including monocytes, macrophages, and dendritic cells play an important role in the initial infection and therefore contribute to its pathogenesis. This is mainly because these cells are critical immune cells responsible for a wide range of both innate and adaptive immune functions.[37]

Macrophages participate in HIV reservoir and are resident in a number of affected tissues, such as the adipose tissue, liver, bone, vascular wall, and brain. If infected macrophages are activated and release both ROS and proinflammatory cytokines inside these tissues, they could participate on local inflammation and related redox-mediated

TABLE 1.1 Evidences of Biomolecule Oxidative Damage and Antioxidant Deficiency in HIV/AIDS Patients

| Place | NI | Evaluation Criteria[a] | References |
|---|---|---|---|
| Grenoble (France) | 43 | P Glutathione, malondialdehyde, total peroxides | Favier et al.[12] |
| Buenos Aires (Argentina) | 20 | P Glutathione, total antioxidant capacity, E superoxide dismutase | Repetto et al.[27] |
| Bonn (Germany) | 102 | E Glutathione, S Glutathione and selenium | Look et al.[29] |
| Toronto (Canada) | 29 | L Glutathione and Cysteine | Walmsley et al.[28] |
| La Habana (Cuba) | 85 | P Glutathione peroxidase (−), E superoxide dismutase (−), P malondialdehyde (+), P hydroperoxide (+) P total antioxidant capacity (−), P glutathione (−) L percent of DNA fragmentation (+) | Gil et al.[16] |
| India | 50 | S Total antioxidant capacity (−), malondialdehyde (+), superoxide dismutase (−), vitamin E and C (−) | Suresh et al.[30] |
| Italy | 26 | P hydroperoxide (+), P Total antioxidant capacity (−), P thiol group (−) | Coaccioli et al.[14] |
| Nigeria | 70 | ↓Vitamin. C, vitamin. E ↑ malondialdehyde, hydroperoxide | Akiibinu et al.[31] |
| Colombia | 45 | ↓ Total antioxidant capacity ↑ malondialdehyde, hydroperoxide | Lagos et al.[32] |
| United States of America | 26 | ↓ Glutathione | Morris et al.[33] |
| *Côte d'Ivoire* | 173 | ↓Vit. C, Vit. E vitamin. C, vitamin. E | Boyvin et al.[34] |

*E*, erythrocytes; *EA*, expired air; *L*, lymphocytes; *NI*, number of individuals; *P*, plasma, *S*, serum; −, diminish; ↔, no change; +, increment.
[a]*Statistical analysis with significant difference respect control group* ($P < .05$).

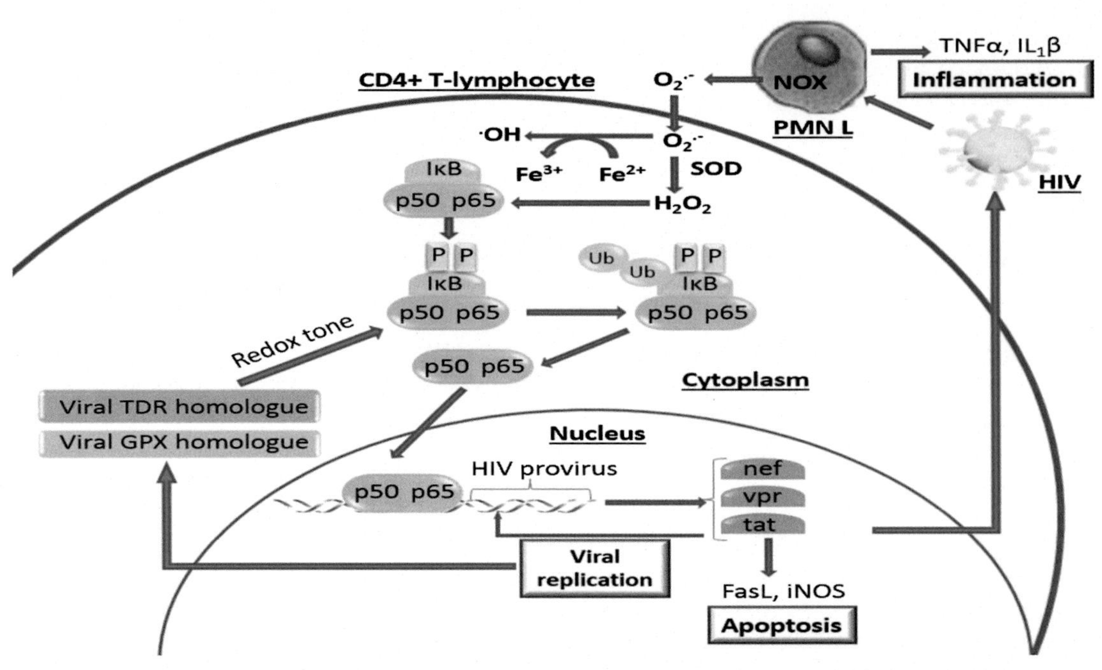


FIGURE 1.1 Molecular interrelation between HIV, T CD4+ lymphocyte, polymorphonuclear cell, inflammation mediators and oxidant-antioxidant species. *FasL*, fas ligand; *GPX*, glutathione peroxidase; *HIV*, human immunodeficiency virus; *IκB*, inhibitor transcription factor; *IL1β*, interleukin 1- β; *iNOS*, inducible -nitric oxide sintase; *nef*, HIV regulatory gene; *NF-κB*, nuclear factor of transcription; *NOX*, NADPH oxidase; *PMN L*, polymorphonuclear lymphocyte; *SOD*, superoxide dismutase; *tat*, HIV regulatory gene; *TDR*, thioredoxin reductase; *TNFα*, tumor necrosis factor-α; *Ub*, ubiquitine; *vpr*, HIV regulatory gene.

signals of transcription factor activation (Fig. 1.1). That activation in turn could mediate other virus replication and infection instauration.[10] Once HIV has established as chronic infection, newly produced virions interact preferentially with cells expressing CD4. CD4+ T cell subset depletion in HIV/AIDS patients is the most dramatic effect of apoptosis mediated by redox abnormalities and induction of Fas/APO-1/CD95 receptor expression. High proportion of lymphocytes expressing Fas was shown to be elevated in HIV-infected individuals. Generally these studies demonstrated that the proportion of Fas-expressing cells increases with disease progression. Increased Fas expression in CD4+ lymphocytes was found by some investigators and others were found in both CD4+ and CD8+ T cells.[38,39]

HIV itself may also cause mitochondrial toxicity and OS contributing to lipoatrophy also during natural infection.[40,41]

These findings could be explained in part by several mechanisms such as chronic inflammatory activation of innate immune system, low intake of antioxidant or their precursors from diet in relation to requirements, malabsorption, enhanced cysteine metabolism in peripheral tissues, downregulation synthesis of antioxidant enzymes by viral protein as Tat, and virally encoded regulatory proteins. All aspects could influence on consequent loss of sulfur group that may account for GSH and antioxidant deficiency during HIV infection.[42,43]

That chronic innate immune activation suppresses functional T cell-mediated adaptive immune responses while sustaining the activated phenotype of T cells with disruption of lymphoid architecture and multifactorial immune suppression, including decreased CD4 and CD8 T cells, which resembles many of the alterations observed during HIV infection.[8,19,21]

Tat-induced protein oxidation is well documented and its effects on lipid peroxidation have also been reported. Recently it has been demonstrated that Tat activates multiple signaling pathways. In one of these, Tat-induced superoxide acts as intermediate, while the other utilizes peroxide as a signal transducer.[39,44]

The expression of HIV-1 in T cells is regulated by the transcription factor NF-κB, under redox-controlled signaling pathways. GSH acts as a free radicals scavenger and is thought to inhibit activation of NF-κB. TRX is related to redox regulation of IκB. NF-κB is involved in the early transcription of HIV-1. Thus, ROS may potentially be involved in the pathogenesis of HIV infection not only through direct effects on cells but also through interactions with NF-κB and activation of HIV replication.[22,44–47]

Some viral proteins interfere with host T cell functions and promote rampant virus replication. Taylor et al. pointed out several regions of HIV-1 with the potential to encode selenoproteins, a GPx viral homologue and a thioredoxin reductase (TDR) homologue.[35] These could be contributing to Se host deficiency. Significance of host Se status is based on the antioxidant properties of amino acid selenocysteine, the catalytic center of selenoenzymes as GPx family.[22] This enzyme regulates biologic oxidative homeostasis by neutralizing metabolically produced ROS. The viral selenoproteins have been suggested could be involved in regulation of NF-κB controlling HIV replication. TDR homologue viral expression could be contributed to activation of NF-κB, while reducing cellular GPx levels via Se sequestration, whereas the viral GPx homologue—as a late-expressed gene—would be expected to deactivate NF-κB by decreasing oxidant tone. Notable another TDR viral homologue action is in the synthesis of deoxyribonucleotides, via ribonucleotide reductase, enhancing stimulation of proviral DNA synthesis activities.[35]. (Fig. 1.1)

Some micronutrients play essential roles in maintaining normal immune function and it may protect immune effector's cells from OS. For HIV patients, it is particularly important to identify metabolic alterations and nutritional deficiencies and determine whether the supplementation will improve clinical outcome.[13,31]

Translation's initiation of the HIV's full length messenger RNA is not only suggested through cap-dependent but also uses an internal ribosome entry site (IRES) located in the 5′untranslated region. This site is the RNA structure region, which directly recruited the 40S ribosomal subunit, at or near an initiation codon. Various models for the 5′UTR encompassing IRES have been proposed based on phylogenetic, chemical mapping and mutagenesis approaches. The wild-type compose includes stem loop. Translation process requires host cell factor called IRES transacting factor. Researches concerning in vitro experiments strongly argue that OS in HIV increases the IRES activity via an effect on stem loop. This data combine chemical and conformational changes caused by OS in favor of viral replication.[48]

The counterpart of NF-κB signaling is the antioxidant gene regulator nuclear factor (erythroid-derived 2)–like 2 (Nrf2), which mediates the Nrf2-antioxidant response elements (ARE) pathway. Nrf2 has an important role in the OS prevention in cells by activating the ARE. Under nonstress conditions, Nrf2 is located at the cytoplasm where it is associated with its inhibitor, Keap-1.[49,50]

Triggering prooxidant conditions, phosphatidylinositol 3-kinase pathway mediates Nrf2 dissociation and its translocation into the nucleus, dissociating from Keap-1. Thereafter, Nrf2 enhances diverse target genes expression that encodes antioxidant compounds synthesis.[49,50]

The Nrf2–ARE axis is active in cells especially relevant for HIV-1 infection, lymphocytes, and macrophages.[44] The role of Nrf2 in HIV-1 infection is not completely understood. Moreover, the Nrf2 complex regulates a wide variety of genes involved in antioxidant defense. As an example, the antioxidant heme oxygenase-1 (HO-1) is regulated by Nrf2. An HIV-1 transgenic rat model reduced expression of Nrf2 and HO-1, leading to increased OS.[4,23]

In previous work, it has been demonstrated also that viral Tat protein liberated by HIV-1–infected cells interferes with calcium homeostasis, activates caspases, and induces mitochondrial generation and accumulation of ROS, important events in the apoptotic cascade of several cell types.[39,51] Those aspects could contribute to the spectrum of malignancies associated with HIV infection.

Also mitochondrial dysfunction and its impact on chronic inflammation during HIV natural evolution have been suggested.[40,51]

High levels of ROS as a consequence of chronic immune system activation by HIV infections could lead to a decline of antioxidant defense molecules and accumulative damage of cellular components generating augmented lipid peroxidation products, oxidized proteins, and altered DNA sequences. Almost redox-implicated enzymes and molecules are physiologically endogenous generated and are involved in detoxification and general metabolism. As a consequence of antioxidant depletion, the detoxification capacity of reactive metabolites is reduced, and this is probably connected to the PO high levels detected and drug side effects in HIV+ patients.[52]

Antioxidant enzymes levels are sensitive to OS. Both increased and decreased levels have been reported in different disease states in which an enhancement of ROS could be a cause or a consequence of the disease.[7,24] Several kinds of molecules contribute to the antioxidant capacity of plasma. The possible interaction among different antioxidants in vivo could also render the measurement of any individual antioxidant less representative of the overall antioxidant status.[20] There is thus experimental evidence that recognizes different metabolic events that occur as a consequence of HIV infection and directly influence the consumption of antioxidant components, thus contributing to the OS.[12,53]

Therefore, some authors consider that increased immune activation and long-term chronic inflammation are major players in the aging process in the general population; it is obvious that these processes are more prevalent in HIV-infected patients than in the general population, even when the infection is well controlled. HIV-infected patients will be more prone to develop, in advance, age-related diseases.[24,42]

## ANTIOXIDANT STATUS AND OXIDATIVE STRESS DURING ANTIRETROVIRAL TREATMENT

Most significant advance in medical management of HIV infection has been the treatment of patients with ART. In latest years, a relevant decline of morbidity and mortality of HIV infection has been observed due to the use of combined therapy named high active antiretroviral therapy (HAART). This treatment can suppress HIV replication to plasma HIV RNA undetectable levels (<50 copies/mL) and improve the immune function in patients, especially CD4 T lymphocytes subsets. Also HAART can successfully prevent AIDS-related morbidity and mortality, resulting in increased life span of HIV-infected patients. In turn, the course of HIV disease has evolved from a universally fatal infection to a manageable chronic illness.[17,54]

However, it has become evident that patients taking otherwise effective ART remain at increased risk of non-AIDS–related morbidity and mortality. Many of these conditions are classically associated with normal aging process but appear to be occurring at an earlier age in HIV-infected persons. Some anti-HIV drug classes are associated with lactic acidosis, hyperlipidemia, glucose intolerance, diabetes mellitus, fat redistribution, wasting, and atherosclerosis. These conditions include premature onset of cardiovascular disease, neurocognitive disease, bone disease, and cancer. These features could be related with OS increased by antiretroviral toxicity, which indicate mitochondria as toxic target (Table 1.2). It could be produced by a common mechanism through mitochondria dysfunction, which contributes to cell senescence and accelerated aging.[52,55,56]

TABLE 1.2 Evidences of Biomolecule Oxidative Damage and Antioxidant Deficiency in Human Immunodeficiency Virus/Acquired Immunodeficiency Syndrome Patients With HAART

| Place | NI | Evaluation Criteria[a] | References |
|---|---|---|---|
| USA | 120 | Three combinations<br>F2 isoprostanes (+) | Hulgan et al.[61] |
| Norway | 20 | Two combinations<br>MDA (−), GSH (−), GSSG (+), VIT E (+), VIT C (−) | Aukrust et al.[58] |
| USA | 164 | Two combinations<br>Syndrome metabolic X, oxidative stress indices (+) associate to PI ± 6 months | Hurwitz et al.[62] |
| Cameroon | 85 | Three combinations<br>Groups sulfhydrils (−), malondialdehyde (TBARS) (+), carbonyls groups (+), albumin (−) y Vit C (−) in plasma | Ngondi et al.[59] |
| USA | 164 | One combination ddI/d4T (lipodystrophy symptoms)<br>75 cases, 71 control<br>F2 isoprostanes (no association) | Hulgan et al.[63] |
| Spain | 245 | Two combinations (NNRTI, PI)<br>total peroxide concentration (+).<br>PI 2-month | Masiá et al.[64] |
| Italy | 86 | Three combinations<br>Serum total antioxidant capacity (d-ROMs) (+) and antioxidant activity (OXY-adsorbent as MPO activity) (−) related to treatment adherence | Mandas et al.[17] |
| India | 100 | Three combinations<br>LPO, GSH ↔ PI 9,2 ± 7 months | Wanchu et al.[65] |
| USA | 194 | Four combinations AZT/d4T/EFZ/NVP<br>Plus 91 control<br>F2 isoprostanes (no association with HIV-1 RNA or HAART) | Redhage et al.[66] |
| Ghana | 228 | ↓Vit. C, Vit. E, SOD, GPx<br>↑MDA | Obirikorang et al.[67] |
| Nigeria | 50 | ↑SOD, MDA | Mgbekem et al.[68] |
| Cuba | 56 | Two combinations<br>MDA (−), GSH (−), PP, HPO, CAT, SOD, PAOP in plasma<br>PI ± 6 months | Gil et al.[69] |
| Brazil | 182 | ↓Vit. E | Itinoseki et al.[70] |
| Nigeria | 50 | ↓GSH, ↑MDA,<br>↓CAT, ↓Vit. C, Vit. E, ↓Vit. A | Abduljalil et al.[55] |

*CAT*, catalase; *GSH*, glutathione; *GPx*, glutathione peroxidase; *HAART*, high active antiretroviral therapy; *MDA*, malondyaldehyde; *NI*, number of individuals; *LPO*, lipoperoxides; *PAOP*, products of advanced oxidized proteins; *PP*, peroxidation potential; *SOD*, superoxide dismutase; *USA*, United States of America; −, diminish; ↔, no change; +, increment.

[a]*Statistical analysis with significant difference* ($P < .05$).

Thus, the emerging picture of HIV-infected patients' health that are being successfully treated with ART therapy in terms of viral suppression is as follows: life span is not normalized by antiretroviral treatment; the risk of age-associated diseases is higher than expected; inflammation remains elevated and CD4+ count often remains low and both predict age-associated events. Also the prevalence of insulin resistance and type II diabetes is increased in treated patients.[18,19]

The DNA pol-$\gamma$, the eukaryotic mtDNA replication enzyme, is inhibited by ART as nucleoside reverse transcriptase inhibitor (NRTI) used commonly during antiviral treatment of HIV/AIDS patients. This inhibition produces mtDNA depletion, which leads to a decreased energy production. This event is cumulative and toxic manifestations increase with duration of exposure.[54]

Hepatic toxicity from ART was reported early in epidemic, and recent reports continue to point out the mitochondria as toxic target and OS as a consequence of therapy.[43,52] Since HAART does not completely eliminate HIV, it is likely that the final outcome of treatment will depend not only on the effective reduction of viral load (VL) but also on recover ability of immune system and residual virus control. Previous studies have suggested a role of OS in HIV replication stimulation, immunodeficiency development, and also in treatment consequences, thus at last contributing to an organism damage vicious cycle.[18]

OS interaction events in disease progression have become intricate in HIV-infected patients with HAART. Virus control with HAART may not, as one might expect, reduce OS levels, on the contrary.[8,53]

Combinations of anti-HIV drugs containing NRTI are generally used during HIV infection as clinical guidelines recommended. Additional adverse effects and/or regimen adherence difficulties have serious consequences such as loss of serum HIV suppression, development of drug-resistant HIV strains, and development of increased probability of opportunistic illness. NRTI is associated with lactic acidosis, hyperlipidemia, glucose intolerance, diabetes mellitus, atherosclerosis, fat redistribution, and wasting syndrome; all of these could be related to increased OS and its toxicity. Phosphorylated-NRTI mitochondrial toxicity may amplify some of the pathophysiologic and phenotypical events in infection.[8,52,54]

Picture of immunosenescence closely resembles observations in patients receiving long-term ART therapy, suggesting that ongoing HIV-related immune dysfunction and inflammation during ART treatment underly premature aging in HIV-infected persons.[25,42]

Collectively, these observations support an emerging relation that posits residual inflammation and suboptimal CD4+ count gains with a hypercoagulable state resulting from several ongoing factors. These include residual viral replication, persistent viral expression, the loss of immunoregulatory cells that should dampen immune activation, increased lymphoid fibrosis, and microbial translocation. Other contributed factors to ongoing inflammation include chronic infection with citomegalovirus, hepatitis C virus, or hepatitis B virus, and thymic dysfunction. How these factors combine to affect CD4+ cell gains, immune function, and early morbidity and mortality is not known.[56,57]

The study and enhancement of surrogate markers of HIV disease progression, including OS indexes, continues to be an important area of research particularly with the advent of therapies that claim to halt or slow the process of immunological decline. Additional markers and combinatorial analysis, which add value to T-CD4+ lymphocyte subset, would therefore be useful, i.e., in the decision of when to start/stop or change therapy.[8,24,26]

Considering factors related to the virus and the treatment, other environmental factors could also prematurely induce aging, such as smoking, sedentary lifestyle, low-nutrition diet and resulting fat gain, or drug use. Even if difficult to do, these factors need to be aggressively taken in count.[4,5]

However, recent studies have indicated a rise in prevalence of HIV-1–associated neurocognitive disorders and related side effects following the era of HAART.[50,52]

Previous studies found ART side effects with simultaneous lowered cellular proliferation and directly affected mitochondrial function in a reversible fashion by decreasing mitochondrial membrane potential and increasing superoxide production. Those events could impact on different biological manner to organisms.[37,40]

ROS elevated levels in HAART previous studies are suggested, thereby provoking OS scale, which has already been well established to occur since HIV infection. Hence, while the oxygen faces a paradox, so does HAART; although VLs may be suppressed, it is at the expense of elevated ROS levels that are known to also activate HIV transcription pathways and promote cell death.[4,40]

Some studies and clinical research found that some ART toxic effect could partially be reversed by previous and concomitant antioxidant treatment. Since antioxidant treatment has contributed to suppress some of prooxidant effects of ART treatment, antioxidants in combination with HAART may impact on others such as neurocognitive disorders, additional opportunistic infections associated with HIV-1 infection, and also could influence on VL.[8,25]

## ANTIOXIDANT STATUS AND OXIDATIVE STRESS DURING ACQUIRED IMMUNODEFICIENCY SYNDROME WITH DIFFERENT CLINICAL CONDITIONS

Redox-sensitive transcription factors and OS are often associated with origin and progression of many human disease states via four critical steps: membrane lipid peroxidation, protein oxidation, DNA damage, and disturbance in reductive equivalents of the cell.[14,42]

The sustained prooxidant microenvironment has been implicated in various diseases and it may exert a pivotal role in HIV-1 disease progression related to appearance of secondary comorbidities, such as cardiovascular diseases, neurological disorders, cancer, diabetes, and aging.[42,57]

Immune activation is increased in HIV-infected patients because of residual HIV infection and other viruses, such as cytomegalovirus reactivation, increased bacterial translocation, and altered gut permeability. Markers of bacterial translocation, such as lipopolysaccharide, and of innate immunity activation, such as sCD14, together with indications of elevated immune activation, have been linked to neurocognitive and cardiovascular comorbidities and to mortality.[23,25,57]

Different mechanisms are involved such as "mitochondrial oxidative stress," which can be exemplified by prooxidants shifting of the thiol/disulphide redox state and damage on glucose tolerance, "inflammatory oxidative conditions," and increased activity of either NOX or XO-induced formation of ROS, or by both.[22,40]

A higher prevalence of age-related morbidities (cardiovascular disease, hypertension, renal failure, bone fracture, and diabetes mellitus) has been reported in HIV-positive individuals. These morbidities occurred at an earlier age, and that polypathology was more frequent than in HIV-negative controls.[56]

Substantial evidence from various researches provides a strong link between antioxidant status and OS, also related to ART effect and OS in HIV infection.[17,58,59] Diverse factors have been associated with an increased production of prooxidants, depletion of antioxidant status, and induction of OS. Despite some contradictory results, evidences produced during the past decade have provided valuable information on biological systems that impact on HIV evolution related to OS condition.[8,56]

Recently, several studies have demonstrated that organisms are capable of initiating an array of regulatory processes in response to OS, including the activation of stress-gene expression and modification of stress-responsive signal transcription pathways. In contrast, there is compelling evidence that these regulatory processes are altered in disease organisms.[7,21,23]

Therefore, stress-induced cellular injury appears to be exaggerated with diseases. This failure to effectively respond to cellular challenge has been postulated to contribute to a reduction in stress tolerance and the development of various pathologies and diseases.[1,3]

Whether HIV itself or the comorbidities associated is most responsible for frailty and other geriatric symptoms is unclear, but it is widely agreed that inflammation related to HIV, other infections, and chronic conditions or lifestyle factors plays a major role in driving more rapid aging and increased morbidity and mortality.[42,56]

## FUTURE PERSPECTIVE

Elucidation of antioxidant status related to oxidative damage in pathogenesis of different diseases has led to a medical revolution that is reassuring a new paradigm of healthcare. Also it had to be taken into consideration that total antioxidant capacity of a tissue or the level of a single class of antioxidants as markers of OS is not sufficient to make inferences about OS without supplemental information on oxidative damage or ROS production.[3,24]

Modern science and research have begun to unravel the molecular components of free-radical biology and biological interrelationships of these components in mediating various disease processes for a better understanding and exploitation in biomedical/clinical sciences.[22,26]

HIV infection results in prolonged continuous stimulation of immunological cells, which is maintained beyond the acute–early phase and throughout the infection evolution. This uncontrolled chronic innate immune activation may lead to a deregulated adaptive immune response, characterized by functionally impaired T cells and increased levels of phenotypic markers of activation.[9] In this context, antioxidants could produce immune modulation and can be used for prophylaxis or therapy of certain diseases along with the mainstream therapy. Several clinical studies and basic research respecting HIV infection have shown a reduction of endogenous enzymes activity in certain pathological condition that could be mitigated by the use of exogenous antioxidants or other alternative intervention.

Supplements of exogenous antioxidants can act directly to quench the free radical or free radical reactions, prevent lipid peroxidation, and also boost the endogenous antioxidant system, and hence deliver the prophylactic or therapeutic activity. Many novel approaches and significant findings have come to light in the last few years and could impact on future management of infection.[53,56]

Sustained interest in the antioxidants use for human disease prevention and management offers opportunities for newer and better therapeutic entities development with either antioxidant activity or antioxidant modulator.[21] Parallel identification, development of natural or synthetic antioxidants and their suitable formulation can provide enormous scope for better treatment of several diseases. On-going studies continue to produce epidemiological evidence suggesting that antioxidant rich foods or antioxidant supplements reduce the risk of chronic disease and promote wellness. Several factors like poor solubility, inefficient permeability, and instability due to storage of food, first pass effect, and gastrointestinal degradation need to be improved. Antioxidants are to be developed as drug targets. Therefore modification in dosage form, physicochemical characteristics, biopharmaceutical properties and pharmacokinetic parameters are important to consider in drug development process. Therapeutic and nutritional fields progression related antioxidants have not only been punctuated by some successes, but by various spectacular failures as well.[2,53] Today, varied diet (natural and healthy food with antioxidant activity) remains the best advice in garnering the benefits of antioxidants and the many other bioactive components available from food. Concurrently, it is also necessary to avoid oxidant sources (cigarette, alcohol, exposure of chemicals, stress, etc.) to keep individuals on healthy condition.[2,5,7]

It is also imperative to reveal the relationship between intake of antioxidants and their dose-dependent functional effects. As clinical evidence emerges and our understanding of genomic differences improves, the specific role of antioxidants with variation of species or genetic differences needs to be identified.[23,60]

The future holds great promise for discoveries of new knowledge about ROS biology and antioxidants and for turning this basic knowledge into practical use for ensuring a healthy life. Developing coordinated research collaborations involving biomedical scientists, phytochemical researchers, nutritionists, and physicians is a critical step evaluating the impact of antioxidants in health and disease for the coming decades.

Since a substantial amount of evidence reveals a role of ROS in inducing OS following HIV infection and OS as a causative factor in the progression of many diseases, including AIDS, a turn of focus should be put on alternatives to suppress the consequently life-threatening disease. Future studies should be undertaken to determine the correct dosages and duration of diverse treatment necessary to curb the adverse effects of HIV infection and its treatment.[8,19,53] Furthermore, comparative studies may serve to identify cofactors that contribute to the development of AIDS. With a better understanding of the cofactors related in disease evolution, there is tremendous hope of improved diagnosis and treatment to perhaps influence again on the course of HIV infection and prevent the onset of AIDS.

## SUMMARY POINTS

- Redox reactions play an accrual role in biological physiology.
- ROS constitute high-energy, reactive molecules that could be generated endogenously and exogenously and can modulate functions and biomolecule damage.
- Antioxidants are endogenous and exogenous sources and accomplish a balance between beneficial and harmful oxidant production.
- OS results from an imbalance between oxidant and antioxidant production.
- Sustained OS contributed to cell and tissue damage is mediated by redox circuits and attempts to cellular structure and function. It could be a component in the pathophysiology concert behind a multitude of diseases as origin or consequence factor.
- Since 1983 HIV natural evolution was related to depletion of antioxidants and increase of biomolecule oxidation.
- The HIV hypothesis of AIDS involved the oxidative biology mechanism. HIV-infected individuals suffer from chronic OS on viral infection.
- ROS-generated oxidant tone is recognized as a mediator of HIV entry to cell, nuclear transcription factor-κB activation encompassed in HIV replication and transcription, and could be related in some type of apoptosis contributing to CD4+ T-cell depletion.
- Whereas many antioxidants are not completely elucidated, their functions and interrelation on redox circuits and their deficiencies have been associated with OS in HIV-positive patients of diverse conditions.
- While many studies indicate antiretroviral therapy to reduce VL to undetectable levels at times, it has revealed to arise accompanied of elevated oxidative and reduced antioxidant indexes.

# References


1. Alfadda A, Sallam R. Reactive oxygen species in health and disease. *J Biomed Biotechnol* 2012:14.
2. Valko M, Leibfritz D, Moncol J, Cronin M, Mazur M, Telser J. Free radicals and antioxidants in normal physiological functions and human disease. *Int J Biochem Cell Biol* 2007;**39**:44–84.
3. Birben E, Murat U, Sackesen C, Erzurum S, Kalayci O. Oxidative stress and antioxidant defense. *WAO J* 2012;**5**:9–19.
4. Kohen R, Nyska A. Oxidation of biological systems: oxidative stress phenomena, antioxidants, redox reactions, and methods for their quantification. *Toxicol Pathol* 2002;**30**:620–50.
5. Devasagayam T, Tilak J, Boloor K, Sane K, Ghaskadbi S. Free radicals and antioxidants in human health: current status and future prospects. *J Assoc Phys India* 2004;**52**:794–804.
6. Schieber M, Chandel N. ROS function in redox signaling and oxidative stress. *Curr Biol* 2014;**24**:453–62.
7. Roberts R, Smith R, Safe S, Szabo C, Tjalkens R. Toxicological and pathophysiological roles of reactive oxygen and nitrogen species. *Toxicology* 2010;**276**:85–94.
8. Aquaro F, Scopelliti F, Pollicita M, Perno F. Oxidative stress and HIV infection: target pathways for novel therapies? *Future HIV Therapie* 2008;**2**:327–38.
9. Boasso A, Shearer G. Chronic innate immune activation as a cause of HIV-1 immunopathogenesis. *Clin Immunol* 2008;**126**:235–42.
10. Elbim C, Pillet S, Prevoste M, Preira A, Girard P. Redox and activation status of monocytes from human immunodeficiency virus-infected patients: relationship with viral load. *J Virol* 1999;**73**:4561–6.
11. Kashou A, Agarwal A. Oxidants and antioxidants in the pathogenesis of HIV/AIDS. *Open Reprod Sci J* 2011;**3**:154–61.
12. Favier A, Sappey C, Leclerc P, Faure P, Micoud M. Antioxidant status and lipid peroxidation in patients infected with HIV. *Chem Biol Interact* 1994;**91**:165–80.
13. Allard J, Aghdassi E, Chau J, Salit I, Walmsley S. Oxidative stress and plasma antioxidant micronutrients in humans with HIV infection. *Am J Clin Nutr* 1998;**67**:143–7.
14. Coaccioli S, Crapa G, Fantera M, Del Giorno R, Lavagna A. Oxidant/antioxidant status in patients with chronic HIV infection. *Clin Ter* 2010;**161**:55–8.
15. Gil L, González I, Díaz M, Bermúdez Y, Hernández D, Abad Y. Evaluación de pacientes pediátricos VIH/sida. *Rev OFIL* 2012;**22**:172–83.
16. Gil L, Martinez G, Gonzalez I, Alvarez A, Molina R. Contribution to characterization of oxidative stress in HIV/AIDS patients. *Pharmacol Res* 2003;**47**:217–24.
17. Mandas A, Lorio E, Congiu M, Balestrieri C, Mereu A. Oxidative imbalance in HIV-1 infected patients treated with antiretroviral therapy. *J Biomed Biotechnol* 2009:7. 749575.
18. Nakagawa F, May M, Phillips A. Life expectancy living with HIV: recent estimates and future implications. *Curr Opin Infect Dis* 2013;**26**:17–25.
19. Dieffenbach C, Fauci A. Thirty years of HIV and AIDS: future challenges and opportunities. *Ann Intern Med* 2011;**154**(11):766–71.
20. Pisoschi A, Pop A. The role of antioxidants in the chemistry of oxidative stress: a review. *Eur J Med Chem* 2015;**97**:55–74.
21. Willcox J, Ash S, Catignani G. Antioxidants and prevention of chronic disease. *Crit Rev Food Sci Nutr* 2004;**44**:275–95.
22. Sen S. Cellular thiols and redox-regulated signal transduction. *Curr Top Cell Regul* 2000;**36**:1–30.
23. Gostner J, Becker K, Fuchs D, Sucher R. Redox regulation of the immune response. *Redox Rep* 2013;**18**:88–94.
24. Jacob K, Hooten N, Trzeciak A, Evans M. Markers of oxidant stress that are clinically relevant in aging and age-related disease. *Mech Ageing Dev* 2013;**134**(3–4):139–57.
25. Reshi M, Su Y, Hong J. RNA viruses: ROS-mediated cell death. *Int J Cell Biol* 2014:2014:16, ID 467452. http://dx.doi.org/10.1155/2014/467452.
26. Kregel K, Zhang H. An integrated view of oxidative stress in aging: basic mechanisms, functional effects, and pathological considerations. *Am J Physiol Regul Integr Comp Physiol* 2007;**292**:18–36.
27. Repetto M, Reides C, Gomez M, Costa M, Griemberg G, Llesuy S. Oxidative stress in blood of HIV patients. *Clin Chim Acta* 1996;**255**:107–17.
28. Walmsley SL, Winn LM, Harrison ML, Uetrecht JP, Wells PG. Oxidative stress and thiol depletion in plasma and peripheral blood lymphocytes from HIV-infected patients: toxicological and pathological implications. *AIDS* November 15, 1997;**11**(14):1689–97. PubMed PMID: 9386803. Epub 1997/12/05. eng.
29. Look M, Rockstroh J, Rao G, Kreuzer K, Barton S, Lemoch H. Serum selenium, plasma glutathione (GSH) and erythrocyte glutathione peroxidase (GSH-Px)-levels in asymptomatic versus symptomatic human immunodeficiency virus-1 (HIV-1)-infection. *Eur J Clin Nutr* 1997;**51**:266–72.
30. Suresh D, Annam V, Pratibha K, Maruti B. Total antioxidant capacity a novel early bio-chemical marker of oxidative stress in HIV infected individuals. *J Biomed Sci* 2009;**16**:61–72.
31. Akiibinu M, Adeshiyan A, Olalekan A. Micronutrients and markers of oxidative stress in symptomatic HIV-positive/aids vigerians: a call for adjuvant micronutrient therapy. *IIOABJ* 2012;**3**:7–11.
32. Lagos G, Cediel V, Villegas S. Especies reactivas de oxigeno y respuesta antioxidante en pacientes VIH positivos y donantes voluntarios de sangre, Pereira, Colombia, 2007–2009. *Revista Médica de Risaralda* 2012;**18**(1):54–64.
33. Morris D, Guerra C, Donohue C, Oh H, Khurasany M, Venketaraman V. Unveiling theMechanisms for decreased glutathione in individuals with HIV infection. *Clin Dev Immunol* 2012;**2012**:1–10.
34. Boyvin L, M'boh G, Ake-Edjeme A, Soumahoro-Agbo M, Séri K, Djaman J. Serum level of two antioxidant vitamins (A and E) in Ivorian (Côte d'Ivoire) people living with human immunodeficiency virus. *Ann Biol Res* 2013;**4**(11):48–54.
35. Cerutti N, Killick M, Jugnarain V, Papathanasopoulos M, Capovilla A. Disulfide reduction in CD4 Domain 1 or 2 is essential for interaction with HIV gp120, which impairs Thioredoxin-driven CD4 dimerization. *JBC* 2014;**289**(15):10455–65. M113.539353.
36. Matthias L, Hogg P. Redox control on the cell surface: implications for HIV-1 entry. *Antioxid Redox Signal* 2003;**5**:133–8.
37. Kameoka M, Kimura T, Ikuta K. Superoxide enhances the spread of HIV-1 infection by cell-to-cell transmission. *FEBS Lett* 1993;**331**(1–2):182–6.
38. Dobmeyer T, Findhammer S, Dobmeyer J, Klein S, Raffel B. Ex vivo induction of apoptosis in lymphocytes is mediated by oxidative stress: role of lymphocyte loss in HIV infection. *Free Radic Biol Med* 1997;**22**:775–85.
39. Buccigrossi V, Laudiero G, Nicastro E, Miele E, Esposito F. The HIV-1 transactivator factor (Tat) induces enterocyte apoptosis through a redox-mediated mechanism. *PLoS One* 2011.

40. Milazzo L, Menzaghi B, Caramma I, Nasi M, Sangaletti O. Effect of antioxidants on mitochondrial function in HIV-1-related lipoatrophy: a pilot study. *AIDS Res Hum Retroviruses* 2010;**26**:1207–14.
41. Miro O, Lopez S, Martínez E, Pedrol E, Milinkovic A. Mitochondrial effects of HIV infection on the peripheral blood mononuclear cells of HIV-infected patients who were never treated with antiretrovirals. *Clin Infect Dis* 2004;**39**:710–6.
42. Capeau J. Premature aging and premature age-related comorbidities in HIV-infected patients: facts and hypotheses. *Clin Infect Dis* 2011;**53**:1127–9.
43. Johnston R, Barre-Sinoussi F. Controversies in HIV cure research. *J Int AIDS Soc* 2012;**15**(16).
44. Zhang H, Sang W, Ruan Z, Wang Y. Akt/Nox2/NF-kappaB signaling pathway is involved in Tat-induced HIV-1 long terminal repeat (LTR) transactivation. *Arch Biochem Biophys* 2011;**505**:266–72.
45. Schreck R, Rieber P, Baeuerle P. Reactive oxygen intermediates as apparently widely used messengers in the activation of the NF-kappa B transcription factor and HIV-1. *EMBO J* 1991;**10**:2247–58.
46. Chandrasekaran V, Taylor E. Molecular modeling of the oxidized form of Nuclear Factor-κB suggests a mechanism for redox regulation of DNA binding and transcriptional activation. *J Mol Graph Model* 2007;**27**:93–107.
47. Zhao L, Olubajo B, Taylor E. Functional studies of an HIV-1 encoded glutathione peroxidase. *Biofactors* 2006;**27**:93–107.
48. Gendron K, Ferbeyre G, Heveker N, Brakier-Gingras L. The activity of the HIV-1 IRES is stimulated by oxidative stress and controlled by a negative regulatory element. *Nucleic Acids Res* 2011;**39**:902–12.
49. Kang K, Lee S, Kim S. Molecular mechanism of nrf2 activation by oxidative stress. *Antioxid Redox Signal* 2005;**7**(11–12):1664–73.
50. Reddy P, Gandhi N, Samikkannu T. HIV-1 gp120 induces antioxidant response element-mediated expression in primary astrocytes: role in HIV-associated neurocognitive disorder. *Neurochem Int* 2012;**61**(5):807–14.
51. Monaghan P, Metcalfe N, Torres R. Oxidative stress as a mediator of life history trade-offs: mechanisms, measurement and interpretation. *Ecol Lett* 2009;**12**:75–92.
52. Deavall D, Martin E, Horner J, Roberts R. Drug-induced oxidative stress and toxicity. *J Toxicol* 2012:13.
53. Rolina D, Lindi M. Reconciling conflicting clinical studies of antioxidant supplementation as HIV therapy: a mathematical approach. *BMC Public Health* 2009;**9**(12).
54. Day B, Lewis W. Oxidative stress in NRTI-induced toxicity. *Cardiovasc Toxicol* 2004;**4**:207–16.
55. Abduljalil M, Liman H, Umar R, Abubakar M. Effect of HIV and HAART on antioxidants markers in HIV positive patients in Sokoto state, Nigeria. *Int J Sci Eng Res* 2015;**6**(2):1069–74.
56. Effros R, Fletcher C, Gebo K. Aging and infectious diseases: workshop on HIV infection and aging: what is known and future research directions. *Clin Infect Dis* 2008;**47**:542–53.
57. Longo-Mbenza B, Longokolo M, Lelo Tshikwela M, Mokondjimobe E, Gombet T. Relationship between younger age, autoimmunity, cardiometabolic risk, oxidative stress, HAART, and Ischemic Stroke in Africans with HIV/aids. *ISRN Cardiol* 2011:897–908.
58. Aukrust P, Muller F, Svardal A, Ueland T, Berge R. Disturbed glutathione metabolism and decreased antioxidant levels in human immunodeficiency virus-infected patients during highly active antiretroviral therapy-potential immunomodulatory effects of antioxidants. *J Infect Dis* 2003;**188**:232–8.
59. Ngondi J, Oben J, Forkah D, Etame L, Mbanya D. The effect of different combination therapies on oxidative stress markers in HIV infected patients in Cameroon. *AIDS Res Ther* 2006;**3**(19).
60. Marcadenti A, Assis C. Dietary antioxidant and oxidative stress: interaction between vitamins and genetics. *J Nutr Health Food Sci* 2015;**3**(1):1–7.
61. Hulgan T, Morrow J, D´Aquila R, Raffanti S, Morgan M. Oxidant stress is increased during treatment of human immunodeficiency virus infection. *Clin Infect Dis* 2003;**37**:1711–7.
62. Hurwitz B, Klaus J, Llabre M, Gonzalez A, Lawrence P, Maher K. Suppression of human immunodeficiency virus type 1 viral load with selenium supplementation. *Arch Int Med* 2007;**167**:148–54.
63. Hulgan T, Hughes M, Sun X, Smeaton L, Terry E, Robbins G, et al. Oxidant stress and peripheral neuropathy during antiretroviral therapy: an aids clinical trials group study. *J Acquired Immune Deficiency Syndr* 2006;**42**:1–5.
64. Masia M, Padilla S, Bernal E, Almenar M, Molina J, Hernandez I. Influence of antiretroviral therapy on oxidative stress and cardiovascular risk: aprospective cross-sectional study in HIV-infected patients. *Clin Ther* 2007;**29**:448–55.
65. Wanchu A, Rana S, Pallikkuth S, Sachdeva R. Oxidative stress in HIV-infected individuals: a cross-sectional study. *AIDS Res Hum Retroviruses* 2009;**25**:1307–11.
66. Redhage L, Shintani A, Haas D, Emeagwali N, Markovic M, Oboho I. Clinical factors associated with plasma F2-isoprostane levels in HIV-infected adults. *HIV Clin Trials* 2009;**10**:181–92.
67. Obirikorang C, Yeboah F, Quaye L. Serum lipid profiling in highly active antiretroviral therapy-naive HIV positive patients in Ghana; any potential risk?. WebmedCentral *Infect Dis* 2010;**1**(10). WMC00987.
68. Mgbekem M, John M, Umoh I, Eyong E, Ukam N, Omotola B. Plasma antioxidant micronutrients and oxidative stress in people living with HIV. *Pakistan J Nutr* 2011;**10**(3):214–9.
69. Gil L, Tarinas A, Hernandez D, Riveron VB, Perez D, Tapanes R. Altered oxidative stress indexes related to disease progression marker in human immunodeficiency virus infected patients with antirretroviral therapy. *J Biomed Aging Pathol* 2011;**1**(1):8–15.
70. Itinoseki K, Rondó P, Luzia L, Souza J, Firmino A, Santos S. Vitamin E concentrations in adults with HIV/AIDS on highly active antiretroviral therapy. *Nutrients* 2014;**6**(9):3641–52.

CHAPTER

# 2

# Oxidative Stress and Tuberculosis–Human Immunodeficiency Virus Coinfection

*Rosario Gravier-Hernández, Lizette Gil-del Valle*

**Pedro Kourí Institute of Tropical Medicine (IPK), La Habana, Cuba**

OUTLINE



## Abstract

Human immunodeficiency virus (HIV) coinfection with *Mycobacterium tuberculosis* represents a major global health challenges worldwide. Oxidative stress (OS) has been reported in HIV, tuberculosis (TB), and HIV/TB patients either playing a role in their pathogenesis or as consequences. OS has been related to increased production of reactive oxygen species (ROS) secondary to phagocyte respiratory burst, malnutrition, and compromised immunity; also it is related to reduced antioxidant status. Immune responses to *M. tuberculosis* and the pathogenesis of TB disease are complex and not completely understood. Antioxidants pathways of *M. tuberculosis* limiting host ROS acting, some oxidative damage involved, cellular function, and how these responses change could mediate pathophysiological situations involved in HIV/TB coinfection are discussed. The goal of this review is therefore to provide a summary of available findings regarding some factors as OS influencing HIV, TB, and coinfection evolution.

**Keywords:** Antioxidants; HIV/TB coinfection; *Mycobacterium tuberculosis*; Oxidative stress; Reactive oxygen species; Tuberculosis.

## List of Abbreviations

**·OH** hydroxyl radical
**AhpC** alkyl hydroperoxidase
**ATT** antituberculosis therapy
**GSH** reduced glutathione
**$H_2O_2$** hydrogen peroxide
**HAART** high active antiretroviral therapy
**HIV** human immunodeficiency virus
**iNOS** inducible nitric oxide synthase
**KatG** catalase - peroxidase - peroxinitite enzyme system
**Lpd** dihydrolipoamide dehydrogenase
***M. tuberculosis*** *Mycobacterium tuberculosis*
**MDA** malondialdehyde



**http://dx.doi.org/10.1016/B978-0-12-809853-0.00002-X**

**MDR** multidrug-resistant *M. tuberculosis*
**MSR** methionine sulfoxide reductases
**NF-kB** nuclear transcription factor
**NO** nitric oxide
$O_2^-$ superoxide anion
**ONOO⁻** peroxynitrite
**OS** oxidative stress
**RNS** reactive nitrogen species
**ROS** reactive oxygen species
**SOD** superoxide dismutase
**SucB** dihydrolipoamide succinyltransferase
**TAS** total antioxidant status
**TB** tuberculosis
**TDR** totally drug-resistant *M. tuberculosis*
**TLRs** Toll-like receptors
**trHbs** truncated hemoglobins
**Trx** thioredoxin
**TrxR** thioredoxin reductase
**WHO** World Health Organization

# INTRODUCTION

At the start of the 21st century, tuberculosis (TB) ranks as the second leading cause of death for an infectious disease worldwide, after the human immunodeficiency virus (HIV). The World Health Organization (WHO) estimated that there were more than 9.6 million new cases of TB each year and 1.5 million TB deaths (1.1 million among HIV-negative people and 0.4 million among HIV-positive people). HIV and TB are a lethal combination, each speeding the other's progress.[1]

TB is a common and often deadly infectious disease caused by *Mycobacterium tuberculosis*. *M. tuberculosis* is a gram-positive acid-fast bacteria transmitted by aerosols of patients infected with pulmonary TB. The outcome of TB infection and disease is highly variable; exposure can be followed by rapid clearance through innate immunity, direct development of active disease, or latent infection that may or may not reactivate up to several decades following initial exposure. Active TB disease comprises a range of presentations, including classical pulmonary TB, and various forms of extrapulmonary disease.[2]

The TB cases have increased as a result of poverty, poor living conditions, malnutrition, HIV coinfection, and inadequate medical care in the developing countries. In addition, inappropriate treatment regimens and patients' poor compliance have led to the appearance of drug-resistant TB.[3]

The therapy of short duration known as directly observed treatment short (DOTS) recommended by the WHO includes drugs as isoniazid, rifampicin, pyrazinamide, streptomycin, and ethambutol. Unfortunately, strategy DOTS stops being the therapeutic option for patients infected with multidrug-resistant *M.tuberculosis* (MDR) defined as TB resistant to at least isoniazid and rifampicin. TB has assumed an even more ominous stance with the emergence of totally drug-resistant (TDR) *M. tuberculosis* strains, which are virtually untreatable. Unfortunately, there has been no new effective vaccine against TB, and in spite of introduction of several new drugs such as bedaquiline, delamanid and a new generation of fluoroquinolones such as gatifloxacin and moxifloxacin, they are yet to be involved in the current routine antituberculosis regimen, though clinical trials for combinatorial therapy along with currently used drugs are underway. The emergence of resistance against these new classes of drugs is also a likely possibility, which will result in the same problems in the future with newer group of drug-resistant strains.[3]

*M. tuberculosis* infection induces reactive oxygen species (ROS) production by activating both mononuclear and polymorphonuclear phagocytes. TB is, therefore, characterized by poor antioxidants defense that exposes to oxidative host tissue damage. *M. tuberculosis* is a recognized facultative intracellular bacterium that replicates and persists within macrophages. To survive, the bacterium must sense and adapt to the oxidative conditions. Several antioxidant defenses including a thick cell wall, small molecule thiols, and protective enzymes are known to help the bacterium withstand the oxidative stress (OS). But generation of ROS may promote tissue injury and inflammation in affected individual. This further contributes to immunosuppression, particularly in those with impaired antioxidant capacity, such as HIV-infected patients. HIV-infected patients' oxidant/antioxidant (redox) balance is also severely disturbed early in the disease. Redox imbalance is related with both infections and it is exacerbated during coinfection.[4]

This review is intended to analyze 70 original investigations and reviews articles focused on the relation of oxidative metabolism and antioxidant status in TB/HIV coinfection.

## OXIDATIVE STRESS IN TUBERCULOSIS INFECTION

Infection starts after inhalation of aerosols containing small numbers of bacilli that have been expulsed to the air by an infected individual. When the bacteria reach the lung alveoli phagocytes, macrophages and dendritic cells take these up. Interaction with innate immune response is the first line of host defense against this pathogen. Pattern recognition receptors such as Toll-like receptors (TLRs) and C-type lectins mediated *M. tuberculosis* initial interactions with host cells. TLRs signals result in the activation of nuclear transcription factor (NF)-kB and, consequently, the transcription of genes producing proinflammatory cytokines and chemokines.[5] On the other hand, both ROS and reactive nitrogen species (RNS) may also play an important role in the suppression of mycobacterial infection. The rapid release of ROS and RNS in response to phagocytic stimuli is referred as respiratory burst/oxidative burst. Both RNS and ROS can be quite harmful to intracellular pathogens. Several mechanisms determine the pathogenesis of TB, such as (1) the generation of ROS and RNS to kill the mycobacteria, (2) the mycobacteria antioxidant mechanisms to avoid death by the OS in phagocytic cells, and (3) the host antioxidant mechanisms to prevent the tissue damage.[6]

One of the bactericidal mechanisms of macrophages is the production of ROS such as superoxide anion ($O_2^-$), hydrogen peroxide ($H_2O_2$), hydroxyl radical, and singlet oxygen. These oxygen species are extremely toxic to microorganisms.[7]

In response to mycobacterial infection, parallel with ROS, another major antimicrobial pathway through inducible nitric oxide synthase (iNOS) leads to increased production of nitric oxide (NO), which further reacts with each other to produce highly reactive (peroxynitrite ($ONOO^-$)) free radicals. Antimycobacterial effects of macrophages have been found in the absence of oxidative burst suggesting the role of other mechanisms. RNS such as NO, nitrate radical ($NO_2$), and nitrate ($NO_3$) are potent effector molecules of macrophage-mediated extracellular and intracellular cytotoxicity against various microorganisms, including the mycobacteria. An early study has also shown that generation of nitrogen species by iNOS, on mouse macrophage activation, was responsible for *M. tuberculosis* growth inhibition and killing inside these cells.[5,8] Their role in human TB infection is, however, controversial. There are experimental data to suggest the putative antimicrobicidal role of NO and related RNS produced by human macrophages as well as the demonstration of high level expression of iNOS in macrophages obtained from bronchoalveolar lavage fluid of patients with active pulmonary TB.[6,9]

### Antioxidant Defense Mechanism of *Mycobacterium tuberculosis*

Bacterial pathogens have developed strategies to counteract production of ROS by the host, which interfere with the synthesis of these products, catabolize, or repair the damage caused by them. A poorly understood aspect of *M. tuberculosis* is the exact mechanism of how it detects and evades the host immune system. To survive inside macrophages, *M. tuberculosis* must withstand ROS produced by phagocyte oxidase (NOX2/gp91phox) and RNS produced by iNOS in the macrophage. The bacterium possesses several antioxidant defenses (Fig. 2.1), including protective enzymes such as catalase - peroxidase - peroxinitritase enzyme system (KatG), superoxide dismutases (SodA and SodC), peroxidase and peroxynitrite reductase complex (AhpC, AhpD, dihydrolipoamide succinyltransferase (SucB), and dihydrolipoamide dehydrogenase (Lpd)), millimolar concentration of mycothiol, and DNA-binding proteins such as Lrs2.[4]

*M. tuberculosis* thioredoxins (Trx) are involved in protection against peroxides,[10] and *trx* genes are transcribed under a variety of OS conditions, suggesting a role in *M. tuberculosis* OS response. The thioredoxin system, consisting of Trx, thioredoxin reductase (TrxR), and NADPH, is a ubiquitous redox pathway found in a wide variety of phyla. Trx is a small redox protein with two redox-active Cys residues in its active site. TrxR catalyzes the reduction of Trx, which when present in its dithiol form is the main disulfide reductase in cells. This redox system has a wide variety of biological functions, including maintaining an intracellular reduced state in the face of an oxidizing extracellular environment.[11] Trx is responsible for maintaining a reducing intracellular environment and regenerating the reduced forms of methionine sulfoxide reductases (MSR) and peroxiredoxins, the redox regulation of enzymes and regulatory proteins by oxidoreduction and the detoxification of ROS. Mtb contains three types of Trx proteins, TrxA, TrxB, and TrxC, along with one TrxR. Trx and TrxR have also been shown to reduce $H_2O_2$ and dinitrobenzenes.[10] In addition, *M. tuberculosis* have thioredoxin-like proteins named disulfide bond oxidoreductase that promote rapid disulfide formation and folding of periplasmic or secreted proteins.[12]

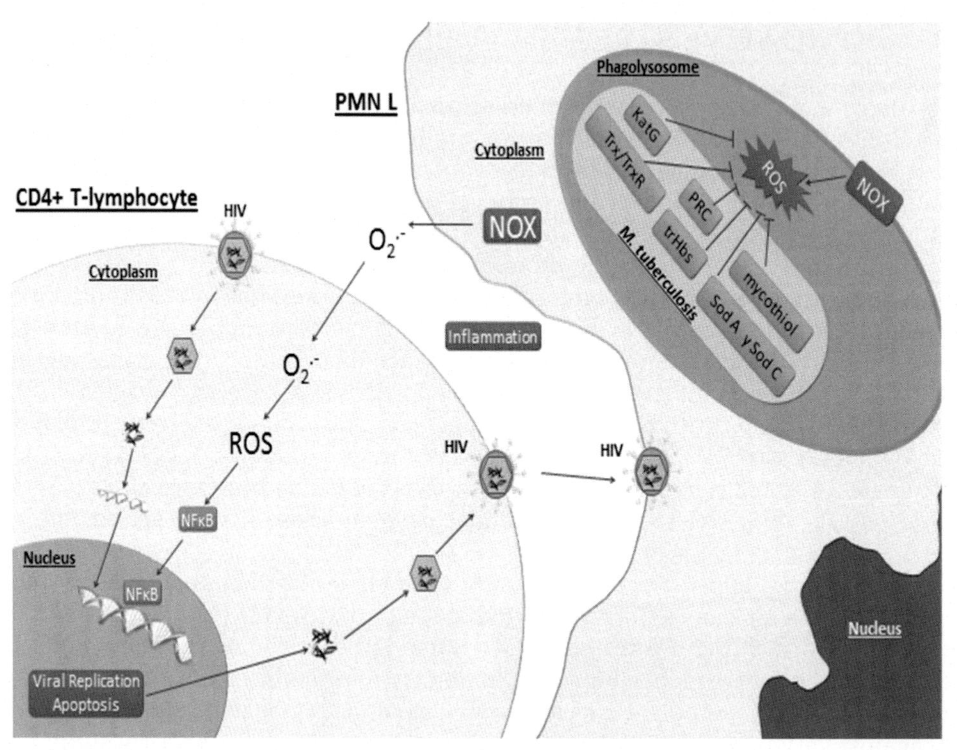


FIGURE 2.1 Redox molecular dynamic including immune system cells, HIV, and *Mycobacterium tuberculosis*. *HIV*, human immunodeficiency virus; *KatG*, catalase; *NF-κB*, nuclear transcription factor; *NOX*, nitric oxide synthase; $O_2^{\cdot -}$, singlet oxygen; *PMNL*, polymorph-nuclear lymphocyte; *PRC*, peroxynitrite reductase complex; *ROS*, reactive oxygen species; *Sod A*, superoxide dismutase A; *Sod C*, superoxide dismutase C; *trHbs*, truncated hemoglobins; *Trx/TrxR*, thioredoxin/thioredoxin reductase.

WhiB proteins were initially thought to have a role as transcriptional regulators; however, most of these proteins were found to be disulfide reductases with properties similar to Trx,[13] involved in *M. tuberculosis* OS response. It was shown that each of the *whiB* family member has a unique transcriptional response pattern to different stresses and two of these, *whiB3* and *whiB6*, were found to be commonly induced by diamide and cumene hydroperoxide.[5,14]

*M. tuberculosis* also has MSR that use NADPH, Trx, and TrxR as the system to reduce methionine sulfoxide to methionine. This pathogen produces two MSR, MsrA and MsrB, which are both required for protection against ROS and RNS.[12]

Alkyl hydroperoxidase (AhpC) is another system used by *M. tuberculosis* to combat OS.[15] *M. tuberculosis* use AhpC to detoxify by reduction alkyl hydroperoxides into less reactive alcohol derivatives. Peroxiredoxins typically use two redox-active Cys residues to reduce their substrates; however, mycobacterial AhpC contains three Cys residues that are directly involved in this catalysis. AhpC was demonstrated to confer protection against both oxidative and nitrosative stress.[15] *M. tuberculosis* Trx and TrxR are not capable of reducing AhpC[10]; AhpD, which is reduced by dihydrolipoamide and Lpd,[16] is needed for the physiological reduction of AhpC. AhpC is linked to Lpd and SucB through AhpD, which acts as an adapter protein.[16] Lpd is a component of three major enzymatic complexes: the pyruvate dehydrogenase complex, the branched amino acid dehydrogenase complex, and the peroxynitrite reductase complex. Thus, the peroxidase activity is uniquely linked to the metabolic state of *M. tuberculosis*.[12] Also, AhpC, AhpD, SucB, and Lpd were found to form a peroxynitrite reductase–peroxidase complex responsible for $ONOO^-$ detoxification.[5,16]

*M. tuberculosis* also has two superoxide dismutases: SodA a Mn–Fe enzyme and SodC a Cu–Zn enzyme. SODs are metalloproteins produced by prokaryotes and eukaryotes to detoxify superoxide radicals. They catalyze the dismutation of $O_2^{\bullet -}$ into $H_2O_2$ and molecular oxygen. SodC was found to be important for bacteria's resistance against exogenous superoxide and $H_2O_2$, as *sodC* mutants had increased susceptibility to these ROS when compared to wild-type.[17] SodC is important for *M. tuberculosis* resistance against ROS produced by activated macrophages. SodA seems to be essential for microorganism growth.[5] It is constitutively expressed under normal conditions and is demonstrated to be a major secretory protein of *M. tuberculosis*.[18] Its expression is also enhanced by $H_2O_2$ exposure and on nutrient starvation.[12,19]

Catalase peroxidases (Kat) are enzyme systems that efficiently protect the bacterium from ROS damage and are used to detoxify $H_2O_2$. *M. tuberculosis* has one enzyme, KatG that shows catalase, peroxidase, and peroxinitrite activities. The *katG* locus is genetically linked to the *furA*. The *katG* has been shown to play a role in the virulence of *M. tuberculosis*. Peroxidase activity seems to be necessary for the activation of isoniazid. Resistance to isoniazid occurs most frequently (40%–50%) through mutations in katG, most commonly Ser-315-Thr. With this mutation, the enzyme cannot activate isoniazid but still protects against host antibacterial ROS.[12]

ROS induce several cellular functions; these include induction of apoptosis through the inhibition of phosphatases and, consequently, activation of signaling molecules such as mitogen-activated protein kinases. *M. tuberculosis* has the ability to inhibit apoptosis, which was recently found to be correlated with the inhibition of ROS accumulation. Therefore, antioxidant enzymes, such as SodA and KatG, seem to be involved in *M. tuberculosis* apoptosis inhibition.[20] By affecting ROS levels, in the host cell, *M. tuberculosis* not only avoids killing but also interferes with other host defense mechanisms against pathogens.[5]

Also this microorganism possesses other nonenzymatic proteins with antioxidant capacity such as mycothiol and truncated hemoglobins (trHbs) between others that complete its antioxidant defenses.[6]

Mycothiol is involved in AhpC reduction and it is present in oxidized–reduced mycothiol (MSSM/2MSH) in millimolar quantities as the major redox buffer. Mycothiol is a low-molecular-weight thiol produced by many members of the actinomycetes, including mycobacteria. It functions like glutathione (GSH), the archetypal redox buffer for superior organism.[12]

trHbs are small heme-binding globin proteins related to, but smaller than, hemoglobin and myoglobin. trHbs are traditionally divided into three classes based on their sequence similarity: trHbN, trHbO, and trHbP. trHbs differ significantly in their sequences and could be substantially different in function, ranging from transport or storage of oxygen to detoxification of ROS and RNS. *M. tuberculosis* has two trHbs: trHbN and trHbO. trHbO has high affinity for $O_2$ because of a high association constant and a low dissociation constant. trHbO can also react with $H_2O_2$ and NO, suggesting a role in detoxification of these two compounds. trHbN was also shown to have potent NO oxidizing activity.[12,21]

In addition, *M. tuberculosis* overexpressing Wag31, a protein that regulates polar cell wall synthesis, was found to be attenuated in human macrophages and had increased susceptibility to $H_2O_2$.[22] Wag31 protects PBP3 protein (involved in synthesis of septal peptidoglycan) from proteolysis induced in OS conditions in mycobacteria. These results suggest that *M. tuberculosis* had also developed mechanisms to preserve cell shape under OS conditions, to survive.[5]

ROS also cause damage in the DNA, which means that *M. tuberculosis* should have efficient DNA repair mechanisms as well. Indeed, several transcriptomic analyses show that DNA repair genes are induced after exposure to $H_2O_2$.[23] DNA repair mechanisms have also been identified as important for microorganism survival during infection in macrophages and lung tissues.[24] *M. tuberculosis* also poses a histone-like protein known as Lsr2. Lsr2 is a DNA-binding protein that likely influences the organization of bacterial chromatin and gene regulation by binding to adenine-thymine-rich segments of DNA. Lsr2 is believed to have roles beyond gene expression. It has been shown to physically protect DNA by binding and shielding it from degradation by ROS.[5,25]

The mechanisms by which *M. tuberculosis* is able to survive and grow within the hostile environment of a host macrophage are likely to be complex and multifactorial. The mechanisms described above show that *M. tuberculosis* OS response includes not only antioxidant enzymes and other molecules that scavenge ROS but also mechanisms involved in the repair of biomolecules oxidative damages. In addition to the ability of the bacteria to influence their microenvironment within the phagosome, the ability of *M. tuberculosis* to produce agents, which counter the toxicity of antimicrobial metabolites produced by activated macrophages, is thought to be important. There are considerable evidences that the production of ROS and/or RNS by macrophages is important in the generation of effective immunity against mycobacteria. Thus, detoxification of ROS and RNS by the bacteria is expected to influence their intracellular survival.[26]

## Oxidative Stress in Tuberculosis-Infected Patients

Although these are important part of the host defense against the mycobacteria, enhanced ROS generation may promote tissue injury and inflammation in affected individual.[27,28]

Redox balance is essential for the normal lung function. Both, an increased oxidant and/or decreased antioxidant, may reverse the physiologic redox balance in favor of oxidants, leading to lung injury[29,30] (Table 2.1). Evidence suggests that increased circulating levels of ROS activity are found in pathogenesis of active pulmonary TB and hence play a role in resultant fibrosis.[27,28,30]

TABLE 2.1 Evidences of Biomolecule Oxidative Damage and Antioxidant Deficiency in Tuberculosis-Infected Patients

| Place | NI | Evaluation Criteria[a] | References |
|---|---|---|---|
| Nigeria | 100 | ↓ Vit. C, Vit. E, and GSH<br>↑ MDA | Nwanjo et al.[34] |
| India | 80 | ↓ Vit. C<br>↑ MDA | Singhahi et al.[32] |
| India | 100 | ↓ TAC<br>↑ MDA | Suresh et al.[36] |
| India | 50 | ↓ GSH and total thiols<br>↑ AOPP | Ramesh et al.[30] |
| Nigeria | 78 | ↓ SOD, CAT, TAC, and GSH<br>↑ MDA | Akiibunu et al.[35] |
| Nigeria | 115 | ↓ Vit. C, Vit. E, Se and Alb<br>↑ MDA | Oyedeji et al.[28] |

*Alb*, albumin; *AOPP*, advanced oxidation protein products; *CAT*, catalase; *GSH*, glutathione; *MDA*, malondialdehyde; *NI*, number of individuals; *Se*, selenium; *SOD*, superoxide dismutase; *Vit.*, vitamin; *TAC*, total antioxidant capacity; ↓, diminish; ↔, no change; ↑, increment.
[a]*Statistical analysis with significant difference respect control group* ($P < .05$).

Pulmonary fibrosis and dysfunction in TB are thought to be a consequence of chronic inflammatory events involving proinflammatory cytokines, activated macrophages, and ROS that stimulate fibroblast proliferation and mononuclear cell DNA damage.[27,31] The granulomatous destruction of the lung tissue itself may cause the liberation of ROS, or, indeed, it may be that the activated macrophages release ROS, which may then cause the local disruption of the essential structure including membrane lipids, deoxyribonucleic acid, and proteins and hence, cause tissue destruction.[32]

Lipid peroxidation, a general mechanism of tissue damage by ROS, is known to be responsible for cell damage and may induce many pathological events.[33] The process of lipid peroxidation is the oxidative conversion of polyunsaturated fatty acids to lipid peroxidation products of which malondialdehyde (MDA) and conjugated dienes are the most widely studied. TB patients have been reported to have increased levels of lipid peroxidation product because of impaired activity of scavenging enzymes.[31,34–36] These levels gradually decreased with antituberculosis therapy (ATT)[32,33,37] (Table 2.1).

If high levels of ROS and RNS are not detoxified, they may cause cellular damage, an important part of which is the oxidation of amino acid residues on proteins, forming protein carbonyls. Protein carbonyl content is a widely used marker of oxidative modification of proteins in inflammatory disease such as TB.[30,38]

The antioxidant homeostasis depends on multiple factors such as antioxidant enzymes, micronutrients, and dietary composition. Dietary antioxidant vitamin status is usually reduced in TB patients (Table 2.1) of lower socioeconomical status, and this could possibly explain the low levels of antioxidant potential of the patients. Moreover, the malnutrition that is commonly associated with patients with TB may further contribute to the impaired antioxidant capacity in these patients, which may result into severe OS that has been reported in TB patients due to malnutrition.[33] Several factors such as low food intake, nutrient malabsorption, inadequate nutrient release from the liver, acute phase response to infection, and inadequate availability of carrier protein may influence circulating antioxidants concentrations.[28,39] Combined effect of these factors causes OS, which is also responsible for TB pathogenesis.[40]

Total antioxidant status (TAS) of an individual is a function of dietary, enzymatic, and other systemic antioxidants. Some researchers observed that TAS in TB patients was lowered significantly as compared to control and level rises during and after ATT (Table 2.2).[29,33,41] Reddy et al. reported that the TAS serum level was significantly low in both treated and untreated TB patients in comparison to normal healthy volunteers.[33] Pawar et al.[40] and Kondaveeti et al.[42] reported that antioxidants supplementation as an adjuvant therapy increases the levels of vitamin C and E as well as TAS and reduces the MDA level (Table 2.2). This suggests that supplementation reduces OS and improves ATT effectiveness and TB patients' outcome.

Micronutrients protect cells from deleterious effects of ROS generated during disease conditions or natural body metabolism. In pulmonary TB patients micronutrients increase effectiveness of ATT, which helps in reducing lesion area in the lungs and improves the outcome of patients. Ascorbic acid and alpha tocopherol act as potents and probably the most important hydrophilic and lipophilic antioxidants, respectively. Reduced levels of vitamin C and E are associated

TABLE 2.2 Evidences of Redox Metabolism Improvement in Tuberculosis-Infected Patients Under Antituberculosis Treatment

| Place | NI | Evaluation Criteria[a] | References |
|---|---|---|---|
| India | 225 | ↓ MDA<br>↑ CAT, SOD, and TAC | Reddy et al.[67] |
| South Africa | 70 | ↑ TAC | Wiid et al.[29] |
| India | 75 | ↓ MDA<br>↑ TAC | Reddy et al.[33] |
| India[b] | 75 | ↓ MDA<br>↑ TAC, Vit. C, and Vit. E | Pawar et al.[40] |
| India | 45 | ↓ MDA<br>↔ TAC | Parchwani et al.[46] |
| India[b] | 200 | ↓ MDA<br>↑ TAC | Kondaveeti et al.[42] |
| Pakistan | 123 | ↓ MDA<br>↑ TAC and SOD | Hashmi et al.[31] |
| Ukraine | 170 | ↓ LPO<br>↑ AOS | Butov et al.[45] |

*AOS*, antioxidants; *CAT*, catalase; *GSH*, glutathione; *LPO*, lipid peroxidation products; *MDA*, malondialdehyde; *NI*, number of individuals; *SOD*, superoxide dismutase; *TAC*, total antioxidant capacity; *Vit.*, vitamin; ↓, diminish; ↔, no change; ↑, increment.
[a]*Statistical analysis with significant difference respect control group* ($P < .05$).
[b]*These patients also have antioxidant supplementation.*

with an impaired immune response.[40] In many studies, concentration of vitamin C, E, and TAS was significantly lower in newly diagnosed TB patients than healthy controls and associated with clinical severity (Table 2.2).[29,40,43] Ascorbic acid plays a major role in pulmonary antioxidant defense. The administration of nutrients such as ascorbic acid and alpha tocopherol has been shown to accelerate TB healing, based on decay cavity closure and negative sputum.[32]

Recently, a significant correlation between high oxidant concentration and low concentration of antioxidants with varying bacillary load as well as severity of disease has been shown; it was also suggested that antioxidants supplementation may prove beneficial as well as may help in fast recovery of TB patients.[6,44–47]

Even following ATT, OS has also been implicated in the pathogenesis of lung fibrosis and lung dysfunction in TB patients.[6,27,37] Furthermore, TB and OS complement each other in development of debilitating complications resulting from this precarious disease. Nutritional supplementation may show effective approach for limiting the oxidative damage and ultimately improving recuperation.[31]

## OXIDATIVE STRESS IN TUBERCULOSIS/HUMAN IMMUNODEFICIENCY VIRUS COINFECTION

The HIV/AIDS epidemic has led to a resurgence of TB. Both reactivation of latent *M. tuberculosis* infection and progressive primary TB are substantially more common in HIV-1–infected subjects.[48]

High levels of OS have been involved in HIV infection. Viral Tat protein induces an enhanced ROS production in HIV-infected patients by mitochondrial generation of $O_2^-$, which in turn may activate NF-κB,[49] thus increasing HIV transcription.[50] Several studies found that both asymptomatic HIV-infected individuals and AIDS patients had higher levels of OS, as indicated by increased plasma metabolites of lipid peroxidation and/or reduced antioxidant levels, compared with healthy controls. Abnormally high levels of ROS because of chronic immune system activation by HIV infections could lead to a decline of antioxidants defense molecules and accumulative damage of cellular components generating augmented lipid peroxidation products, oxidized proteins, and altered DNA sequences. Almost redox-implicated enzymes and molecules physiologically endogenous generated are involved in detoxification and general metabolism. Persistent OS has a dramatic impact on immunological, clinical, and nutritional status.[51] In addition, several studies relate OS with high active antiretroviral therapy (HAART) associated to mitochondrial dysfunction.[52–54] OS-mediated cell damage results in part via ROS reactions mainly due to NRTI. This antiviral class has been implicated as a cause of several insidious and sometimes irreversible chronic toxicities. Reports about HAART influences on OS indexes noticed a decrease in the antioxidant system, an increase in damaged molecules, and a failure to repair oxidative damage.[55] Ngondi et al. assessed the effect of different HAART in OS parameters and found an increase in lipid oxidation while antioxidants decrease. They showed significant differences in lineal association between some parameters as sulphydryls and albumin with CD4+ T lymphocytes count and with viral load.[52]

HIV and TB are severe global dual epidemics. Coinfection with HIV and *M. tuberculosis* increases disease progression of both diseases. Higher HIV viral loads are observed in coinfection. The high levels of inflammation and immune activation may create an optimal cytokine milieu for HIV replication. *M. tuberculosis* infection is linked to lower HIV-specific CD8+ and CD4+ T cell polyfunctionality. *M. tuberculosis* load may contribute to the loss of T-cell function. Decreased T-cell function may be a contributing factor to increased HIV disease progression in coinfection as functionally defective HIV-specific T cells may be less able to control HIV.[48]

The natural course of TB has been exacerbated by the HIV-manipulated immunological reaction, affecting macrophage function, cytokine production, and failure to contain initial or latent *M. tuberculosis* infection and disruption of granuloma. HIV induces primary or reactivated TB through the killing of CD4+ T cells within granulomas. CD4+ T cells play a major role in controlling the virulence of *M. tuberculosis* inside and outside the granulomas. HIV infection induces some functional changes in those cells decreasing their ability to contain *M. tuberculosis*. In addition, HIV replication is increased at sites of *M. tuberculosis* infection leading to an exacerbated pathological process.[56]

HIV is a powerful risk factor for development of all forms of TB including MDR and TDR-TB; for this reason, MDR and TDR-TB are often associated with higher mortality rates in HIV infected when compared with the HIV noninfected.[57]

TB and HIV act in deadly synergy. HIV infection increases the risk of TB infection on exposure, progression from latent infection to active TB, risk of death if not timely treated for both infections, and risk of recurrence even if successfully treated. Correspondingly, TB is the most common opportunistic infection and cause of mortality among people living with HIV, difficult to diagnose and treat owing to challenges related to comorbidity, pill burden, cotoxicity, and drug interactions.[57,58] Numerous MDR-TB outbreaks have been documented in HIV-positive individuals as a result of the depressed immune system and high susceptibility to infection. The prevalent hypothesis is that HIV infection favors the transmission of multidrug-resistant strains of *M. tuberculosis*.[57,59]

OS has been shown to enhance HIV replication, to induce the production of several inflammatory cytokines and to promote lymphocyte apoptosis and T-cell dysfunction and could therefore contribute to increased viral replication and progression of immunodeficiency in patients dually infected with HIV and TB.[43]

The induction of ROS production by *M. tuberculosis* contributes to immunosuppression, particularly in those patients with impaired antioxidant capacity, such as HIV-infected patients. Moreover, the malnutrition, which is commonly present in patients with TB, can add to the impaired antioxidant capacity in these patients.[33]

van Lettow reports that micronutrient malnutrition and wasting are more severe in adults with pulmonary TB who have higher HIV load (Table 2.3). The association between high-plasma HIV load and nutrient deficiencies was strongest for the major plasma carotenoids and selenium.[60]

Coinfected patients had higher concentrations of MDA than did HIV-seronegative TB patients, possibly reflecting increased OS in the former group (Table 2.3). In addition, ongoing TB infection has a greater impact on antioxidant

TABLE 2.3 Evidences of Biomolecule Oxidative Damage and Antioxidant Deficiency in Tuberculosis/Human Immunodeficiency Virus Coinfected Patients

| Place | NI | Evaluation Criteria[a] | References |
|---|---|---|---|
| Rwanda | 94 | ↓ Vit. A | Rwangabwoba et al.[68] |
| The United States | 44 | ↓ Se | Shor-Posner et al.[69] |
| Ethiopia | 195 | ↓ Vit. C, Vit. E, Vit. A, and GSH<br>↑ MDA | Madebo et al.[43] |
| Malawi | 801 | ↓ Se, total carotenoids, α tocopherol | van Lettow et al.[60] |
| The United States | 26 | ↓ GSH | Guerra et al.[63] |
| Nigeria | 100 | ↓ GSH, CAT, and Vit. C<br>↑ MDA | Awodele et al.[50] |
| India | 100 | ↓ GSH and SOD<br>↑ MDA | Gouripur et al.[70] |

*CAT*, catalase; *GSH*, glutathione; *MDA*, malondialdehyde; *NI*, number of individuals; *Se*, selenium; *SOD*, superoxide dismutase; *TAC*, total antioxidant capacity; *Vit.*, vitamin; ↓, diminish; ↔, no change; ↑, increment.
[a]*Statistical analysis with significant difference respect control group* ($P < .05$).

status on HIV infection. Madebo et al. reported that concentrations of antioxidant vitamins and of several thiol compounds were lower and concentrations of MDA were higher (Table 2.3). These findings further support a link between OS, TB, and HIV infection.[43] Awodele et al. reported that there is lower antioxidant potential and higher lipid peroxidation in HIV–TB coinfected patients as compared to the HIV patients on HAART and the seronegative patients (Table 2.3).[50]

Combination of enhanced OS and decreased concentrations of several antioxidants may have important pathogenic consequences in HIV-infected TB patients.[43]

GSH levels are decreased in patients with HIV-1 infection, and this decrease causes increased risk of developing TB in HIV-infected patients. This decrease would be associated with reduced capacity of monocytes to kill intracellular *M. tuberculosis*. Pulmonary TB patients also had significant lower blood total thiols.[30] GSH plays a major role in the maintenance of the cellular redox state. GSH scavenges peroxide species, which can be harmful to the cells. GSH also plays a role in the normal function of the immune system. Of particular interest is the ability of GSH to enhance the activation of lymphocytes that play a major role in the pathology of HIV infection. GSH has been demonstrated to be depleted in HIV-positive individuals.[61] Low GSH levels have been shown to result in the activation of NF-kB, which is necessary for active transcription of the HIV provirus.[62] Depleted GSH levels have also been shown to play a role in the apoptosis of CD4+T cells.[63] GSH also facilitates the control of growth of intracellular *M. tuberculosis* in macrophages and has direct antimycobacterial activity.[64,65] Furthermore, GSH in combination with IL-2 and IL-12 augments natural killer cell functions to control *M. tuberculosis* infection.[63,66] The deficiency of antioxidants may markedly increase OS, possibly adversely affecting the immune response and predisposing to drug toxicity.[43]

Diverse interactive pathway interconnecting HIV, TB, and immune system cells contributes to redox imbalance. OS has been recognized as an important player in diseases development. Antioxidant supplementation has been used to influence in better outcome and avoiding chronic inflammatory consequences. Future studies that completely characterize the biochemistry interaction of HIV, TB, host cells, and ROS need to be performed. Also therapeutic research related to these areas has to be encouraged.

## SUMMARY POINTS

- HIV natural evolution, TB infection, and coinfection were related with both depletion of antioxidants and increase of biomolecules oxidation.
- Endogenous ROS generated during immune activation can modulate functions and biomolecule damage in HIV/TB coinfection.
- Antioxidants are endogenous and exogenous sources, and it could be affected during coinfection.
- OS could be a cause and/or a consequence during HIV/TB coinfection.
- Sustained inflammation during coinfection could be related to OS condition.
- Whereas many ROS are host generated to kill microorganism, *M. tuberculosis* have specific antioxidant mechanism as protective enzymes with antioxidant activity, mycothiol, trHbs, and DNA-binding proteins between others to avoid damage.
- Antiretroviral and ATT used in coinfected patients could also affect redox balance.

## References


1. WHO. *Global tuberculosis report*. 2014. Available from: http://www.who.int/tb/publications/globalreport/en/.
2. Coscolla M, Gagneux S. Consequences of genomic diversity in *Mycobacterium tuberculosis*. *Semin Immunol* 2014;**26**:431–44.
3. Neel R, Gandhi N, Sarita S, Jason R, Andrews V, Vella A, et al. HIV co-infection in multidrug- and extensively drug-resistant tuberculosis results in high early mortality. *Am J Respir Crit Care Med* 2010;**181**:80–6.
4. Brugarolas P, Movahedzadeh F, Wang Y, Zhang N, Bartek I, Gao Y, et al. The oxidation-sensing regulator (MosR) is a new redoxdependent transcription factor in *Mycobacterium tuberculosis*. *J Biol Chem* 2012;**287**(45):37704–12.
5. Elviro O. *Mycobacterium tuberculosis host adaptation and evolution reflected by defense mechanisms against oxidative stress*. [PhD]. Universidad de Lisboa; 2012.
6. Verma I, Jindal S, Ganguly N. Oxidative stress in tuberculosis. In: Ganguly N, editor. *Studies on respiratory disorders, oxidative stress in applied basic research and clinical practice*. New York: Springer Science+Business Media; 2014. p. 101–14.
7. Bartos M, Falkinham J, Pavlik I. Mycobacterial catalases, peroxidases, and superoxide dismutases and their effects on virulence and isoniazid-susceptibility in mycobacteria – a review. *Vet Med* 2004;**49**(5):161–70.
8. Chan J, Xing Y, Magliozzo R, Bloom B. Killing of virulent *Mycobacterium tuberculosis* by reactive nitrogen intermediates produced by activated murine macrophages. *J Exp Med* 1992;**175**:1111–22.

9. Nicholson S, Bonecini-Almeida M, Silva J, Nathan C, Xie Q. Inducible nitric oxide synthase in pulmonary alveolar macrophages from patients with tuberculosis. *J Exp Med* 1996;**183**(5):2293–302.
10. Zhang Z, Hillas P, Ortiz de Montellano P. Reduction of peroxides and dinitrobenzenes by *Mycobacterium tuberculosis* thioredoxin and thioredoxin reductase. *Arch Biochem Biophys* 1999;**363**:19–26.
11. Hall G, Shah M, McEwan P, Laughton C, Stevens M, Westwell A, et al. Structure of *Mycobacterium tuberculosis* thioredoxin C. *Acta Cryst* 2006;**62**:1453–7.
12. Kumar A, Farhana A, Guidry L, Saini V, Hondalus M, Steyn A. *Redox homeostasis in mycobacteria: the key to tuberculosis control?*. Cambridge University Press; 2011. 39 p.
13. Alam M, Garg S, Agrawal P. Studies on structural and functional divergence among seven WhiB proteins of *Mycobacterium tuberculosis* H37Rv. *The FEBS Journal* 2009;**276**:76–93.
14. Geiman D, Raghunand T, Agarwal N, Bishai W. Differential gene expression in response to exposure to antimycobacterial agents and other stress conditions among seven *Mycobacterium tuberculosis* whiB-like genes. *Antimicrob Agents Chemotherapy* 2009;**50**:2836–41.
15. Master S, Springer B, Sander P, Boettger E, Deretic V, Timmins G. Oxidative stress response genes in *Mycobacterium tuberculosis*: role of ahpC in resistance to peroxynitrite and stage-specific survival in macrophages. *Microbiology* 2002;**148**:3139–44.
16. Bryk R, Lima C, Erdjument-Bromage H, Tempst P, Nathan C. Metabolic enzymes of mycobacteria linked to antioxidant defense by a thioredoxin-like protein. *Science* 2002;**295**:1073–7.
17. Piddington D, Fang F, Laessig T, Cooper A, Orme I, Buchmeier N. Cu,Zn superoxide dismutase of *Mycobacterium tuberculosis* contributes to survival in activated macrophages that are generating an oxidative burst. *Infect Immunity* 2001;**69**:4980–7.
18. Andersen P, Askgaard D, Ljungqvist L, Bennedsen J, Heron I. Proteins released from *Mycobacterium tuberculosis* during growth. *Infect Immun* 1991;**59**:1905–10.
19. Betts J, Lukey P, Robb L, McAdam R, Duncan K. Evaluation of a nutrient starvation model of *Mycobacterium tuberculosis* persistence by gene and protein expression profiling. *Mol Microbiol* 2002;**43**:717–31.
20. Miller J, Velmurugan K, Cowan M, Briken V. The type I NADH dehydrogenase of *Mycobacterium tuberculosis* counters phagosomal NOX2 activity to inhibit TNF-alpha-mediated host cell apoptosis. *PLoS Pathog* 2010;**6**:1000864.
21. Ouellet H, Ranguelova K, Labarre M, Wittenberg J, Wittenberg B, Magliozzo R, et al. Reaction of *Mycobacterium tuberculosis* truncated hemoglobin O with hydrogen peroxide: evidence for peroxidatic activity and formation of protein-based radicals. *J Biol Chem* 2007;**282**:7491–503.
22. Mukherjee P, Sureka K, Datta P, Hossain T, Barik S, Das K, et al. Novel role of Wag31 in protection of mycobacteria under oxidative stress. *Mol Microbiol* 2009;**73**:103–19.
23. Voskuil M, Bartek I, Visconti K, Schoolnik G. The response of *mycobacterium tuberculosis* to reactive oxygen and nitrogen species. *Front Microbiology* 2011;**2**:105.
24. Cappelli G, Volpe E, Grassi M, Liseo B, Colizzi V, Mariani F. Profiling of *Mycobacterium tuberculosis* gene expression during human macrophage infection: upregulation of the alternative sigma factor G, a group of transcriptional regulators, and proteins with unknown function. *Res Microbiology* 2006;**157**:445–55.
25. Colangeli R, Haq A, Arcus V, Summers E, Magliozzo R, McBride A, et al. The multifunctional histone-like protein Lsr2 protects mycobacteria against reactive oxygen intermediates. *Proc Natl Acad Sci USA* 2009;**106**:4414–8.
26. Springer B, Master S, Sander P, Zahrt T, Mcfalone M, Song J, et al. Silencing of oxidative stress response in *Mycobacterium tuberculosis*: expression patterns of ahpC in virulent and avirulent strains and effect of ahpC inactivation. *Infect Immun* 2001;**39**(10):5967–73.
27. Jack C, Jackson M, Hind C. Circulating makers of free radical activity in patients with pulmonary tuberculosis. *Tuber Lung Dis* 1994;**75**:132–7.
28. Oyedeji S, Adesina A, Oke O, Oguntuase N, Esan A. Oxidative stress and lipid profile status in pulmonary tuberculosis patients in South Western Nigeria. *Greener J Med Sci* 2013;**3**(6):228–32.
29. Wiid I, Seaman T, Hoal E, Benade A, Van Helden P. Total antioxidant levels are low during active TB and rise with anti-tuberculosis therapy. *IUBMB Life* 2004;**56**(2):101–6.
30. Ramesh SK, Amareshwara M, Sameer RM. Study of protein oxidation and antioxidants status in pulmonary tuberculosis patients. *Int J Pharma Bio Sci* 2011;**2**(3):104–9.
31. Hashmi M, Ahsan B, Shah S, Khan M. Antioxidant capacity and lipid peroxidation product in pulmonary tuberculosis. *Al Ame en J Med S C I* 2012;**5**(3):313–9.
32. Singhahi R, Arora D, Singh R. Oxidative stress and ascorbic acid levels in cavitary pulmonary tuberculosis. *J Clin Diagn Res* 2010;**4**:3437–41.
33. Reddy Y, Murthy S, Krishna D, Prabhakar M. Oxidative metabolic changes in pleural fluid of tuberculosis patients. *Bangladesh J Pharmacol* 2009;**4**:69–72.
34. Nwanjo H, Oze G. Oxidative imbalance and non-enzymic antioxidant status in pulmonary tuberculosis infected subjects. *Carcinogenic Potential Pakistan J Nutr* 2007;**6**(6):590–2.
35. Akiibinu M, Ogunyemi E, Shoyebo E. Levels of oxidative metabolites, antioxidants and neopterin in Nigerian pulmonary tuberculosis patients. *Eur J Gen Med* 2011;**8**(13):213–8.
36. Suresh D, Annam V, Pratibha K. Immunological correlation of oxidative stress markers in tuberculosis patients. *Int J Biol Med Res* 2010;**1**(4):185–7.
37. Kwiatkowska S, Piaseika G, Zieba M, Piotrows K, Nowak D. Increased serum concentrations of conjugated dienes and malonyldialdehyde in patients of pulmonary tuberculosis. *Respir Med* 1999;**93**(4):272–6.
38. Kulkarni A, Madrasi N. Relationship of nitric oxide and protein carbonyl in tuberculosis. *Indian J Tuberc* 2008;**55**:138–44.
39. Akiibinu M, Ogunyemi O, Arinola O. Assessment of antioxidants and nutritional status of pulmonary tuberculosis patients in Nigeria. *Eur J Gen Med* 2008;**5**(4):208–11.
40. Pawar B, Suryakar A, Khandelwal A. Effect of micronutrients supplementation on oxidative stress and antioxidant status in pulmonary tuberculosis. *Biomed Res* 2011;**22**(4):455–9.
41. Kowalski J, Janiszewska-Drobińska B, Pawlicki L, Ceglinski T, Irzmanski R. Plasma antioxidative activity in patients with pulmonary tuberculosis. *Pol Merkur Lekarski* 2004;**16**(92):119–22.
42. Kondaveeti S, Annam V, Suresh D. Oxidative stress index as a novel biochemical marker in tuberculosis; with therapeutic benefit of antioxidant supplementation. *BMC Infect Dis* 2012;**12**:66.

43. Madebo T, Lindtjorn B, Aukrust P, Berge R. Circulating antioxidants and lipid peroxidation products in untreated tuberculosis patients in Ethiopia. *Am J Clin Nutr* 2003;**78**:117–22.
44. Mohod K, Dhok A, Kumar S. Status of oxidants and antioxidants in pulmonary tuberculosis with varying bacillary load. *J Exp Sci* 2011;**2**(6):35–7.
45. Butov D, Kuzhko M, Kuznetsova I, Grinishina O, Maksimenko O, Butova T, et al. Dynamics of oxidant-antioxidant system in patients with multidrugresistant tuberculosis receiving anti-mycobacterial therapy. *J Pulm Respir Med* 2013;**3**(5):1–3.
46. Parchwani D, Singh S, Patel D. Total antioxidant status and lipid peroxides in patients with pulmonary tuberculosis. *Natl J Community Med* 2011;**2**(2):225–8.
47. Dalvi S, Patil V, Ramraje N. The roles of glutathione, glutathione peroxidase, glutathione reductase and the carbonyl protein in pulmonary and extra pulmonary tuberculosis. *J Clin Diagn Res* 2012;**6**(9):1462–5.
48. Mulu A, Kassu A, Huruy K, Ameni G. Tuberculosis - human immunodeficiency virus coinfection: bidirectional effect. *Pharmacologyonline* 2008;**2**:301–18.
49. Schreck R, Rieber P, Baeuerle P. Reactive oxygen intermediates as apparently widely used messengers in the activation of the NF-κB transcription factor and HIV-1. *EMBO J* 1991;**10**:2247–58.
50. Awodele O, Olayemi S, Nwite J, Adeyemo T. Investigation of the levels of oxidative stress parameters in HIV and HIV-TB co-infected patients. *J Infect Dev Ctries* 2012;**6**(1):79–85.
51. Guaraldi G, Orlando G, Zona S, Menozzi M, Carli F, Garlassi E, et al. Premature age-related comorbidities among HIV-infected persons compared with the general population. *Clin Infect Dis* 2011;**53**(11):1120–6.
52. Ngondi J, Oben J, Forkah D, Etame L, Mbanya D. The effect of different combination therapies on oxidative stress markers in HIV infected patients in Cameroon. *AIDS Res Ther* 2006;**3**:19–26.
53. Masiá M, Padilla S, Bernal E. Influence of antiretroviral therapy on oxidative stress and cardiovascular risk: a prospective cross-sectional study in HIV-infected patients. *Clin Ther* 2007;**29**:448–55.
54. Caron M, Boccara F, Lagathu C, Antoine B, Cervera P, Bastard J, et al. Adipose tissue as a target of HIV-1 antiretroviral drugs. Potential consequences on metabolic regulations. *Curr Pharm Des* 2010;**16**(30):3352–60.
55. Aukrust P, Muller F, Svardal A, Ueland T, Berge R, Froland S. Disturbed glutathione metabolism and decreased antioxidants levels in human immunodeficiency virus-infected patients during highly active antiretroviral therapy-potential immunomodulatory effects of antioxidants. *J Infect Dis* 2003;**188**:232–8.
56. Geldmacher C, Ngwenyama N, Schuetz A, Petrovas C, Reither K, Heeregrave E, et al. Preferential infection and depletion of *Mycobacterium tuberculosis*–specific CD4 T cells after HIV-1 infection. *J Exp Med* 2010;**207**:2869–81.
57. Girma G. Prevalence of multidrug-resistanttuberculosis and convergence of MDR-TB and HIV infection. *Glob J Microbiol Res* 2015;**3**(1):117–25.
58. Isaakidis P, Varghese B, Mansoor H, Cox H, Ladomirska J, Saranchuk P. Adverse events among HIV/MDR-TB coinfected patients receiving antiretroviral and second line anti-TB treatment in Mumbai, India. *PLoS One* 2012;**7**:40781.
59. McCray E, Onorato I. In: Bastian I, Portaels F, editors. *The interaction of human immunodeficiency virus and multidrug-resistant Mycobacterium tuberculosis*. Netherlands: Kluwer Academic Publishers; 2000.
60. van Lettow M, Harries A, Kumwenda J, Zijlstra E, Clark T, Taha T, et al. Micronutrient malnutrition and wasting in adults with pulmonary tuberculosis with and without HIV co-infection in Malawi. *BMC Infect Dis* 2004;**4**(1):1–8.
61. Buhl R, Jaffe H, Holroyd K, Wells F, Mastrangeli A. Systemic glutathione deficiency in symptom-free HIV seropositive individuals. *Lancet* 1989;**2**:1294–8.
62. Staal F, Roederer M, Herzenberg L. Intracellular thiols regulate activation of nuclear factor kB and transcription of human immunodeficiency virus. *Proc Natl Acad Sci USA* 1990;**87**:9943–7.
63. Guerra C, Morris D, Sipin A, Kung S, Franklin M, Gray D, et al. Glutathione and adaptive immune responses against *Mycobacterium tuberculosis* infection in healthy and HIV infected individuals. *PLoS One* 2011;**6**(12):28378.
64. Venketaraman V, Dayaram Y, Amin A, Ngo R, Green R, Talaue M, et al. Role of glutathione in macrophage control of Mycobacteria. *Infect Immun* 2003;**71**(4):1864–71.
65. Venketaraman V, Dayaram Y, Talaue M, Connell N. Glutathione and Nitrosoglutathione in macrophage defense against *M. tuberculosis*. *Infect Immun* 2005;**73**:1886–9.
66. Millman A, Salman M, Dayaram Y, Connell N, Venketaraman V. Natural killer cells, glutathione, cytokines and innate immunity against *Mycobacterium tuberculosis*. *J Interferon Cytokine Res* 2008;**28**:1–13.
67. Reddy Y, Murthy S, Krishna D, Prabhakar M. Role of free radicals and antioxidants in tuberculosis patients. *Indian J Tuberculosis* 2004;**51**:213–8.
68. Rwangabwoba J, Fischman H, Semba R. Serum vitamin A levels during tuberculosis and human immunodeficiency virus infection. *Int J Tuberc Lung* 1998;**2**(9):771–3.
69. Shor-Posner G, Miguez M, Pineda L, Rodriguez A, Ruiz P, Castillo G, et al. Impact of selenium status on the pathogenesis of mycobacterial disease in HIV-1-infected drug users during the Era of Highly anctive antiretroviral therapy. *J Acquir Immune Defic Syndr* 2002;**26**:169–73.
70. Gouripur T, Desai P, Vani A, Gouripur K, Patil V. Comparison of lipid peroxidation product and enzymatic antioxidants in newly diagnosed pulmonary tuberculosis patients with and without human deficiency virus infection. *Int J Pharm Bio Sci* 2012;**3**(3):391–7.

CHAPTER

# 3

# Dysfunctional High-Density Lipoprotein in Relation to Oxidative Stress and Human Immunodeficiency Virus

*Vasiliki D. Papakonstantinou*[1], *Theodoros Kelesidis*[2]

[1]Department of Science Nutrition-Dietetics, Athens, Greece; [2]David Geffen School of Medicine at UCLA, Los Angeles, CA, United States

OUTLINE



## Abstract

HIV-related cardiovascular disease (CVD) is a main cause of morbidity and mortality among HIV-infected persons, despite potent antiretroviral therapy (ART). However, the mechanisms of HIV-related CVD are unclear. Chronic HIV infection is a state of oxidative stress driven by several mediators including HIV per se, bacterial translocation, ART, mitochondrial dysfunction, and inflammation. Oxidative stress contributes to oxidation of lipoproteins. These oxidized lipoproteins carry oxidized lipids that have pleotropic proinflammatory and immunomodulatory effects. Oxidized low-density lipoprotein is the main oxidized lipoprotein that has been implicated in pathogenesis of atherosclerosis. However, in states of systemic inflammation and oxidative stress (such as chronic HIV infection), high-density lipoprotein (HDL) also gets oxidized and loses its antiinflammatory and antioxidant function. This oxidized HDL (ox-HDL) becomes proinflammatory, prooxidant, and dysfunctional and can also promote atherogenesis. However, the role of ox-HDL in chronic HIV infection remains poorly understood. This chapter summarizes available scientific evidence regarding the role of ox-HDL in HIV immunopathogenesis.

**Keywords:** (ox)HDL; (ox)LDL; Cardiovascular disease; Human immunodeficiency virus; Oxidation; Oxidative stress.

## List of Abbreviations

**ABCA-1** ATP-binding cassette-1
**AIDS** Acquired immune deficiency syndrome
**ART** Antiretroviral therapy
**Cav-1** Caveolin-1
**CIMT** Carotid intima-media thickness
**CVD** Cardiovascular disease
**GPx** Glutathione peroxidase



HIV/AIDS
http://dx.doi.org/10.1016/B978-0-12-809853-0.00003-1

**HDL** High-density lipoprotein
**HIV** Human immunodeficiency virus
**IL** Interleukin
**LCAT** Lecithin–cholesterol acyltransferase
**LDL** Low-density lipoprotein
**LPS** Lipopolysaccharide
**Ox-** Oxidized-
**PAF** Platelet-activating factor
**PAF-AH** PAF-acetylhydrolase
**PI** Protease inhibitor
**PLs** Phospholipids
**PON** Paraoxonase
**ROS** Reactive oxygen species
**TNF-α** Tumor necrosis factor-alpha

## INTRODUCTION

HIV-related cardiovascular disease (CVD) is a main cause of morbidity and mortality among HIV population,[1] despite the great success of antiretroviral therapy (ART) to diminish the appearance of AIDS and increase the life span of HIV-infected persons.[2] Indeed, the risk of atherosclerotic coronary heart disease and other cardiac abnormalities is increased among HIV-infected persons.[3] The exact molecular mechanism is not clear, but immune activation, inflammation, ART, and viremia have been implicated (Fig. 3.1).[4] Elucidating the mechanisms of HIV-related CVD will pave the way to develop new strategies for prevention and treatment.

Chronic HIV infection is a state of chronic oxidative stress.[5] Oxidative stress derives from an imbalance of homeostasis between prooxidant and antioxidant systems, which causes the generation of reactive oxygen species (ROS).[6] ROS are highly involved in HIV infection,[7] can affect cellular responses,[8] and can lead to cellular apoptosis.[9] However, the mechanisms by which chronic HIV infection induces oxidative stress are unclear. There is a plethora of theories involving HIV per se,[10] bacterial translocation,[11] ART,[12] mitochondrial dysfunction,[13] and inflammation.[14]

Oxidative stress contributes to formation of oxidized lipids that have pleotropic proinflammatory and immunomodulatory effects. However, oxidized lipids per se are also prooxidant.[15] Generally, oxidized lipids release both ROS and complex products, which induce monocyte migration and exert cytotoxic effects. The usual types of oxidized lipids that are produced are simple hydroxy fatty acids, $NO_2$ fatty acids, oxidized cholesterol species, various isoprostanes, lipid aldehydes, and oxidized phospholipids (ox-PLs) (Fig. 3.2).[16] Phospholipids (PLs) are one of the main types of lipids and major components of cell membranes that have an important role in cell biochemistry. Lipid peroxidation can derive either from enzymatic (lipoxygenases, cyclooxygenases, and cytochrome P450) or nonenzymatic (ROS and free radicals) mechanisms. Generally, enzymatic formation of ox-PLs is strictly regulated while the nonenzymatic formation of ox-PLs is poorly controlled. The oxidation of PLs has several physiological and pathological effects (via receptor and nonreceptor mechanisms) on endothelial cells, smooth muscle cells, platelets, and macrophages (Fig. 3.1). Ox-PLs can act as proinflammatory mediators by stimulating extravasation and activation of

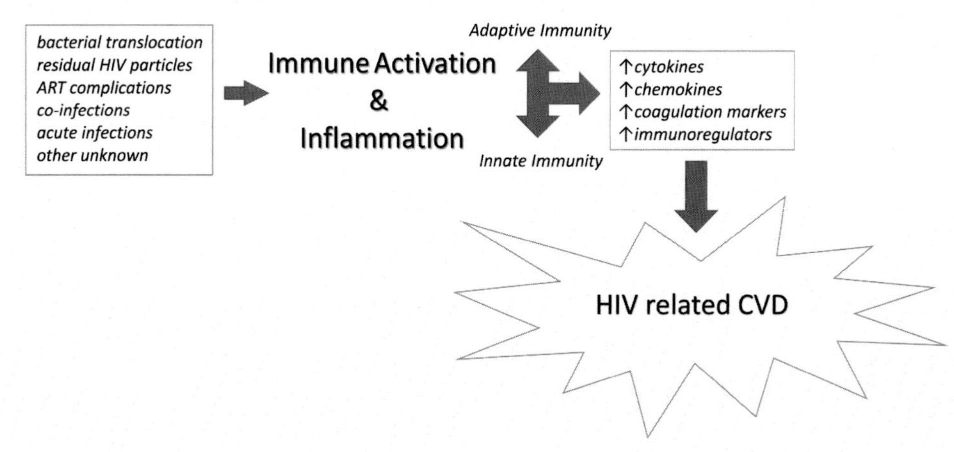


FIGURE 3.1 The causes of HIV-related cardiovascular disease. *ART*, antiretroviral therapy; *CVD*, cardiovascular disease; *HIV*, human immunodeficiency virus.

monocytes, upregulation of proinflammatory cytokines, enhanced thrombogenesis, and increased oxidative stress. However, they also act as antiinflammatory mediators via their role in endothelial barrier function, their protective effects against endotoxin (lipopolysaccharide)-induced inflammation, and through numerous antiinflammatory mechanisms.[17,18]

Lipoproteins are rich in PLs containing polyunsaturated fatty acids, which are highly prone to oxidative modification, forming oxidized lipoproteins. The modification of low-density lipoprotein (LDL) to oxidized LDL (ox-LDL) is far more studied than the oxidation of high-density lipoprotein (HDL). The oxidation of LDL is mediated by the products of lipid peroxidation found in plasma and in the subendothelial space. Ox-LDL is responsible for the recruitment of inflammatory mediators, such as platelet-activating factor (PAF), migration of monocytes into the subendothelial space during acute inflammatory processes, foam production, and atherogenesis (Fig. 3.3).[19,20] Moreover, ox-LDL induces apoptosis of vascular endothelial cells through mitochondrial pathways.[21]

HDL counteracts effects of ox-LDL and holds cardioprotective role, as it removes cholesterol and PLs from cells through reverse cholesterol transport.[22] Additionally, normal HDL protects against LDL-induced cytotoxicity,[23] removes ROS from LDL, inhibits the oxidation of PLs in LDL, and attenuates effects of ox-LDL and inflammatory mediators such as tumor necrosis factor-alpha and interleukin-1 (Fig. 3.4).[24,25] HDL has also immunomodulatory role in monocytes, macrophages, dendritic cells, T cells, and B cells (Fig. 3.5).[26] Generally, the beneficial role of HDL against CVD is due to apoA-I protein and the enzymes it carries such as platelet-activating factor acetylhydrolase (PAF-AH), paraoxonase (PON), lecithin–cholesterol acyltransferase, glutathione peroxidase.[27,28] Addition of apoA-I or apoA-I peptide mimetic or normal HDL or PON to human artery wall cells can prevent LDL oxidation. However, HDL from patients with coronary artery disease fails to protect LDL from oxidation[24,25] and can also induce proinflammatory and proatherogenic properties, amplifying vascular inflammation.[29] In particular, during an acute inflammatory response the protein and lipid composition of HDL is altered.[30] PON and PAF-AH activities are reduced while ceruloplasmin, serum amyloid A, and apo-J levels are increased. In these cases, it seems that HDL becomes dysfunctional, loses its antioxidant and antiinflammatory abilities, and promotes lipid peroxidation and LDL oxidation.[31] These activities are associated with impaired reverse cholesterol transport,

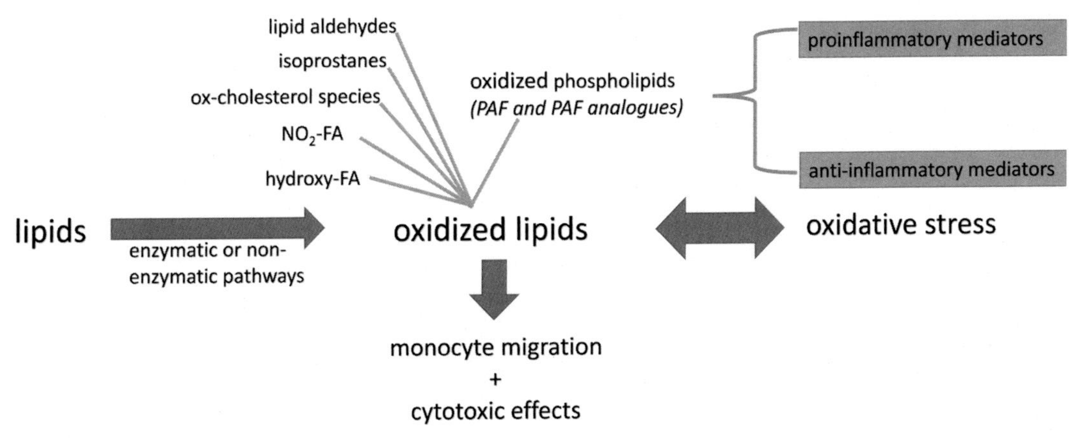


FIGURE 3.2 The cross-talk between oxidative stress and lipids contributes to formation of oxidized lipids that have pleotropic proinflammatory and immunomodulatory effects. *PAF*, platelet-activating factor.

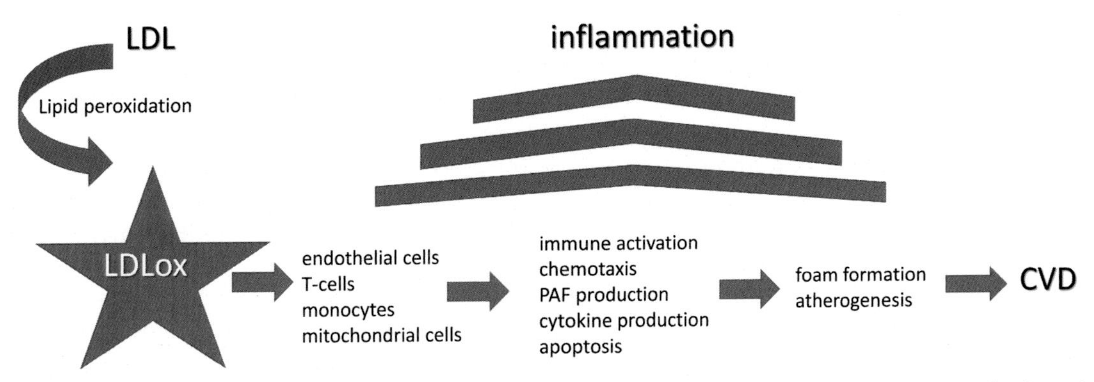


FIGURE 3.3 Proatherogenic effects of oxidized low-density lipoprotein (ox-LDL). *CVD*, cardiovascular disease; *PAF*, platelet-activating factor.

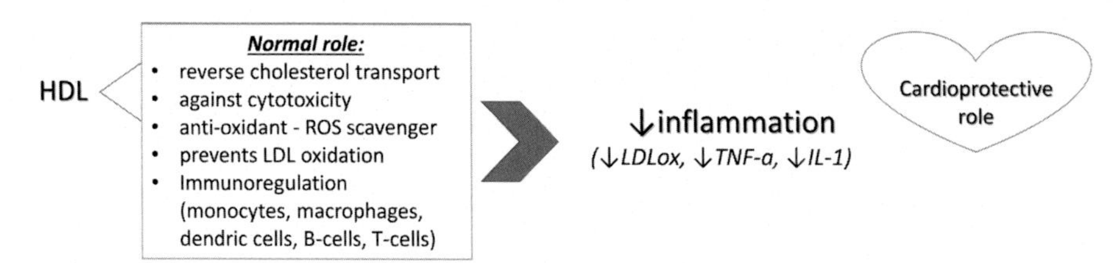


FIGURE 3.4 Antiinflammatory and antiatherogenic effects of high-density lipoprotein (HDL). *IL*, interleukin; *LDL*, low-density lipoprotein; *ox-LDL*, oxidized low-density lipoprotein; *ROS*, reactive oxygen species; *TNF-α*, Tumor necrosis factor-alpha.

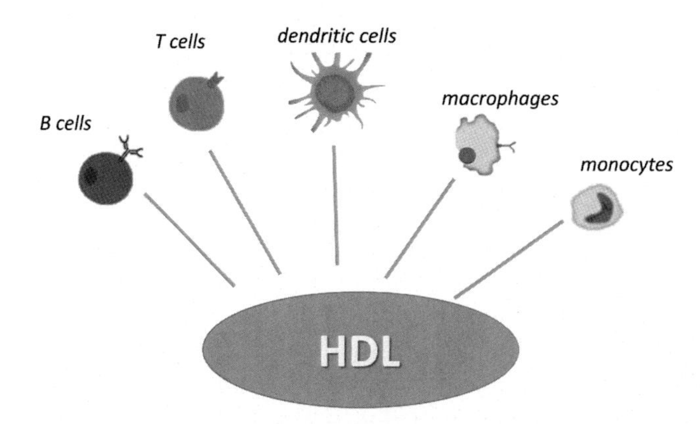


FIGURE 3.5 High-density lipoprotein (HDL) mediated effects on immune cells.

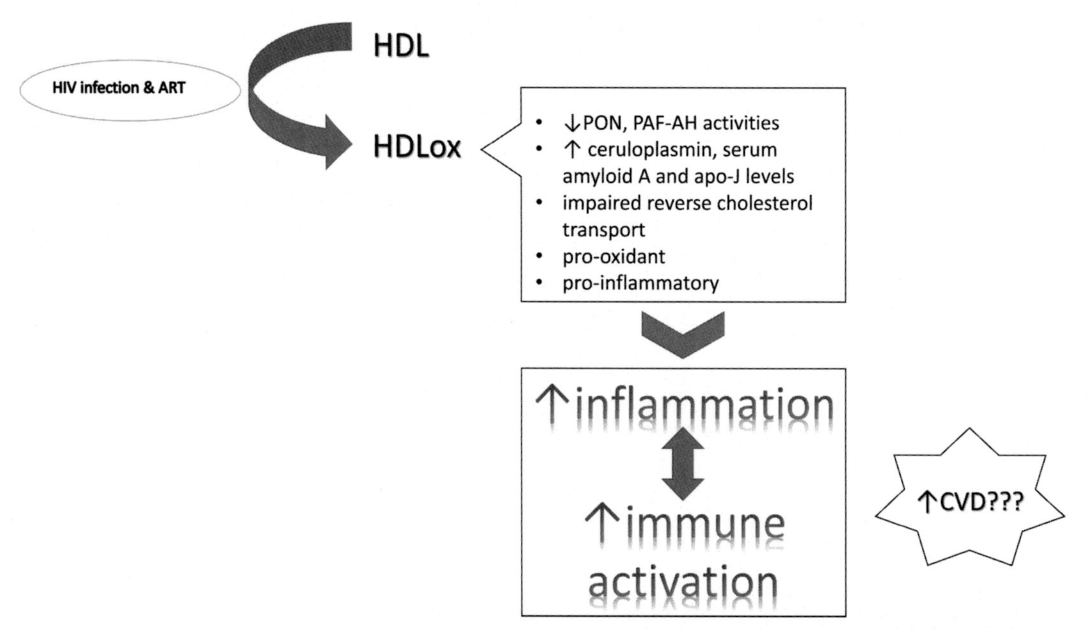


FIGURE 3.6 Prooxidant, proinflammatory, proatherogenic, and immunomodulatory effects of oxidized high-density lipoprotein (ox-HDL). *ART*, antiretroviral therapy; *CVD*, cardiovascular disease; *PAF-AH*, platelet-activating factor acetylhydrolase; *PON*, paraoxonase.

meaning that dysfunctional HDL cannot transfer cholesterol successfully and therefore atherogenic lipoproteins and oxidized lipids accumulate in the cells (Fig. 3.6).[32] Dysfunctional HDL and CVD have been identified in several inflammatory conditions as systemic lupus erythematosus, type II diabetes mellitus, metabolic syndrome, Crohn's disease, and rheumatoid arthritis.[19]

HIV infection is a state of dyslipidemia,[33] immune activation, and inflammation. However, a significant reduction of HDL antioxidant/antiinflammatory activities has been found.[34] Additionally, HIV-infected patients were found to have increased HDL redox activity compared to matched healthy volunteers with comparable HDL levels.[35] Also, oxidized lipoproteins of HIV-infected subjects were associated with markers of inflammation and immune activation.[36] Thus, dysfunctional HDL appears to have an important role in (1) HIV-related CVD, (2) HIV-related inflammation, and (3) HIV-related immune dysfunction. Therefore, the study of dysfunctional HDL may be promising for reduction of CVD in HIV population.

## OXIDIZED LIPOPROTEINS AND IMMUNE SYSTEM IN HUMAN IMMUNODEFICIENCY VIRUS

HIV-1 infection is characterized by chronic inflammation and immune activation. Potent ART can normalize CD4+ counts, diminish viral load, and prolong the life expectancy of HIV patients. However, residual inflammation and immune activation despite potent ART may lead to increased morbidity and mortality in HIV-infected persons. Although several markers such as PAF, cytokines/chemokines, and plasma markers of immune activation correlate with HIV-related comorbidities, the exact pathogenesis of HIV-related comorbidities is unknown.[37]

Generally, untreated HIV patients carry an atherogenic lipid profile, with high total cholesterol, triglycerides and LDL, low HDL, and insulin resistance.[38,39] As a consequence of oxidative stress, lipids and PLs get oxidized and obtain novel biological properties including immunomodulatory effects.[40] Therefore, lipoproteins, LDL and HDL, which carry the oxidized lipids, get modified and contribute to the pathogenesis of many diseases as CVDs.[41] Interestingly, HIV patients who receive ART show higher values in markers of oxidative stress compared to healthy subjects or naïve patients because of increased ROS production and alterations in antioxidant systems.[42] In chronic HIV infection, oxidized lipoproteins may contribute to immune activation and can induce T-cell activation. Oxidized lipoproteins correlated with several plasma biomarkers of inflammation and coagulation in ART-treated HIV patients.[36] Interestingly, ox-LDL and oxidized HDL (ox-HDL) display differential immunoregulatory effects during the first years of ART initiation.

Generally, ox-LDL is a proinflammatory lipoprotein and is far more studied than ox-HDL. HIV-infected subjects have increased levels of plasma ox-LDL despite potent ART.[43] LDL binds endotoxin, which is present in HIV-infected subjects and may also contribute to immune activation.[44] Ox-LDL correlated to several markers of immune activation, while in vitro stimulation with ox-LDL may expand inflammatory monocytes.[45] In addition, rosuvastatin administration decreased levels of ox-LDL and markers of monocyte activation.[46]

On the other hand, HDL is an antiinflammatory lipoprotein that generally prevents the oxidation of PLs. Although HDL may have a protective role in HIV-infected individuals,[47] numerous protein and lipid compositional changes in HDL occur in inflammatory states such as HIV infection.[48] Ox-HDL may also affect monocytes and dendritic cell function during other inflammatory conditions apart from HIV infection.[49] Interestingly, the lipid content of HDL particles is more prone to oxidation compared to LDL when human plasma is exposed to ROS.[50] It is known that during HIV infection, HDL becomes dysfunctional; however, the molecular changes and mechanisms that promote antiinflammatory HDL conversion to proinflammatory HDL are currently unknown. Therefore in HIV patients HDL displays impaired antioxidant activity and increased lipid peroxide content.[34,43] In contrast to ox-LDL, ox-HDL levels are decreased after initiation of ART. In addition, ox-HDL correlated positively to several markers of inflammation and immune activation.[36,51] Thus, oxidized lipoproteins contribute to HIV-reported immune activation.[36]

## HUMAN IMMUNODEFICIENCY VIRUS HIGH-DENSITY LIPOPROTEIN FUNCTION AND CARDIOVASCULAR DISEASE

HDL is the most powerful independent negative predictor of CVD. The structure and function of HDL may be able to predict atherosclerosis in chronic HIV infection. HIV-infected subjects have dysfunctional HDL, which is decreased after initiation of ART in ART-naïve HIV-infected subjects with low cardiovascular risk.[52,53] There are mixed results regarding associations of ox-HDL and subclinical atherosclerosis in HIV-infected patients with low cardiovascular risk on potent ART. We have previously described a statistically significant positive relationship between HDL redox activity and the percentage of noncalcified coronary atherosclerotic plaque.[35] On the other hand, in other studies there are no clear correlations between dysfunctional HDL and subclinical atherosclerosis progression.[43] It is possible that ox-HDL may have differential role on carotid intima-media thickness (CIMT) compared to calcified plaque.[52] Larger prospective studies that determine both calcified plaque and CIMT are needed to define the role of ox-HDL in HIV-related CVD. However, the role of ox-HDL in HIV-related CVD remains largely unknown.

Ox-HDL is proinflammatory, prooxidant, and has impaired functions such as reduced cholesterol efflux and impaired antiinflammatory and antioxidant functions.[54] Inside the arterial wall, the accumulated cholesterol is deposited in the macrophages and thus leads to atherosclerosis. Reverse cholesterol transport is a physiological pathway, which transports cholesterol from the macrophages to the liver for excretion in the bile. ATP-binding cassette-1 (ABCA-1) eases the efflux of cellular cholesterol and mediates the maturation and enlargement of HDL particles. The cholesterol efflux process is an important component of the reverse cholesterol transport, which requires both active transporters as ABCA-1 and acceptor lipoprotein particles. Studies have shown that the reverse cholesterol transport is impaired in HIV patients not only from the ART (such as protease inhibitors), or the high inflammation levels, but also from the virus. HIV-1 and/or its proteins are responsible for changes in cholesterol efflux capacity during HIV infection, and ART can restore cholesterol efflux capacity.[55] HIV protein Nef can induce the degradation of ABCA-1 and therefore reverse the cholesterol efflux from macrophages into foam cells.[39] However, there is a counterbalance mechanism between Nef and caveolin-1, a scaffolding protein and component of caveolae. Normally caveolin-1 is implicated in a plethora of physiological cell processes such as cell cycle regulation, signal transduction, endocytosis, cholesterol trafficking, and efflux and thus holds an important role in cholesterol homeostasis. However, during HIV infection the viral protein Tat induces Cav-1 expression. The excess of Cav-1 acts favorably for the host as it manages to bind to Nef and stimulate cholesterol efflux. Subsequently, as cholesterol is removed from the cells, atherosclerosis is avoided and thereby HIV infectivity is reduced as well.[56] The activation of cholesterol efflux suppresses both HIV replication and infectivity.[57] Therefore, the interplay between HIV, lipoproteins, and cholesterol efflux may contribute to HIV-related CVD.

## CONCLUSIONS

HIV-related CVD is a main cause of morbidity and mortality among HIV-infected persons, despite potent ART. However, the mechanisms of HIV-related CVD are unclear. Chronic HIV infection is a state of oxidative stress that may contribute to oxidation of lipoproteins. These oxidized lipoproteins carry oxidized lipids that have pleotropic proinflammatory and immunomodulatory effects. Ox-LDL is the main oxidized lipoprotein that has been implicated in pathogenesis of atherosclerosis. However, in states of systemic inflammation and oxidative stress (such as chronic HIV infection), HDL also gets oxidized and loses its antiinflammatory and antioxidant function. This ox-HDL becomes proinflammatory, prooxidant, and dysfunctional and can also promote atherogenesis. However, the role of ox-HDL in chronic HIV infection remains poorly understood. In contrast to ox-LDL, ox-HDL levels are decreased after initiation of ART. In addition, in both ART-naïve and ART-treated HIV-infected patients, ox-HDL correlated positively to several markers of inflammation and immune activation.[36,51] Since HDL is the most powerful independent negative predictor of CVD, the structure and function of HDL may be able to predict atherosclerosis in chronic HIV infection. HIV-infected subjects have dysfunctional HDL, which is decreased after initiation of ART in ART-naïve HIV-infected subjects with low cardiovascular risk. There are mixed results regarding associations of ox-HDL and subclinical atherosclerosis in HIV-infected patients with low cardiovascular risk on potent ART. Overall complex interactions between HIV-1, ART, cholesterol metabolism, immune system, and lipoproteins may contribute to HIV-related CVD. More studies are needed to determine the role of ox-HDL in atherosclerosis and immune activation associated with HIV infection.

## SUMMARY POINTS

- HIV-related CVD remains the main morbidity and cause of death among HIV population.
- The exact molecular mechanism of HIV-related CVD is not clear, but oxidative stress and oxidized lipoproteins may contribute to pathogenesis of HIV-related CVD.
- Chronic HIV infection is a state of oxidative stress that may contribute to formation of oxidized lipoproteins that carry oxidized lipids with pleotropic proinflammatory and immunomodulatory effects.
- Ox-LDL is the main oxidized lipoprotein that has been implicated in pathogenesis of atherosclerosis.
- In states of systemic inflammation and oxidative stress (such as chronic HIV infection), high-density lipoprotein gets oxidized (ox-HDL), loses its antiinflammatory and antioxidant function, becomes proinflammatory, prooxidant, and dysfunctional, and can also promote atherogenesis.
- In both ART-naïve and ART-treated HIV-infected patients, ox-HDL correlated positively to several markers of inflammation and immune activation.
- Since HDL is the most powerful independent negative predictor of CVD, the structure and function of HDL may be able to predict atherosclerosis in chronic HIV infection.

- There are mixed results regarding associations of ox-HDL and subclinical atherosclerosis in HIV-infected patients with low cardiovascular risk on potent ART.
- Overall complex interactions between HIV-1, oxidative stress, ART, cholesterol metabolism, immune system, and lipoproteins may contribute to HIV-related CVD.
- More studies are needed to determine the role of ox-HDL in HIV-related atherosclerosis and immune activation.

## References


1. Currier JS. Update on cardiovascular complications in HIV infection. *Top HIV Med* 2009;**17**(3):98–103.
2. Deeks SG, Lewin SR, Havlir DV. The end of AIDS: HIV infection as a chronic disease. *Lancet* 2013;**382**(9903):1525–33.
3. Currier JS, Lundgren JD, Carr A, et al. Epidemiological evidence for cardiovascular disease in HIV-infected patients and relationship to highly active antiretroviral therapy. *Circulation* 2008;**118**(2):e29–35.
4. Guaraldi G, Orlando G, Zona S, et al. Premature age-related comorbidities among HIV-infected persons compared with the general population. *Clin Infect Dis* 2011;**53**(11):1120–6.
5. Pace GW, Leaf CD. The role of oxidative stress in HIV disease. *Free Radic Biol Med* 1995;**19**(4):523–8.
6. Rahal A, Kumar A, Singh V, et al. Oxidative stress, prooxidants, and antioxidants: the interplay. *Biomed Res Int* 2014;**2014**:761264.
7. Schreck R, Rieber P, Baeuerle PA. Reactive oxygen intermediates as apparently widely used messengers in the activation of the NF-kappa B transcription factor and HIV-1. *EMBO J* 1991;**10**(8):2247–58.
8. Louboutin JP, Strayer D. Role of oxidative stress in HIV-1-associated neurocognitive disorder and protection by gene delivery of antioxidant enzymes. *Antioxidants (Basel)* 2014;**3**(4):770–97.
9. Kannan K, Jain SK. Oxidative stress and apoptosis. *Pathophysiology* 2000;**7**(3):153–63.
10. Sacktor N, Haughey N, Cutler R, et al. Novel markers of oxidative stress in actively progressive HIV dementia. *J Neuroimmunol* 2004;**157**(1–2):176–84.
11. Dubourg G, Lagier JC, Hue S, et al. Gut microbiota associated with HIV infection is significantly enriched in bacteria tolerant to oxygen. *BMJ Open Gastroenterol* 2016;**3**(1):e000080.
12. Hulgan T, Morrow J, D'Aquila RT, et al. Oxidant stress is increased during treatment of human immunodeficiency virus infection. *Clin Infect Dis* 2003;**37**(12):1711–7.
13. Ma R, Yang L, Niu F, Buch S. HIV tat-mediated induction of human brain microvascular endothelial cell apoptosis involves endoplasmic reticulum stress and mitochondrial dysfunction. *Mol Neurobiol* 2016;**53**(1):132–42.
14. Baliga RS, Chaves AA, Jing L, Ayers LW, Bauer JA. AIDS-related vasculopathy: evidence for oxidative and inflammatory pathways in murine and human AIDS. *Am J Physiol Heart Circ Physiol* 2005;**289**(4):H1373–80.
15. Yoshida Y, Umeno A, Akazawa Y, Shichiri M, Murotomi K, Horie M. Chemistry of lipid peroxidation products and their use as biomarkers in early detection of diseases. *J Oleo Sci* 2015;**64**(4):347–56.
16. Ayala A, Munoz MF, Arguelles S. Lipid peroxidation: production, metabolism, and signaling mechanisms of malondialdehyde and 4-hydroxy-2-nonenal. *Oxid Med Cell Longev* 2014;**2014**:360438.
17. Ashraf MZ, Kar NS, Podrez EA. Oxidized phospholipids: biomarker for cardiovascular diseases. *Int J Biochem Cell Biol* 2009;**41**(6):1241–4.
18. Aldrovandi M, O'Donnell VB. Oxidized PLs and vascular inflammation. *Curr Atheroscler Rep* 2013;**15**(5):323.
19. Navab M, Reddy ST, Van Lenten BJ, Anantharamaiah GM, Fogelman AM. The role of dysfunctional HDL in atherosclerosis. *J Lipid Res* 2009;**50**(Suppl.):S145–59.
20. Demopoulos CA, Karantonis HC, Antonopoulou S. Platelet activating factor- a molecular link between atherosclerosis theories. *Eur J Lipid Sci Technol* 2003;**105**(11):705–16.
21. Vindis C, Elbaz M, Escargueil-Blanc I, et al. Two distinct calcium-dependent mitochondrial pathways are involved in oxidized LDL-induced apoptosis. *Arterioscler Thromb Vasc Biol* 2005;**25**(3):639–45.
22. Oram JF, Yokoyama S. Apolipoprotein-mediated removal of cellular cholesterol and phospholipids. *J Lipid Res* 1996;**37**(12):2473–91.
23. Hessler JR, Robertson Jr AL, Chisolm 3rd GM. LDL-induced cytotoxicity and its inhibition by HDL in human vascular smooth muscle and endothelial cells in culture. *Atherosclerosis* 1979;**32**(3):213–29.
24. Navab M, Hama SY, Anantharamaiah GM, et al. Normal high density lipoprotein inhibits three steps in the formation of mildly oxidized low density lipoprotein: steps 2 and 3. *J Lipid Res* 2000;**41**(9):1495–508.
25. Navab M, Hama SY, Cooke CJ, et al. Normal high density lipoprotein inhibits three steps in the formation of mildly oxidized low density lipoprotein: step 1. *J Lipid Res* 2000;**41**(9):1481–94.
26. Catapano AL, Pirillo A, Bonacina F, Norata GD. HDL in innate and adaptive immunity. *Cardiovasc Res* 2014;**103**(3):372–83.
27. Watson AD, Berliner JA, Hama SY, et al. Protective effect of high density lipoprotein associated paraoxonase. Inhibition of the biological activity of minimally oxidized low density lipoprotein. *J Clin Invest* 1995;**96**(6):2882–91.
28. Watson AD, Navab M, Hama SY, et al. Effect of platelet activating factor-acetylhydrolase on the formation and action of minimally oxidized low density lipoprotein. *J Clin Invest* 1995;**95**(2):774–82.
29. Fogelman AM. When good cholesterol goes bad. *Nat Med* 2004;**10**(9):902–3.
30. Rohrer L, Hersberger M, von Eckardstein A. High density lipoproteins in the intersection of diabetes mellitus, inflammation and cardiovascular disease. *Curr Opin Lipidol* 2004;**15**(3):269–78.
31. Ansell BJ, Navab M, Hama S, et al. Inflammatory/antiinflammatory properties of high-density lipoprotein distinguish patients from control subjects better than high-density lipoprotein cholesterol levels and are favorably affected by simvastatin treatment. *Circulation* 2003;**108**(22):2751–6.
32. Navab M, Ananthramaiah GM, Reddy ST, et al. The double jeopardy of HDL. *Ann Med* 2005;**37**(3):173–8.
33. El-Sadr WM, Mullin CM, Carr A, et al. Effects of HIV disease on lipid, glucose and insulin levels: results from a large antiretroviral-naive cohort. *HIV Med* 2005;**6**(2):114–21.

34. Kelesidis T, Yang OO, Currier JS, Navab K, Fogelman AM, Navab M. HIV-1 infected patients with suppressed plasma viremia on treatment have pro-inflammatory HDL. *Lipids Health Dis* 2011;**10**:35.
35. Zanni MV, Kelesidis T, Fitzgerald ML, et al. HDL redox activity is increased in HIV-infected men in association with macrophage activation and non-calcified coronary atherosclerotic plaque. *Antivir Ther* 2014;**19**(8):805–11.
36. Kelesidis T, Jackson N, McComsey GA, et al. Oxidized lipoproteins are associated with markers of inflammation and immune activation in HIV-1 infection. *AIDS* 2016;**30**(17):2625–33.
37. Kelesidis T, Papakonstantinou V, Detopoulou P, et al. The role of platelet-activating factor in chronic inflammation, immune activation, and comorbidities associated with HIV infection. *AIDS Rev* 2015;**17**(4):191–201.
38. Grunfeld C, Pang M, Doerrler W, Shigenaga JK, Jensen P, Feingold KR. Lipids, lipoproteins, triglyceride clearance, and cytokines in human immunodeficiency virus infection and the acquired immunodeficiency syndrome. *J Clin Endocrinol Metab* 1992;**74**(5):1045–52.
39. Asztalos BF, Matera R, Horvath KV, et al. Cardiovascular disease-risk markers in HIV patients. *J AIDS Clin Res* 2014;**5**(7).
40. Hansson GK, Hermansson A. The immune system in atherosclerosis. *Nat Immunol* 2011;**12**(3):204–12.
41. Tsimikas S, Miller YI. Oxidative modification of lipoproteins: mechanisms, role in inflammation and potential clinical applications in cardiovascular disease. *Curr Pharm Des* 2011;**17**(1):27–37.
42. Sharma B. Oxidative stress in HIV patients receiving antiretroviral therapy. *Curr HIV Res* 2014;**12**(1):13–21.
43. Kelesidis T, Tran TT, Stein JH, et al. Changes in inflammation and immune activation with atazanavir-, raltegravir-, darunavir-based initial antiviral therapy: ACTG 5260s. *Clin Infect Dis* 2015;**61**(4):651–60.
44. Hossain E, Ota A, Karnan S, et al. Lipopolysaccharide augments the uptake of oxidized LDL by up-regulating lectin-like oxidized LDL receptor-1 in macrophages. *Mol Cell Biochem* 2015;**400**(1–2):29–40.
45. Zidar DA, Juchnowski S, Ferrari B, et al. Oxidized LDL levels are increased in HIV infection and may drive monocyte activation. *J Acquir Immune Defic Syndr* 2015;**69**(2):154–60.
46. Hileman CO, Turner R, Funderburg NT, Semba RD, McComsey GA. Changes in oxidized lipids drive the improvement in monocyte activation and vascular disease after statin therapy in HIV. *AIDS* 2016;**30**(1):65–73.
47. Calmy A, Montecucco F, James RW. Evidence on the protective role of high-density lipoprotein (HDL) in HIV-infected individuals. *Curr Vasc Pharmacol* 2015;**13**(2):167–72.
48. Siegel MO, Borkowska AG, Dubrovsky L, et al. HIV infection induces structural and functional changes in high density lipoproteins. *Atherosclerosis* 2015;**243**(1):19–29.
49. Skaggs BJ, Hahn BH, Sahakian L, Grossman J, McMahon M. Dysfunctional, pro-inflammatory HDL directly upregulates monocyte PDGFRbeta, chemotaxis and TNFalpha production. *Clin Immunol* 2010;**137**(1):147–56.
50. Bowry VW, Stanley KK, Stocker R. High density lipoprotein is the major carrier of lipid hydroperoxides in human blood plasma from fasting donors. *Proc Natl Acad Sci USA* 1992;**89**(21):10316–20.
51. Duprez DA, Neuhaus J, Kuller LH, et al. Inflammation, coagulation and cardiovascular disease in HIV-infected individuals. *PLoS One* 2012;**7**(9):e44454.
52. Naqvi TZ, Lee MS. Carotid intima-media thickness and plaque in cardiovascular risk assessment. *JACC Cardiovasc Imaging* 2014;**7**(10):1025–38.
53. Gillard BK, Raya JL, Ruiz-Esponda R, et al. Impaired lipoprotein processing in HIV patients on antiretroviral therapy: aberrant high-density lipoprotein lipids, stability, and function. *Arterioscler Thromb Vasc Biol* 2013;**33**(7):1714–21.
54. Navab M, Reddy ST, Van Lenten BJ, Fogelman AM. HDL and cardiovascular disease: atherogenic and atheroprotective mechanisms. *Nat Rev Cardiol* 2011;**8**(4):222–32.
55. Lo J, Rosenberg ES, Fitzgerald ML, et al. High-density lipoprotein-mediated cholesterol efflux capacity is improved by treatment with antiretroviral therapy in acute human immunodeficiency virus infection. *Open Forum Infect Dis* 2014;**1**(3):ofu108.
56. Lin S, Nadeau PE, Wang X, Mergia A. Caveolin-1 reduces HIV-1 infectivity by restoration of HIV Nef mediated impairment of cholesterol efflux by apoA-I. *Retrovirology* 2012;**9**:85.
57. Morrow MP, Grant A, Mujawar Z, et al. Stimulation of the liver X receptor pathway inhibits HIV-1 replication via induction of ATP-binding cassette transporter A1. *Mol Pharmacol* 2010;**78**(2):215–25.

CHAPTER

# 4

# Aging With HIV and Oxidative Stress

*Ilaria Motta, Andrea Calcagno*

**Department of Medical Sciences, University of Torino, Torino, Italy**

OUTLINE



## Abstract

The proportion of patients aging with HIV is increasing and they present special characteristics of frailty and premature aging that differ from general population even with suppressive antiretroviral therapy (ART). Hyperproduction of reactive oxygen and nitrogen species and the consequent oxidative stress (OS) contribute to this phenomenon. The chapter firstly reviews the basic knowledge of aging process pointing out the role of OS in the mechanisms of genomic instability, telomere shortening, epigenetic alterations, loss of proteostasis, mitochondrial dysfunction, cellular senescence, and stem cell exhaustion. Secondly, an insight on the impact of long-term HIV infection on OS is presented focusing on HIV pathogenesis, the effect of chronic use of antiretrovirals, and the resulting chronic immune activation and inflammation. A complex scenario originates for HIV-infected patients and many aspects still need to be elucidated. Additional studies and additional measures to ART are warranted to improve health of this specific population.

**Keywords:** ART; Cellular senescence; Frailty; HIV; Immune activation; Inflammaging; Oxidative stress; Premature aging; RNS; ROS.

## List of Abbreviations

**ART** antiretroviral therapy
**CVD** cardiovascular disease
**HBV** hepatitis B
**HCV** hepatitis C
**NO** nitric oxide
**OS** oxidative stress
**RNS** reactive nitrogen species
**ROS** reactive oxygen species



*HIV/AIDS*
http://dx.doi.org/10.1016/B978-0-12-809853-0.00004-3

# INTRODUCTION

## What Is Oxidative Stress?

Oxidative stress (OS) may be the result of an imbalance between the production of reactive oxygen species (ROS) and the ability to detoxify oxygen reactive intermediates or repair the resulting damage. The term ROS includes the superoxide anion ($O_2^-$), the hydrogen peroxide radical ($H_2O_2$) and other free radicals, and oxygen anions. These are produced predominantly in mitochondria by enzymes such as nicotinamide adenine dinucleotide phosphate hydrogen (NADPH) oxidases that use NADPH to reduce $O_2$. Exogenous sources are radiations, atmospheric pollutants, and chemicals. In addition to ROS the cells produce RNS (reactive nitrogen species) derived from NO (nitric oxide, produced by NOS, nitric oxide synthase) and superoxide. Types of RNS are peroxynitrite ($ONOO^-$), nitrogen dioxide ($NO_2$), and dinitrogen trioxide ($N_2O_3$). RNS along with ROS cause extensive cellular damage depending by their relative rates of formation and removal. Cell function is optimal at reducing environment and, under physiological condition, an efficient defensive system of antioxidants prevents oxidative damage "eliminating" ROS/RNS. At the cellular level families of enzymes such as superoxide dismutases, catalases, and peroxidases play an important role in the conversion of ROS to oxygen and water.[1] Nonenzymatic antioxidants are also important in removing free radicals and included in this group are the lipid-soluble antioxidants vitamin E and beta-carotene and water-soluble antioxidants as vitamin C and glutathione (GSH). In pathologic states, the inability to eliminate free radicals causes the shift of the balance in favor of oxidation; this process is called "OS" and affects major cellular components, including proteins, lipids, and DNA. When the antioxidant defenses are depleted we are subject to low level but chronic OS that contributes to cellular damage, mutations, and senescence. Moreover, the OS contributes to the pathogenesis of diseases (cancer, renal disease, neurodegeneration, vasculopathy) and to the normal process of aging.[2–4] (Fig. 4.1) Long-term HIV infection is associated with premature aging and frailty

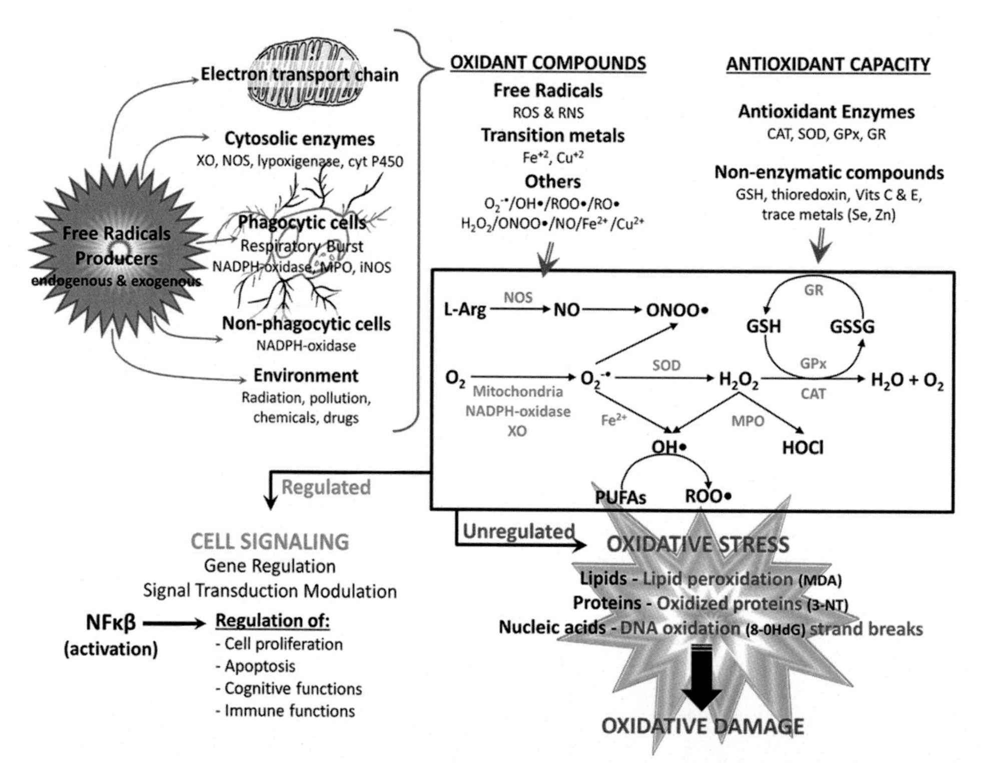


FIGURE 4.1 **Reactive species participate in normal cellular function or in pathological mechanisms depending on their overproduction.** The balance between oxidants compounds and antioxidant defense determines the end result. Optimal physiologic levels leads to beneficial effects, with ROS and RNS acting as second messengers in intracellular signaling cascades (modulation of gene regulation and signal transduction pathways, mainly by activation of NFκB), regulating several physiological functions. However, when overproduction of ROS/RNS is higher than the antioxidant system, the equilibrium status favors oxidant versus antioxidant reactions, leading to oxidative stress, in which ROS/RNS have harmful effects, because of their reaction with various macromolecules (lipids, proteins and nucleic acids), contributing to cellular and tissue oxidative damage, and the development of age-related impairments. *CAT*, catalase; *Fe*, iron ions; *GPx*, glutathione peroxidase; *GR*, glutathione reductase; *GSH*, reduced glutathione; *$H_2O_2$*, hydrogen peroxide; *HOCl*, hypochlorous acid; *iNOS*, inducible nitric oxide synthase; *MD*, Amalondialdehyde; *MPO*, myeloperoxidase; *ONOO•*, peroxynitrite anion; *RNS*, reactive nitrogen species; *RO•*, alkoxyl radical; *ROS*, reactive oxygen species; *SOD*, superoxide dismutase; *XO*, xanthine oxidase. *Reproduced from von Bernhardi R, Eugenín-von Bernhardi L, Eugenín J. Microglial cell dysregulation in brain aging and neurodegeneration.* Front Aging Neurosci *2015;7:124, with permission from Frontiers and authors, 2015.*

phenotype. The role of OS has been partially investigated and a state of chronic immune inflammation and activation results with detrimental consequences. This is considered both a major issue in HIV pathogenesis and in the elderly and aged-HIV population are at an increasing risk of comorbidities and disease progression.

## AGING PROCESS AND OXIDATIVE STRESS

The aging process per se is harder to study than diseases since it is inevitable. In the elderly several functional changes in organs' function are observed and configure a slow decline: pathological aging is, by contrast, characterized by a high incidence of comorbidities, frailty, and the acceleration of the normal aging process.

In 1956, Denham Harman proposed the "Free Radical Theory of Aging" and can be considered the basis of reactive oxygen biochemistry. By 1980, oxyradicals were accepted as biological entities, though their significance for aging and disease remained generally unappreciated. Around 1980 protein oxidation in aging tissue was studied, giving rise to serious consideration of OS as a pathological factor.[5]

Better understanding of the signaling pathways and cellular events involved in aging shows that these are characteristic of many chronic degenerative diseases (neurodegeneration, cardiovascular disease, chronic obstructive pulmonary disease).[6] Aging can affect several tissues and processes resulting in complex functional changes. For examples, microglia undergo several age-related changes that contribute to the generation of a chronic mild inflammatory environment, including an increased production of inflammatory cytokines and the production of ROS. These changes have been linked to the appearance of cognitive deficits and the onset of chronic neurodegenerative diseases.[7] (Fig. 4.2)

Common mechanisms have now been identified, and López-Otín and colleagues described nine biological hallmarks of the aging process: genomic genetic instability, telomere shortening, epigenetic alteration, loss of proteostasis, deregulated nutrient sensing, mitochondrial dysfunction, cellular senescence, stem cell exhaustion, and altered intercellular communication.[8] (Fig. 4.3)

Many of these pathways are driven by chronic OS and its role in each of them is further discussed below.

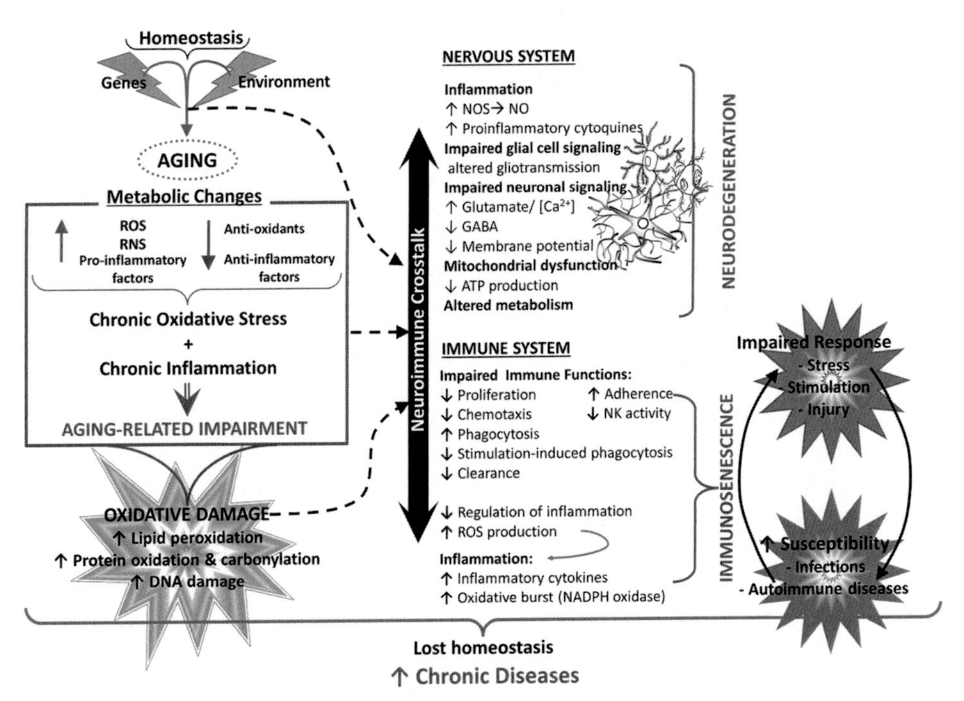


FIGURE 4.2 **Aging of the nervous and immune system and the neuroimmune cross talk.** Healthy aging of the nervous and immune systems depends both on genetic and environmental factors. Aging is associated with a state of low-grade chronic oxidative stress and inflammation. The central nervous system (CNS) and the immune system are especially vulnerable to oxidative damage, which contributes to oxidative stress and inflammation. Age-related changes in the immune function, known as immunosenescence, results in increased susceptibility to infections and cancer, inflammation and autoimmune diseases. In the CNS, oxidative stress has a negative impact on function, leading to mitochondrial dysfunction and impaired energetic metabolism, altered neuronal, and glial signaling. *NO*, Nitric oxide; *NOS*, nitric oxide synthase; *RNS*, reactive nitrogen species; *ROS*, reactive oxygen species. *Figure as originally published in von von Bernhardi R, Eugenín-von Bernhardi L, Eugenín J. Microglial cell dysregulation in brain aging and neurodegeneration.* Front Aging Neurosci 2015;7:124, *with permission from Frontiers and authors.*

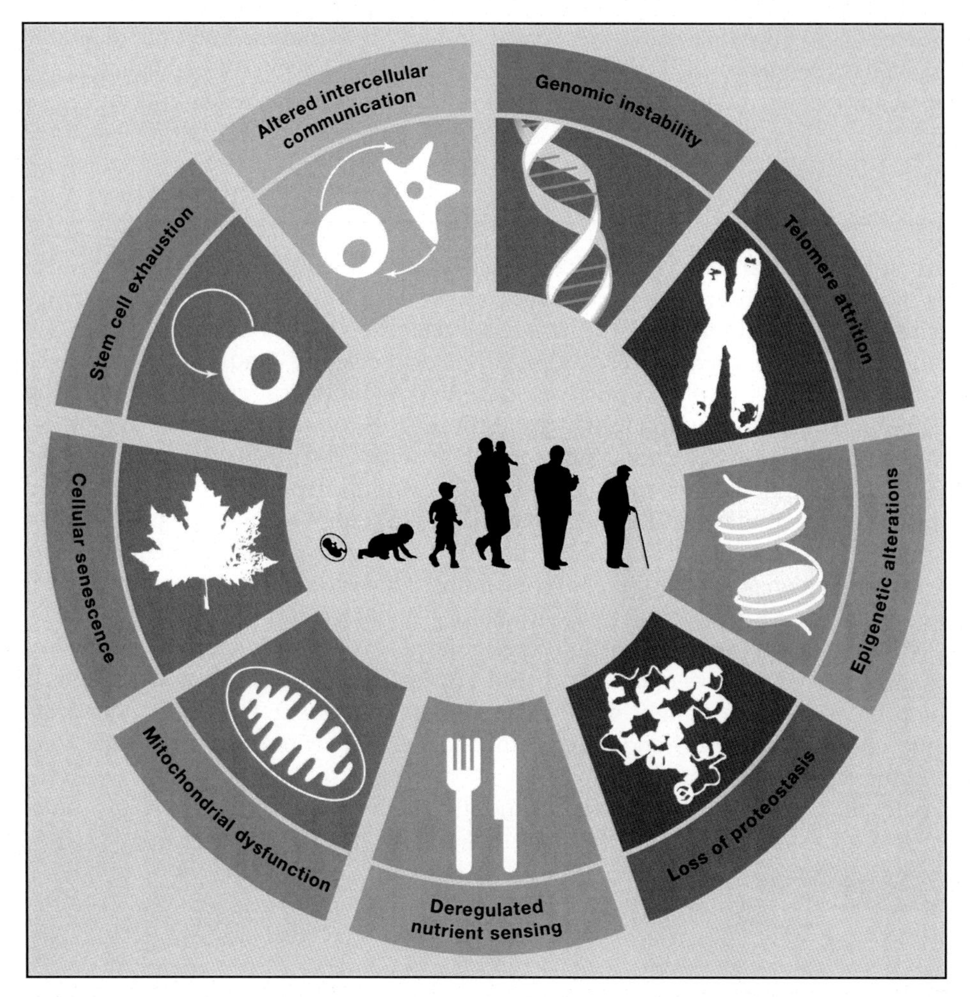


FIGURE 4.3 **The hallmarks of aging.** The scheme enumerates the nine hallmark of aging: genomic instability, telomere attrition, epigenetic alterations, loss of proteostasis, deregulated nutrient sensing, mitochondrial dysfunction, cellular senescence, stem cell exhaustion, and altered intercellular communication. *Figure as originally published in López-Otín C, Blasco MA, Partridge L, Serrano M, Kroemer G. The hallmarks of aging.* Cell *2013;***153***:1194–1217, with permission from Elsevier.*

## Genomic Instability

Genomic instability can be caused by OS response and can lead to tumor progression in case of DNA damage and mutation in cancer-related genes. NO overproduction and lipid peroxidation results in intermediates that can react with DNA basis to form exocyclic DNA adducts and have been found to have a role in carcinogenesis. Also response to infections can cause genomic instability. On activation mainly macrophages produce ROS and NO, which at lower concentrations act as signaling molecules. However, increased production of these molecules by macrophages is considered to be pathological because of interference with signaling pathways related to apoptosis, cell cycle arrest, and the ability to cause severe damage to DNA and proteins. Indeed, several studies have reported that high levels of NO and ROS production in response to bacterial or viral infections weaken cellular immunity by inducing genomic instability due to DNA damage.[9]

## Telomere Shortening

Telomere length during aging process is subject to shortening and factors such as inflammation and OS increase the rate of telomere attrition, leading to telomere dysfunction-mediated cellular senescence and accelerating the aging process.

Telomere exhaustion explains the limited proliferative capacity of some types of in vitro cultured cells, the so-called replicative senescence or Hayflick limit. In 1961, Hayflick and Moorhead noted that human diploid cells lines could be cultured for only a limited number of passages. This gave rise to the telomerase theory of aging and the Hayflick limit. Telomere shortening is associated with a number of age-related diseases such as osteoarthritis, atherosclerosis, coronary heart disease, and atrial fibrillation.[10]

## Epigenetic Alteration

Epigenetic alteration refers to changes in gene expression that are achieved without affecting the DNA sequence at the base-pair level. They involve DNA methylation, chromatin remodeling, and noncoding RNA. OS is associated with different type of DNA damage, included epigenetic alteration. For example, human colon cancerogenesis is mediated by the silencing of CDX1 gene and aberrant hypermethylation of tumor suppressor gene promoter regions, induced by $H_2O_2$ treatment in T84 cells, causes gene silencing and the progression to a malignant phenotype. Analog mechanisms have been described in hepatocellular carcinoma.[11]

## Loss of Proteostasis

Proteostasis is the process to maintain the homeostasis of the proteome, the building, and turnover of proteins. The proteostasis consists in chaperone-mediated folding, proteasomal degradation, and autophagy. Autophagy is the process for degradation of long-lived proteins and damaged organelles in lysosomes. This is essential for the cells to maintain the pool of stem cells in quiescent state (G0-reversible quiescent state) and prevent senescence. Failure of autophagy is "physiological" in aged cells resulting in toxic cell waste accumulation, mitochondrial dysfunction, and OS. Increased ROS drive satellite cell senescence and findings from studies in mice showed that ROS inhibition prevents senescence.

Sarcopenia, the age-related loss of skeletal muscle mass and function, is a clear example of this regenerative impairment in human satellite cells. Comparing to young cells, old cells present higher levels of defective proteins (showed by p62) and mitochondrial accumulation associated with ROS increase. The causal role of impaired autophagy in cellular senescence was supported by the use of rapamycin. Rapamycin, an inhibitor of mTOR, mimics cellular starvation by blocking signals required for cell growth and proliferation, restoring the abnormal mitochondrial content and reducing level of ROS.[12]

## Mitochondrial Dysfunction

Mitochondria deteriorate with age, losing respiratory activity, accumulating damage to their DNA (mitochondrial DNA (mtDNA)), and producing excessive amounts of ROS.

Dysfunctional mitochondria are engulfed by autophagosomes and transported to lysosome to be degraded and recycled by cell. High levels of ROS and the consequent produced OS are associated with high levels of p62 that act as a trigger for mitochondrial autophagy, called mithophagy.[13]

Mitochondrial dysfunction has a profound impact on the aging process, accelerating aging in mammals, but it is less clear whether improving mitochondrial function, for example, through mitohormesis, can extend life span in mammals, although suggestive evidence in this sense already exists. Mitohormesis occurs when low levels of OS induced by either caloric restriction, exercise, or other stimuli trigger an adaptive response that improves overall stress resistance. Mild OS may in fact promote longevity and metabolic health through mechanisms that increase endogenous antioxidant defenses and involved inhibition of glycolysis, impairment of insulin-like signaling, and certain mutations in mitochondrial electron transport chain components.

So ROS can generate detrimental oxidative damage but have been found to be physiologically vital for signal transduction, gene regulation, and redox regulation, among others, implying that their complete elimination would be harmful.

The redox stress hypothesis proposes that aging-associated functional losses are primarily caused by a progressive prooxidizing shift in the redox state of the cells, which leads to the overoxidation of redox-sensitive protein thiols and the consequent disruption of the redox-regulated signaling mechanisms.[14]

## Cellular Senescence

Cellular senescence is a stable and long-term loss of proliferative capacity.[15] The senescence growth arrest is, by definition, limited to mitotic tissues. The principal contributor is the DNA damage response pathway, which can be activated by stresses such as ROS, certain DNA lesions, telomere attrition, oncogene activation, and tumor suppressor gene inactivation. Cellular senescence is now recognized as an important tumor suppressive mechanism that prevents cancer progression in vivo. Senescent cells are usually removed by immune surveillance and phagocytosis but seem to accumulate with age. This accumulation may be the result of their increased production as well as the result of decreased immune clearance. Chronic immune activation will cause immunosenescence, commonly defined as the functional decline of the adaptive immune system with age. Lymphocytes and macrophages are largely affected during immunosenescence, and the continuous age-related, antigenic stress provokes a variety of changes: the clonal expansion of T and B cells, the decrease and the exhaustion of naive T cells, promoting the reduction of the "immunological space". Moreover, the response to new antigens will be reduced by this process. In general, data on immunosenescence indicate that changes occurring over time might be considered the result of global reshaping, where the immune system continuously looks for possible stable points for optimal functioning. Possibly, this phenomenon is more general, likely involving the stress response in every tissue and organ.

Unlike apoptosis, senescent cells are still metabolically active with altered secretory phenotypes (senescence-associated secretory phenotype, SASP). SASPs are associated with the increased production of inflammatory cytokines during the aging process, which likely contributes to the chronic inflammation connected with aging.

This process is called **inflammaging** and is an example of altered intercellular communication. Macrophages are the major target of this process and some authors named this "macrophaging".[12,16] (Fig. 4.4)

Mitochondrial oxidative metabolism regulates macrophage polarization, T cell activation, differentiation, and memory cell formation. Thus, mitochondria not only sustain immune cell phenotypes but also are necessary for establishing immune cell phenotype and function. In a proinflammatory state this is accomplished by mitochondria shifting from producing ATP via oxidative metabolism to producing building blocks for macromolecule synthesis via anapleurosis and glutaminolysis. The shift from catabolism to anabolism is critical to affect cell expansion, production of inflammatory mediators, and immune cell fate commitments.

Chronic inflammation is associated with normal aging but is a major risk factor for age-related diseases such as dementia, cardiovascular diseases, arthritis, and cancer. Healthy centenarians typically present lower plasma inflammatory levels than frail centenarians.[17]

### Stem Cell Exhaustion

Aging limits the ability of stem cells to differentiate, self-renew, and respond to environmental factors.

Moderate levels of ROS are needed to maintain stem cell self-renewal, but OS, in particular high ROS levels, due to stress and inflammation, induce stem cell differentiation to short-term repopulating cells and enhanced motility.[18]

Chronic inflammation can directly or indirectly disrupt stem cell function: inflammatory mediators can drive stem cells differentiation (interleukin (IL)-1, IL-6, IL-17, IL-27, TNF-α[19–21]), while proteases and destructive immune cell activities (for example, by thickening the basal lamina around muscle satellite cells by extracellular matrix deposition, impeding satellite cell function) can eliminate stem cells niches.

Consequences of the exhaustion of hematopoietic stem cells, mesenchymal stem cells, and intestinal epithelial stem cells are anemia, osteoporosis and atherosclerosis, and decreased intestinal function, respectively.[17] (Fig. 4.4)

A substantial body of evidence now implicates OS in the pathogenesis of atherosclerosis. ROS generated during cellular metabolic pathways modify polyunsaturated fatty acids and lipoproteins, resulting in proinflammatory and atherogenic particles such as oxidized low-density lipoprotein. In addition, high levels of ROS may contribute to endothelial dysfunction through nitric oxide inactivation.

## ANTIAGING MOLECULES (SIRTUINS AND α-KLOTHO)

During aging process there is also a reduction in antiaging molecules, such as sirtuins and Klotho, which further accelerates the aging process. Understanding these molecular mechanisms has identified several novel therapeutic targets, and some drugs have already been developed to potentially slow the aging process, as well as lifestyle interventions, such as diet and physical activity (see Part 2).

Aging promotes endothelial senescence and this is associated with atherosclerosis and is considered one of the major risk factors for cardiovascular diseases (CVDs). In fact, atherosclerotic plaques show features of cellular senescence as shown by reduced cell proliferation, irreversible growth arrest, apoptosis, elevated DNA damage, epigenetic modifications, and telomere shortening.

### The Sirtuins

The sirtuins comprise a highly conserved family proteins present in virtually all species from bacteria to mammals. Sirtuins are members of the highly conserved class III histone deacetylases, and seven sirtuin genes (sirtuins 1–7)

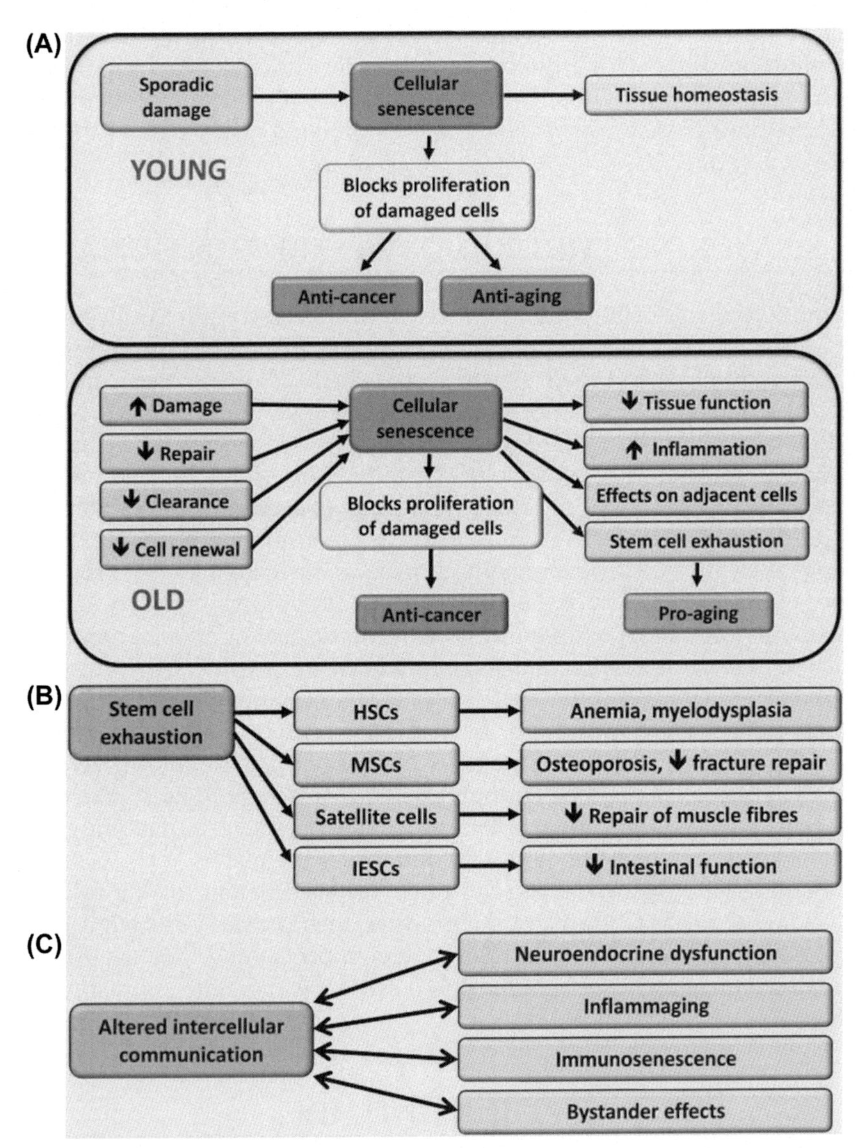


FIGURE 4.4 **Cellular Senescence, Stem Cell Exhaustion and Altered Intercellular Communication.** (A) Cellular senescence. In young organisms, cellular senescence prevents the proliferation of damaged cells, thus protecting from cancer and contributing to tissue homeostasis. In old organisms, the pervasive damage and the deficient clearance and replenishment of senescent cells result in their accumulation, and this has a number of deleterious effects on tissue homeostasis that contribute to aging. (B) Stem cell exhaustion. Consequences of the exhaustion of hematopoietic stem cells (HSCs), mesenchymal stem cells (MSCs), satellite cells, and intestinal epithelial stem cells (IESCs) are exemplified. (C) Altered intercellular communication. Examples of altered intercellular communication associated with aging. *HSCs*, Hematopoietic stem cells; *IESCs*, intestinal epithelial stem cells; *MSCs*, mesenchymal stem cells, satellite cells. *Figure as originally published in López-Otín C, Blasco MA, Partridge L, Serrano M, Kroemer G. The hallmarks of aging.* Cell *2013;***153***:1194–1217, with permission from Elsevier.*

have been identified and characterized in mammals. Sirtuins' activity is linked to metabolic control, apoptosis, cell survival, development, inflammation, and healthy aging. They act as sensors of energy and redox status in cells; they modulate the activity of key metabolic enzymes as well as regulate transcription of metabolic genes. Sirtuins are intimately linked to the cellular response to OS and play a role in modulating it. Sirtuins also have been studied intensively as potential antiaging and age-related diseases targets in several human diseases including cancer, diabetes, inflammatory disorders, and neurodegenerative diseases.[22]

## α-Klotho

α-Klotho is predominantly produced in the kidneys and its expression is downregulated by chronic kidney disease–associated uremic toxins, OS, and enhanced activity of the renin-angiotensin-aldosterone system. The gene was discovered in 1997 as an antiaging gene. Lower α-Klotho concentrations are associated with progressive chronic kidney disease, higher prevalence of cardiovascular disease, arterial stiffness, and vascular calcification. However, a small intervention study with antioxidant molecules (pravastatin and vitamin E) did not be able to show changes in plasma klotho levels.[23] Other findings describe that plasma klotho levels were inversely associated with subclinical carotid atherosclerosis in HIV-infected patients receiving combined antiretroviral therapy (ART).[24]

## MEASURING THE OXIDATIVE STRESS IN HUMANS

Further research on the pivotal role of OS on aging process is needed and many studies are on going. Heterogenous results and different biomarkers used to measure OS make it more difficult to compare analysis and results.

The determination of free radicals and their effects is challenging. Free radicals can be determined directly by physical methods such as electron-spin resonance spectrometry or chemiluminescence, but their low concentrations and their very short half-lives make these methods of limited use. Another approach is the determination of by-products derived from reaction of free radicals with biological compounds such as DNA, proteins, and lipids. As polyunsaturated fatty acids are critical targets of free radicals, it is usual to evaluate lipid oxidation by-products level. The chemical rearrangement of fatty acyl radicals usually leads to the formation of conjugated dienes, lipid hydroperoxides, and malondialdehyde (MDA). However, methods used to evaluate these by-products are known to be nonspecific and insensitive, and the evaluation of these parameters after induction of OS is limited. Prostaglandin-like compounds (F2-isoprostanes (F2-IsoPs)) produced by nonenzymatic peroxidation of arachidonic acid are apparently formed in situ, esterified to phospholipids, and subsequently released, possibly by a phospholipase. In contrast with lipid hydroperoxides, they are chemically stable end products of lipid peroxidation, and measurement of their levels in plasma might thus represent a sensitive and specific method for detection of in vivo lipid oxidative damage.[25]

Studies quantifying F2-IsoP concentration as an index of OS both in vitro and in vivo have confirmed that the concentration of F2-IsoPs in plasma is a reliable, noninvasive measurement of lipid peroxidation. Increased F2-IsoP concentrations have been associated with known risk factors for atherosclerosis, including long-term cigarette smoking, polygenic hypercholesterolemia, diabetes, and hyperhomocysteinemia.[26]

In conclusion, these and similar data have pointed out a reconsideration for the role of ROS in aging. ROS is a stress-elicited survival signal finalized to offset the gradual deterioration associated with aging. With age progression ROS levels raise to preserve survival until an inversion occurs and they eventually aggravate, instead of mitigate, the age-associated damage. This new vision could settle the apparent contrasting evidence concerning the positive, negative, or neutral age-related ROS effects.

## AGING AND HIV

Despite proven efficacy of ART in terms of reduction of morbidity and mortality, life expectancy of treated patients is shorter than the general population and is dependent on the age when ART is started.[27]

Consequently, the questions that arise are do HIV-positive persons age prematurely? And if is so OS has a role in it?

### Comorbidity and Frailty

Recent studies showed that age-related diseases are more common in HIV-positive patients than in the general population. Multimorbidity is the norm among those aging with HIV. In one analysis, 65% of HIV-infected individuals between 50 and 59 years of age had at least one comorbid diagnosis, and 7% had a major medical comorbidity, a substance use disorder, and a psychiatric diagnosis.[28]

Guaraldi et al.[29] found that CVD, hypertension, bone fractures, renal failure, and diabetes mellitus had higher incidence in HIV population and HIV-specific factors (lower nadir CD4 cell count and longer exposure to ART) were independent predictors of age-related diseases.

Other studies show that the incidence of cancer, liver disease, and neurocognitive impairment are also higher in HIV-positive patients than in age-matched uninfected people.

The apparent paradox based on the findings that early ART prevents noninfectious comorbidities related to lower CD4 cell count, and the above-reported increased prevalence of age-related diseases secondary to longer exposure of ART[24] has been partially resolved by results from the SMART study where is shown that HIV disease presents higher risk of several non-AIDS complications and that ART reduces the risk of these events.[30]

Moreover, the prevalence of polypathologies in HIV-positive patients was approximately the same to the prevalence observed in members of general population who were 10–15 years older.

In general population the simultaneous presence of CVD, hypertension, diabetes mellitus, renal failure, and bone fracture represents a frailty phenotype and has been associated with aging. Frailty has been recognized as a distinct clinical syndrome where comorbidities are an etiologic risk factor and disability and increased mortality an outcome. Therefore, frailty phenotype could be considered a potential surrogate marker of aging process.[31] Characteristics of frailty are involuntary weight loss, sarcopenia, weakness, poor endurance, feeling exhausted, slow gait speed, and low physical activity.[32]

In HIV-positive patients a HIV-specific aging phenotype has been described: earlier onset of comorbidities usually presented at later stage in general population, higher risk of non-AIDS-related morbidity and mortality despite ART defined a premature aging in this population. It is now frequently considered that HIV-infected patients are aging prematurely, considered to be old when >50 years of age, compared with 65 years of age in the general population.[33]

Although factors involved in premature aging have not been adequately addressed, critical factors resulting in higher degree of frailty include long-term drug toxicity, lifestyle factors, higher chronic systemic inflammation, persistent viral replication, immunosenescence, and microbial translocation.[34]

## Role of Antiretroviral Therapy in Oxidative Stress

ART toxicity seems to activate the mechanisms associated with frailty phenotype. Existing reports showed that elevated levels of F2-IsoP (index of OS, see above for details) were associated with antiretroviral agents use and suppressed viral replication.[26]

Undoubted benefits from ART were stated from many clinical trials,[30] but few studies based on earlier recommended regimen regarding the influence of ART in the process of premature aging gave contradictory information and so far there is no consensus among researchers. Data showed that ART-induced mitochondrial toxicity may have a central role in accelerated aging. An accumulation of mtDNA mutations increases mitochondrial OS and shortens mitochondrial average life span and is primarily induced by the class of NRTIs (nucleoside reverse transcriptase inhibitors) (see chapter 1.2) and PIs (protease inhibitors) used to boost this effect by increasing intracellular concentrations of NRTIs. Moreover, thymidine analogs have additionally been shown to cause cellular senescence through an increase in OS and induction of mitochondrial dysfunction in human fibroblast cell lines and in subcutaneous adipose tissue from ART-treated patients.

PIs themselves induce prelamin A accumulation inhibiting *ZMPSTE24*, which follows genomic instability and cellular senescence.[35] Metabolic perturbations such as lipodystrophy phenotype and insulin resistance are associated with systemic inflammation[36] and with the long-term use of PIs. These antiretroviral class agents directly trigger the production of ROS, and the chronic use seems to enhance the consequent detrimental effects. The exact mechanism is not completely understood, but data showed that they increase NADPH oxidase (NOX) and decrease Cu/ZnSOD activity elevating cytosolic ROS. Intracellular ROS (from cytosol and intermitochondrial membrane space) are able to inhibit respiratory chain complexes. A decreased electron transporter chain activity will generate more reactive species perpetrating the cycle. An increase production of superoxide will enhance mitochondrial ROS levels. So far there is no evidence that PIs directly inhibit respiratory chain complexes.

Decrease cytochrome *c* oxidase 4 (COX4) gene expression and increase NOX activity (through elevated mRNA levels of NOX subunits $p22^{phox}$, $p40^{phox}$, $p47^{phox}$, and $p67^{phox}$) are some of the long-term transcriptional effects. PI-induced ROS is then related to cardio metabolic disfunction mediated by ubiquitin-proteasome system[37] (Fig. 4.5).

Furthermore some data report that ART is involved in immunosenescence and chronic inflammation by changing phenotype of senescent cells and secreting proinflammatory cytokines (see above).[38]

## Role of Lifestyle in Oxidative Stress

Regarding lifestyle factors, we know from large cohort studies[39,40] that the prevalence in HIV-positive patients of alcohol and illicit drugs consuming, smoking, and coinfections with hepatitis B (HBV)/hepatitis C (HCV) is higher than in general population and this further worsen the frailty of this population. For example, consumption of high quantities of ethanol increases MDA and reduces GSH and plasma vitamin E levels and partially nullifies the antioxidant capacity of red wine rich in polyphenolic compounds.[41]

Methamphetamine users present higher levels of OS due to dysregulation of the dopaminergic system, hypertermia, apoptosis, and neuroinflammation. Methamphetamine abuse is correlated with increased viral replication,

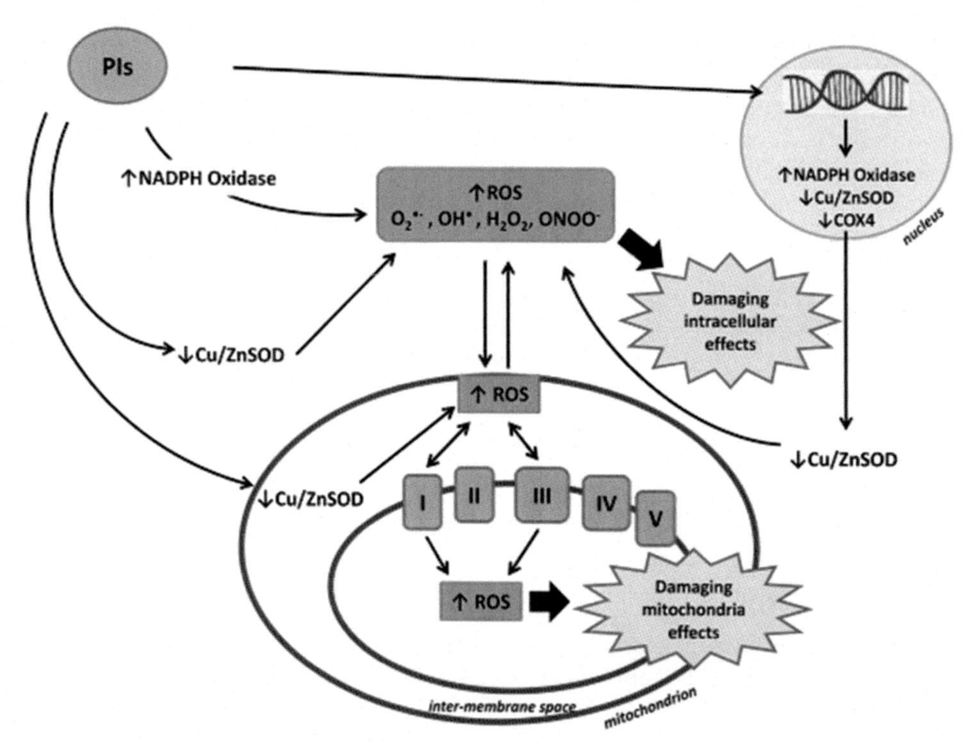


FIGURE 4.5 **HIV-protease inhibitors (PIs) and oxidative stress.** HIV PIs elicit a variety of effects at the cellular level. For example, PIs increase NADPH oxidase (NOX), and decrease Cu/ZnSOD and COX4 gene expression. This would be expected to increase and decrease NOX and Cu/ZnSOD activities, respectively. Moreover, lower COX4 levels may impair mitochondrial respiratory capacity. PIs can also directly affect enzyme activities of Cu/ZnSOD and NOX in the cytosol, resulting in ROS generation with damaging intracellular effects. Mitochondrial Cu/ZnSOD activity is also downregulated and thus increases the generation of ROS within the intermembrane space. Higher ROS levels may inhibit complexes I and III of the ETC, thereby increasing matrix and cytosolic ROS levels. In this manner, mitochondria are also damaged. *COX4*, cytochrome *c* oxidase 4; *Cu/ZnSOD*, copper/zinc superoxide dismutase; $H_2O_2$, hydrogen peroxide; $O_2^{\cdot}$, Superoxide free radical; *NADPH*, nicotinamide adenine dinucleotide phosphate hydrogen; *OH•*, hydroxyl radical; *ONOO⁻*, peroxynitrite; *PI*, protease inhibitor; *ROS*, reactive oxygen species. *Figure as originally published in Reyskens KM, Essop MF. HIV protease inhibitors and onset of cardiovascular diseases: a central role for oxidative stress and dysregulation of the ubiquitin-proteasome system.* Biochim Biophys Acta 2014;**1842**:256–68, *with permission from Elsevier.*

enhanced Tat-mediated neurotoxicity and neurocognitive impairment. Moreover, in periphery inflammatory response and expression of HIV coreceptors seem to be increased.[42]

In HCV-related chronic hepatitis reduced levels of GSH are found in serum and liver, an increased ROS production by mitochondria and the interference of HCV with NrF/ARE pathway are reported and many more mechanisms explain the importance of ROS for the onset of HCV-associated pathogenesis.[43]

## Role of HIV in Oxidative Stress

Excess inflammation has been postulated as a potential driver of frailty pathogenesis through mitochondrial dysfunction, which subsequently results in increased OS.

Observational studies have shown that frailty is associated with several proinflammatory biomarkers, including IL-6, C-reactive protein, and TNF-α.[44]

To have a more comprehensive understanding of the association between OS and premature aging in HIV population, multiple HIV pathogenesis aspects have to be considered in our analysis.

HIV induces production of ROS through the regulatory protein Tat and the envelope glycoprotein gp120 contributing to immunopathogenesis of HIV.

A dose-dependent increase in OS and decrease in intracellular GSH have been observed in immortalized blood-brain barrier endothelial rat cells treated with Tat and gp120, explaining also the etiopathogenic role of HIV in the development of HIV-related dementia.

HIV replicates in a highly oxidized environment and cells infected with HIV generate a large amount of ROS/RNS and this is accompanied by deregulation of superoxide dismutase, catalase, and GSH. ROS have the ability to start a cascade with activation of nuclear factor (NF)-kb that controls the transcription of genes for HIV replication leading to increase of it. ART, suppressing HIV, partially reestablishes antioxidant ability, however, as previous stated, PIs increase ROS production in several cells including macrophages.

HIV-activated macrophages via TNF-α release, and activated polymorphonuclear leukocytes, also contribute to the generation and accumulation of ROS that can be accompanied by a deficient antioxidant ability and increased cell death.[45]

Free radicals have been associated with decrease of immune cell proliferation, loss of immune function, followed by apoptosis, loss of T cell response, T helper imbalance, and a premature exhaustion of patient's immune system, called immunosenescence (see above) characterized by continuous stimulation of immune system and low grade of inflammation in untreated HIV infection.[46]

Consequently, there is a shift versus oxidative status due to excessive consumption of antioxidant molecules to protect cells against ROS-induced damage, which contributes to further enhance the prooxidative status and suggests that OS contributes to the progression of HIV from asymptomatic stages to AIDS.

Biomarkers of OS have been associated with higher mortality from all causes in HIV population. Masia and colleagues[47] reported measurement of F2-IsoPs and MDA (free radical-induced peroxidation products) in 54 HIV-positive patients with matched controls. Higher levels of F2-IsoPs and more marginally MDA were associated with death and independent of the HIV transmission category, CD4 cell count, HIV viral load, and subclinical inflammation measured with C-reactive protein.

Chronic immune activation and ROS are associated with loss of intestinal mucosal integrity that allows bacterial translocation and increased plasma levels of lipopolysaccharide and soluble CD14 that have been found to correlate with HIV mortality.[48] Intestinal mucosa is one of the first sites to be interested in the infection, and early T regulatory response may suppress HIV-specific immunity, allowing HIV to persist.

Vpr, an HIV protein that contribute to HIV-1 pathogenesis, has been proven to increase ROS and HIF-1 (a biomarker of OS) expression in human microglial cells, leading to mitochondrial dysfunction and OS. OxPC, formed from phospholipids in response to OS, was identified in atherosclerotic lesions to be induced by rVpr (recombinant Vpr), in a ROS-dependent manner, because it is reverted by the addition of N-acetyl-L-cysteine, an ROS scavenger molecule.

Dysfunction of antioxidant enzymes in monocytes and cerebrospinal fluid has been found to lead to neuronal loss and neurocognitive impairment. Nef, another regulatory protein, induces ROS production in human astrocytes and promotes their rapid death, having a role in HIV-associated dementia insurgence. In some HIV-positive patients, despite effective ART, blood-brain barrier impairment has been found to be associated with markers of neurological damage (tau and phosphorylated tau), and this finding might support the role of blood-brain barrier in HIV-associated neurocognitive impairment and possibly mediated by OS and ROS.[49]

Depression is the most common psychiatric disorder in HIV/AIDS population and neurooxidative stress has been found to be associated with depression as shown by Rivera-Rivera et al.[50] In a small cohort of HIV patients levels of inflammatory cytokines (IL-15, IP-10, IL-12 p40/p70) correlated with levels of depression symptoms. Antioxidant activity was measured and conserved catalase, and superoxide dismutase activity was compared between two groups and a higher activity of catalase in depressed patients was recorded but was not significant.

Unexpectedly lower plasma levels of lipid peroxidation (MDA and 8-isoprostane) were found in depressed patients. This can partially be explained by the increase activity of catalase in depressed patients or higher levels of adherence to ART in nondepressed patients and the consequent increased in OS, as previously reported. From other studies in non-HIV patients oxidative DNA damage seems to be the first step of neurodegenerative process and aging.[51]

Further studies specific in HIV-positive population are warranted to confirm the role of cytokines and OS in the pathogenesis of depression and other central nervous system diseases and the therapeutical implication of agents that inhibit ROS.

In conclusion chronic immune activation, OS, and HIV infection are interconnected and play an important role in disease progression.[52] Immune activation is manifested in many ways including ROS, elevated levels of proinflammatory cytokines, and chemokines. ROS and the subsequent OS are associated with chronic immune activation during viral replication, immune dysfunction, programmed cell death, and neurological damage, all considered as major determinants of HIV-disease progression (Fig. 4.6).

With the availability of ART, HIV infection passed from being a death sentence to a complex chronic disease that needs a multidisciplinary approach. A high immune activation burden is present even in long-term treated patient, and OS has a role in this scenario.[53]

Some contradictory points still have to be elucidated by future studies and investigations. Dietary restriction, physical exercise, and supplements with antioxidants and vitamins showed to have effects in reducing oxidative damage, but more comprehensive research is needed in specific elderly HIV population (see Part 2).

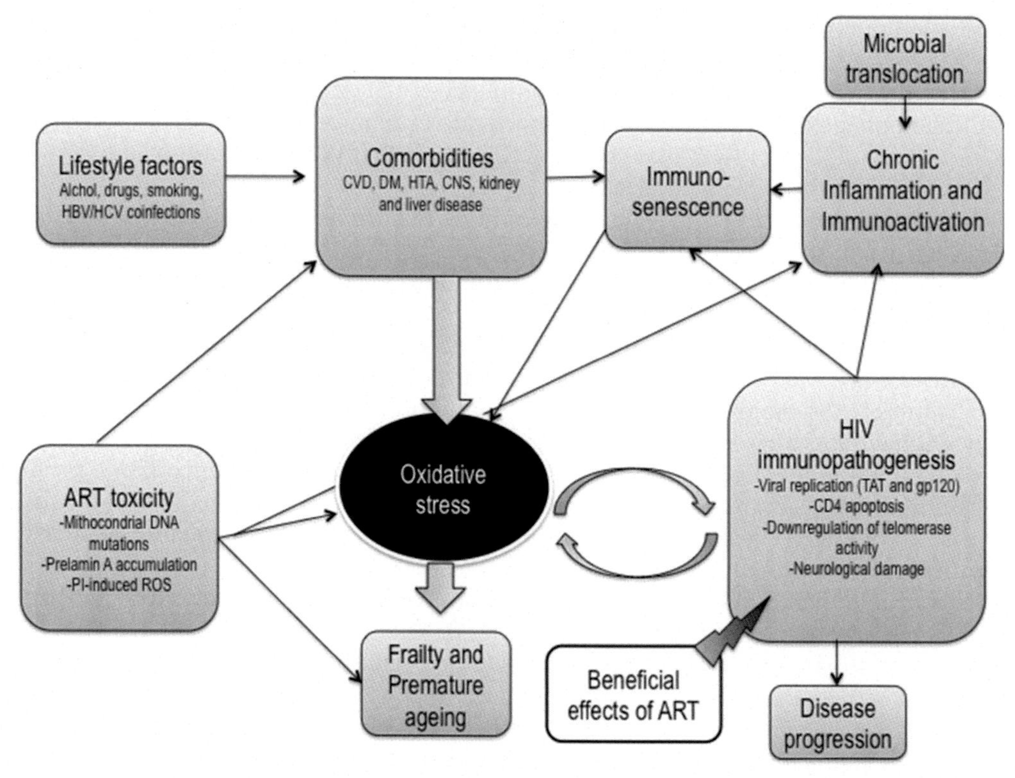


FIGURE 4.6 **Interconnection between HIV, aging, and oxidative stress.** See text for details. *ART*, antiretroviral therapy; *CNS*, central nervous system; *CVD*, cardiovascular disease; *DM*, diabetes mellitus; *HTA*, arterial hypertension; *PI*, protease inhibitor; *ROS*, Reactive oxygen species.

## SUMMARY POINTS

- During the aging process hyperproduction of reactive oxygen species (ROS) is connected with mechanisms of genomic instability, telomere shortening, epigenetic alterations, loss of proteostasis, mitochondrial dysfunction, cellular senescence, and stem cell exhaustion.
- HIV-positive patients present non-AIDS-related diseases, which are typical related with aging. Frailty phenotype is a potential surrogate marker of premature aging in HIV population.
- Chronic immune activation, inflammation, and immunosenescence are strictly related with the release of ROS and have a direct impact on HIV pathogenesis and the development of comorbidities.
- Antiretroviral therapy has undoubted benefits, but available data show that chronic use is related with production of ROS and consequently with OS.
- A reconsideration for the role of ROS in aging is necessary, and further investigation specific in aged-HIV population is needed to design strategies of intervention to reduce the burden of chronic diseases, immune activation, and inflammation.

## References


1. Fridovich I. Oxygen toxicity: a radical explanation. *J Exp Biol* 1998;**201**:1203–9.
2. Griendling KK, FitzGerald GA. Oxidative stress and cardiovascular injury: Part I: basic mechanisms and in vivo monitoring of ROS. *Circulation* 2003;**108**:1912–6.
3. Frei B. Reactive oxygen species and antioxidant vitamins: mechanisms of action. *Am J Med* 1994;**97**:5S–13S. discussion 22S – 28S.
4. Diaz MN, Frei B, Vita JA, Keaney JF. Antioxidants and atherosclerotic heart disease. *N Engl J Med* 1997;**337**:408–16.
5. Ortuño-Sahagún D, Pallàs M, Rojas-Mayorquín AE. Oxidative stress in aging: advances in proteomic approaches. *Oxid Med Cell Longev* 2014;**2014**:573208.
6. Barnes PJ. Mechanisms of development of multimorbidity in the elderly. *Eur Respir J* 2015;**45**:790–806.
7. von Bernhardi R, Eugenín-von Bernhardi L, Eugenín J. Microglial cell dysregulation in brain aging and neurodegeneration. *Front Aging Neurosci* 2015;**7**:124.
8. López-Otín C, Blasco MA, Partridge L, Serrano M, Kroemer G. The hallmarks of aging. *Cell* 2013;**153**:1194–217.

9. Mohanty S, Dal Molin M, Ganguli G, et al. *Mycobacterium tuberculosis* EsxO (Rv2346c) promotes bacillary survival by inducing oxidative stress mediated genomic instability in macrophages. *Tuberculosis (Edinb)* 2016;**96**:44–57.
10. Aubert G, Lansdorp PM. Telomeres and aging. *Physiol Rev* 2008;**88**:557–79.
11. Zhang R, Kang KA, Kim KC, et al. Oxidative stress causes epigenetic alteration of CDX1 expression in colorectal cancer cells. *Gene* 2013;**524**:214–9.
12. García-Prat L, Martínez-Vicente M, Perdiguero E, et al. Autophagy maintains stemness by preventing senescence. *Nature* 2016;**529**:37–42.
13. Onyango IG, Dennis J, Khan SM. Mitochondrial dysfunction in Alzheimer's disease and the rationale for bioenergetics based therapies. *Aging Dis* 2016;**7**:201–14.
14. Sohal RS, Orr WC. The redox stress hypothesis of aging. *Free Radic Biol Med* 2012;**52**:539–55.
15. Zhang J, Rane G, Dai X, et al. Ageing and the telomere connection: an intimate relationship with inflammation. *Ageing Res Rev* 2016;**25**:55–69.
16. Franceschi C, Capri M, Monti D, et al. Inflammaging and anti-inflammaging: a systemic perspective on aging and longevity emerged from studies in humans. *Mech Ageing Dev* 2007;**128**:92–105.
17. Freund A, Orjalo AV, Desprez P-Y, Campisi J. Inflammatory networks during cellular senescence: causes and consequences. *Trends Mol Med* 2010;**16**:238–46.
18. Ludin A, Gur-Cohen S, Golan K, et al. Reactive oxygen species regulate hematopoietic stem cell self-renewal, migration and development, as well as their bone marrow microenvironment. *Antioxid Redox Signal* 2014;**21**:1605–19.
19. Gopinath SD, Rando TA. Stem cell review series: aging of the skeletal muscle stem cell niche. *Aging Cell* 2008;**7**:590–8.
20. Huang H, Kim HJ, Chang E-J, et al. IL-17 stimulates the proliferation and differentiation of human mesenchymal stem cells: implications for bone remodeling. *Cell Death Differ* 2009;**16**:1332–43.
21. Seita J, Asakawa M, Ooehara J, et al. Interleukin-27 directly induces differentiation in hematopoietic stem cells. *Blood* 2008;**111**:1903–12.
22. Conti V, Corbi G, Simeon V, et al. Aging-related changes in oxidative stress response of human endothelial cells. *Aging Clin Exp Res* 2015;**27**:547–53.
23. Adema AY, van Ittersum FJ, Hoenderop JG, et al. Reduction of oxidative stress in chronic kidney disease does not increase circulating α-klotho concentrations. *PLoS One* 2016;**11**:e0144121.
24. Jeong SJ, Song JE, Kim SB, et al. Plasma klotho levels were inversely associated with subclinical carotid atherosclerosis in HIV-infected patients receiving combined antiretroviral therapy. *AIDS Res Hum Retroviruses* 2013;**29**:1575–81.
25. Feillet-Coudray C, Tourtauchaux R, Niculescu M, et al. Plasma levels of 8-epiPGF2alpha, an in vivo marker of oxidative stress, are not affected by aging or Alzheimer's disease. *Free Radic Biol Med* 1999;**27**:463–9.
26. Hulgan T, Morrow J, D'Aquila RT, et al. Oxidant stress is increased during treatment of human immunodeficiency virus infection. *Clin Infect Dis* 2003;**37**:1711–7.
27. Life expectancy of individuals on combination antiretroviral therapy in high-income countries: a collaborative analysis of 14 cohort studies. *Lancet (London, England)* 2008;**372**:293–9.
28. Justice AC. HIV and aging: time for a new paradigm. *Curr HIV/AIDS Rep* 2010;**7**:69–76.
29. Guaraldi G, Orlando G, Zona S, et al. Premature age-related comorbidities among HIV-infected persons compared with the general population. *Clin Infect Dis* 2011;**53**:1120–6.
30. El-Sadr WM, Lundgren JD, Neaton JD, et al. CD4+ count-guided interruption of antiretroviral treatment. *N Engl J Med* 2006;**355**:2283–96.
31. Kooij KW, Wit FW, Schouten J, et al. HIV infection is independently associated with frailty in middle-aged HIV type 1-infected individuals compared with similar but uninfected controls. *AIDS* 2016;**30**:241–50.
32. Fried LP, Tangen CM, Walston J, et al. Frailty in older adults: evidence for a phenotype. *J Gerontol A Biol Sci Med Sci* 2001;**56**:M146–56.
33. Capeau J. Premature aging and premature age-related comorbidities in HIV-infected patients: facts and hypotheses. *Clin Infect Dis* 2011;**53**:1127–9.
34. Calcagno A, Nozza S, Muss C, et al. Ageing with HIV: a multidisciplinary review. *Infection* 2015;**43**:509–22.
35. Brand MD. The sites and topology of mitochondrial superoxide production. *Exp Gerontol* 2010;**45**:466–72.
36. Lagathu C, Eustace B, Prot M, et al. Some HIV antiretrovirals increase oxidative stress and alter chemokine, cytokine or adiponectin production in human adipocytes and macrophages. *Antivir Ther* 2007;**12**:489–500.
37. Reyskens KM, Essop MF. HIV protease inhibitors and onset of cardiovascular diseases: a central role for oxidative stress and dysregulation of the ubiquitin-proteasome system. *Biochim Biophys Acta* 2014;**1842**:256–68.
38. Smith RL, de Boer R, Brul S, Budovskaya Y, van Spek H. Premature and accelerated aging: HIV or HAART? *Front Genet* 2012;**3**:328.
39. Helleberg M, May MT, Ingle SM, et al. Smoking and life expectancy among HIV-infected individuals on antiretroviral therapy in Europe and North America. *AIDS* 2015;**29**:221–9.
40. Ingle SM, May MT, Gill MJ, et al. Impact of risk factors for specific causes of death in the first and subsequent years of antiretroviral therapy among HIV-infected patients. *Clin Infect Dis* 2014;**59**:287–97.
41. Da Silva MS, Rudkowska I. Novel functional foods for optimal oxidative status in healthy ageing. *Maturitas* 2016. http://dx.doi.org/10.1016/j.maturitas.2016.04.001. Published online April 4.
42. Silverstein PS, Shah A, Weemhoff J, Kumar S, Singh DP, Kumar A. HIV-1 gp120 and drugs of abuse: interactions in the central nervous system. *Curr HIV Res* 2012;**10**:369–83.
43. Medvedev R, Ploen D, Hildt E. HCV and oxidative stress: implications for HCV life cycle and HCV-associated pathogenesis. *Oxid Med Cell Longev* 2016;**2016**:9012580.
44. Hubbard RE, O'Mahony MS, Savva GM, Calver BL, Woodhouse KW. Inflammation and frailty measures in older people. *J Cell Mol Med* 2009;**13**:3103–9.
45. Salmen S, Berrueta L. Immune modulators of HIV infection: the role of reactive oxygen species. *J Clin Cell Immunol* 2012:03. http://dx.doi.org/10.4172/2155-9899.1000121.
46. Aquaro S, Scopelliti F, Pollicita M, Perno CF. Oxidative stress and HIV infection: target pathways for novel therapies? *Futur HIV Ther* 2008;**2**:327–38.
47. Masiá M, Padilla S, Fernández M, et al. Oxidative stress predicts all-cause mortality in HIV-infected patients. *PLoS One* 2016;**11**:e0153456.

48. Sandler NG, Wand H, Roque A, et al. Plasma levels of soluble CD14 independently predict mortality in HIV infection. *J Infect Dis* 2011;**203**:780–90.
49. Calcagno A, Atzori C, Romito A, et al. Blood brain barrier impairment is associated with cerebrospinal fluid markers of neuronal damage in HIV-positive patients. *J Neurovirol* 2016;**22**:88–92.
50. Rivera-Rivera Y, García Y, Toro V, et al. Depression correlates with increased plasma levels of inflammatory cytokines and a dysregulated oxidant/antioxidant balance in HIV-1-infected subjects undergoing antiretroviral therapy. *J Clin Cell Immunol* 2014:5. http://dx.doi.org/10.4172/2155-9899.1000276.
51. Trushina E, McMurray CT. Oxidative stress and mitochondrial dysfunction in neurodegenerative diseases. *Neuroscience* 2007;**145**:1233–48.
52. Appay V, Kelleher AD. Immune activation and immune aging in HIV infection. *Curr Opin HIV AIDS* 2016;**11**:242–9.
53. Regidor DL, Detels R, Breen EC, et al. Effect of highly active antiretroviral therapy on biomarkers of B-lymphocyte activation and inflammation. *AIDS* 2011;**25**:303–14.

CHAPTER

# 5

# Antioxidants in Breast Milk of Lactating Mothers with HIV

*Sheu K. Rahamon*[1], *Abdulfatah A. Onifade*[2], *Olatunbosun G. Arinola*[2]

[1]Edo University, Iyamho, Edo State, Nigeria; [2]University of Ibadan, Ibadan, Nigeria

OUTLINE



## Abstract

Antioxidants play important roles in growth, development, detoxification, and effective immune responses. It is reported that the serum concentrations of antioxidants are altered in people living with human immunodeficiency virus (HIV). However, there is the dearth of information on the effects of HIV on the quality of breast milk, especially its antioxidants components. The breast milk of lactating mothers with HIV contains all the antioxidants that are normally found in human milk, although there are slight alterations. Breast milk concentrations of copper, iron, total antioxidant potential, riboflavin, vitamin $B_6$, vitamin C, and folate are lower in lactating mothers with HIV compared with HIV-negative mothers. Other antioxidant trace elements and vitamins are in comparable concentrations. It appears that shedding of HIV is not only the problem with breastfeeding but also a reduction in breast milk concentrations of certain antioxidants that are vital for infant growth and survival.

**Keywords:** Antioxidants; Breast milk; Essential trace metals; Lactating mothers with HIV; Vitamins.

## List of Abbreviations

**Cu** Copper
**DDT** Dichloro-diphenyl-trichloroethane
**DHA** Dehydroascorbic acid


http://dx.doi.org/10.1016/B978-0-12-809853-0.00005-5

**DNA** Deoxyribonucleic acid
**EGF** Epidermal growth factor
**Fe** Iron
**FIL** Feedback inhibitor of lactation
**FRAP** Ferric reducing ability of plasma
**HIV** Human immunodeficiency virus
**Ig** Immunoglobulin
**IGF** Insulin-like growth factor
**IL** Interleukin
**Mn** Manganese
**NADPH** Nicotinamide adenine dinucleotide phosphate (reduced)
**NF-kB** Nuclear factor kappa B
**NGF** Nerve growth factor
**PCBs** Polychlorinated biphenyls
**RNA** Ribonucleic acid
**Se** Selenium
**sIgA** Secretory immunoglobulin A
**SOD** Superoxide dismutase
**TAC** Total antioxidant capacity
**TGF** Transforming growth factor
**TNF-α** Tumor necrosis factor-alpha
**TRAP** Total radical trapping antioxidant parameter
**UNICEF** United Nations Children's Fund
**WHO** World Health Organization
**Zn T-4** Zinc transporter-4
**Zn** Zinc

# INTRODUCTION

Breast milk is a dynamic fluid containing numerous bioactive compounds that are necessary for the survival of the newborn.[1] Its unique significance to the growth and overall health of newborn, especially in the first 6 months of life, is well established. In 1989, the World Health Organization and the United Nations Children's Fund[2] made a joint declaration that the ideal food for infants is breast milk and cannot be equaled by artificial alternatives. This declaration was predicated on the unique properties of breast milk, which protect infants from morbidity and mortality that could result from infection and chronic diseases at a critical time when the immune system is still naïve.

# COMPOSITION OF HUMAN MILK

The normal constituents of human breast milk are numerous (Table 5.1). The first bioactive compounds recognized in human milk were antibodies as transfer of immunity from mother to infant through breast milk was reported as early as 1892.[3] Other components include lactobacilli (*Bifidobacterium bifidum*), secretory immunoglobulin A, and other immunoglobulin classes, leucocytes, lactoferrin, lysozyme, haptocorrin, triglyceride, α-lactalbumin, soluble tumor necrosis factor-alpha receptors, interleukin (IL)-1 receptor antagonist, IL-10 transforming growth factor β, protease inhibitors, hormones, complement factors, antitrypsin, adiponectin, and glycans among others.[4–7] In addition, there are numerous nutritional factors in the breast milk. These factors include macronutrients such as protein and fat, essential trace elements such as selenium (Se), zinc (Zn), copper (Cu), iron (Fe), and vitamins such as vitamins A, B, C, D, E, and K (Table 5.2).

There are myriad of factors that influence the concentration of breast milk constituents. These include stage of lactation (colostral, transitional, mature, and involutional), time of the day, breastfeeding routine, parity, age, regional differences, season of the year, time since last meal, gestational age of the newborn, and other maternal factors such as nutrition. This shows that the composition of breast milk is not uniform, and hence, the daily intake of milk constituents by the breastfed infant is dependent on the differences in the breast milk composition. It is of importance to note that variations in concentration of breast milk constituents are not inversely related to the breast milk volume.

TABLE 5.1 Some Nonnutritional Components of Breast Milk

| Components With Antimicrobial Activity | Components With Inflammatory Functions |
|---|---|
| Immunoglobulins (secretory IgA, IgG, IgM) | Interleukins 1, 6, 8, and 10 |
| Lactoferrin | Tumor necrosis factor-alpha |
| Lysozyme | Tumor growth factor-beta |
| Lactoperoxidase | Alpha-1-antitrypsin |
| Alpha-lactalbumin | |
| Complement factors | |
| **Components With Transport and Absorption Activities** | **Digestive Enzymes** |
| Beta-casein | Lipase |
| Folate-binding protein | Amylase |
| Insulin-like growth factor–binding protein | Esterase |
| Alpha-lactalbumin | |
| **Growth Factors** | **Other Possible Components** |
| Insulin-like growth factors | Viruses |
| Epidermal growth factor | Polychlorinated biphenyls, |
| Nerve growth factors | Dichloro-diphenyl-trichloroethane |
| Transforming growth factor | DNA, RNA |
| | Drugs |

*DNA*, deoxyribonucleic acid; *RNA*, ribonucleic acid.
*This Table was adapted from the reports of Prentice A. Constituents of human milk.* Food Nutr Bull *1996;***17***(4); Lönnerdal B. Nutritional and physiologic significance of human milk proteins.* Am J Clin Nutr *2003;***77***(6):1537S–43S.*

TABLE 5.2 Some Nutritional Factors in Human Milk

| Macronutrients | Micronutrients |
|---|---|
| Carbohydrate (mainly lactose) | ***Vitamins*** |
| Energy | Water-soluble vitamins |
| Protein | (vitamin C, riboflavin, niacin, |
| ***Fat*** | and pantothenic acid) |
| **Total fat** | Fat-soluble vitamins |
| Saturated fatty acids | (vitamins A, D, E, and K) |
| Monounsaturated fatty acids | ***Essential trace elements*** |
| Polyunsaturated fatty acids | Copper (Cu) |
| Linoleic acid | Iron (Fe) |
| Alpha-linolenic acid | Zinc (Zn) |
| Arachidonic acid | Selenium (Se) |
| Docosahexaenoic acid | Manganese (Mn) |
| *Trans* fatty acids | |

## HUMAN IMMUNODEFICIENCY VIRUS SHEDDING INTO BREAST MILK

It is well known that shedding of human immunodeficiency virus (HIV) into breast milk is a feature in mothers with HIV. Approximately 40% of milk samples from HIV-infected mothers contain HIV RNA, but it is likely that only a small percentage of the HIV RNA in breast milk represents infective virions.[8] Postnatal transmission of the virus is thus possible via breastfeeding. Although several attempts, such as flash-heating and pasteurization, are made to eliminate HIV from breast milk, it is still apparent that breastfeeding by mothers living with HIV is unsafe for the newborn. This prompted the advice that HIV-infected mothers who have uninterrupted access to safe breast milk substitutes should avoid breastfeeding. In developing countries, however, this advice against breastfeeding is met with numerous challenges. First, there is a general cultural belief that a child who is not breastfed will not be kind to the mother when he/she grows up. Second, the usual problem of discrimination against people living with HIV is that most mothers, especially the younger and the less educated ones, find it difficult to cope with the discrimination specifically in public places when the need to feed their babies arises. Third, the widespread poverty in the developing countries is another concern. Usually, artificial substitutes are given free of charge to the mothers living with HIV in most health centers; some mothers still find it difficult to transport themselves to the health centers when they exhaust the formula given to them. This compels some mothers to breastfeed their babies pending when they will be able to access the formula again. This mixed-feeding system increases the chance of HIV transmission to the child.

Based on these aforementioned challenges, one would have expected that there will be heightened research on the quality of breast milk in mothers living with HIV, but hitherto, studies on the impact of HIV infection on components of breast milk, especially the antioxidants, are relatively scarce.[9] Efforts in this line are probably militated against by the assumption that all babies breastfed by mothers living with HIV have the risk of being infected. Also, the inaccessibility of formula by mothers living with HIV appears to affect mothers in the developing countries, where often time little attention is given to quality research more than their counterpart in the developed world.

## ANTIOXIDANTS AND HUMAN IMMUNODEFICIENCY VIRUS INFECTION

To achieve effective breastfeeding, breast milk is expected to provide adequate amounts of vital micronutrients, including the vitamins and essential trace elements with antioxidants properties. Generally, antioxidants are substances that prevent the damaging effects of free radicals emanating from aerobic metabolism and/or the actions of xenobiotics among others. They play important roles in growth, development, detoxification, and effective immune responses. It is known that micronutrients are extremely essential in maintaining quality health and development; their need, however, increases when there is trauma or infection such as HIV. Up until now, there is still the dearth in knowledge on the effects of infection on these micronutrients as against the general knowledge on optimal requirements.[10]

Although there are important factors determining the antioxidants component of breast milk and plasma, alteration in plasma levels of antioxidants in people living with HIV is usually attributed to increases in nutrient expenditure, loss of appetite, nutrient malabsorption, urinary loss, redistribution from plasma to tissues, and complex metabolic alterations that result in weight loss and wasting.[11,12] Therefore, deficiencies of vitamins and minerals such as vitamins A, B, C, and E and essential trace metals such as Se, Fe, manganese (Mn), Cu, and Zn have been reported as common observations in people living with HIV.[12,13] Since there is a cross-talk between the blood and the breast milk, a similar deficiency pattern cannot be ruled out in the breast milk of mothers living with HIV. However, very little is known about the mechanisms that control the secretion of micronutrients into human milk.

## BREAST MILK LEVELS OF CU AND FE

Copper (Cu) is an essential trace element that is usually bound to caeruloplasmin and to a lesser extent amino acid and low-molecular chelators. Its concentration in human breast milk is about 20%–25% of that in the serum.[14] Usually, the amount of Cu provided by the breast milk is insignificant with regard to meeting the Cu requirement of the newborn. Hence, newborn term infants have ample stores of Cu primarily in the liver, and these are mobilized in early life.[15] In humans, breast milk Cu concentration declines during lactation. In mature breast milk of mothers living with HIV, Cu level has been reported to be lower compared with mothers not living with HIV.[16] This reduction is thought to be due to low prolactin as a result of poor/nonsuckling. Suckling increases milk Cu secretion due to its direct relationship with circulating prolactin level.[17] In contrast, caeruloplasmin, a Cu transport protein, is higher in the breast milk of lactating mothers with HIV compared with HIV-negative mothers.[9]

The concentration of iron (Fe) in breast milk is low in relation to serum Fe and to estimated Fe requirements of infants. Its concentration in breast milk is about 20%–30% of serum Fe;[14] this relatively low concentration of Fe in human milk is thought to be a form of protection for the infants against Fe toxicity since term infants are born with ample stores that can be utilized in the first 6 months of life.

Serum Fe is almost exclusively bound to transferrin, but the major Fe-binding protein in human milk is lactoferrin and to a lesser extent xanthine oxidase. Usually, Fe is shielded from the body fluids and stored by ferritin to prevent ionic Fe from producing oxidative damage. In infection, there is an acute phase response that increases synthesis of ferritin, thereby causing uptake of Fe with a resultant fall in plasma Fe level.[18]

During lactation, breast milk Fe level declines as lactation progresses. This decline is usually functional as it parallels decrease in transferrin receptor and ferroportin expression, and not due to tissue depletion.[19] In mothers living with HIV, breast milk Fe level is reported to be lower than that in the breast milk of HIV-negative mothers (Table 5.3). In contrast, breast milk transferrin level was found to be elevated in mothers living with HIV compared with HIV-negative mothers.[9] If it were to be in the serum, this picture is that of Fe deficiency anemia. Usually, Fe deficiency anemia is indicated when there is low Fe concentration with elevated transferrin concentration.

TABLE 5.3 Breast Milk Concentrations of Antioxidant Trace Elements and Total Antioxidant Potential in Human Immunodeficiency Virus–Lactating Mothers and Human Immunodeficiency Virus–Negative Mothers[16]

| Trace Elements | HIM | HNM |
|---|---|---|
| Zn (μg/dL) | 146 ± 20 | 148 ± 16 |
| Cu (μg/dL) | 68 ± 5[a] | 71 ± 5 |
| Fe (μg/dL) | 66 ± 6[a] | 71 ± 6 |
| Se (μg/L) | 64 ± 6 | 65 ± 3 |
| Mn (μg/dL) | 61 ± 7 | 64 ± 5 |

*Cu*, copper; *Fe*, iron; *HIM*, *HIV*-infected mothers; *HNM*, *HIV*-negative mothers; *Mn*, manganese; *Se*, selenium; *Zn*, zinc.
[a]*Significantly different at* P < .05.

TABLE 5.4 Correlation Between Plasma and Breast Milk Concentrations of Antioxidant Trace Elements and TAP in Human Immunodeficiency Virus–Lactating Mothers and Human Immunodeficiency Virus–Negative Mothers[16]

| | HIM | | HNM | |
|---|---|---|---|---|
| Trace Elements | r-values | *P*-values | r-values | *P*-values |
| Cu (μg/dL) | −0.355 | 0.135 | −0.093 | 0.626 |
| Zn (μg/dL) | 0.042 | 0.866 | −0.249 | 0.184 |
| Se (μg/L) | −0.112 | 0.647 | 0.299 | 0.108 |
| Fe (μg/dL) | −0.022 | 0.928 | 0.233 | 0.215 |
| Mn (μg/dL) | 0.496 | 0.043[a] | −0.157 | 0.406 |
| TAP (μmolTE/L) | −0.111 | 0.651 | 0.000 | 0.999 |

*Cu*, copper; *Fe*, iron; *HIM*, *HIV*-infected mothers; *HNM*, *HIV*-negative mothers; *Mn*, manganese; *Se*, selenium; *TAP*, total antioxidant potential; *TE*, trolox equivalent; *Zn*, zinc.
[a]*Significantly different at* P<.05.

Therefore, aside from the danger of shedding HIV in the breast milk, there is even a nutritional danger of Fe deficiency in breastfed infants of lactating mothers with HIV. Unfortunately, infants fed with formula usually have poor Fe status, in spite of the fact that formula contains up to three times more Fe than does breast milk.[20] Thus, infants depend more on breast milk Fe to meet their needs.

Domelleof et al.[21] reported that no correlation exists between human milk Fe and iron-status variables in the serum. This is not different in mothers living with HIV as no significant correlation was observed between the plasma Fe level and that of the breast milk (Table 5.4).[16] This thus indicates that there is a need to ensure optimal Fe supplement in infants (and not maternal Fe supplementation) of mothers living with HIV who are breastfed or those fed with treated (flash-heated or pasteurized) breast milk.

## BREAST MILK LEVEL OF ZN

During infection, there is increased synthesis of metallothionein, leading to the uptake of Zn into the liver, thereby causing a fall in plasma Zn level. Similarly, plasma albumin usually falls during acute phase response, thereby causing a fall in plasma Zn level since albumin is its binding protein.[10]

Unlike Fe and Cu, serum Zn concentration is considerably lower than that in the breast milk.[14] There is an indication that mechanisms ensuring mammary gland Zn uptake as well as subsequent secretion into the breast milk are so effective that the increased level of Zn persists for the first several months of lactation.[14] Every day, more than 0.5–1.0 mg of Zn is taken up by the mammary gland and secreted into the breast milk. This amount is almost twice that of Zn transferred across the placenta to the fetus during late pregnancy; this illustrates the remarkable capacity of the mammary gland to transport Zn.[21] It has been shown that breast milk Zn concentration is similar in women with low Zn status and those who receive daily supplements, thereby showing that homeostasis of breast milk Zn transfer is tightly regulated.[22]

During human lactation, the concentration of breast milk Zn reduces, whereas plasma Zn level increases.[23] This has been attributed to relocalization of Zn transporter-4. No difference has been reported between the concentrations of Zn in the breast milk of mothers living with HIV and mothers without HIV. Rahamon et al.[16] showed that the breast milk Zn levels are similar in both groups and there was no correlation between breast milk and plasma Zn levels. Also, albumin, a Zn carrier, which was lower in plasma of the mothers living with HIV, was not affected in the breast milk. This indicates that metabolic changes associated with HIV might not affect the mammary gland Zn uptake as well as its secretion into the breast milk in mothers living with HIV. However, some women, irrespective of HIV status, are known to have abnormally low breast milk Zn level, hence, placing their infants at the risk of transient neonatal Zn deficiency. Infants of such women who are breastfed without any form of Zn supplementation usually experience severe eczema and decreased growth by 2–3 months of age. This is even exacerbated in premature infants due to their lower Zn stores at birth. In this group of women, breast milk Zn concentration cannot be increased by maternal Zn supplementation.[24]

## BREAST MILK LEVELS OF SE AND MN

Selenium (Se) is an essential component of about 25 selenoproteins, which are involved in protection against oxidative damage. It plays a vital role in a number of physiological processes such as immune response and viral suppression. In HIV, there is Se deficiency, which is related to faster disease progression and to mortality, especially in children. Its supplementation has been shown to stimulate glutathione peroxidase activity, thereby reducing nuclear factor kappa B activation in HIV-1–infected cell lines.[25]

In infant nutrition, there is a particular emphasis on Se because breast milk is its only source during this most rapid period of growth.[26] Breast milk Se level is dependent on maternal intake, which depends on where she lives, the soil on which what she consumes are grown, and supplementation. During lactation, Se level in the breast milk decreases significantly through the phases of lactation.

Like other essential trace elements, there is presently little information on the impact of HIV infection on human breast milk Se concentration. Rahamon et al.[16] reported that the breast milk Se level, as well as plasma Se level, of lactating mothers living with HIV was not different from that of HIV-negative mothers. This probably indicates that Se concentration in the breast milk of mothers living with HIV might be optimal when there is adequate plasma concentration.

Manganese (Mn) is present in minute quantity in the body. It is a component of the antioxidant enzyme superoxide dismutase, which is the principal antioxidant enzyme in the mitochondria. Neonates have been found to be in negative Mn balance after birth. Interestingly, human breast milk also contains a very low level of Mn. This might be a physiological measure as early life exposure to elevated Mn level is associated with poor neurodevelopment and cognitive performance resulting from the ability of developing nerve cells to highly express receptors for manganese transport protein and the immaturity of the liver, which impedes its bile elimination system. Although there is higher Mn content in most formula, Mn from human breast milk is still believed to be more available than that from the formula.[27]

Similar to Se and Zn, breast milk concentration of Mn does not appear to be affected by HIV infection. The Mn concentration in the breast milk of mothers living with HIV is similar to that of mothers without HIV. However, there seems to be a positive correlation between plasma and breast milk levels of Mn in mothers living with HIV.[16] This suggests that their plasma Mn level probably dictates the quantity that is found in the breast milk.

## VITAMINS IN HUMAN MILK

Other vital micronutrients required for health and growth are vitamins. They are organic compounds required in minute quantities in the body, especially for their catalytic functions. All of the types of vitamins are present in human milk in nutritionally significant concentration except vitamin K.

Adequate amount of vitamins in the breast milk is essential for infant's health. Generally, the type and concentrations of vitamins in the breast milk is dependent on the maternal type and concentrations. This underscores the simple fact that adequate intake by mothers is a prerequisite for adequate concentration in the breast milk. As with the essential trace elements in the breast milk, there is lack of information on the effect of infections such as HIV on the quality and quantity of vitamins in the breast milk of mothers living with HIV.

## VITAMINS A AND B IN HUMAN MILK

Vitamin A is a micronutrient that is usually obtained from animal-derived foods as preformed vitamin A or as provitamin A carotenoids, which is converted to retinol after absorption. It is required for growth, immune function, and bone remodeling among others. Due to its numerous functions, its excess or deficiency impacts negatively on human health, especially in children. For example, vitamin A deficiency is associated with childhood blindness, diarrhea, respiratory illnesses, depressed immune function, and increased risk of mortality. In contrast, excess intake, resulting from acute intoxication or chronic ingestion, can cause hypervitaminosis A.[28]

Vitamin A in the breast milk is well absorbed by children as breast milk also contains lipase that helps in vitamin A digestion. Adequate transfer of vitamin A from mother to child is very essential for child survival. Usually, mother-to-child transfer of vitamin A occurs during gestation and lactation. After birth, infants of well-nourished mothers have small vitamin A reserve, which gets augmented as soon as lactation begins. Although vitamin A belongs to the group of breast milk components that are affected by maternal concentration, its concentration varies during the phase of lactation with colostrum having the highest concentration. Mother's milk continues to be the best delivery source of vitamin A to the infants, and hence, diseases such as xerophthalmia are practically not found in fully breastfed infants.[29]

The B vitamins are water-soluble vitamins performing myriads of important functions such as carboxylation, decarboxylation, transamination, acylation, oxidation, reduction, methyl group transfer, and energy utilization among others. They are made up of eight vitamins consisting thiamine ($B_1$), riboflavin ($B_2$), niacin ($B_3$), pantothenic acid ($B_5$), pyridoxine ($B_6$), biotin ($B_7$), folate ($B_9$), and cobalamin ($B_{12}$). Although B vitamins are usually grouped together, each of the various types performs unique but synergistic functions. Depending on the type, B vitamins are commonly found in cereals and foods of animal and plant origins. Their deficiencies are associated with a number of diseases mainly due to the resulting impairment in their numerous functions.

All B vitamins are found in the breast milk and their concentration, except folate, largely depends on maternal status as maternal depletion can rapidly and substantially reduce their secretion into the breast milk.[30] This explains why diseases such as beriberi and epileptiform convulsions are found in exclusively breastfed infants of mothers with certain B vitamins deficiency. During postpartum period, the concentrations of B vitamins in the breast milk change. However, exclusively breastfed infants between 0 and 6 months derive adequate B vitamins from breast milk alone.

## VITAMINS C AND D IN HUMAN MILK

Vitamin C are molecules with antiscorbutic properties. These include ascorbic acid and its oxidized form, dehydroascorbic acid. It is a water-soluble vitamin found in many fruits and vegetables. Many of the biologic properties of vitamin C are linked to its antioxidant properties as it can donate a hydrogen atom to form a radical that is reduced back to ascorbate by glutathione or reduced nicotinamide adenine dinucleotide phosphate. Vitamin C is essential to the health and development of infants due to its role in the synthesis of collagen, tendons, and bone with its deficiency usually manifesting as scurvy. To prevent scurvy, therefore, it is necessary to consume vitamin C regularly since it is water soluble and cannot be stored in the body.[31]

Vitamin C is an essential nutrient in human milk, which is dependent on maternal dietary intake. At birth, vitamin C concentration in infant's blood is higher than that in the maternal blood.[32] Its concentration in breast milk is also higher than in maternal blood, but the breast milk concentration declines progressively as the lactation progresses. Irrespective of the lactation phase, breastfed infants of vitamin C-replete mothers are still able to meet the recommended daily intake during the first 6 months of life.

Vitamin D, a fat-soluble vitamin, can be obtained from diets and can be synthesized endogenously. It performs an essential function in bone metabolism as its deficiency results in impaired bone metabolism manifesting as rickets in children and osteomalacia in adults. Its deficiency in children has also been associated with increased risk of respiratory tract infections.

During gestation, there is placental transfer of vitamin D to the fetus. This transfer is dependent on maternal vitamin D status as maternal supplementation increases the breast milk concentration. The concentration of vitamin D in breast milk of healthy lactating mothers is low and insufficient to prevent vitamin D deficiency in exclusively breastfed infants. Therefore, intake of vitamin D from breast milk and adequate sunlight exposure are required to maintain optimal concentration to avoid rickets after birth.[33]

## VITAMINS E AND K IN HUMAN MILK

Vitamin E is a fat-soluble vitamin with established roles in antioxidant defense system and cellular respiration. It is mainly obtained from oils and fats while meats, fruits, and vegetables contribute little quantity of vitamin E. There is a synergistic interaction between vitamins C and E in lipoperoxidation reduction as they help regenerate the reduced forms of each other.

Breast milk continues to be the most important source of vitamin E to newborn as placental transfer of vitamin E is usually limited. It becomes even more important in premature and low-birth-weight infants who are characterized by poor placenta transfer and limited adipose tissue.

Breast milk concentration of vitamin E decreases as the milk becomes mature. This change in concentration is independent of parity, anthropometric nutritional status, socioeconomic status, and habitual dietary intake of vitamin E by the mother.[34] Pockets of report have, however, shown that maternal plasma level of vitamin E determines its content in the transitional milk but not the colostrum. This, thus, suggests that vitamin E uptake by the mammary gland involves distinct transfer mechanisms.[35]

Vitamin K is another member of the fat-soluble vitamins family. It plays important roles in blood clotting and bone metabolism. Margarines, plant oils, milk products, vegetables, and eggs are common sources of vitamin K. There is limited placenta transfer of vitamin K during gestation and haemorrhagic disease of the newborn manifests when there is poor placenta transfer.

Concentration of vitamin K in breast milk tends to decrease over the lactation period. Early breast milk and mature milk have low vitamin K contents, which cannot meet the recommended intake for infants especially in the first 6 months of life.[10] Unfortunately, supplementation with vitamin K is usually not recommended in exclusively breast-fed infants as their liver is still immature and very high doses of vitamin K can elevate serum bilirubin concentration.[36]

Reduced plasma concentrations of vitamins, especially the fat-soluble class, are found in pregnant women living with HIV. This necessitated the strong suggestion for multivitamin supplementation in this group of people. Even with supplementation, however, HIV disease especially at the advanced stage can suppress vitamin release from their stores. For example, vitamin A release from the liver is suppressed by HIV, thereby leading to its low plasma level despite adequate liver stores.[37] Adequate maternal vitamin pool as well as multivitamin supplementation especially with vitamins B, C, and E reduces adverse pregnancy outcomes and possibly mother-to-child transfer of HIV.

## VITAMINS IN THE BREAST MILK OF HUMAN IMMUNODEFICIENCY VIRUS–LACTATING MOTHERS

Presently, there is little information on the concentrations of vitamins in the breast milk of mothers living with HIV. A South African study showed that lactating mothers with HIV have low serum concentrations of retinol, folate cobalamin, and alpha-tocopherol.[38] In addition, serum alpha-tocopherol reduces progressively during the first 6 months after delivery. This indicates that the breast milk concentrations of these vitamins could also be low since the breast milk concentration of most vitamins, especially the water-soluble vitamins, largely depend on maternal status. As in serum, there is the dearth of information on vitamin concentrations in the breast milk of lactating HIV mothers. In a study assessing the effects of flash-heat treatment on vitamin concentrations in the breast milk of lactating mothers with HIV in South Africa, it was observed that the median concentrations of riboflavin, $B_6$, vitamin C, and folate in unheated breast milk were lower compared with concentrations in well-nourished women (Table 5.5).[30,39,40] Vitamin A was, however, an exception as its median concentration was higher in the breast milk of lactating mothers with HIV compared with the reported concentration in well-nourished women.[39,40] This single report is not enough to conclude on the vitamin concentrations in the breast milk of lactating mothers with HIV, and hence, more studies are suggested.

## TOTAL ANTIOXIDANT CAPACITY IN BREAST MILK OF HUMAN IMMUNODEFICIENCY VIRUS–LACTATING MOTHERS

The practical impossibility of measuring all antioxidants in samples of interest leads to the concept of total antioxidant capacity (TAC). This concept assesses the reductant properties of antioxidants against free radicals or the delayed production of a measurable free radical. There are two common methods for assessing TAC; these include the total radical trapping antioxidant parameter assay or ferric reducing ability of plasma assay.[10] As interesting as

TABLE 5.5 Breast Milk Concentrations of Vitamins in Human Immunodeficiency Virus–Lactating Mothers and Human Immunodeficiency Virus–Negative Well-Nourished Mothers

| Vitamins | HIM[a39] | Well-Nourished Women[40] |
|---|---|---|
| Riboflavin | 0.01 mg/L | 0.35 mg/L |
| $B_6$ | 0.04 mg/L | 0.13 mg/L |
| A | 637.00 μg/L | 485.00 μg/L |
| C | 27.00 mg/L | 50.00 mg/L |
| Folate | 19.45 μg/L | 81.00 μg/L[30] |

*HIM, HIV*-infected mothers.
[a]*Median concentration.*
*This Table was adapted, with slight modification, from the reports of Allen LH. B vitamins in breast milk: relative importance of maternal status and intake, and effects on infant status and function.* Adv Nutr *2012;3(3):362–69; Allen LH. Multiple micronutrients in pregnancy and lactation: an overview.* Am J Clin Nutr *2005;81(5):1206–12S; Israel-Ballard KA, Abrams BF, Coutsoudis A, Sibeko LN, Cheryk LA, Chantry CJ. Vitamin content of breast milk from HIV-1-infected mothers before and after flash-heat treatment.* J Acquir Immune Defic Syndr *2008;48(4):444–49.*

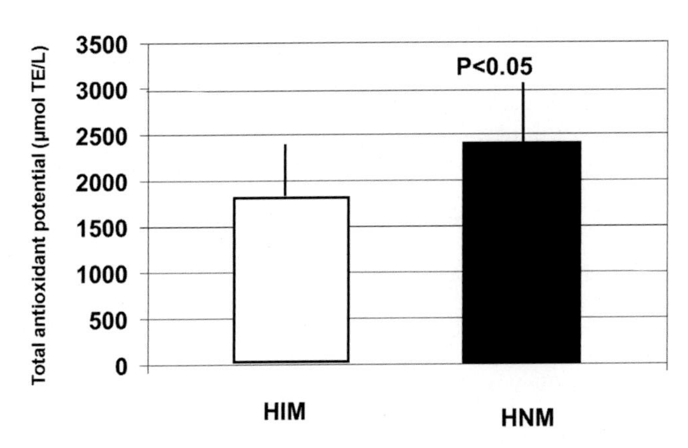


FIGURE 5.1 Breast milk concentrations of total antioxidant potential (μmolTE/L) in human immunodeficiency virus–lactating mothers and human immunodeficiency virus–negative mothers.[16] *HIM, HIV*-infected mothers; *HNM, HIV*-negative mothers.

this concept might be, it is usually affected by endogenous antioxidants such as albumin and uric acid whose levels are affected by the prevailing clinical condition. For example, acute phase response or change in renal function can alter TAC result without reflecting the changes in antioxidant vitamins and trace elements concentration. However, this problem can be overcome by calculating the antioxidant gap.

Effect of HIV infection on the TAC in breast milk is poorly understood. A Nigerian study showed that TAC was significantly lower in the breast milk of lactating mothers with HIV compared with HIV-negative mothers (Fig. 5.1). A similar, but insignificant, reduction was observed in the maternal plasma. This reduction in the total antioxidants concentration in the breast milk of lactating mothers with HIV is thought to be due to low concentrations of antioxidant trace metals such as Cu and Fe in their milk.

## CONCLUSION

From the few available reports on antioxidants in breast milk of lactating mothers with HIV, it appears that shedding of HIV is not only the problem with breastfeeding but also a reduction in breast milk concentrations of certain antioxidants that are vital for infant growth and survival. Therefore, multiantioxidants supplementation could be beneficial in infants of lactating mothers with HIV who are exclusively breastfed or fed with flash-heated or pasteurized breast milk.

## SUMMARY POINTS

- The breast milk of lactating mothers with HIV contains all the dietary antioxidants that are normally found in human milk.
- Breast milk concentrations of copper (Cu), iron (Fe), total antioxidant potential, riboflavin, vitamin $B_6$, vitamin C, and folate are lower in lactating mothers with HIV compared with HIV-negative mothers.
- Other antioxidant trace elements and vitamins are in comparable concentrations in lactating mothers with HIV and HIV-negative mothers.
- Shedding of HIV might not only be the problem with breastfeeding but also a reduction in breast milk concentrations of certain antioxidants that are vital for infant growth and survival.
- Multiantioxidants supplementation could be beneficial in infants of lactating mothers with HIV who are exclusively breastfed or fed with flash-heated or pasteurized breast milk.
- Due to the little available information, there is an urgent need for more information on breast milk concentrations of antioxidants in lactating mothers with HIV.

## References


1. World Health Organization. Collaborative study team on the role of breastfeeding on the prevention of infant mortality. Effect of breastfeeding on infant and child mortality due to infectious diseases in less developed countries: A pooled analysis. *Lancet* 2000;**355**:451–5.
2. World Health Organization (WHO) and United Nations Children's Fund (UNICEF). *Joint statement protecting, promoting and supporting breast-feeding: The special role of maternity services*. 1989. Geneva.
3. Newburg DS. Oligosaccharides in human milk and bacterial colonization. *J Pediatr Gastroenterol Nutr* 2000;**30**:S8–17.
4. Hanson L. Comparative immunological studies of the immune globulins of human milk and blood serum. *Int Arch All Appl Immunol* 1961;**18**:241–67.
5. Buescher ES. Anti-inflammatory characteristics of human milk: how, where, why. *Adv Exp Med Biol* 2001;**501**:207–22.
6. Newburg DS, Walker WA. Protection of the neonate by the innate immune system of developing gut and of human milk. *Pediatr Res* 2007;**61**(1):2–8.
7. Prentice A. Constituents of human milk. *Food Nutr Bull* 1996;**17**(4).
8. Lewis P, Nduati R, Kreiss JK, John GC, Richardson BA, Mbori-Ngacha D, Ndinya-Achola J, Overbaugh J. Cell-free human immunodeficiency virus type 1 in breast milk. *J Infect Dis* 1998;**177**(1):34–9.
9. Rahamon SK, Arinola GO. Immunoglobin classes and acute phase proteins in the breast milk and plasma of Nigerian HIV-infected lactating mothers. *Eur J Gen Med* 2012;**9**(4):241–6.
10. Shenkin A, Baines M, Fell GS, Lyons TDG. Vitamins and trace elements. In: Burtis C, Ashwood E, Bruns D, editors. *Tietz textbook of clinical chemistry and molecular diagnostics*. Missouri: Elsevier Saunders; 2006. p. 1075–164.
11. Shenkin A. Trace elements and inflammatory response: implications for nutritional support. *Nutrition* 1995;**11**(1 Suppl.):100–5.
12. Piwoz EG, Bentley ME. Women's voices, women's choices: the challenge of nutrition and HIV/AIDS. *J Nutr* 2005;**135**(4):933–7.
13. Arinola OG, Adedapo KS, Kehinde AO, Olaniyi JA, Akiibinu MO. Acute phase proteins, trace elements in asymptomatic human immunodeficiency virus infection in Nigerians. *Afr J Med Med Sci* 2004;**33**(4):317–22.
14. Lönnerdal B. Trace element transport in the mammary gland. *Annu Rev Nutr* 2007;**27**:165–77.
15. Olivares M, Araya M, Uauy R. Copper homeostasis in infant nutrition: deficit and excess. *J Pediatr Gastroenterol Nutr* 2000;**31**:102–11.
16. Rahamon SK, Arinola GO, Akiibinu MO. Total antioxidant potential and essential trace metals in the breast milk and plasma of Nigerian human immunodeficiency virus-infected lactating mothers. *J Res Med Sci* 2013;**18**(1):27–30.
17. Kelleher SL, Lönnerdal B. Mammary gland copper transport is stimulated by prolactin through alterations in Ctr1 and Atp7A localization. *Am J Physiol Regul Integr Comp Physiol* 2006;**291**(4):R1181–91.
18. Cousins RJ, Leinart AS. Tissue-specific regulation of zinc metabolism and metallothionein genes by interleukin 1. *FASEB J* 1988;**2**:2884–90.
19. Leong WI, Lönnerdal B. Iron transporters in rat mammary gland: effects of different stages of lactation and maternal iron status. *Am J Clin Nutr* 2005;**81**(2):445–53.
20. Pizarro F, Yip R, Dallman PR, Olivares M, Hertrampf E, Walter T. Iron status with different infant feeding regimens: relevance to screening and prevention of iron deficiency. *J Pediatr* 1991;**118**(5):687–92.
21. Domellöf M, Lönnerdal B, Dewey KG, Cohen RJ, Hernell O. Iron, zinc, and copper concentrations in breast milk are independent of maternal mineral status. *Am J Clin Nutr* 2004;**79**(1):111–5.
22. King JC. Enhanced zinc utilization during lactation may reduce maternal and infant zinc depletion. *Am J Clin Nutr* 2002;**75**(1):2–3.
23. Dewey KG, Lönnerdal B. Milk and nutrient intake of breast-fed infants from 1 to 6 months: relation to growth and fatness. *J Pediatr Gastroenterol Nutr* 1983;**2**(3):497–506.
24. Atkinson SA, Whelan D, Whyte RK, Lönnerdal B. Abnormal zinc content in human milk. Risk for development of nutritional zinc deficiency in infants. *Am J Dis Child* 1989;**143**(5):608–11.
25. Sappey C, Legrand-Poels S, Best-Belpomme M, Favier A, Rentier B, Piette J. Stimulation of glutathione peroxidase activity decreases HIV type 1 activation after oxidative stress. *AIDS Res Hum Retroviruses* 1994;**10**(11):1451–61.
26. Zachara BA, Pilecki A. Selenium concentration in the milk of breast-feeding mothers and its geographic distribution. *Environ Health Perspect* 2000;**108**(11):1043–6.
27. Chan WY, Bates Jr JM, Raghib MH, Rennert OM. Bioavailability of manganese in milk studied in in vitro and in vivo systems. In: Schramm VL, Wedler FC, editors. *Manganese in metabolism and enzyme function*. New York: Academic Press; 1986. p. 17.

28. Haskell MJ, Brown KH. Maternal vitamin A nutriture and the vitamin A content of human milk. *J Mammary Gland Biol Neoplasia* 1999;**4**(3):243–57.
29. Stoltzfus RJ, Underwood BA. Breast-milk vitamin A as an indicator of the vitamin A status of women and infants. *Bull World Health Organ* 1995;**73**(5):703–11.
30. Allen LH. B vitamins in breast milk: relative importance of maternal status and intake, and effects on infant status and function. *Adv Nutr* 2012;**3**(3):362–9.
31. Francis J, Rogers K, Brewer P, Dickton D, Pardini R. Comparative analysis of ascorbic acid in human milk and infant formula using varied milk delivery systems. *Int Breastfeed J* 2008;**3**:19.
32. Macy I, Kelly HJ. Human and Cow's milk in infant nutrition. In: Kon SK, Cowie AT, editors. *Milk the mammary gland and its secretion*, vol. 2. New York, NY, USA: Academic Press; 1961. p. 265–304.
33. Dawodu A, Tsang RC. Maternal vitamin D status: effect on milk vitamin D content and vitamin D status of breastfeeding infants. *Adv Nutr* 2012;**3**(3):353–61.
34. Lima MS, Dimenstein R, Ribeiro KD. Vitamin E concentration in human milk and associated factors: a literature review. *J Pediatr (Rio J)* 2014;**90**(5):440–8.
35. da Silva Ribeiro KD, Lima MS, Medeiros JF, de Sousa Rebouças A, Dantas RC, Bezerra DS, Osório MM, Dimenstein R. Association between maternal vitamin E status and alpha-tocopherol levels in the newborn and colostrum. *Matern Child Nutr* 2016. http://dx.doi.org/10.1111/mcn.12232.
36. Lucey JF, Doln RG. Hyperbilirubinemia of newborn infants associated with the parenteral administration of a vitamin K analogue to the mothers. *Pediatrics* 1959;**23**(3):553–60.
37. Mehta S, Fawzi W. Effects of vitamins, including vitamin A, on HIV/AIDS patients. *Vitam Horm* 2007;**75**:355–83.
38. Papathakis PC, Rollins NC, Chantry CJ, Bennish ML, Brown KH. Micronutrient status during lactation in HIV-infected and HIV-uninfected South African women during the first 6 months after delivery. *Am J Clin Nutr* 2007;**85**(1):182–92.
39. Israel-Ballard KA, Abrams BF, Coutsoudis A, Sibeko LN, Cheryk LA, Chantry CJ. Vitamin content of breast milk from HIV-1-infected mothers before and after flash-heat treatment. *J Acquir Immune Defic Syndr* 2008;**48**(4):444–9.
40. Allen LH. Multiple micronutrients in pregnancy and lactation: an overview. *Am J Clin Nutr* 2005;**81**(5):1206S–12S.

CHAPTER

# 6

# Oxidative Stress in HIV in Relation to Metals

*Zephy Doddigarla[1], Lingidi J. Lakshmi[1], Jamal Ahmad[2], Muhammad Faisal[2]*

[1]Mayo Institute of Medical Sciences, Barabanki, India; [2]Aligarh Muslim University, Aligarh, India

OUTLINE



## Abstract

Trace metals have been associated with adverse clinical outcomes during human immunodeficiency virus infection (HIV infection), and new studies are emerging, which suggest that micronutrient supplementation may help reduce morbidity and mortality during HIV infection. Naturally, when an organism enters into a host, numerous changes take place due to the unfavorable presence of the organism since, the inherent quality (immunity) of the host intensifies the modulation of immune system. Such immune responses are active over time in a host, and alter general immune mechanisms especially antioxidant systems, which over time affect the whole body. In the process, depletes tissues trace metal reserve resources in order to counter reactive oxygen species generated by the virus in the host. To maintain the integrity of immune system, therefore, it is very essential to manage trace metal concentrations in HIV-infected individual. Thus, concentrations of trace metals play a vital role in the progression of the disease.

**Keywords:** Human immunodeficiency virus (HIV); Nrf-2-antioxidant responsive elements (Nrf-2-ARE); Nuclear factor erythroid2-related factor 2 (Nrf-2); Nuclear factor kappa-light-cascade-enhancer of activated B cells (NF-κB); Reactive oxygen species (ROS); Tumor necrosis factor-α (TNF-α).

## List of Abbreviations

**ATP** Adenosine triphosphate
**CD-4** Cluster of differentiation-4
**DNA** Deoxyribonucleic acid
**dNTPs** Deoxyribonucleotide tripohosphates
**ETC** Electron transport chain
**HIV** Human immunodeficiency virus



*HIV/AIDS*
http://dx.doi.org/10.1016/B978-0-12-809853-0.00006-7

**NADPH oxidase** Nicotinamide adenine dinucleotide phosphate oxidase
**NF-κB** Nuclear factor kappa-light-cascade-enhancer of activated B cells
**NOS** Nitric oxide synthases
**HIF-1α** Hypoxic inducible factor-1alpha
**HIV LTR** HIV long terminal repeats
**IκB** Inhibitory kappa kinase beta
**Nrf-2** Nuclear factor erythroid2-related factor 2
**Nrf-2-ARE** Nrf-2-antioxidant responsive elements
**ROS** Reactive oxygen species
**SOD** Superoxide dismutase
**TNF-α** Tumor necrosis factor-α

## INTRODUCTION

Long-term relationship between an organism and a host, one of the things, turns out. For instance, an organism may evolve to become less harmful to its host, or a host may evolve to cope with the unfavorable presence of that organism.[1] But, none of the things turn out in case of human immunodeficiency virus infection (HIV infection), for instance, neither organism (HIV) evolve to become less harmful to its host (human) or host (human) evolve to cope with the unfavorable presence of that organism (HIV). Since the inherent quality (immunity) of the host is not potent enough to cope with or to kill the virus.[2] Meanwhile, the host modulates robust immunological defenses to prevent the organism to enter or to kill the organism.[2–4] In the process to eliminate the HIV infection from the human body, the immunological responses like antioxidants and trace metals reserve resources gets depleted. This depletion of antioxidants and trace metals availability is due to the counter action of the continuous production of reactive oxygen species (ROS) due to the presence of organism in the host.[5,6] In addition, as these immunological responses are active over time in a host, therefore, they further affect antioxidant systems directly[2–4] and trace metal concentrations indirectly. For example, trace metals including zinc, copper, manganese, selenium, cobalt, nickel, molybdenum, and chromium are essential factors for antioxidants and also for immunological functions in humans.[7–9] On the other hand, literature also reports that some trace metals enhance severity of the HIV infection with metals such as iron and copper.[7–12] Moreover, surveys consistently have shown that trace metals decrease in HIV-infected individuals when compared to healthy controls.[12–14] Thus, concentrations of trace metals play a vital role in the HIV disease progress. The present chapter emphasizes how oxidative stress contributes to trace elements deficiency and thereafter the pathological aspects that arise from the metals concentration in the host immune structure due to HIV.

## REACTIVE OXYGEN SPECIES AND HUMAN IMMUNODEFICIENCY VIRUS ENTRY

The HIV gains entry into the immune cells by either alpha-chemokine receptors or chemokine receptor 5 receptors.[2] After HIV entry into the immune cells, initially the virus remains in a quiescent state.[3] A unique feature of the retroviral life cycle is it weakens the host immune system making the host unable to cope with or kill the virus.[3] Interestingly, this process happens prior to HIV activation to make favorable conditions for its survival. Let us see how HIV weakens the immune system.

Normally, ROS, including superoxide anion, hydrogen peroxide, hydroxyl radical, hypochlorous acid, hydroperoxide radical, nitric oxide, and peroxynitrite, are present in the human body.[5,6] The superoxide produce is during electron leakage from the mitochondrial electron transfer chain and during enzymatic reactions, including nicotinamide adenine dinucleotide phosphate (NADPH) oxidase, xanthine oxidase, and aldehyde oxidase (Fig. 6.1). Conversion of superoxide to hydrogen peroxide is through the enzyme superoxide dismutase (SOD). Meanwhile, the enzyme that generates hypochlorous acid is myeloperoxidase (Fig. 6.1). Physiologically, nitric oxide is an important vasodilator and produced by enzyme nitric oxide synthases (NOS). Inducible-NOS, endothelial-NOS, and neuronal-NOS are three isoforms of NOS present in the human body. Peroxynitrite forms when nitric oxide reacts with superoxide. Of all ROS, hydroxyl radical and peroxynitrite are the most reactive.

In a healthy human being, low or moderate levels of ROS principally play a critical role in growth, reproduction, and replication. Previous studies have shown the balance of these species is vital for various physiological reactions.[7,8] The disproportionate increase of ROS is associated with pathogenesis of many infections.[5–8] Likewise, increase in ROS has been shown to involve with HIV pathogenesis.[3,4] The factors that lead to ROS increase depend on the nutritional status of trace metals and also the optimum functioning of the host immune system.

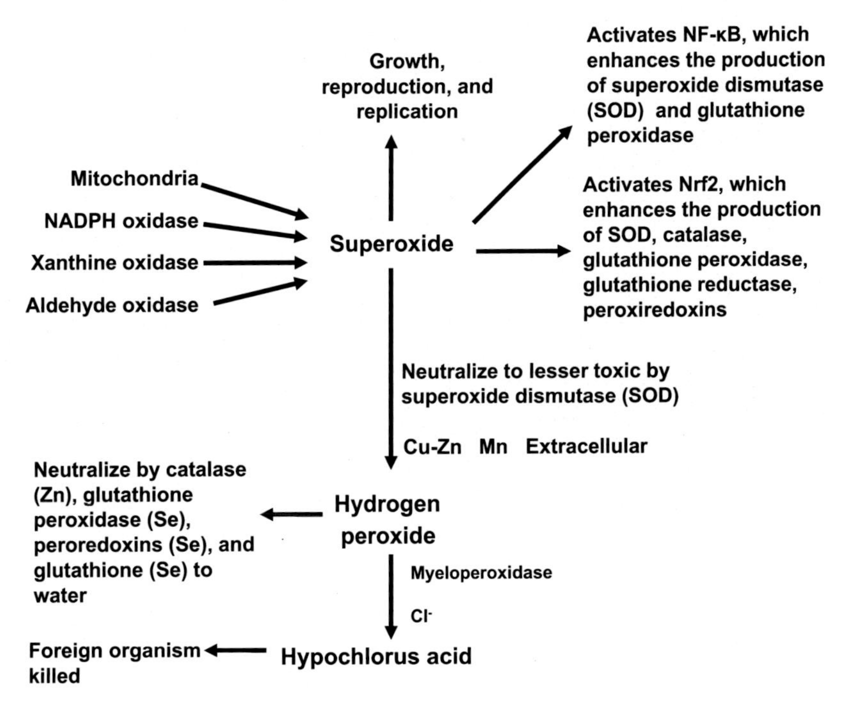


FIGURE 6.1 Sources of reactive oxygen species and their combat mechanism. *Cl*, chlorine; *Cu–Zn*, copper–zinc; *Mn*, manganese; *NF-κB*, nuclear factor kappa-light-cascade-enhancer of activated B cells; *Nrf-2*, nuclear factor erythroid2-related factor 2; *Se*, selenium; *Zn*, zinc.

## EFFECT OF REACTIVE OXYGEN SPECIES ON TRACE METALS

Although, HIV remains quiet in the initial stage of its entry into the host, the physiological fighting mechanisms of host try to remove the infection. This is done by generating superoxide radicals (Fig. 6.2). Kashou et al.[15] reported uptake of oxygen initiates superoxide production and subsequent activation of the NADPH oxidase. Consistently, studies have observed superoxide in peripheral blood mononuclear cells in the primary stage of HIV infection. Studies have also shown generation of superoxide radical in lymphocytes and macrophages in the primary stage of HIV infection.[5,6,16] This production of superoxide is enhanced as generation of superoxide could not cope with the HIV infection. Further, HIV Tat protein triggers ROS production via NADPH oxidases, by induction of spermine oxidase and by mitochondrial dysfunction.[17,18] HIV Vpr protein is another important factor in enhancing ROS production.[19] It was shown that Vpr causes mitochondrial dysfunction and ROS production in mitochondria.[19,20] In accordance, a study has observed constitutive production of hydrogen peroxide by neutrophils in the early stages.[21] This ascertains the fact that there is increased production of SOD during the primary stage of HIV infection. Basically, SOD exists in three isoforms such as copper–zinc-SOD (in cytoplasm), manganese-SOD (mitochondrial matrix), and extracellular-SOD (outside cytoplasm).[22] And also molybdenum is a component of pterin coenzyme essential for xanthine oxidase and aldehyde oxidase activities.[7,8] From the earlier studies, the activities of SOD, NADPH oxidase, and xanthine oxidase have been observed to increase in HIV-infected individuals.[5,6,15,16] In view of this, therefore, the need for trace elements such as zinc, copper, manganese, and molybdenum is enhanced. The need of trace metals is consequently adjusted through the reserve stores (Fig. 6.2). On the contrary, superoxide and hydrogen peroxide production are enhanced because the generation of superoxide could not cope with or kill the HIV infection.

## EFFECT OF IRON AND COPPER ON OTHER TRACE METALS (FIG. 6.3)

Next, the increase in hydrogen peroxide gives rise to hydroxyl ion (Fig. 6.3) since, the hydrogen peroxide undergoes Fenton and Haber–Weiss reactions in the presence of iron and copper metals, generating hydroxyl radical.[9] Increase in hydroxyl radicals is harmful to the cell. Given that, the hydroxyl radical oxidizes almost any molecule in its proximity including phospholipids, proteins, and deoxyribonucleic acid (DNA).[5,6] In accordance to studies[5,6], a report has observed accumulation of malondialdehyde, and 4-hydroxynonenol as lipid peroxidation products, 8-oxoguanine as oxidized nucleic acid bases, and increased protein carbonyl content, which is manifested as protein damage.[23] Further, in agreement with the literature, HIV-infected subjects have been shown to excrete higher quantities of malondialdehyde into their urine, which reflects increased levels in lipid peroxidation.[24] Interestingly, damage to the membrane can predict three things in HIV-infected subjects. Firstly, damage to the membrane alters membrane structure and activates acute phase proteins. As zinc is also useful to maintain membrane integrity and if zinc deficiency is prominent at this stage, then it intensifies the affliction of membrane architecture. Subsequently, these changes favor impairment in trace metals absorption. For example, the presence of acute phase protein, i.e., ceruloplasmin in the serum enhances the need for copper and iron since ceruloplasmin is the transport protein of copper and iron. Pasupathi et al.[12] observed significant plasma ceruloplasmin increase in HIV-infected subjects compared to healthy control.

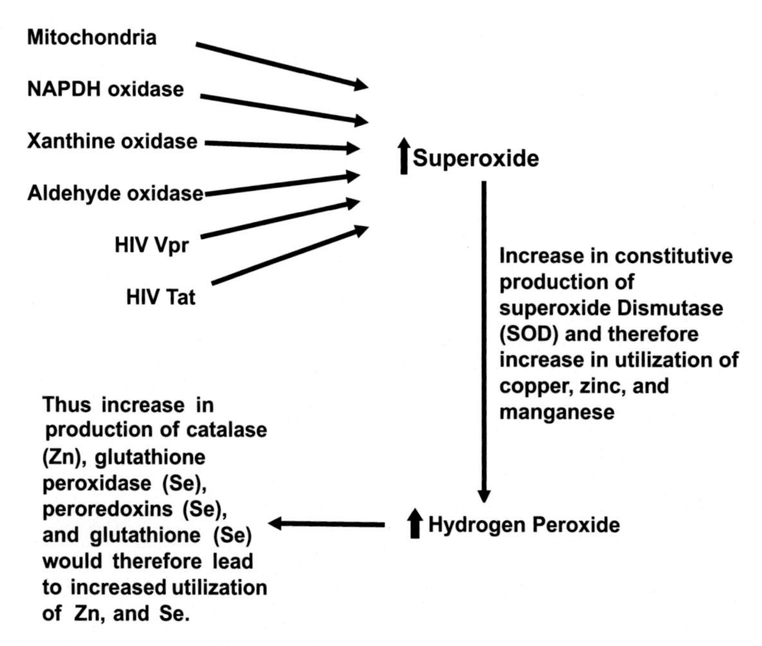


FIGURE 6.2 Sources of reactive oxygen species in human immunodeficiency virus–infected (HIV-infected) cell. The diagram also shows how the need for trace metals is increased in HIV-infected cell. *HIV*, human immunodeficiency virus; *Se*, selenium; *Zn*, zinc.

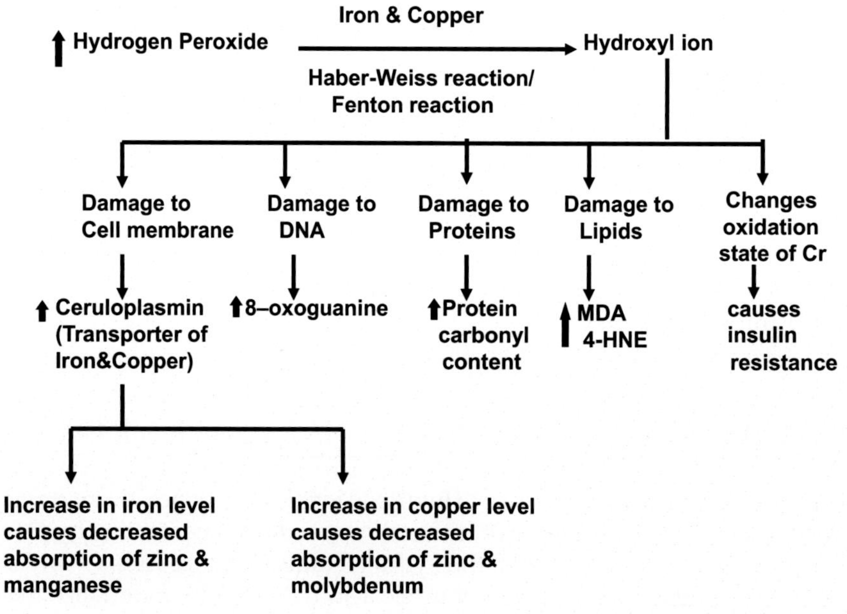


FIGURE 6.3 This diagram shows the effect of reactive oxygen species on iron and copper absorption and subsequent impairment in other trace metal absorption. *4-HNE*, 4-hydroxynonenol; *Cr*, chromium; *DNA*, deoxyribonucleic acid; *MDA*, malonaldialdehyde.

In addition, HIV-infected patients show increased levels of copper in blood along with ceruloplasmin.[13] The findings[10,12,13] suggest that increase in ceruloplasmin may increase the absorption of copper and iron from the intestines.

Secondly, the increase in ROS changes the oxidation state of the metal and can hamper metal absorption. Transition elements are found in multiple oxidation states.[25] The reactivity of these metals increases because of low ionization energy. Since ROS are highly reactive and electron deficient, there is a high possibility when they come into contact with transition metals to alter oxidation states of transition metals. A study has reported that ROS react with redox-cycling metal ions and result in an increased free radical formation.[26] Stohs et al.[27] postulated that intracellular chromium$^{+3}$ can be reduced to chromium$^{+2}$ generating hydroxyl radical. Thus, change in oxidation state decreases bioavailability of transition metal such as zinc and chromium.

Thirdly, the competition between the trace metals is also another factor that should be taken into consideration during HIV infection. It is observed in the literature that when supplemented with zinc, copper absorption is affected.[7] Zinc and copper will competitively inhibit each other's absorption.[8] Zinc is therapeutically given to reduce copper absorption.[6] Studies have also shown that in HIV-infected individuals, zinc and copper are inversely proportional to each other.[12–14] The inverse relationship between zinc and copper was associated with increased mortality.[14] Moreover, in the literature[7,8] it is reported that copper is effective in removing molybdenum from the body. On contrary, it is also reported that molybdenum induces copper deficiency.[8] Interestingly, the present review has reported increase in copper concentration in HIV infection. Therefore, increase in the absorption of copper may account for molybdenum deficiency in retrovirus-infected patients. Unfortunately, no report has stated molybdenum levels in HIV-infected patients.

## TRACE METALS AND LYMPHOCYTES (FIG. 6.4)

The HIV also shows a high degree of specialization toward reproducing at a faster rate than their host.[2] This takes place along with trace metal utilization. With this effect, virus life cycle involves the death of cluster of differentiation-4 (CD-4) cell, to exit the present CD-4 cell and to enter the next, evolve to be more virulent or even alter the behavior or other enzymatic properties of the host immune system to make it more vulnerable.[2,28] For instance, increasing the iron concentration in the host hastens the pavement for the oxidized environment since hydrogen peroxide is converted to hydroxyl radical in the presence of iron. And also, DNA is highly susceptible to hydroxyl radicals. DNA fragmentation is a part of programmed cell death process called apoptosis.[3] Apoptosis can be induced when there is irreparable damage done to DNA, and it is commenced by nucleic acid damage by ROS in the case of HIV infection.[29,30] Studies provide evidence that infection-induced ROS contributes to CD-4 cells depletion by increasing rate of apoptosis.[3,30] Studies have also shown that T-lymphocytes exhibit increased rate of apoptosis than lymphocytes

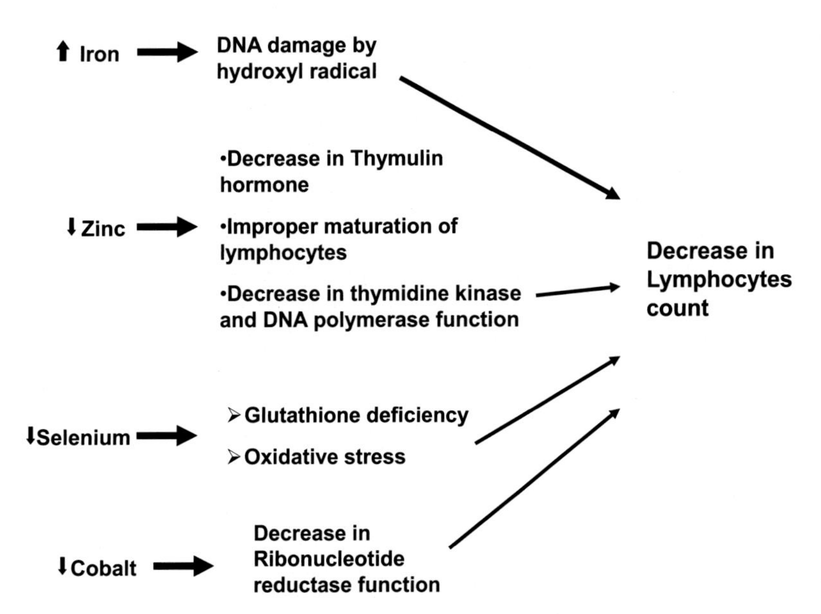


FIGURE 6.4 Hypothetical representation, which shows the effect of trace metals on lymphocytes. Increase in iron level and decrease in levels of zinc, selenium, and cobalt affect lymphocytes count in HIV-infected individual. *DNA*, deoxyribonucleic acid; *HIV*, human immunodeficiency virus.

from control subjects.[30,31] Banjoko et al.[10] observed increased iron levels in HIV-infected individuals compared with healthy controls. The study also observed a negative correlation between levels of iron and CD-4 cells in HIV-infected individuals.[10] In addition, higher levels of iron are associated with increased infection and replication.[11] Interestingly, the literature[7,8] says iron hampers the absorption of manganese. It is also reported in the literature that iron inhibits absorption of zinc. Therefore, the increase in iron levels may decrease the absorption of zinc and manganese and also account for zinc and manganese deficiencies in HIV-infected patients. However, the deficiency of manganese in the primary stages of HIV-infection subjects has not fully elucidated.

Long-term relationship of HIV-1 tends to reduce host biological fitness by utilizing the trace metals resources, thus initiating trace metal deficiencies and hampering host survival. This makes the first move by the virus to modify host immunological behavior toward trace metal utilization. The deficiency leads to progressive decrease of CD-4 cells. Trace metal zinc plays a vital role in maintenance of normal T cells and other immune systems.[32] T-lymphocytes function declined significantly when there is no intake of zinc for several weeks.[33] Lymphocytes devoid of zinc showed a depressed T cell-dependant antibody function. Studies have shown apoptosis can be induced to cells of lymphoid and myeloid origins with zinc deficiency.[33–36] Saha et al.[36] suggested that deficiency of zinc leads to T-lymphocyte dysfunction. Another study observed thymic function and lymph node atrophy in zinc deficiency.[37] Moreover, zinc is also cofactor for thymulin hormone, which is responsible for the maturation and differentiation of T-lymphocytes.[38] Zinc is an important cofactor for DNA synthesis enzymes such as thymidine kinase and DNA polymerase.[35] Therefore, zinc deficiency will definitely hamper synthesis of T-lymphocytes, and zinc deficiency is prominently seen in HIV-infected individuals.

One other possible mechanism underlying lymphocyte death is the requirement for cobalt for DNA synthesis.[7] Ribonucleotide reductase enzyme is required for synthesis of deoxyribonucleotide tripohosphates (dNTPs) in host cell.[6,7] Inhibition of this enzyme activity depletes the pool of dNTPs, which in turn inhibits host DNA synthesis. This may occur in the nonreplicating host cell. Ribonucleotide reductase needs cobalt as a coenzyme for its function.[6–8] As a preventive measure, there is possibility of HIV to decrease the absorption of cobalt from the intestine. But, till date no such decrease has been reported in HIV-infected individuals. However, cobalt deficiency seems to affect natural killer cells activity in some studies. Further, cobalt deficiency has shown to suppress macrophage phagocytosis.[8] Also, trace metals (zinc, copper, manganese, and molybdenum) start to get utilized in a larger quantity from the stores.

The other key aspect to manage the inherent quality of the virus to increase their own sustenance (fitness) is by exploiting hosts for resources necessary for virus survival and replication. Moreover, to maintain the precision of an organism to sustain the host environment, the organism has to make its own genome competent. While the precision of virus replication is dependent on a number of variables, the most important is the proofreading activity.[39,40] As in the case of retroviruses, proofreading activity is absent. Nevertheless, the presence of balanced dNTPs and the presence of reverse transcriptase enzyme concentrations are subject to matter for proofreading.[39] Besides this, the inherent properties of the organism such as protecting self from error-prone processes and also to limit self from the error-inducers are equally important.[40] Since, trace metals are vital for the enzymes required for replication, their concentrations likewise matter for their enzyme function. Ribonucleic acid polymerase and reverse transcriptase are zinc-dependant enzymes.[39] In addition, the HIV reverse transcriptase also requires divalent cations for enzyme activities. Therefore, HIV reverse transcription process can be altered if the concentration of reverse transcriptase or divalent cations or dNTPs is disbalanced. In a study, HIV replication has been modulated by changes in different components of the retrotranscription reaction.[40] For example, in vitro substitution of magnesium by manganese cations has been shown to increase misincorporation of dNTPs and to alter substrate specificity.[39] Treatment with manganese to HIV-infected cells resulted in increase in the mutant and mutation frequencies, respectively.[40] As discussed earlier, HIV manipulates host resources intelligently. The survival instinct is an innate property of the organism, which is to protect itself from error-prone processes and by limiting itself from the error-inducers. Since, zinc and magnesium are also required for HIV replication, zinc present in the host is directed to virus replication instead of host DNA synthesis, whereas magnesium, which is essential for adenosine triphosphate (ATP) synthesis, is not altered as there is an increase need for ATP synthesis. Likewise, manganese that can harm HIV replication, its absorption, is hampered. In place of zinc, copper is directed to host cells, thus indirectly aggravating the ROS production.

## DEFICIENCY OF TRACE METALS LEADS TO OXIDATIVE STRESS

Neutralization of ROS especially hydrogen peroxide and hydroxyl ions requires thiol-containing substances.[41–46] Thiol-dependent antioxidant system mainly consists of glutathione, thioredoxins, glutaredoxins, and peroxiredoxins. Neutralization of hydrogen peroxide is done by multiple enzymes such as glutathione peroxidases, thioredoxins, and peroxiredoxins and also by catalase. The enzymes (glutathione peroxidases, thioredoxins, and peroxiredoxins) are also responsible for protecting lipid peroxides, thus protecting lipids from oxidative damage. More importantly, thiol-dependant enzymes act synergistic with each other to manage reduce state of each enzyme. Further, thiol-dependent enzymes maintain intracellular glutathione level, which is capable of preventing damage to cellular components due to free radicals, peroxides, and lipid peroxides. The ratio between reduced and oxidized forms of glutathione designates intracellular reduce environment. The thiol quantity is directly proportional to the reduced glutathione levels in the cell. Trace metal selenium is the cofactor for glutathione peroxidase.[41]

Concentrations of intracellular glutathione in the peripheral mononuclear cells and lymphocytes were found to be reduced in asymptomatic HIV-seropositive individuals compared to healthy individuals.[42] Cirelli et al.[43] evaluated serum selenium levels in HIV-infected patients at different stages. Symptom-free HIV-infected individuals had lower levels of selenium compared with healthy controls. In addition, HIV-infected individuals often have lower concentrations of acid-soluble thiol, an important marker of antioxidant activity in the blood.[44] The other findings in HIV-infected patients are lower plasma levels of selenium in HIV-infected patients when compared with plasma levels of selenium in healthy controls.[45] In addition, the study observed a positive relation between CD-4 and selenium levels.[46] A wide variety of laboratory and clinical abnormalities in various species have been described with trace metal deficiency. Moreover, as the HIV infection progresses the concentration of zinc and copper varies between the two.

Trace metal deficiency most often occurs proportionately when metal utilization is increased.[8] This aspect takes place during HIV infection as the need for the antioxidants to combat the ROS is increased. Further, when these factors combine along with poor nutrition intake, trace element deficiency deepens. Several studies observed increased in lipid peroxidation and decreased in SOD, catalase, and glutathione peroxidase in HIV-infected subjects compared to healthy controls.[12,47,48] Thus, the stage when there is antioxidant forces depletion denotes the reflection of oxidative stress in the HIV-infected individual.

## DEFICIENCY OF ZINC FURTHER LEADS TO PROTEIN DEFICIENCY

Zinc is an essential ingredient in zinc-binding finger-loop domains.[49] The presence of zinc in the domains is necessary for site-specific binding to DNA and for gene expression.[49] The zinc ion stabilizes folding of the domain.[49] Further, zinc is necessary for synthesis of protein molecules and for maintenance of normal transport proteins. Transport protein such as albumin binds zinc to major extent. The other proteins that transport zinc are $\alpha_2$-macroglobulin, transferrin, and prealbumin.[50] Serum levels of albumin, transferrin, and prealbumin decrease in zinc-deficient patients. Thus, zinc deficiency might occur as the metal availability might be used for viral mRNA replication rather than for the host DNA replication.

Metallothionein is a low molecular weight binding protein, which contains high quantity of zinc.[6] Plasma metallothionein concentrations decrease with zinc deficiency, as zinc is an important inducer of metallothionein synthesis.[7] In human cells, expression of metallothionein genes is regulated by zinc levels in the body.[51]

## OXIDATIVE STRESS ENHANCES ACTIVATION OF HUMAN IMMUNODEFICIENCY VIRUS AND FURTHER ENHANCES ANTIOXIDANTS DEPLETION

Oxidative stress is generated as a consequence from over production of oxygen-derived species throughout different phases of the HIV life cycle, irrespective of normal levels of antioxidant system, as it weighs more toward the oxidant system,[14] subsequently making antioxidant system over used in combating the oxidants or making it vulnerable to low levels by virtue of antioxidant production. As a result, exhaustion of endogenous antioxidant moieties with concomitant production of ROS takes place.

The oxidized environment that is generated causes a favorable platform for HIV to get activated (Fig. 6.5). The pivotal event is the activation of nuclear factor kappa-light-cascade-enhancer of activated B cells (NF-κB) transcription factor.[52] Hydrogen peroxide is the main ROS that is responsible for HIV activation.[53] And also the depletion of reduced glutathione, which is the main intracellular antioxidant, is a crucial step for the activation of HIV. Depletion of glutathione activates hypoxic inducible factor-1alpha (HIF-1α). Zinc deficiency also activates HIF-1α. This factor further contributes to ROS formation through dysfunctional mitochondrial activity. ROS induces dissociation of complex between NF-κB and inhibitory kappa kinase beta (IκB) and results in the translocation of NF-κB to the nucleus.[52] Inside the nucleus it interacts with *cis*-acting units present in HIV long terminal repeats (HIV LTR).[54] After a brief period of time, HIV LTR gets activated. Activation of HIV LTR was shown to be the main requirement of

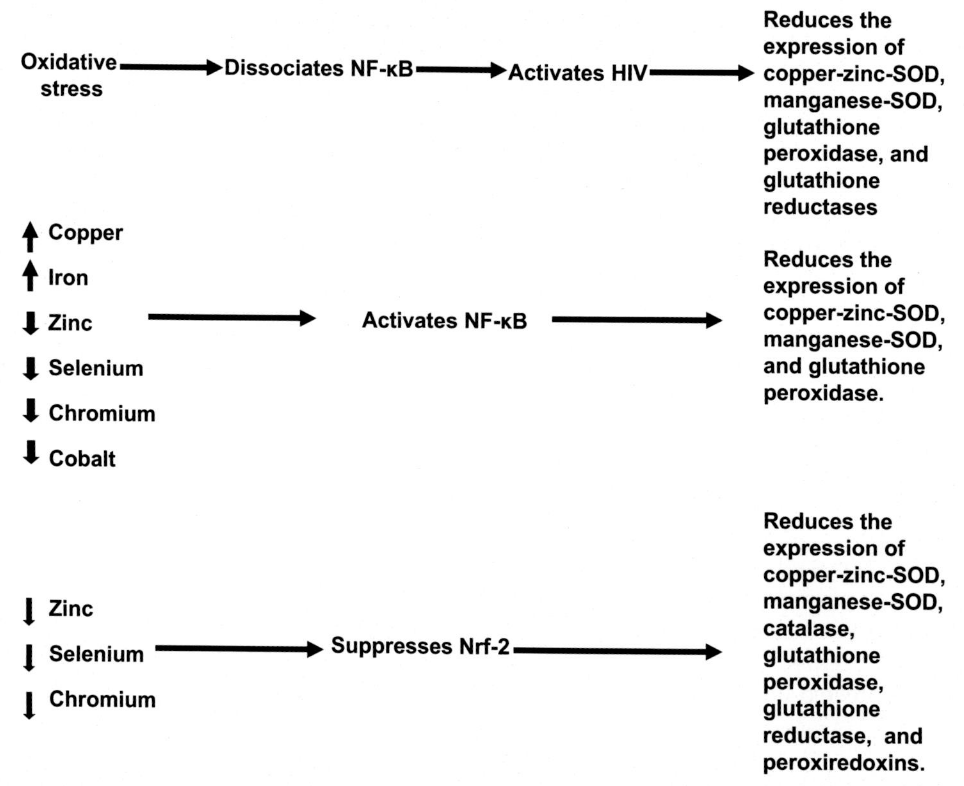


FIGURE 6.5 The figure depicts how oxidative stress activates human immunodeficiency virus (HIV), and also antioxidant status regulated and maintained by oxidative stress, nuclear factor kappa-light-cascade-enhancer of activated B cells (NF-κB), and nuclear factor erythroid2-related factor 2 (Nrf2). Further, the effect of trace metals concentrations in modulating transcription factors. *SOD*, superoxide dismutase.

HIV-1 replication.[54] NF-κB is responsible for the activation of HIV LTR. Tumor necrosis factor (TNF)-α is also another cytokine that activates HIV-1 LTR.[55]

The hallmark of HIV infection is progressive suppression of antioxidant enzymes expression. After HIV activation, the HIV Tat protein suppresses the expression of manganese-SOD through inhibition of binding of Sp1 and Sp3 transcription factor to manganese-SOD gene promoter.[56] In another study, it was shown that Tat protein of HIV-1 represses expression of manganese-SOD in HeLa cells.[57] In addition, studies done in HIV-transgenic rats demonstrate a decrease in copper–zinc-SOD expression.[58] Tat downregulates the expression of glutathione-synthesizing enzymes, such as glutathione peroxidase and glutathione reductase, leading to a decrease in glutathione content.[2,5]

Incorporation of viral proteins into the nucleus may hamper the utilization of ATPs that are required for the transcription of antioxidant enzymes, defusing the robust design of the cell and hampering the homeostasis of immunologic response toward the removal of HIV virus. Vpr causes the suppression of ATP biosynthesis in mitochondria, thus triggering the depletion of GSH as two molecules of ATP are required for every molecule of glutathione moiety.[13] Thus, the activation of HIV further enhances oxidative stress in the host.

## DEFICIENCY OF TRACE METALS ACTIVATES TRANSCRIPTION FACTORS

Redox balance is controlled by a battery of transcriptional factors including NF-κB and nuclear factor erythroid2-related factor 2 (Nrf-2) (Fig. 6.5). They regulate redox status through modulating ROS-generating enzymes and antioxidant enzymes. Further, ROS activate and amplify nuclear transcription factors, further establishing lysogeny.

NF-κB has a central role in inflammatory response. Numerous studies have characterized NF-κB as an important redox-sensitive transcription factor. Low concentration of ROS could activate NF-κB activity.[59] Glutathione, as discussed earlier, is a major intracellular thiol, when present in normal quantity, prevents the translocation of NF-κB to nucleus.[60] Several antioxidant enzymes including SOD and glutathione peroxidase contain NF-κB–binding sites[61] (Fig. 6.1). Therefore, thiol deficiency increases the nuclear translocation of NF-κB. Studies that observed low antioxidant capacity also observed high ROS and activation of NF-κB.

Persichini et al.[62] demonstrated that copper activated the transcription factor NF-κB in the liver and lung tissues of rats. In addition, they reported that the effect was mediated by oxidative stress with production of TNF-α and induction of NOS and nitrites. These results suggest that a physiopathological status, characterized by hypercupremic situations, may lead to the onset of inflammation through production of ROS and activation of NF-κB.

It is known that HIV infection is regulated of its expression by NF-κB through redox-controlled signal transduction pathways. Sappey et al.[63] demonstrated that iron induced cytotoxic and activating effects of hydrogen peroxide. These effects were observed in lymphocyte-infected cells by HIV. Concomitantly, NF-κB activation by hydrogen peroxide, when treated with deferoxamine (DFO), was observed to be decreased in the DFO-treated lymphocytes-infected cells with HIV. DFO is an iron-chelating agent. In addition, the study also suggests that other chemically unrelated iron chelators also provide protection against cytotoxicity, NF-κB activation, and HIV-1 activation in lymphocytic cells challenged with hydrogen peroxide.

Makropoulos et al.[64] have investigated the effects of selenium, an integral component of glutathione peroxidase, on NF-κB activation. In selenium-deprived Jurkat and ESb-L T-lymphocytes, supplementation of selenium led to a substantial increase of glutathione peroxidase activity. Analysis revealed that NF-κB activation in response to TNF-α was significantly inhibited. Likewise, a study reported that the HIV-1 LTR showed a dose-dependent inhibition of NF-κB–controlled gene expression by selenium. The effects of selenium were specific for NF-κB, and also data suggest that selenium supplementation may be used to modulate the expression of NF-κB target genes and HIV-1. In addition, Faure et al.[65] observed activation of NF-κB measured in peripheral blood monocytes can be reduced by selenium supplementation, confirming its importance in the activation of NF-κB.

Sahin et al.[66] in their report concluded that chromium may play a protective role in cerebral antioxidant defense system by reducing inflammation through NF-κB inhibition. In addition, they suggested that histidinate form of chromium was superior to picolinate form of chromium in reducing NF-κB expression in the brain of diabetic rats. NF-κB and TNF-α are known to induce insulin resistance[67] (Fig. 6.2). Aghdassi et al.[68], in a report on a randomized, double-blind, placebo-controlled trial on HIV-infected patients, observed improvement in insulin resistance after chromium supplementation.

Oh et al.[69] have evaluated the effects of cobalt chloride on TNF-α–induced inflammation and ROS in renal tubular epithelial cells. TNF-α–treated renal tubular epithelial cells showed an increase in the nuclear translocation of phosphorylated NF-κB and NF-κB transcriptional activity and a decrease in IκBα protein expression. These changes were restored by cobalt chloride. On treatment with cobalt chloride, they observed reduced generation of ROS induced by TNF-α. Thus, this study suggests that cobalt chloride has a protective effect on TNF-α–induced inflammation through the inhibition of NF-κB and ROS in renal tubular epithelial cells.

## EFFECT OF TRACE METALS ON NUCLEAR FACTOR ERYTHROID2-RELATED FACTOR 2

Nrf-2, a basic leucine zipper transcription factor, is encoded by the Nrf-2 gene in humans.[70] Nrf-2 is stabilized when it is in complex with keap1 welch protein. Nrf-2 is also able to induce the expression of antioxidant enzymes and major antioxidant proteins (glutathione). In addition, Nrf-2 is also known to regulate the expression of glutathione peroxidase, peroxiredoxinases, and thioredoxinases.[70] Moreover, evidence exists to support that Nrf-2 plays a major role in regulating cell differentiation and proliferation. In vivo, HIV-1 suppresses the Nrf-2 pathway.[71]

To explore associations between plasma selenium level and NRF-2–regulated cytoprotective genes expression, an observational study was conducted, which reported Nrf-2 mRNA level was positively correlated with peroxiredoxin and glutathione peroxidase gene expression.[72] On the other hand, plasma selenium level was significantly inversely associated with peroxiredoxin and glutathione peroxidase. In addition, Sahin et al.[66] suggested that chromium supplementation was effective in increasing Nrf-2 expression in the brain of diabetic rats. Thus, the findings may suggest a possible link between plasma selenium and chromium levels and cytoprotective response at gene level.

Ishida et al.[73] studied the effect of Nrf-2–antioxidant responsive elements (Nrf-2-ARE) on zinc transporters. Zinc transporters are solute carrier family members. To date, zinc transporters and zinc transporters–like proteins have been identified. Zinc transporters control intracellular zinc levels by effluxing zinc from the cytoplasm into the extracellular fluid, intracellular vesicles, and organelles; zinc transporters–like proteins also contribute to control intracellular zinc levels with influxing zinc into the cytoplasm. Further, they observed that the Nrf-2-ARE signal transduction pathway induces the expression of zinc transporters. As the availability of zinc is deficient in HIV-infected person, therefore, there is a possibility that Nrf-2-ARE cascade may be impaired, substantially incurring reduced zinc level inside the cell.

NF-κB and Nrf-2 cross-talk each other[74] as such they are inversely proportional to each other in physiological conditions.

## EPILOGUE

- It is concluded that trace metals are very essential for optimum functioning of immune system and their dysregulation leads to the enhancement of pathogenesis of HIV infection.
- To maintain the integrity of immune system, therefore, it is very essential to manage trace metal concentrations in HIV-infected individual.
- Trace metals have been associated with adverse clinical outcomes during HIV infection, which suggest that micronutrient supplementation may help reduce morbidity and mortality during HIV infection.[75]
- Trace elements should not be considered as direct antiretroviral agents but rather as potentially immune-modulator compounds that can achieve the effect through indirect mechanisms, possibly through inactivation of triggering of HIV gene expression, such as NF-kB and TNF-α.

## SUMMARY POINTS

- Entry of HIV enhances ROS generation.
- Increase in ROS initiates trace metals depletion, including zinc, manganese, and molybdenum during initial stage of HIV infection.
- Levels of iron and copper have an effect on other trace metals concentrations in the body.
- Trace metals are very essential for optimum functioning of immune system and their dysregulation leads to the enhancement of pathogenesis of HIV infection.
- Deficiency of trace metals affects the levels of lymphocytes.
- Oxidative stress enhances activation of HIV and further enhances antioxidants depletion.
- Further deficiency of trace metals and depletion of antioxidants activate transcription factors.
- To maintain the integrity of immune system, therefore, it is very essential to manage trace metal concentrations in HIV-infected individual.
- To conclude, trace metals should not be considered as direct antiretroviral agents but rather as potentially immune-modulator compounds that can achieve the effect through indirect mechanisms, possibly through inactivation of triggers of HIV gene expression, such as cytokines.

## References


1. Combes C. *The art of being a parasite*. University of Chicago Press; 2005.
2. Rosenberg ZF, Fauci AS. The immunopathogenesis of HIV infection. *Adv Immunol* 1989;**47**:377–431.
3. Baruchel S, Wainberg MA. The role of oxidative stress in disease progression in individuals infected by the human immunodeficiency virus. *J Leukocyte Biol* 1992;**52**(1):111–4.
4. a Pace GW, Leaf CD. The role of oxidative stress in HIV disease. *Free Radic Biol Med* 1995;**19**(4):523–8.
   b Bayr H. Reactive oxygen species. *Crit Care Med* 2005;**33**(12):S498–501.
5. Dedon PC, Tannenbaum SR. Reactive nitrogen species in the chemical biology of inflammation. *Arch Biochem Biophys* 2004;**423**(1):12–22.
6. a *Vasudevan textbook of biochemistry for medical students*. 7th ed. Jaypee Brothers, Reprint 2013; 1995.
   b Vasudevan DM, Sreekumari S, Vaidyanathan K. Mineral metabolism and abnormalities 514–523. Free Rad Antioxid:433–438
7. Chatterjea MN, Shinde R. *Textbook of Medical Biochemistry*. 7th ed. Jaypee Brothers; 1993. p. 570–94. Reprint 2008.
8. Chaturvedi UC, Shrivastava R, Upreti RK. Viral infections and trace elements: a complex interaction. *Curr Sci* 2004;**87**(11):1536–54.
9. Liochev SI, Fridovich I. The Haber-Weiss cycle—70 years later: an alternative view. *Redox Rep* 2002;**7**(1):55–7.
10. Banjoko SO, Oseni FA, Togun RA, Onayemi O, Emma-Okon BO, Fakunle JB. Iron status in HIV-1 infection: implications in disease pathology. *BMC Clin Pathol* 2012;**12**(1):1.
11. Chang HC, Bayeva M, Taiwo B, Palella Jr FJ, Hope TJ, Ardehali H. Short communication: high cellular iron levels are associated with increased HIV infection and replication. *AIDS Res Hum Retroviruses* 2015;**31**(3):305–12.
12. Pasupathi P, Ramchandran T, Sindhu PJ, Saranavan G, Bakthavathsalam G. Enhanced oxidative stress markers and antioxidant imbalance in HIV infection and AIDS patients. *J Sci Res* 2009;**1**(2):370–80.
13. Lai H, Lai S, Shor-Posner G, Ma F, Trapido E, Baum MK. Plasma zinc, copper, copper: zinc ratio, and survival in a cohort of HIV-1-infected homosexual men. *J Acquir Immune Defic Syndr* 2001;**27**(1):56–62.
14. Doddigarla Z, Bhaskar MV, Lingidi JL, Ashraf R. Evaluation of zinc, copper and oxidative stress in HIV seropositive cases. *HIV AIDS Rev* 2013;**12**(3):79–81.

15. Kashou A, Agarwal A. Oxidants and antioxidants in the pathogenesis of HIV/AIDS. *Open Reprod Sci J* 2011;**3**(1).
16. Matés JM, Segura JA, Alonso FJ, Márquez J. Intracellular redox status and oxidative stress: implications for cell proliferation, apoptosis, and carcinogenesis. *Arch Toxicol* 2008;**82**(5):273–99.
17. Gu Y, Wu RF, Xu YC, Flores SC, Terada LS. HIV Tat activates c-Jun amino-terminal kinase through an oxidant-dependent mechanism. *Virology* 2001;**286**(1):62–71.
18. Mastrantonio R, Cervelli M, Pietropaoli S, Mariottini P, Colasanti M, Persichini T. HIV-tat induces the Nrf-2/are pathway through NMDA receptor-elicited spermine oxidase activation in human neuroblastoma cells. *PLoS One* 2016;**11**(2):e0149802.
19. Vilhardt F, Plastre O, Sawada M, Suzuki K, Wiznerowicz M, Kiyokawa E, Trono D, Krause KH. The HIV-1 Nef protein and phagocyte NADPH oxidase activation. *J Biol Chem* 2002;**277**(44):42136–43.
20. Macreadie IG, Thorburn DR, Kirby DM, Castelli LA, de Rozario NL, Azad AA. HIV-1 protein Vpr causes gross mitochondrial dysfunction in the yeast *Saccharomyces cerevisiae*. *FEBS Lett* 1997;**410**(2–3):145–9.
21. Elbim C, Prevot MH, Bouscarat F, Franzini E, Chollet-Martin S, Hakim J, Gougerot-Pocidalo MA. Polymorphonuclear neutrophils from human immunodeficiency virus-infected patients show enhanced activation, diminished fMLP-induced L-selectin shedding, and an impaired oxidative burst after cytokine priming. *Blood* 1994;**84**(8):2759–66.
22. Zelko IN, Mariani TJ, Folz RJ. Superoxide dismutase multigene family: a comparison of the CuZn-SOD (SOD1), Mn-SOD (SOD2), and EC-SOD (SOD3) gene structures, evolution, and expression. *Free Radic Biol Med* 2002;**33**(3):337–49.
23. Deresz LF, Lazzarotto AR, Manfroi WC, Gaya A, Sprinz E, Oliveira ÁRD, Dall'Ago P. Oxidative stress and physical exercise in HIV positive individuals. *Revista Brasileira de Medicina do Esporte* 2007;**13**(4):275–9.
24. Sönnerborg A, Carlin G, Åkerlund B, Jarstrand C. Increased production of malondialdehyde in patients with HIV infection. *Scand J Infect Dis* 1988;**20**(3):287–90.
25. Figgis BN, Lewis J, Wilkins RG. *Modern coordination chemistry*. New York: Interscience; 1960. p. 403.
26. Kornweitz H, Burg A, Meyerstein D. Plausible mechanisms of the Fenton like reactions, M= Fe (II) and Co (II), in the presence of RCO2- Substrates: are OH. Radicals formed in the process? *J Phys Chem A* 2015;**119**(18):4200–6.
27. Stohs SJ, Bagchi D. Oxidative mechanisms in the toxicity of metal ions. *Free Rad Biol Med* 1995;**18**(2):321–36.
28. Hunt PW, Brenchley J, Sinclair E, McCune JM, Roland M, Shafer KP, Hsue P, Emu B, Krone M, Lampiris H, Douek D. Relationship between T cell activation and CD-4+ T cell count in HIV-seropositive individuals with undetectable plasma HIV RNA levels in the absence of therapy. *J Infect Dis* 2008;**197**(1):126–33.
29. Meyaard L, Otto SA, Jonker RR, Mijnster MJ, Keet RP, Miedema F. Programmed death of T cells in HIV-1 infection. *Science* 1992;**257**(5067):217–9.
30. Gougeon ML, Lecoeur H, Dulioust A, Enouf MG, Crouvoiser M, Goujard C, Debord T, Montagnier L. Programmed cell death in peripheral lymphocytes from HIV-infected persons: increased susceptibility to apoptosis of CD-4 and CD8 T cells correlates with lymphocyte activation and with disease progression. *J Immunol* 1996;**156**(9):3509–20.
31. Oyaizu N, Adachi Y, Hashimoto F, McCloskey TW, Hosaka N, Kayagaki N, Yagita H, Pahwa S. Monocytes express Fas ligand upon CD-4 cross-linking and induce CD-4+ T cells apoptosis: a possible mechanism of bystander cell death in HIV infection. *J Immunol* 1997;**158**(5):2456–63.
32. Failla ML. Trace elements and host defense: recent advances and continuing challenges. *J Nutr* 2003;**133**(5):1443S–7S.
33. Sugarman B. Zinc and infection. *Rev Infect Dis* 1983;**5**(1):137–47.
34. Elmes ME. Apoptosis in the small intestine of zincdeficient and fasted rats. *J Pathol* 1977;**123**(4):219–23.
35. Riordan JF. Biochemistry of zinc. *Med Clin N Am* 1976;**60**(4):661–74.
36. Saha AR, Hadden EM, Hadden JW. Zinc induces thymulin secretion from human thymic epithelial cells in vitro and augments splenocyte and thymocyte responses in vivo. *Int J Immunopharmacol* 1995;**17**(9):729–33.
37. Dardenne M, Boukaiba N, Gagnerault MC, Homo-Delarche F, Chappuis P, Lemonnier D, Savino W. Restoration of the thymus in aging mice by in vivo zinc supplementation. *Clin Immunol Immunopathol* 1993;**66**(2):127–35.
38. Bach JF, Dardenne M, Pleau JM. Biochemical characterisation of a serum thymic factor. *Nature* 1977;**266**:55–7.
39. Auld DS, Kawaguchi H, Livingston DM, Vallee BL. RNA-dependent DNA polymerase (reverse transcriptase) from avian myeloblastosis virus: a zinc metalloenzyme. *Proc Natl Acad Sci* 1974;**71**(5):2091–5.
40. Vartanian JP, Sala M, Henry M, Wain-Hobson S, Meyerhans A. Manganese cations increase the mutation rate of human immunodeficiency virus type 1 ex vivo. *J Gen Virol* 1999;**80**(8):1983–6.
41. Rotruck JT, Pope AL, Ganther HE, Swanson AB, Hafeman DG, Hoekstra W. Selenium: biochemical role as a component of glutathione peroxidase. *Science* 1973;**179**(4073):588–90.
42. Walmsley SL, Winn LM, Harrison ML, Uetrecht JP, Wells PG. Oxidative stress and thiol depletion in plasma and peripheral blood lymphocytes from HIV-infected patients: toxicological and pathological implications. *AIDS* 1997;**11**(14):1689–97.
43. Cirelli A, Ciardi M, de Simone C, Sorice F, Giordano R, Ciaralli L, Costantini S. Serum selenium concentration and disease progress in patients with HIV infection. *Clin Biochem* 1991;**24**(2):211–4.
44. de Quay B, Malinverni R, Lauterburg BH. Glutathione depletion in HIV-infected patients: role of cysteine deficiency and effect of oral N-acetylcysteine. *AIDS* 1992;**6**(8):815–20.
45. Look MP, Rockstroh JK, Rao GS, Kreuzer KA, Barton S, Lemoch H, Sudhop T, Hoch J, Stockinger K, Spengler U, Sauerbruch T. Serum selenium, plasma glutathione (GSH) and erythrocyte glutathione peroxidase (GSH-Px)-levels in asymptomatic versus symptomatic human immunodeficiency virus-1 (HIV-1)-infection. *Eur J Clin Nutr* 1997;**51**(4):266–72.
46. Baum MK, Shor-Posner G, Lai S, Zhang G, Lai H, Fletcher MA, Sauberlich H, Page JB. High risk of HIV-related mortality is associated with selenium deficiency. *JAIDS* 1997;**15**(5):370–4.
47. Delmas-Beauvieux MC, Peuchant E, Couchouron A, Constans J, Sergeant C, Simonoff M, Pellegrin JL, Leng B, Conri C, Clerc M. The enzymatic antioxidant system in blood and glutathione status in human immunodeficiency virus (HIV)-infected patients: effects of supplementation with selenium or beta-carotene. *Am J Clin Nutr* 1996;**64**(1):101–7.
48. Suresh DR, Annam V, Pratibha K, Prasad BM. Total antioxidant capacity–a novel early bio-chemical marker of oxidative stress in HIV infected individuals. *J Biomed Sci* 2009;**16**(1):1.
49. Prasad AS. Discovery of human zinc deficiency and studies in an experimental human model. *Am J Clin Nutr* 1991;**53**(2):403–12.

50. Kiilerich S, Christiansen C. Distribution of serum zinc between albumin and α2-macroglobulin in patients with different zinc metabolic disorders. *Clin Chim Acta* 1986;**154**(1):1–6.
51. Kaegi JH, Schaeffer A. Biochemistry of metallothionein. *Biochemistry* 1988;**27**(23):8509–15.
52. Wang X, Hai C. Novel insights into redox system and the mechanism of redox regulation. *Mol Biol Rep* 2016;**43**(7):607–28.
53. Schreck R, Rieber P, Baeuerle PA. Reactive oxygen intermediates as apparently widely used messengers in the activation of the NF-kappa B transcription factor and HIV-1. *EMBO J* 1991;**10**(8):2247.
54. Hiscott J, Kwon H, Génin P. Hostile takeovers: viral appropriation of the NF-kB pathway. *J Clin Invest* 2001;**107**(2):143–51.
55. Duh EJ, Maury WJ, Folks TM, Fauci AS, Rabson AB. Tumor necrosis factor alpha activates human immunodeficiency virus type 1 through induction of nuclear factor binding to the NF-kappa B sites in the long terminal repeat. *Proc Natl Acad Sci* 1989;**86**(15):5974–8.
56. Marecki JC, Cota-Gomez A, Vaitaitis GM, Honda JR, Porntadavity S, Clair DKS, Flores SC. HIV-1 Tat regulates the SOD2 basal promoter by altering Sp1/Sp3 binding activity. *Free Radic Biol Med* 2004;**37**(6):869–80.
57. Flores SC, Marecki JC, Harper KP, Bose SK, Nelson SK, McCord JM. Tat protein of human immunodeficiency virus type 1 represses expression of manganese superoxide dismutase in HeLa cells. *Proc Natl Acad Sci* 1993;**90**(16):7632–6.
58. Kline ER, Kleinhenz DJ, Liang B, Dikalov S, Guidot DM, Hart CM, Jones DP, Sutliff RL. Vascular oxidative stress and nitric oxide depletion in HIV-1 transgenic rats are reversed by glutathione restoration. *Am J Physiol Heart Circ Physiol* 2008;**294**(6):H2792–804.
59. Kabe Y, Ando K, Hirao S, Yoshida M, Handa H. Redox regulation of NF-κB activation: distinct redox regulation between the cytoplasm and the nucleus. *Antioxid Redox Signal* 2005;**7**(3–4):395–403.
60. Allan ME, Storey KB. Expression of NF-κB and downstream antioxidant genes in skeletal muscle of hibernating ground squirrels, *Spermophilus tridecemlineatus*. *Cell Biochem Funct* 2012;**30**(2):166–74.
61. Kaur P, Kaur G, Bansal MP. Tertiary-butyl hydroperoxide induced oxidative stress and male reproductive activity in mice: role of transcription factor NF-κB and testicular antioxidant enzymes. *Reprod Toxicol* 2006;**22**(3):479–84.
62. Persichini T, Percario Z, Mazzon E, Colasanti M, Cuzzocrea S, Musci G. Copper activates the NF-κB pathway in vivo. *Antioxid Redox Signal* 2006;**8**(9–10):1897–904.
63. Sappey C, Boelaert JR, Legrand-Poels S, Forceille C, Favier A, Piette J. Iron chelation decreases NF-k B and HIV type 1 activation due to oxidative stress. *AIDS Res Hum Retroviruses* 1995;**11**(9):1049–61.
64. Makropoulos V, Brüning T, Schulze-Osthoff K. Selenium-mediated inhibition of transcription factor NF-κ B and HIV-1 LTR promoter activity. *Arch Toxicol* 1996;**70**(5):277–83.
65. Faure P, Ramon O, Favier A, Halimi S. Selenium supplementation decreases nuclear factor-kappa B activity in peripheral blood mononuclear cells from type 2 diabetic patients. *Eur J Clin Invest* 2004;**34**(7):475–81.
66. Sahin K, Tuzcu M, Orhan C, Gencoglu H, Ulas M, Atalay M, Sahin N, Hayirli A, Komorowski JR. The effects of chromium picolinate and chromium histidinate administration on NF-κB and Nrf-2/HO-1 pathway in the brain of diabetic rats. *Biol Trace Element Res* 2012;**150**(1–3):291–6.
67. Hotamisligil GS, Peraldi P, Budavari A, Ellis R. IRS-1-mediated inhibition of insulin receptor tyrosine kinase activity in TNF-alpha-and obesity-induced insulin resistance. *Science* 1996;**271**(5249):665.
68. Aghdassi E, Arendt BM, Salit IE, Mohammed SS, Jalali P, Bondar H, Allard JP. In patients with HIV-infection, chromium supplementation improves insulin resistance and other metabolic abnormalities: a randomized, double-blind, placebo controlled trial. *Curr HIV Res* 2010;**8**(2):113–20.
69. Oh SW, Lee YM, Kim S, Chin HJ, Chae DW, Na KY. Cobalt chloride attenuates oxidative stress and inflammation through NF-κB inhibition in human renal proximal tubular epithelial cells. *J Korean Med Sci* 2014;**29**(Suppl. 2):S139–45.
70. Moi P, Chan K, Asunis I, Cao A, Kan YW. Isolation of NF-E2-related factor 2 (Nrf-2), a NF-E2-like basic leucine zipper transcriptional activator that binds to the tandem NF-E2/AP1 repeat of the beta-globin locus control region. *Proc Natl Acad Sci* 1994;**91**(21):9926–30.
71. Aleksunes LM, Manautou JE. Emerging role of Nrf-2 in protecting against hepatic and gastrointestinal disease. *Toxicol Pathol* 2007;**35**(4):459–73.
72. Reszka E, Wieczorek E, Jablonska E, Janasik B, Fendler W, Wasowicz W. Association between plasma selenium level and NRF-2 target genes expression in humans. *J Trace Elem Med Biol* 2015;**30**:10.
73. Ishida T, Takechi S. Nrf-2-ARE-dependent alterations in zinc transporter mRNA expression in HepG2 cells. *PLoS One* 2016;**11**(11):e0166100.
74. George LE, Lokhandwala MF, Asghar M. Novel role of NF-κB-p65 in antioxidant homeostasis in human kidney-2 cells. *Am J Physiol Renal Physiol* 2012;**302**(11):F1440–6.
75. Semba RD, Tang AM. Micronutrients and the pathogenesis of human immunodeficiency virus infection. *Br J Nutr* 1999;**81**(03):181–9.

SECTION 2

# ANTIOXIDANTS AND HIV/AIDS

CHAPTER

# 7

# HIV and Gender Differences in Diet: A Focus on Antioxidants

*Ajibola I. Abioye*[1,2], *Wafaie W. Fawzi*[3]

[1]Rhode Island Hospital, Providence, RI, United States; [2]Brown University, Providence, RI, United States; [3]Harvard T.H. Chan School of Public Health, Boston, MA, United States

OUTLINE



## Abstract

Male human immunodeficiency virus (HIV) patients experience faster disease progression and earlier mortality, even while receiving antiretroviral therapy. The role of dietary differences by gender is unclear. Understanding the gender differences in the diet of HIV patients is critical to guide the design of effective nutritional interventions and eliminate disparities in survival and the experience of clinical outcomes among HIV patients. Evidence from large population studies suggest that the intake of vitamins B and C and zinc, important in slowing HIV disease progression, may be lower among males than females. Studies assessing the intake of the other antioxidants have yielded inconsistent findings. This pattern is also reflected in the intake of specific food items by gender. These gender differences in dietary intake may be explained by a skills gap in food preparation by gender and society's expectations concerning masculinity. The design of interventions to address these disparities should bear in mind gender-based nutrition goals among males and females.

**Keywords:** Antioxidants; Diet; HIV; Masculinity; Sex; Vitamins.

## List of Abbreviations

**AIDS** Acquired Immunodeficiency Syndrome
**ART** Antiretroviral therapy
**CRP** C-Reactive protein
**CSB** Corn soya blend
**DDS** Dietary Diversity Score
**FANTA** Food and Nutrition Technical Assistance III Project
**HIV** Human Immunodeficiency Virus
**MDD-W** Minimum Dietary Diversity Score for Women



HIV/AIDS
http://dx.doi.org/10.1016/B978-0-12-809853-0.00007-9

## INTRODUCTION

Understanding what HIV-infected patients eat, how it influences their health, and how their diet may differ by sex may enable better design of nutritional components of HIV care and treatment programs toward improving survival, treatment outcomes, nutritional status, and of patients and their overall quality of life.[1] Human diet varies by gender, and gender is a key factor in the epidemiology of HIV.[2,3] HIV is the most common cause of death among adults aged 15–49 years living in developing countries, accounting for 10% of deaths among males and 15% among females in that age-group.[4] While the burden of HIV in developed countries is not as much, gender is still an important factor in the epidemiology, with substantial risk of infection and poorer outcomes among homosexual men and in the United States, African American males.[5,6]

While sex is a biological classification, gender is a social construct that captures how individuals are identified based on how they speak, walk, dress, interact, and function in society.[7] The studies on which this chapter is based focus on sex differences but draw on gender to explain some of the findings.

## GENDER DIFFERENCES IN HUMAN IMMUNODEFICIENCY VIRUS CLINICAL OUTCOMES

### Male Human Immunodeficiency Virus Patients Have Worse Outcomes Than Females

Among HIV-infected adults, there is considerable evidence from over 100 primary studies and a metaanalysis that disease progression and mortality differ by gender, especially in a setting of antiretroviral therapy, and male HIV-infected patients experience worse outcomes.[3]The results of a recent metaanalysis examining these differences are summarized in Table 7.1. Specifically, all-cause mortality and disease progression occur faster among males than females. The drivers of these differences are not fully understood. It appears that patient's age, antiretroviral use, disease severity, and country income classification are important factors. Men typically present to care and treatment centers with more advanced disease, lower CD4 counts, and higher viral loads.[3,8,9] On commencing care, men are less likely to perceive and report HIV-related illnesses. For instance, incident tuberculosis was reported to be greater among males in a large urban Tanzanian cohort of 67,686 HIV-infected patients.[10] Hepatitis B coinfection has also been reported to be more common among males in the same setting.[11] Finally, males also adhere poorly to care.[8]

The pathogenesis of HIV disease has been shown to differ by gender. Female HIV patients mount a better response to HIV infection, with greater activation of T-cells, production of interferon-α, and expression of interferon-

TABLE 7.1 Results From Metaanalysis of Gender Differences in Human Immunodeficiency Virus Clinical Outcomes

| Outcome | RR (95% CI) | Heterogeneity ($I^2$ *P*-value) | Heterogeneity and Subgroup Analysis |
|---|---|---|---|
| All-cause mortality | 1.23 (1.17, 1.29) | 89%, 0.001 | Related to patient's age, country income classification, HAART use; Relationship only seen in patients below 50 years of age, receiving HAART and from low and middle income countries; Stronger relationship with lower CD4 count and reported adherence to ART ≥90% |
| AIDS-related mortality | 1.02 (0.82, 1.30) | 46%, 0.047 | Related to age; Decreased risk of AIDS-related mortality among patients below 50 years of age |
| Disease progression | 1.11 (1.02, 1.21) | 50%, 0.01 | Related to country income classification; Relationship only seen in patients living in low and middle income countries |
| Immunologic failure | 1.19 (0.97, 1.47) | 79%, 0.002 | None identified |
| Change in CD4 count per year (per mm$^3$) | −5 (−14, 3) | 95%, <0.001 | None identified |
| Virologic failure | 1.26 (0.99, 1.61) | 65%, 0.01 | Related to age; Risk of virologic failure decreases compared to females |

The $I^2$ statistic is the proportion of the variation in the outcome assessment that is explained by the considered factor. HAART, Highly active antiretroviral therapy. Modified from *Abioye A, et al. Are there differences in disease progression and mortality among male and female HIV patients on antiretroviral therapy? A metaanalysis of observational cohorts.* AIDS Care *2015;**27(12)**:1468–86.*

stimulated genes among females than male counterparts.[12] Studies have also shown how the concentrations of inflammatory biomarkers may differ by gender.[12–14] Although the directions of these relationships have not been consistent, there have been reported relationships with cytokines and biologic markers known to predict early mortality, cardiovascular disease, and poorer response to treatment, especially among males compared to females.[12–14]

Genetic and environmental factors may also account for these differences. Estrogen concentrations differ by gender, and estrogen modifies inflammatory responses. Production of interferon-α is greater as a result, with resultant attenuation of viral replication among females.[15] It has even been suggested that estrogen receptor signaling may represent a potential therapeutic target in HIV care.[16] Gender differences in HIV disease are, however, seen before the onset of puberty, with lower viral loads among girls compared to boys.[17] Human and rhesus monkey models have shown major histocompatibility complex class I markers and a X chromosomal locus to be related to progression to AIDS, with heterozygous female carriers showing significantly slower CD4 decline and lower viral load compared to homozygous females.[18]

## DIFFERENTIAL EFFECTS OF ANTIOXIDANTS

There is no evidence that antioxidants modify HIV pathogenesis or treatment outcomes differently among males compared to females. Any effect modification likely occurs through differential genetic transcription, phenotypic expression, or hormonal regulation of antioxidant effects. No effect modification of the relationship of diet quality and HIV outcomes by gender was observed among patients in a Tanzanian cohort of 2038 HIV-infected adults followed up for 50 months, after adjusting for multiple potential confounders including adherence to antiretroviral therapy.[19] No similar effect modification was observed in a randomized controlled trial of lipid-based supplements (with high vs. regular dose vitamins) in Lusaka, Zambia, after 12 weeks on antiretroviral therapy (ART).[20]

A study among HIV-infected adults in Nepal suggested that the effect of dietary B-vitamins on the concentrations of an inflammatory biomarker may differ by gender. While significant relationships were observed between C-reactive protein (CRP) concentrations with niacin and cobalamin intake among men but not women, significant differences were noted with riboflavin among women. Pyridoxine intake was associated with CRP among men and women to significantly different extents.[16] It is probable, however, that these interactions with gender were observed due to much lower intake levels among females, rather than a true lack of effect. Though estrogen is related to CRP concentration in serum, it is plausible that appropriate adjustment for energy intake may attenuate these interactions remarkably.

While studies in the general population have established that males and females have different dietary preferences, the role of dietary practices in the gender differences in HIV outcomes has only been sparsely researched. The subsequent sections of this chapter will summarize what is known on the subject and highlight research gaps.

## DIFFERENCES IN INTAKE OF VITAMINS, MINERALS, AND ANTIOXIDANTS

### Vitamins B, C, E, and Zinc

Observational and interventional studies have established the value of intake of vitamins B, C, E, and zinc in preventing immunologic failure, slowing disease progression, and reducing the risk of mortality among HIV patients.[21,22] Baum et al. reported a fourfold decrease in the risk of immunologic failure and diarrhea following zinc supplementation among North American HIV–infected adults.[21] Substantially reduced risks of progression to stage III or IV disease, AIDS-related mortality or all-cause mortality among Tanzanian HIV-infected adults who received multivitamin supplements—B, C, and E, compared to others who received placebo have previously been reported[22].

Few studies have directly examined gender differences in dietary intake among HIV-infected adults. A cross-sectional analysis of dietary data among 2038 HIV-infected adults at the time of ART initiation observed significant gender differences in the dietary pattern of HIV patients.[2] Gender differences in the intake of micronutrients and vitamins in this cohort are summarized in Table 7.2. Male HIV-infected adults were less likely to meet the dietary reference intake of vitamins B and zinc, micronutrients known to modify the risk of HIV disease progression and mortality. Intake of vitamin E was extremely low and did not differ by gender. Intake of vitamin C was lower among males but not statistically significant. This study was set in a large HIV treatments program in Dar es Salaam, Tanzania. Similar gender differences were found by colleagues among 497 patients in neighboring Kenya.[23] Taken together with the

TABLE 7.2 Gender Differences in Intake of Micronutrients and Vitamins

| | RDA | | % Meeting RDA | | | |
|---|---|---|---|---|---|---|
| **Nutrient Adequacy** | **Females** | **Males** | **Females** | **Males** | **RR (95% CI)** | ***P*-value** |
| Vitamin A (μg/d) | 700 | 900 | 39.8 | 39.5 | 0.96 (0.82, 1.13) | 0.64 |
| Thiamin (mg/d) | 1.1 | 1.2 | 76.7 | 71.6 | 0.98 (0.93, 1.03) | 0.46 |
| Riboflavin (mg/d) | 1.1 | 1.3 | 69.9 | 53.1 | 0.82 (0.74, 0.92) | 0.0006 |
| Niacin (mg/d) | 14 | 16 | 22.8 | 10.2 | 0.39 (0.27, 0.55) | <0.0001 |
| Vitamin B6 (mg/d) | 1.3 | 1.3 | 73.2 | 77.2 | 1.03 (0.98, 1.09) | 0.22 |
| Vitamin B12 (μg/d) | 2.4 | 2.4 | 46.5 | 49.9 | 1.05 (0.92, 1.20) | 0.48 |
| Pantothenate (mg/d) | 5 | 5 | 22.2 | 23.8 | 0.97 (0.77, 1.24) | 0.83 |
| Vitamin C (mg/d) | 75 | 90 | 83.5 | 77.3 | 0.95 (0.90, 1.01) | 0.08 |
| Vitamin D (μg/d) | 15 | 15 | 0.4 | 0.9 | 2.91 (1.21, 7.00) | 0.02 |
| Vitamin E (mg/d) | 15 | 15 | 0.5 | 0.6 | 2.04 (0.29, 14.46) | 0.48 |
| Folic acid (μg/d) | 400 | 400 | 21.7 | 24.4 | 1.08 (0.85, 1.38) | 0.53 |

Modified from *Abioye, AI, et al. Gender differences in diet and nutrition among adults initiating antiretroviral therapy in Dar es Salaam, Tanzania.* AIDS Care *2015;***27***(6):706–15.*

findings that demonstrated sex differences in survival, these findings suggested a possible mediating influence of diet on the relationship of gender and survival among people living with HIV.

Studies from the United States and United Kingdom also reported gender differences in dietary intake, although the direction of the association differed from that reported by the African studies.[24,25] A cross-sectional analysis of dietary data from 679 HIV-infected adults enrolled in the Nutrition for Healthy Living study in Massachusetts and Rhode Island, USA, reported significantly poorer intake of vitamins B, C, E, and zinc among females. The authors, however, found no significant association between micronutrient intake and clinical outcomes including CD4 count and progression to AIDS. While 72% of participants were using at least one antiretroviral drug, patients in the Tanzanian study were at ART initiation. Although it is unclear whether dietary intake was energy-adjusted and how, energy adjustment is unlikely to explain the difference in the direction of the association in this study compared to the Tanzanian studies. The studies also assessed dietary intake differently—food frequency questionnaires in the Tanzanian studies, diet records in the US studies. Klassen et al. report on 196 HIV patients in London, UK, and observed no significant differences in the intake of vitamin C, although energy intake was greater among males.[25] Sociological and cultural differences are likely to explain these differences in dietary patterns by gender in Tanzanian and Kenya versus in the United Kingdom and the United States.

### Other Vitamins, Micronutrients, and Antioxidants

There has been recent interest in exploring the possible role of vitamin D in HIV disease progression and mortality. Observational studies have suggested a possible influence of blood levels of vitamin D (cholecalciferol) on the risk of incident pulmonary tuberculosis, wasting, disease progression, and mortality among HIV patients,[26,27] and an ongoing clinical trial may establish its utility as a supplement for inclusion in HIV treatment programs.[28] Studies examining the association of serum vitamin D with gender have yielded inconsistent results. While vitamin D concentrations were higher among females in Maceio, Brazil,[29] the concentrations were lower among females in Brussels, Belgium,[30] Kathmandu, Nepal,[31] and Catania, Italy.[32] No gender differences were observed among patients in Dar es Salaam, Tanzania,[26] Ochoa, Spain,[33] Nijmegen, Netherlands,[34] and in four US cities.[35] The extent to which these studies adjusted for physical activity and exposure to sunlight is unclear. Intake of vitamin D was greater among males in the Tanzanian cohort[2] (Table 7.3).

Studies from developing and developed countries have reported gender differences in the intake of dietary iron among HIV-infected adults, with worse intake levels among females.[2,36] Recommended iron intake is substantially lower for men, and it is well known that dietary iron intake among females does not meet recommended levels. Animal protein, especially meat, is the major source of iron, and its intake was lower among females in the

TABLE 7.3 Gender Differences in Intake of Minerals

| | RDA | | % Meeting RDA | | | |
|---|---|---|---|---|---|---|
| **Nutrient Adequacy** | **Females** | **Males** | **Females** | **Males** | **RR (95% CI)** | ***P*-value** |
| Sodium (g/d) | 1.5 | 1.5 | 99.9 | 99.8 | 1.00 (0.89, 1.13) | 0.97 |
| Potassium(μg/d)[2] | 4.7 | 4.7 | 99.9 | 100 | | |
| Manganese (mg/d) | 2.3 | 2.3 | 98.4 | 94.1 | 0.97 (0.91, 1.04) | 0.42 |
| Magnesium(mg/d) | 320 | 420 | 28.1 | 19.4 | 1.05 (0.83, 1.32) | 0.7 |
| Copper (mg/d)[2] | 900 | 900 | 0 | 0 | | |
| Calcium (mg/d) | 1000 | 1000 | 0.6 | 0.3 | 0.32 (0.06, 1.76) | 0.19 |
| Iron (mg/d) | 18 | 8 | 5.7 | 82.7 | 14.29 (11.42, 17.89) | <0.0001 |
| Zinc | 8 | 11 | 7.9 | 0.3 | 0.05 (0.01, 0.20) | <0.0001 |
| Phosphorus | 700 | 700 | 91.5 | 91.4 | 0.98 (0.95, 1.01) | 0.12 |

Modified from *Abioye, AI, et al. Gender differences in diet and nutrition among adults initiating antiretroviral therapy in Dar es Salaam, Tanzania.* AIDS Care *2015;***27***(6):706–15.*

TABLE 7.4 Gender Differences in Intake of Food Groups

| **Food Group** | **Mean Intake (Servings per Day)** | | | |
|---|---|---|---|---|
| | **Female** | **Male** | **Mean Difference** | ***P*-value** |
| Vegetables | 3.63 | 3.25 | −0.47 (−0.86, −0.07) | 0.02 |
| Cereals | 3.78 | 3.51 | −0.21 (−0.44, 0.001) | 0.05 |
| Sweet | 1.44 | 1.31 | −0.18 (−0.32, −0.04) | 0.01 |
| Oil | 1.62 | 1.58 | −0.09 (−0.25, 0.07) | 0.28 |
| Fish | 0.92 | 0.91 | 0.01 (−0.08, 0.10) | 0.75 |
| Dairy | 0.29 | 0.28 | −0.02 (−0.08, 0.04) | 0.56 |
| Eggs | 0.25 | 0.26 | 0.02 (−0.04, 0.07) | 0.58 |
| Pulses | 1.24 | 1.31 | 0.06 (−0.08, 0.20) | 0.38 |
| Fruits | 3.13 | 3.21 | 0.05 (−0.27, 0.38) | 0.75 |
| Tuber | 1.39 | 1.48 | 0.08 (−0.09, 0.25) | 0.34 |
| Alcoholic beverage | 0.01 | 0.01 | −0.005 (−0.02, 0.01) | 0.5 |
| Nonalcoholic beverage | 1.07 | 1.18 | 0.05 (−0.06, 0.21) | 0.35 |
| Meat | 0.55 | 0.7 | 0.14 (0.06,0.21) | 0.001 |
| Dietary diversity score | 10.99 | 11.07 | 0.06 (−0.08, 0.20) | 0.42 |

Modified from *Abioye, AI, et al. Gender differences in diet and nutrition among adults initiating antiretroviral therapy in Dar es Salaam, Tanzania.* AIDS Care *2015;***27***(6):706–15.*

Tanzanian cohort.[2] See Tables 7.3 and 7.4 for gender differences in the intake of minerals and food groups from the Tanzanian cohort. Iron deficiency among HIV patients is associated with severe anemia, declining CD4 counts, and increased risk of AIDS-related and all-cause mortality.[2,37] In a prospective cohort of 48,068 nonpregnant HIV-infected adults in Dar es Salaam, Tanzania, although female gender was associated with increased risk of anemia, there was no difference in the risk of iron deficiency anemia.[38] Conversely, high iron stores are potentially associated with poorer outcomes among HIV-infected adults due to activity of oxidants.[39,40]

Studies that examined concentrations of retinol-binding protein (vitamin A) concentrations among Korean HIV patients on antiretroviral therapy[41] and with ascorbate (vitamin C) and tocopherol (vitamin E) among US adults[42] found no gender differences.

## DIFFERENCES IN INTAKE OF FOOD GROUPS

There were gender differences in the intake of foods and food groups among patients in the Tanzanian studies.[2] While females were more likely to eat vegetables, cereals, and sweets in their diet, males were more likely to have meat and tubers. These differences were reflected in the intake of micronutrients and antioxidants. Examining intake of food groups, however, lends itself better to clinical and public health recommendations. Although there were substantial gender differences in the total calorie intake among HIV patients in a small Ghanaian study (n = 50), there were no significant differences in the intake of grains, fruits, and vegetables[43] (Table 7.4).

The relationship of diet quality with the prospective risk of HIV-related treatment outcomes was assessed among 2038 HIV-infected adults initiating antiretroviral therapy in Tanzania.[19] Outcomes considered included all-cause mortality, wasting, severe anemia, pulmonary tuberculosis, and immunologic failure. Significant dose-related relationships were observed with some of these outcomes. For every additional food group an individual introduces to their diet, their risk of all-cause mortality and severe weight loss (≥10% of baseline weight) during follow-up reduces. Intake of up to nine food groups daily was associated with 56% reduced risk of mortality and 41% reduced risk of severe weight loss during follow-up compared to intake of ≤4 food groups. Diet quality was assessed using Food and Nutrition Technical Assistance III Project's Dietary Diversity Score.[44] Usual intake below 1 serving per day of the food group was regarded as trivial and did not contribute to the score. Food groups considered include dairy, eggs, meat, fish, vegetable, fruits, pulses/legumes, beverage, cereal, tubers, sweets, and oils. Similar results were obtained with the minimum dietary diversity scores, which is scaled in favor of intake of vegetables and fruits.[45] By providing adequate amounts and variety of macronutrients and micronutrients, usual intake of good quality diet slows HIV disease progression, reduces the risk of AIDS-related mortality, likely through improved immunologic responses.

The possibility of effect modification of the relationship of diet quality and HIV outcomes by gender was assessed in the Tanzanian cohort and found no significant association. This absence of significant interaction suggests a role for differential intake of antioxidants, rather than a differential effect, by gender, in explaining the gender differences observed. Interventions that increase the intake of a diverse diet will therefore likely yield protective effects regardless of gender.

### Basis of Gender Differences

Studies evaluating the mechanism underlying gender differences in diet have employed quantitative and qualitative approaches. Qualitative measurement approaches allow culturally appropriate examination of the subject. They may also shed light on the sociologic factors that may influence dietary practices.

The assigning of roles within the family by gender, especially in African societies, fosters a skills gap that adversely affects males. Males are brought up differently than women. Young females are more likely to be taught how to prepare meals because it is expected that this would be a considerable part of their responsibility as they grow up into adulthood and especially if they choose to start a family.[46] Every male, on the other hand, is expected to grow up to marry a woman who would provide his meals. Males, especially if unmarried, are therefore likely to have poor quality diets simply because they lack the necessary skills to make good enough meals. Irrespective of HIV status, females are more likely to be judged by their body weight, especially in Western societies.[46] They therefore take greater interest in eating healthy. Among HIV patients, these have great implications for disease progression and survival.

Expectations concerning masculinity in global societies may also be adversely affecting whether and how men care for themselves. Males are thought to be the stronger sex, and individual males are expected to strive to protect this façade of strength. This begins early in life with male infants in African societies being cared for differently than females.[47] Adult males are less likely to report experience of clinical symptoms[8] or even drug-related adverse events.[48] They are less likely to get tested for HIV or disclose their HIV status because of the potential shame and emasculation with which the weakness of HIV is associated.[49] Males are also less likely to utilize HIV care and treatment and adhere to antiretroviral therapy.[49] It has also been shown that these norms influence dietary practices in the general population in Western societies.[50] It seems reasonable to infer that these expectations of masculinity may also influence whether a male patient accesses nutritional counseling services as part of HIV care and the extent to which they adhere to the health-care provider's recommendations, even in developing countries.

Substantial gender differences in dietary intake likely begin early in life. From a prospective cohort of HIV-exposed infants in a Rwandan supplementary feeding program, researchers detected gender differences in feeding patterns as early as 6 months of age.[47] Mothers of these infants were provided with corn soya blend as an acceptable, feasible, affordable, sustainable, and safe option for replacement feeding, and nutritional status and feeding practices were assessed. Among male infants, breastfeeding initiation was delayed, breastfeeding cessation was earlier, and nutritional status was poorer, compared to females, during follow-up. Researchers inferred that behavioral attitudes modifying the allocation of food favors girls in Rwanda. Societal norms likely drive these behavioral attitudes. Male infants are thought to be stronger and requiring less care. Interventions to address these disparities will need to factor in the entrenched nature of the norms driving these factors.

It is unlikely that gender differences discussed above may be explained by measurement errors. Though dietary intake is often based on self-report, the consistent finding of these differences when diet records, 24-h recalls, and food frequency questionnaires were employed limits the likelihood of biased assessment. Any errors are likely to be random, attenuating the effects seen, rather than changing the directions. Few studies have employed objective biomarker-based approaches to assess antioxidant intake among HIV patients, and these have not examined gender differences. Results from studies reporting on blood vitamin D levels have been inconsistent[26,29–32,35,33] and are threatened by residual confounding.

## Interventions

From the foregoing, we have shown that there are significant gender differences in disease progression, treatment outcomes, and survival among HIV-infected adults, which are partly explained by dietary intake of micronutrients and antioxidants. These dietary patterns become entrenched very early in life and are likely driven by societal norms. Appropriately intervening in this regard is likely to effectively eliminate these gender differences because differential intake of antioxidants, rather than a differential effect by gender, drives these differences.

There are suggestions that female HIV patients lose their initial immunologic advantage after a few years on treatment and have similar or worse outcomes.[51] Although this requires further investigation, interventions to sustainably improve disparities in outcomes would need to include males and females.

Regardless of HIV status, implementation of nutrition interventions is typically successful only if target populations accept the interventions. Interventions that may successfully address the gender differences in dietary intake and treatment outcomes among people living with HIV must be sensitive to the underlying biologic and sociologic factors. A qualitative study among 2764 adolescents from four European countries suggests that girls, unlike boys, are more likely to accept strategies that promote healthy eating—including counseling, teaching sessions, and reduced cost of healthy eating options.[52] No gender differences were reported, however, in the preference for interventions that discourage unhealthy eating such as placing warning labels or increasing the price of or ban of unhealthy snacks and drinks. Understanding which interventions males and females prefer in each setting may improve effectiveness of the interventions. The mass media contributes to perpetuating masculine stereotypes in modern settings and can be useful for improving the acceptability of interventions and correcting unhealthy stereotypes.[50]

Although interventions to increase nutrition knowledge among HIV-infected adults may be helpful, they are unlikely to suffice. Most HIV-infected adults appear to understand the critical role of nutrition in the pathogenesis of the disease, but other factors affect the translation of that knowledge into practice. Anand and Puri examined nutritional knowledge among 400 HIV-infected adults attending care at a hospital in New Delhi, India, and observed no significant gender differences.[53] Involving junior cadre health workers and family members may be helpful to improve nutrition knowledge and communication.[54] Behavioral practices such as alcohol intake and cigarette smoking differ by gender, even among HIV patients[25] and probably reflect a general attitude toward health in general.[46] A comprehensive nutrition program should include smoking cessation and alcohol intake reduction programs. Counseling and other supportive interventions based on an understanding of the goals that are more likely to successfully motivate males to take up healthy eating, compared to females, may be effective. Men may be more likely to modify their dietary behavior if they are convinced this will help them manage their weight, prevent diseases and have more energy for daily functioning.[55]

Improving the intake of dietary antioxidants among males and females will likely decrease the gender disparities discussed above and improve the health and quality of life of people living with HIV.

## SUMMARY POINTS

- Male HIV patients experience faster disease progression and earlier mortality, even while receiving antiretroviral therapy.
- Evidence from large population studies suggest that the intake of vitamins B and C and zinc, important in slowing HIV disease progression, may be lower among males than females.
- Studies assessing the intake of the other antioxidants have yielded inconsistent findings.
- For example, while studies assessing serum concentrations of vitamin A have observed no gender differences, studies of serum concentrations of vitamin D have reported higher concentrations among males, higher concentrations among females, or no differences by gender
- These patterns are also reflected in the intake of specific food items by gender.
- These gender differences in dietary intake may be explained by a skills gap in food preparation by gender and society's expectations concerning masculinity. The design of interventions to address these disparities should bear in mind gender-based nutrition goals among males and females.

## References

1. WHO. *Nutrition requirements for people living with HIV/AIDS: report of a technical consultation*. Geneva: World Health Organization; 2003.
2. Abioye AI, Isanaka S, Liu E, et al. Gender differences in diet and nutrition among adults initiating antiretroviral therapy in Dar es Salaam, Tanzania. *AIDS Care* 2015;**27**(6):706–15.
3. Abioye A, Soipe A, Salako A, et al. Are there differences in disease progression and mortality among male and female HIV patients on antiretroviral therapy? A meta-analysis of observational cohorts. *AIDS Care* 2015;**27**(12):1468–86.
4. IHME. *GBD compare*. Seattle, WA: Institute for Health Metrics and Evaluation (IHME), University of Washington; 2015.
5. CDC. Diagnoses of HIV infection in the United States and dependent areas, 2011, HIV Surveillance Report, 2013, vol. 23. Centers for Disease Control and Prevention; 2014.
6. ECDC. *HIV/AIDS surveillance in Europe 2011*. Stockholm: European Centre for Disease Prevention Control/WHO Regional Office for Europe; 2012.
7. Lorber J, Farrell SA. *The social construction of gender*. Newbury Park, CA: SAGE publications; 1991.
8. CDC. *Differences between HIV-infected men and women in antiretroviral therapy outcomes-six African countries, 2004–2012*. MMWR Morbidity and mortality weekly report. Atlanta, GA: Centers for Disease Control and Prevention; 2013. p. 945.
9. Jiang H, Yin J, Fan Y, et al. Gender difference in advanced HIV disease and late presentation according to European consensus definitions. *Scientific Reports* 2015;**5**:14543.
10. Liu E, Makubi A, Drain P, et al. Tuberculosis incidence rate and risk factors among HIV-infected adults with access to antiretroviral therapy. *AIDS* 2015;**29**(11):1391–9.
11. Hawkins C, Christian B, Ye J, et al. Prevalence of hepatitis B co-infection and response to antiretroviral therapy among HIV-infected patients in Tanzania. *AIDS* 2013;**27**(6):919–27.
12. Addo MM, Altfeld M. Sex-based differences in HIV type 1 pathogenesis. *J Infect Dis* 2014;**209**(Suppl. 3):S86–92.
13. Krebs SJ, Slike BM, Sithinamsuwan P, et al. Sex differences in soluble markers vary before and after the initiation of antiretroviral therapy in chronically HIV-infected individuals. *AIDS* 2016;**30**(10):1533–42.
14. Mangili A, Ahmad R, Wolfert RL, et al. Lipoprotein-associated phospholipase A2, a novel cardiovascular inflammatory marker, in HIV-infected patients. *Clin Infect Dis* 2014;**58**(6):893–900.
15. Griesbeck M, Scully E, Altfeld M. Sex and gender differences in HIV-1 infection. *Clin Sci (London, Engl 1979)* 2016;**130**(16):1435–51.
16. Poudel-Tandukar K, Chandyo RK, Dietary B. Vitamins and serum C-reactive protein in persons with human immunodeficiency virus infection: the positive living with HIV (POLH) study. *Food Nutr Bull* 2016.
17. Ruel TD, Zanoni BC, Ssewanyana I, et al. Sex differences in HIV RNA level and CD4 cell percentage during childhood. *Clin Infect Dis* 2011;**53**(6):592–9.
18. Siddiqui RA, Sauermann U, Altmuller J, et al. X chromosomal variation is associated with slow progression to AIDS in HIV-1-infected women. *Am J Hum Genet* 2009;**85**(2):228–39.
19. Abioye AI, Isanaka S, Madzorera I, et al. *Dietary diversity is associated with decreased mortality and severe weight loss among HIV-infected Tanzanian adults initiating antiretroviral therapy*. Unpublished. 2016.
20. Filteau S, PrayGod G, Kasonka L, et al. Effects on mortality of a nutritional intervention for malnourished HIV-infected adults referred for antiretroviral therapy: a randomised controlled trial. *BMC Med* 2015;**13**:17.
21. Baum MK, Lai S, Sales S, Page JB, Campa A. Randomized, controlled clinical trial of zinc supplementation to prevent immunological failure in HIV-infected adults. *Clin Infect Dis* 2010;**50**(12):1653–60.
22. Fawzi WW, Msamanga GI, Spiegelman D, et al. A randomized trial of multivitamin supplements and HIV disease progression and mortality. *N Engl J Med* 2004;**351**(1):23–32.
23. Onyango AC, Walingo MK, Mbagaya G, Kakai R. Assessing nutrient intake and nutrient status of HIV seropositive patients attending clinic at Chulaimbo Sub-District hospital, Kenya. *J Nutr Metab* 2012;**2012**:306530.
24. Kim JH, Spiegelman D, Rimm E, Gorbach SL. The correlates of dietary intake among HIV-positive adults. *Am J Clin Nutr* 2001;**74**(6):852–61.
25. Klassen K, Goff LM. Dietary intakes of HIV-infected adults in urban UK. *Eur J Clin Nutr* 2013;**67**(8):890–3.

26. Sudfeld CR, Wang M, Aboud S, Giovannucci EL, Mugusi FM, Fawzi WW. Vitamin D and HIV progression among Tanzanian adults initiating antiretroviral therapy. *PLoS One* 2012;**7**(6):e40036.
27. Sudfeld CR, Giovannucci EL, Isanaka S, et al. Vitamin D status and incidence of pulmonary tuberculosis, opportunistic infections, and wasting among HIV-infected Tanzanian adults initiating antiretroviral therapy. *J Infect Dis* 2012:jis693.
28. Fawzi W, Sudfeld CR, Wang M, et al. *Trial of vitamin D in HIV progression*. ClinicalTrialsGov: National Institutes for Health; 2013. p. NCT01798680.
29. Canuto JM, Canuto VM, de Lima MH, et al. Risk factors associated with hypovitaminosis D in HIV/aids-infected adults. *Arch Endocrinol Metab* 2015;**59**(1):34–41.
30. Theodorou M, Serste T, Van Gossum M, Dewit S. Factors associated with vitamin D deficiency in a population of 2044 HIV-infected patients. *Clin Nutr (Edinburgh, Scotland)* 2014;**33**(2):274–9.
31. Poudel-Tandukar K, Poudel KC, Jimba M, Kobayashi J, Johnson CA, Palmer PH. Serum 25-hydroxyvitamin d levels and C-reactive protein in persons with human immunodeficiency virus infection. *AIDS Res Human Retroviruses* 2013;**29**(3):528–34.
32. Pinzone MR, Di Rosa M, Celesia BM, et al. LPS and HIV gp120 modulate monocyte/macrophage CYP27B1 and CYP24A1 expression leading to vitamin D consumption and hypovitaminosis D in HIV-infected individuals. *Eur Rev Med Pharmacol Sci* 2013;**17**(14):1938–50.
33. Cervero M, Agud JL, Garcia-Lacalle C, et al. Prevalence of vitamin D deficiency and its related risk factor in a Spanish cohort of adult HIV-infected patients: effects of antiretroviral therapy. *AIDS Res Hum Retroviruses* 2012;**28**(9):963–71.
34. Van Den Bout-Van Den Beukel CJ, Fievez L, Michels M, et al. Vitamin D deficiency among HIV type 1-infected individuals in the Netherlands: effects of antiretroviral therapy. *AIDS Res Hum Retroviruses* 2008;**24**(11):1375–82.
35. Dao CN, Patel P, Overton ET, et al. Low vitamin D among HIV-infected adults: prevalence of and risk factors for low vitamin D Levels in a cohort of HIV-infected adults and comparison to prevalence among adults in the US general population. *Clin Infect Dis* 2011;**52**(3):396–405.
36. Gross R, Bellamy SL, Ratshaa B, et al. Effects of sex and alcohol use on antiretroviral therapy outcomes in Botswana: a cohort study. *Addiction (Abingdon, England)* 2016.
37. O'Brien ME, Kupka R, Msamanga GI, Saathoff E, Hunter DJ, Fawzi WW. Anemia is an independent predictor of mortality and immunologic progression of disease among women with HIV in Tanzania. *J Acquir Immune Defic Syndr* 2005;**40**(2):219–25.
38. Petraro P, Duggan C, Spiegelman D, et al. Determinants of anemia among human immunodeficiency virus-positive adults at care and treatment clinics in dar es Salaam, Tanzania. *Am J Trop Med Hyg* 2016;**94**(2):384–92.
39. Crist MB, Melekhin VV, Bian A, et al. Higher serum iron is associated with increased oxidant stress in HIV-infected men. *J Acquir Immune Defic Syndr* 2013;**64**(4):367–73.
40. Gordeuk VR, Delanghe JR, Langlois MR, Boelaert JR. Iron status and the outcome of HIV infection: an overview. *J Clin Virol* 2001;**20**(3):111–5.
41. Han SH, Chin BS, Lee HS, et al. Serum retinol-binding protein 4 correlates with obesity, insulin resistance, and dyslipidemia in HIV-infected subjects receiving highly active antiretroviral therapy. *Metab Clin Exp* 2009;**58**(11):1523–9.
42. Stephensen CB, Marquis GS, Jacob RA, Kruzich LA, Douglas SD, Wilson CM. Vitamins C and E in adolescents and young adults with HIV infection. *Am J Clin Nutr* 2006;**83**(4):870–9.
43. Wiig K, Smith C. An exploratory investigation of dietary intake and weight in human immunodeficiency virus-seropositive individuals in Accra, Ghana. *J Am Diet Assoc* 2007;**107**(6):1008–13.
44. Kennedy G, Ballard T, Dop M. *Guidelines for measuring household and individual dietary diversity*. Nutrition and Consumer Protection Division, Food and Agriculture Organization of the United Nations; 2010.
45. Herforth A. *Access to adequate nutritious food: new indicators to track progress and inform action*. Oxford University Press; 2015.
46. Wang WC, Worsley A, Hunter W. Similar but different. Health behaviour pathways differ between men and women. *Appetite* 2012;**58**(2):760–6.
47. Condo JU, Gage A, Mock N, Rice J, Greiner T. Sex differences in nutritional status of HIV-exposed children in Rwanda: a longitudinal study. *Trop Med Int Health* 2015;**20**(1):17–23.
48. Squires KE, Bekker LG, Eron JJ, et al. Safety, tolerability, and efficacy of raltegravir in a diverse cohort of HIV-infected patients: 48-week results from the REALMRK Study. *AIDS Res Hum Retroviruses* 2013;**29**(6):859–70.
49. Nyamhanga TM, Muhondwa EP, Shayo R. Masculine attitudes of superiority deter men from accessing antiretroviral therapy in Dar es Salaam, Tanzania. *Glob Health Action* 2013;**6**:21812.
50. Gough B. 'Real men don't diet': an analysis of contemporary newspaper representations of men, food and health. *Soc Sci Med* 2007;**64**(2):326–37.
51. Mosha F, Muchunguzi V, Matee M, et al. Gender differences in HIV disease progression and treatment outcomes among HIV patients one year after starting antiretroviral treatment (ART) in Dar es Salaam, Tanzania. *BMC Public Health* 2013;**13**:38.
52. Stok FM, de Ridder DT, de Vet E, et al. Hungry for an intervention? Adolescents' ratings of acceptability of eating-related intervention strategies. *BMC Public Health* 2016;**16**:5.
53. Anand D, Puri S. Nutritional knowledge, attitude, and practices among HIV-positive individuals in India. *J Health, Popul Nutr* 2013;**31**(2):195–201.
54. Rodas-Moya S, Kodish S, Manary M, Grede N, de Pee S. Preferences for food and nutritional supplements among adult people living with HIV in Malawi. *Public Health Nutr* 2016;**19**(4):693–702.
55. Caperchione CM, Vandelanotte C, Kolt GS, et al. What a man wants: understanding the challenges and motivations to physical activity participation and healthy eating in middle-aged Australian men. *Am J Mens Health* 2012;**6**(6):453–61.

C H A P T E R

# 8

# Knowledge, Attitude, and Practices of HIV Positive Adults in India: Rural–Urban Differences

*Deepika Anand, Seema Puri*
**University of Delhi, New Delhi, India**

O U T L I N E



## Abstract

The Millennium Development Goals era witnessed reduction in human immunodeficiency virus/acquired immunodeficiency virus (HIV/AIDS) epidemic across the globe and now the Sustainable Development Goal 3 commits to end the AIDS epidemic by 2030. There are enough research and data available on the strong relationship between nutrition and HIV infection. However, data on the nutritional knowledge, attitude, and practices (KAP) of people living with HIV/AIDS (PLHIV) are still scanty. Studying KAP of PLHIV is important as an appropriate level of knowledge and attitude toward nutrition is an important factor that might influence the actual nutrient intake. A number of national and international agencies have developed education modules on improving nutritional practices among PLHIV to be used in resource constrained settings, though the extent to which these are used and the impact of these has not been evaluated in detail. India is a country with diverse cultures and there is stark demographic difference in its rural and urban setting. Thus, this chapter throws light on the nutritional KAP specifically highlighting the difference (if it exists) between the rural and urban population affected by HIV.

**Keywords:** HIV; India; Nutritional KAP; Rural; Urban.

## List of Abbreviations

**AIDS** Acquired Immunodeficiency Virus
**ART** Antiretroviral Treatment
**ARV** Antiretroviral
**HIV** Human immunodeficiency Virus
**KAP** Knowledge, Attitude, and Practices
**NACO** National AIDS Control Organization
**NACP** National AIDS Control Program



*HIV/AIDS*
http://dx.doi.org/10.1016/B978-0-12-809853-0.00008-0

**NSS** National Service Scheme
**PLHIV** People living with HIV/AIDS
**RDA** Recommended Dietary Allowance
**REE** Resting Energy Expenditure
**SECRT** Sex Education, Counseling, Research Training/Therapy
**UNAIDS** United Nations AIDS Organization
**UTA** Universities Talk AIDS Program
**UTs** Union Territories

# INTRODUCTION

India has a five-decade long history of nutrition-specific programs delivering nutrition services to the community including HIV-affected populations. With the stabilization of the HIV epidemic and with better availability and use of antiretroviral treatment (ART) medications, India is now facing newer challenges. With better treatment options available, there is an emerging need for nutritional care of people living with HIV/AIDS (PLHIV) to provide them with a better quality of life. This chapter touches on the epidemic of virus in the country, relationship between nutrition and HIV including nutritional requirements, government initiatives in the country and throws light on the rural–urban differences in the nutritional knowledge, attitude, and practices (KAP) of adult PLHIV.

# MAIN TEXT: PREVALENCE OF HUMAN IMMUNODEFICIENCY VIRUS INFECTION

## Global

According to the recent United Nations AIDS Organization report (2013), there were 35 million [33.2 million–37.2 million] people living with HIV at the end of the year 2013. Agewise statistics indicate that there are 3.2 million [2.9 million–3.5 million] children younger than 15 years living with HIV and 4 million [3.6 million–4.6 million] young people 15–24 years old living with HIV, 29% of whom are adolescents aged 15–19 years. There are 16 million [15.2 million–16.9 million] women aged 15 years and older living with HIV; 80% live in sub-Saharan Africa. Globally, 15 countries account for nearly 75% of all people living with HIV[1]. The report also mentions that the number of people who are newly infected with HIV is continuing to decline in most parts of the world—a decline of 38% from 2001. In case of children, in the year 2013, 240 000 [210 000–280 000] children were newly infected with HIV, 58% lower than 2002. Also, the AIDS-related deaths have fallen by 35% since 2005, when the highest number of deaths was recorded. In the past 3 years alone, AIDS-related deaths have fallen by 19%, which represents the largest annual decline in the past 10 years. Almost 12.9 million people were receiving antiretroviral therapy globally at the end of 2013. The percentage of people living with HIV who are not receiving antiretroviral therapy has been reduced from 90% [90%–91%] in 2006 to 63% [61%–65%] in 2013. But on the other side, 22 million, or 3 of 5 people living with HIV are still not accessing antiretroviral therapy. The number of children receiving antiretroviral therapy is appallingly low—a mere 24% [22%–26%] i.e., three of four children living with HIV or 76% [74%–78%] are not receiving HIV treatment.[1]

## India

According to the National AIDS Control Organization recent estimates[2] the number of people living with HIV/AIDS in India is 21.17 lakhs (17.11 lakhs–26.49 lakhs) in 2015 compared to 22.26 lakhs (18.00 lakhs–27.85 lakhs) in 2007. Children (<15 years) account for 6.54%, while two-fifth (40.5%) of total HIV infections are among females. India is estimated to have around 86 (56–129) thousand new HIV infections in 2015, showing 66% decline in new infections from 2000 to 32% decline from 2007, the year sets as baseline in the National AIDS Control Program-IV (NACP-IV). Children (<15 years) accounted for 12% (10.4 thousand) of total new infections, while the remaining (75.9 thousand) new infections were among adults (15+years). Between 2000 and 2015, new HIV infections dropped from 2.51 lakhs to 86 thousand, a reduction of 66% against a global average of 35%. The adult (15–49 age-group) HIV prevalence at national level has continued its steady decline from an estimated peak of 0.38% in 2001–03 through 0.34% in 2007 and 0.28% in 2012 to 0.26% in 2015[2]. With the revision of ART guidelines for treatment initiation at CD4 count <350 cells/mm$^3$

[1] South Africa (18%); Nigeria (9%); India (6%); Kenya (5%); Mozambique (4%); Uganda (4%); United Republic of Tanzania (4%); Zimbabwe (4%); United States (4%); Zambia (3%); Malawi (3%); China (2%); Ethiopia (2%); Russian Federation (2%); and Brazil (2%).

from 2012 onward, the projected need for treatment of children with HIV is estimated at around 0.86 lakhs (70,000–108,000). While the treatment coverage has increased from 2007 to 2011, approximately 34% of the total estimated need for children needing treatment, actually received ART in 2011. The treatment coverage for adults increased in 2011, when nearly 52% of the estimated 785,000 (681,000–872,000) PLHIV needing treatment were receiving ART.[3]

### Statewise Picture

Among the states/union territories (UTs), in 2015, undivided Andhra Pradesh and Telangana have the highest estimated number of PLHIV (3.95 lakhs) followed by Maharashtra (3.01 lakhs), Karnataka (1.99 lakhs), Gujarat (1.66 lakhs), Bihar (1.51 lakhs), and Uttar Pradesh (1.50 lakhs). These seven states together account for two thirds (64.4%) of total estimated PLHIV. Rajasthan (1.03 lakhs), Tamil Nadu (1.43 lakhs), and West Bengal (1.29 lakhs) are other states with estimated PLHIV numbers of 1 lakh or more.[2] Besides these states, Maharashtra, Chandigarh, Tripura, and Tamil Nadu have shown estimated adult HIV prevalence greater than the national prevalence (0.26%), while Odisha, Bihar, Sikkim, Delhi, Rajasthan, and West Bengal have shown an estimated adult HIV prevalence in the range of 0.21%–0.25%. All other states/UTs have levels of adult HIV prevalence below 0.20%.[2]

## NUTRITION-SPECIFIC GOVERNMENT INITIATIVES IN THE COUNTRY

Consolidating the gains made during NACP-III, the NACP-IV (2012–17) was launched to accelerate the process of reversal and to further strengthen the epidemic response in India through a cautious and well-defined integration process over the period 2012–17. The objectives of NACP-IV are to reduce new infections and provide comprehensive care and support to all PLHIV and treatment services for all those who require it. Under the care, support, and treatment service package of NACP-IV, there is a provision for the nutritional and psychosocial support through community and support centers. Also, the pediatric centers of excellence setup in the country focus on the provision of counseling for adherence to ART and also on nutrition linked issues. As part of continuous capacity building efforts, technical guidelines, and training modules have been developed. Specific to nutrition there are guidelines for providing nutritional care and support for adults living with HIV and AIDS[4] and nutritional guidelines for HIV exposed and infected children (0–14 years of age).[5]

## NUTRITION AND HUMAN IMMUNODEFICIENCY VIRUS

Scientific evidence from across the world has revealed that macronutrients and micronutrients are critical for fighting HIV infection because they are required by the immune system and major organs to attack infectious pathogens including HIV. It is well proven that good nutrition helps to strengthen the immune system and reduce the severity and impact of opportunistic infections in PLHIV. It is known that an immune dysfunction as a result of HIV/AIDS leads to malnutrition, and this in turn leads to further immune dysfunction with accelerated progression to AIDS.[6] Nutritional problems have been shown to be significant and contribute to poor health and death in HIV/AIDS patients. Weight loss, lean tissue depletion, lipoatrophy, loss of appetite, diarrhea, and the hypermetabolic state each increases risk of death. NACO has given nutritional guidelines for adult as well as for pediatric HIV population in India. The adult nutritional guidelines are summarized in Table 8.1.

TABLE 8.1 Nutritional Requirements for Adult PLHIV[4]

| | |
|---|---|
| Energy | During asymptomatic stage—energy needs are increased by 10% over accepted levels for otherwise healthy people. During symptomatic stage—20%–30% increase is recommended |
| Protein | 1 g protein per kg body weight per day is recommended. The protein intake should not be greater than 12%–15% of the total calories. |
| Fat | About 20%–30% of the total energy should be provided by fat. At least 50% of fat intake should consist of vegetable oils rich in EFA and the amount of saturated, monounsaturated, and polyunsaturated fat should be <7%, >10%, and <10%, respectively, of the total calories. |
| Micronutrients | A daily dose of micronutrients providing "1 RDA" vitamins and minerals may be supplemented if the available diets are not balanced and do not contain variety of animal sourced foods, fruits, and vegetables. However, the guidance on supplementation of micronutrients with regard to the amount and duration is not available. |
| Water | 2 L of water per day is recommended. Safe drinking water should be roller-boiled for 3 min, cooled, and stored in clean, covered containers. |

Studies consistently show that micronutrients particularly serum antioxidant vitamins and minerals decrease, while oxidative stress increases during AIDS progression. The optimization of nutritional status, intervention with foods, and supplements, including these nutrients and other bioactive food components, is needed to maintain the immune system.[7]

It is well established that the nutritional needs of PLHIV are influenced by several factors such as age, physical activity level, clinical stage, viral load, etc., to name a few. Adequate intake of macro- and micronutrients is necessary to maintain good nutritional status. The need is primarily to maintain a balance of food from each food group and to consume a variety within each food group. This explains the need for having good nutritional knowledge, positive attitude, and adherence to good nutritional practices by the PLHIVs.

## NUTRITIONAL KNOWLEDGE, ATTITUDE, AND PRACTICES

Adequate nutritional knowledge has been shown to be effective in the management of HIV infections and has an influence on the health outcomes of infected individuals.[8] Low socioeconomic status may be treated as a confounding factor as it is an independent predictor of HIV transmission and mortality, even after controlling for confounders such as age, disease stage, and access to health care.[9–11] There is paucity of information on the effect of having an HIV positive status on nutritional knowledge.[9,12]

Dietary intake may be influenced positively by nutritional counseling, by raising awareness about needed quantities of food to meet increased demand and adequate dietary diversity. In multiple studies, counseling has been shown to increase intake and improve weight and fat mass and may also increase fat-free mass and lean body mass.[13–15] The goal of nutritional counseling is to improve the quality of the diet to reach required amounts of energy, protein, and micronutrients as well as to increase intake to meet changes in resting energy expenditure. Indeed, counseling can increase intake itself[13,16] and can be employed with relatively few resources to accomplish dietary diversity, which is associated with improved micronutrient intake and improved growth in children.[17–20] Counseling remains the first-line therapy in most programs for mild and moderate malnutrition.[21] Adequate nutrition and food is recognized along with nutrition interventions and psychological support to break the silence in the fight against HIV/AIDS.[7,22–24] A study by Segal-Isaacson (2006)[25] reported that health education sessions by a therapist improved the dietary patterns of disadvantaged HIV-infected women in America.

In India, several programs have operated for improving KAP regarding the disease in HIV positive patients. The "Universities Talk AIDS" program (UTA) run by the Indian government as a strategy to prevent HIV/AIDS by informing students and encouraging discussion about healthy human sexuality. UTA works through the National Service Scheme, began on an experimental basis in 1991 in 59 universities.[26] The Family Planning Association of India's Sex Education, Counseling, Research Training/Therapy project is spearheading the incorporation of sex education into its family planning activities.[27] The project's focus is on educated urban youth aged 15–29 years—a sector that will become India's future parents and leaders but has been neglected by the government and voluntary organizations. With respect to nutrition, NACP-III (2007–12) recognized nutrition as an important intervention for the care and treatment of HIV-infected individuals, and nutritional assessment and counseling were made essential components under NACP.

This section discusses the level of KAP of adult PLHIV in India highlighting differences (if any) in the rural and urban settings. The urban area sample was selected from New Delhi region (n = 400), while the rural sample was taken from Odisha state (n = 186). The purpose of conducting a KAP study among PLHIV was to understand what the affected people know about nutrition; how they perceive the interaction between nutrition, the disease and their health and how do they behave. Research has shown that both macronutrient and micronutrient deficiencies contribute to immune dysfunction and can lead to progression of disease.[28–31] Consumption of proper nutrients, which can be enhanced by knowledge of importance of good nutrition for the PLHIV and proper dietary practices, can support an already-compromised immune system.[30]

### Key Sociodemographic Indicators for the Urban Sample

The urban sample comprised of 400 adults (21–59 years old), which included 245 males (61.3%), 144 females (36%) and 11 transgender (2.7%). The study sample mainly had subjects in the age-group of 21–30 (46.3%) years followed by 31–40 years (36.5%). As for education, 30.7% reported no formal education (illiterate group), 15.2% had primary level education, 44% secondary, and 10% had college or above education. Around 39% of the respondents reported that they had an HIV-infected family member staying with them. At the time of interview, 38% of the sample was unemployed, 29% was into some kind of business, 20.5% was salaried population, 9% was involved in heavy work such as cultivation or construction work, and 3.5% reported to be students or domestic help. The rate of unemployment increased from the time of detection of virus to the time of interview. The major reasons reported for the same were ill health and their HIV status. Around 50% of the sample reported their annual family income to be less than Rs. 30,000 i.e., less than Rs. 2500 per family per month.

## Key Demographic Indictors for Rural Sample

The majority of rural sample (males = 120; females = 66) belonged to the age-group of 31–40 years (54%). Around 40% of the sample had primary education (n = 77); 26% (n = 51) had secondary education, 25% (n = 49) was illiterate, and only 9% of sample had higher education i.e., college and above.

A large section of the rural sample was employed as construction workers (37%) and around 33% of them were unemployed primarily due to health reasons and around 20% were involved in some kind of small-scale business. The income profile shows that the majority of subjects (44%) had an annual family income less than Rs. 20,000.

The KAP questionnaire had a knowledge section, which was designed to test the knowledge of PLHIV on HIV and nutrition in HIV. The questions covered topics such as risk factors for HIV, symptoms of HIV, diagnosis of HIV, treatment of HIV, nutrient requirement in HIV, and healthy foods to improve health. The attitude section gauged the prevailing attitudes and beliefs regarding the HIV infection. The statements covered issues such as demography of HIV, precautions to be practiced, treatment, healthy food choices, and healthy dietary practices during the infection. The practices section assessed the dietary practices of the population with regard to HIV infection including snacking, food group consumption, exercise, fluid intake, and medication. After summing up the scores of the individual sections, the final KAP score was obtained. The KAP score ranged from 15 to 75. The subjects were classified according to their scores obtained in each section. The scoring pattern is showed in Table 8.2.

Table 8.3 shows the mean scores of the rural and urban sample. The results indicate that the rural population had significantly lower nutritional knowledge as compared to the urban sample ($P > .05$), but there were no significant differences in the attitude and practices or the total KAP scores between the two groups. Higher KAP scores by the urban group could be attributed to the better education status and greater exposure to the information pertaining to HIV while living in a large city.

The mean knowledge scores for both urban and rural sample belonged to moderate category. This may be explained on the basis that both the groups were exposed to the nutritional counseling provided at the ART centers. The sample had to meet the counselor during their monthly visits to the ART center for collecting their ARV medications. Both the groups had poor attitude scores, which may be explained that mere knowledge that they are HIV positive makes their attitude toward disease and importance of nutrition negative. The practices score was again in the moderate category for both the groups indicating that the sample was practicing certain good dietary practices, but there is a scope for improvement especially in areas pertaining to food safety and hygiene. Also, the overall

TABLE 8.2 Scoring Pattern for Knowledge, Attitude, and Practices

| | Knowledge | Attitude | Practices | Total KAP |
|---|---|---|---|---|
| Poor | 0–5 | 15–26 | 0–5 | 15–35 |
| Moderate | 6–10 | 27–35 | 6–10 | 36–55 |
| Good | 11–15 | 36–45 | 11–15 | 56–75 |

TABLE 8.3 Knowledge, Attitude, and Practices (KAP) Scores of Sample

| | Score Range | Urban (n = 400) | Rural (n = 186) |
|---|---|---|---|
| Knowledge[a] | 0–15 | 8.3 ± 2.2 | 7.4 ± 1.9 |
| Attitude | 15–45 | 34.4 ± 3.7 | 34.1 ± 3.7 |
| Practices | 0–15 | 8.1 ± 2.3 | 8.1 ± 2.1 |
| Total KAP | 15–75 | 50.8 ± 5.5 | 49.7 ± 5.1 |

[a]*Differences significant as tested by t test (P < .05).*

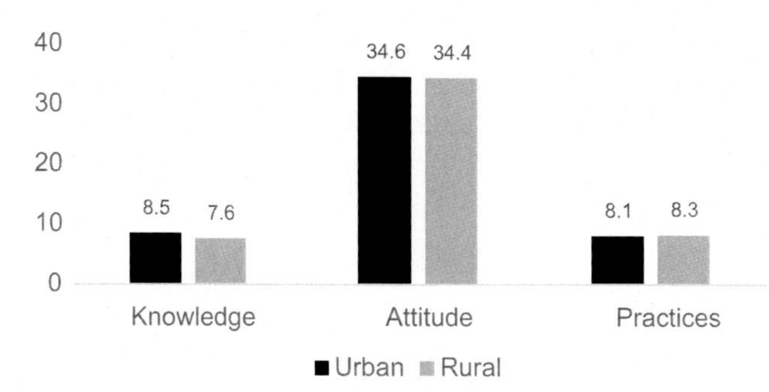


FIGURE 8.1 Knowledge, attitude, and practices scores of males.

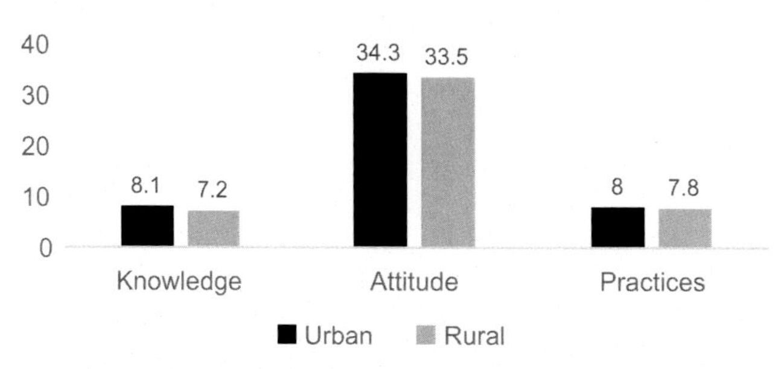


FIGURE 8.2 Knowledge, attitude, and practices scores of females.

KAP score for both urban and rural population was poor indicating that there is an urgent need for sensitizing and bringing about change in their dietary practices and making their attitude more positive toward the importance of nutrition during HIV infection. Though the basic knowledge about nutrition is present in these individuals but their attitude requires positive reinforcement and dietary practices need to be improved through rigorous and continuous interventions.

Further, genderwise analysis of KAP scores (Figs. 8.1 and 8.2) shows that the rural sample, both males and females, scored low on knowledge and attitude as compared to their urban counterparts. Rural men scored higher on practices as compared to the urban counterparts ($P > .05$).

There are a number of KAP studies done among PLHIV in India,[32–35] but still there is paucity of data specifically addressing the issue of nutrition. The studies primarily focus on KAP about the disease rather than the role and importance of nutrition during the infection. Considering that the HIV/AIDS picture in terms of prevalence, treatment, and awareness has changed over last three decades, the need is now to improve their quality of life where good nutritional status plays a major role. Efforts and focus should be on not merely improving the nutritional knowledge rather should more be on bringing about the behavior change.

## SUMMARY POINTS

- PLHIV (both rural and urban) possesses knowledge about HIV and nutrition but there is scope for improvement, and there is also a need to further increase their nutrition awareness levels. This may require constant delivery of nutrition linked messages through different media, i.e., visits to ART centers, broadcasting messages on mass media or awareness campaigns to name a few.
- Community level interventions should focus on converting the existing knowledge of individuals into healthy practices. This would again require constant counseling wherein PLHIVs should be made aware about the importance of nutrition in maintaining good quality of life. They should be encouraged to adopt healthy and safe eating practices to strengthen their immunity and stay protected from infections.
- The entire KAP dimension (in both urban and rural settings) needs to be strengthened by continuous and rigorous interventions, which aim at bringing about behavior change. Especially to bring about behavior change in rural population, use of simple, and easy to understand messages in local language should be adopted.
- As the treatment for the disease has penetrated in the interiors of the country and majority of the adult and pediatric population is now availing the ART regimen; nutrition counseling should also aim to bring about positive behavior and attitude change rather than merely communicating nutrition messages.

## References


1. UNAIDS. *UNAIDS GAP report*. 2013.
2. NACO. *India HIV estimations*. Technical Report. National AIDS Control Organization and National Institute of Medical Statistics, ICMR, Ministry of Health and Family Welfare, Government of India; 2015.
3. NIMS, NACO. *Technical report – India, HIV estimates*. 2012.
4. NACO. *National guidelines for providing nutritional care and support for adults living with HIV and AIDS*. National AIDS Control Organization, Ministry of Health and Family Welfare, Government of India; 2012.
5. NACO. *Nutrition guidelines for HIV exposed and infected children (0-14 years)*. Ministry of Health and Family Welfare, Government of India; 2012.
6. Oguntibeju OO, Van den Heever WM, Van Schalkwyk FE. The interrelationship between nutrition and the immune system in HIV infection: a review. *Pak J Biol Sci* December 15, 2007;**10**(24):4327–38.
7. Suttajit M. Advances in nutrition support for quality of life in HIV+/AIDS. *Asia Pac J Clin Nutr* 2007;**16**(1):318–22.
8. Masuku SK, Lan SJ. Nutritional knowledge, attitude, and practices among pregnant and lactating women living with HIV in the Manzini region of Swaziland. *J Health Popul Nutr* June 2014;**32**(2):261–9.
9. Odimayo MS, Olanrewaju WI, Omilabu SA, et al. Prevalence of rotavirus-induced diarrhea among children under 5 years in Ilorin, Nigeria. *J Trop Pediatr* 2008;**54**(5):343–6.
10. Nasidi A, Harry TO. *The epidemiology of HIV/AIDS in Nigeria*. Paperback ed. 2006. p. 17–35.
11. Anabwani G, Navario P. Nutrition and HIV/AIDS in sub-Saharan Africa: an overview. *Nutrition* January 2005;**21**(1):96–9.
12. Quach L, Mayer K, McGarvey ST, Lurie MN, Do P. Knowledge, attitudes, and practices among physicians on HIV/AIDS in Quang Ninh, Vietnam. *AIDS Patient Care STDS* May 2005;**19**(5):335–46.
13. de Luis D, Aller R, Bachiller P, Gonzalez-Sagrado M, de Luis J, Izaola O, Terroba C, Cuellar L. Isolated dietray counselling program versus supplement and dietary counselling in patients with human immunodeficiency virus infection. *Med Clin (Barc)* 2003;**120**(15):565–7.
14. de Luis Roman DA, Bachiller P, Izaola O, Romero E, Martin J, Arranz M, Eiros Bouza JM, Aller R. Nutritional treatment for acquired immunodeficiency virus infection using an enterotropic peptide-based formula enriched with n-3 fatty acids: a randomized prospective trial. *Eur J Clin Nutr* 2001;**55**:1048–52.
15. Schwenk A, Steuck H, Kremer G. Oral supplements as adjunctive treatment to nutritional counseling in malnourished HIV-infected patients: randomized controlled trial. *Clin Nutr* 1999;**18**:371–4.
16. Zambelli A, Comi D, Meraviglia P, Cargnel A. Effect of two different diets on AIDS patients' nutritional status. XI International Conference on AIDS. *Int AIDS Soc* 1996;**11**:293.
17. Moursi MM, Arimond M, Dewey KG, Treche S, Ruel MT, Delpeuch F. Dietary diversity is a good predictor of the micronutrient density of the diet of 6- to 23-month-old children in Madagascar. *J Nutr* 2008;**138**:2448–53.
18. Kennedy GL, Pedro MR, Seghieri C, Nantel G, Brouwer I. Dietary diversity score is a useful indicator of micronutrient intake in nonbreast-feeding Filipino children. *J Nutr* 2007;**137**:472–7.
19. Steyn NP, Nel JH, Nantel G, Kennedy G, Labadarios D. Food variety and dietary diversity scores in children: are they good indicators of dietary adequacy? *Public Health Nutr* 2006;**9**:644–50.
a Stover, Bollinger. *The economic impact of AIDS. The policy project*. 1999.
20. Penny ME, Creed-Kanashiro HM, Robert RC, Narro MR, Caulfield LE, Black RE. Effectiveness of an educational intervention delivered through the health services to improve nutrition in young children: a cluster-randomised controlled trial. *Lancet* 2005;**365**:1863–72.
21. Sztam KA, Fawzi WW, Duggan C. Macronutrient supplementation and food prices in HIV treatment. *J Nutr* 2009. http://jn.nutrition.org/content/140/1/213S.full.pdf+html.
22. Sherlekar S, Udipi SA. Role of nutrition in the management of HIV infection/AIDS. *J Indian Med Assoc* 2002;**100**(6):385–90.
23. Dey SK, Pal NK, Set I. Nutrition status of HIV seropositive subjects and counselling - scenario in urban Calcutta and rural North Bengal, India. *Int Conf AIDS* 2000:9–14.
24. Thuita FM, Mirie W. Nutrition in the management of acquired immunodeficiency syndrome. *East Afr Med J* 1999;**76**(9):507–9.
25. Segal-Isaacson CJ, Tobin JN, Weiss SM, Brondolo E, Vaughn A, Wang C, et al. Improving dietary habits in disadvantaged women with HIV/AIDS: the SMART/EST women's project. *AIDS Behav* 2006;**10**(6):659–70.
26. Bhatt SD, Dhoundiyal NC. Country watch: India. *AIDS STD Health Promot Exch* 1997;**3**:11–2.
27. Watsa M. Sexuallity counselling in India. *Plan Parent Chall* 1993;**2**:25–7.
28. Tang AM. Weight loss, wasting, and survival in HIVpositive patients: current strategies. *AIDS* 2003;**13**:23–7.
29. Tang AM, Graham NM, Kirby AJ, McCall LD, Willett WC, Saah AJ. Dietary micronutrient intake and risk of progression to acquired immunodeficiency syndrome (AIDS) in human immunodeficiency virus type 1 (HIV-1)-infected homosexual men. *Am J Epidemiol* 1993;**138**:937–51.
30. Walsh CM, Dannhauser A, Joubert G. Impact of a nutrition education programme on nutrition knowledge and dietary practices of lower socioeconomic communities in the Free State and Northern Cape. *SAJCN* 2003;**16**:89–95.
31. Steinhart CR. HIV-associated wasting in the era of highly active antiretroviral therapy (HAART): a practice- based approach to diagnosis and treatment. *AIDS Read* 2001;**11**:557–69.
32. Sanjay S, et al. Evaluation of impact of health education regarding HIV/AIDS on knowledge and attitude among persons living with HIV/AIDS. *Indian J Comm Med* 2003;**28**(1):30–3.
33. Ahmed M, Gaash B. Awareness of HIV/AIDS in a remotely located conservative district of J&K (kargil). *Indian J Comm Med* 2002;**27**(1):12–8.
34. Gaash B, et al. Knowledge, Attitude and belief on HIV/AIDS among female senior secondary students in Srinagar district of Kashmir. *Health Popul Perspect Iss* 2003;**2**(3):101–9.
35. Paul D, Gopalakrishnan N. Knowledge regarding modes of transmission and prevention of sexually transmitted diseases including HIV/AIDS among child development project officers. *Indian J Comm Med* 2001;**26**(3):141–4.

CHAPTER

# 9

# Antioxidants in HIV in Africa: Supplements, Local Diet, and Education

Germaine S. Nkengfack Nembongwe[1], Heike Englert[2]

[1]University of Dschang, Dschang, Cameroon; [2]Muenster University of Applied Sciences, Muenster, Germany

OUTLINE



## Abstract

Since the discovery of human immunodeficiency virus (HIV), causative agent of acquired immunodeficiency syndrome (AIDS) in 1981, approximately 25 million lives have been claimed, especially in sub-Saharan Africa. HIV causes chronic immune activation, provoking oxidative stress, and reducing antioxidant level in the body. Studies confirm that regular intake of medication and supplements can slow down disease progression. However, such medications and nutritional supplements are expensive and often inaccessible to a majority, who are from the lower social class. Recent studies show that a health affirming lifestyle and an optimal intake of antioxidants through local diets is also able to reduce oxidative stress, increase levels of endogenous antioxidative enzymes (glutathione, superoxide dismutase, and catalase). It is therefore intimating to promote the use of local diets, with its high and available content in bioactive agents, through education. Educating patients on adequate nutrition will help improve health outcomes and make this approach sustainable.

**Keywords:** Antioxidant; Africa; Diet; Education; Oxidative stress; Supplements.



HIV/AIDS
http://dx.doi.org/10.1016/B978-0-12-809853-0.00009-2

## List of Abbreviations

**AIDS** Acquired Immunodeficiency Syndrome
**ART** Antiretroviral therapy
**CD** Cluster of Differentiation
**DNA** Deoxyribonucleic Acid
**Gp 120** glycoprotein 120
**GSH** Glutathione
**HAART** Highly Active Antiretroviral Therapy
**HIV** Human Immunodeficiency Virus
**IL-1, IL-6** Interleukin 1, Interleukin 6
**LDL** Low-density lipoprotein
**LGSH** Liposomal Glutathione
**MCP-1** Monocyte Chemotactic Protein-1
**NAD+** Nicotinamide Adenine Dinucleotide
**NNRTIs** nonnucleoside reverse transcriptase inhibitors
**NRTIs** nonreverse transcriptase inhibitors
**PARP** Poly ADP ribose polymerase
**PI** Protease Inhibitors
**PLWHA** people living with HIV/AIDS
**RANTES** Regulated on activation, normal T cell expressed, and secreted
**RDA** Required daily allowance
**Tat** Transactivator of Transcription
**Th1** Type-1 helper
**Th2** Type-2 helper
**TNF-α** Tumor Necrosis Factor
**UNAIDS** Joint United Nations Program on HIV/AIDS
**WHO** World Health Organization

# INTRODUCTION

According to estimates from Joint United Nations Program on HIV/AIDS (UNAIDS), approximately 36.7 million people are living with human immunodeficiency virus (HIV) globally, of this, 24 million live in sub-Saharan Africa.[1] Important research efforts to reduce the burden of HIV on infected individuals lead to the discovery of antiretroviral therapy (ART) and later on highly active antiretroviral therapy (HAART). Thanks to this therapy, the rate of HIV/AIDS related mortality has consistently reduced. Report from UNAIDS show that 1.1 million people died of acquired immunodeficiency syndrome (AIDS) in 2015 in sub-Saharan Africa. This represents a 45% decrease since its peak in 2005.[1] However, ART/HAART cannot cure HIV/AIDS but can slow down the rate of disease progression.[2] Besides that, the use of HAART is associated with many side effects such as vomiting, nausea, fatigue; insulin resistance, etc., which may worsen the individual's health condition.[3] It is well known that various factors play a role in the progression of the HIV infection. One of the most important factors is oxidative stress. Oxidative stress does not only occur due to the presence of the HIV virus but also during the HAART therapy, exacerbating oxidative stress in infected people.[4] To counteract the effects of oxidative stress, aside HAART, alternative methods are used to reduce the burden of HIV on infected individuals and slow down disease progression. Evidence exists that optimum nutrition through adequate supply of antioxidants, through supplements and/or local diet can improve clinical conditions in patients and minimize free radical production and oxidative stress.[5] Another important strategy to enable people living with HIV/AIDS (PLWHA) improve their antioxidant status is to educate them on appropriate nutritional habits. This can be done in support groups or in nutrition-related projects for PLWHA.[6]

# OXIDATIVE STRESS IN HUMAN IMMUNODEFICIENCY VIRUS

Oxidative stress, caused by a redox imbalance between the oxidative and antioxidant system, is thought to play an important role in the progression of HIV. Oxidative stress is the result of enhanced productions of reactive oxygen species (ROS), causing progressive oxidative damage and cell injury.[7] ROS are highly reactive compounds with a half-life varying between nanoseconds and a few seconds. Also, ROS are capable of reacting with nonradicals to produce new radicals, which can in turn attack polysaccharides, DNA, and proteins and can lead to cell death. ROS originates endogenously mainly in the mitochondria by the oxidative production of energy

$H_2O_2$ + 2GSH ---------(CAT)-----> $2H_2O$ + GSSG

GSSG + NADPH +$H^+$ -------------> 2GSH + $NADP^+$

2 LOOH + 2 GSH ------(GPX)---------> 2 LOH + $H_2O$ + GSSG

$2O_2^{\cdot}$ + $2H^+$------(SOD)-------> $O_2$ + $H_2O_2$

$2H_2O_2$------- (CAT)-------> $H_2O$ + $O_2$.

FIGURE 9.1 Some important equations.

and through the metabolism of arachidonic acid. Exogenous sources of ROS are, for example, rays (UV rays and related therapies), smoking cigarette, and air pollution (ozone, NO, exhaust fumes, industrial waste, food waste, pesticides and heavy metals, medication).[8] ROS include superoxide, hydrogen peroxide ($H_2O_2$), hydroxyl free radicals such as nitrogen intermediates, nitric oxide (NO), and peroxynitrite. Oxidative damage may take place in all biomolecules (lipids, proteins, carbohydrate, and nucleic acids) in the body impairing cell functions, leading to cell death, and progression of diseases such as HIV/AIDS.[9] A direct consequence of this redox imbalance is the depletion of endogenous antioxidants such as superoxide dismutase (SOD), catalase (CAT), and glutathione (GSH), through the production of large amounts of inflammatory products such as superoxide, peroxynitrite, and NO (Fig. 9.1).

Also, HIV infection is known to enhance expression of proinflammatory cytokines such as interleukin 1, interleukin 6 (IL-1, IL-6), RANTES, TNF-α, and monocyte chemotactic protein-1,[10] which in turn enhance replication of HIV. In particular, ROS has the capacity of starting a cascade as second messengers, provoking the activation of NF-kb (a transcriptional promoter of protein that are involved in the inflammatory and acute-phase response), which controls the transcription of gens for HIV replication. NF-kb is a transcription factor that is normally bound in the cells with I-kb in the cytoplasm in its active form. During oxidative stress, it begins to bind DNA and in this way increase replication.[11] Further, ROS does not only affect the antioxidative status directly but also indirectly. ROS, such as peroxynitrite and peroxide, which increase under deficiency of arginine and selenium, a well-known antioxidant can cause NAD+/niacin depletion via DNA damage and PARP, which can later on lead to tryptophan oxidation for compensatory niacin synthesis. The tryptophan deficiency resulting from tryptophan oxidation—which is the first step in the de novo niacin synthesis pathway—is known to impair T-cell function, leading to immunological tolerance.[12]

## ANTIOXIDANTS IN SUPPLEMENTS AND LOCAL DIETS FOR PEOPLE LIVING WITH HIV/AIDS IN AFRICA

Antioxidants are substances produced by the human body in small quantities, which are capable of inhibiting or even blocking oxidative stress. Studies support the fact that dietary antioxidants play an important role in protecting the body from many diseases such as cancer, atherosclerosis, neurogenerative diseases, diabetes, etc.[13] Antioxidants are known to counteract the effects of oxidant and evidence exists on their positive effects on CD4cells and viral load of infected people.[14] There exist two major groups of antioxidants. The enzymatic endogenous antioxidants (e.g., SOD, GSH, CAT, etc.), which can protect the body against ROS and the nonenzymatic exogenous antioxidants such as vitamin A, β-carotene, vitamin C, vitamin E, selenium, and zinc. Exogenous antioxidants are essential micronutrients that cannot be produced by the body thus are available in food especially in fruits and vegetables. To recognize antioxidants in nature, a general rule of the thumb is to select intensely colored fruit or vegetables since most exogenous antioxidants are pigments (e.g., anthocyanin) found in plants.[8] Specific African food with high antioxidants content include leguminous plants (black beans, red beans, red kidney beans), green leafy vegetables (*Amaranth* sp. (pigweed), *Vigna unguiculata* (Cowpea), *Solanum retroflexum* (Nightshade), *Brassica carinata* (Kale), *Cucurbita pepo, Cucurbita Moschata and Cucurbita maxima* (leaves of pumpkin), spinach, carrots; fruits (tomatoes, papaya, grapefruits, mangoes, palm kernels, squash, avocado, passion fruit, ananas, melon varieties, *Dacryodes edulis* (African plum), red cabbage, etc.; spices such as peppermint, hot chili pepper, oregano, lemon balm are also good sources of antioxidants, even in small quantities.[8,61] Antioxidants are also available as capsules, tablets, and syrups, for sale as nutritional supplements. However, be it in the local diet or as supplements, antioxidants are known for their important role in neutralizing free radicals on their own or by acting in conjunction with the enzymatic system.[15,16] Meanwhile, it is important to note that in spite of its function in reducing oxidative stress and strengthening immune function, supplements are very expensive, especially for a majority in sub-Saharan Africa, home for about two-third of the PLWHA worldwide.[1] Besides the fact that supplements confer benefits in HIV patients, they are associated with side effects. Evidence exists that the supplementation of vitamin A and E is linked with increase production of immunosuppressive chemokines (CCR5), reducing CD4 counts and increasing viral load.[17,18] Another study precise that vitamin A supplementation increases the risk of mother-to-child transmission (MTCT) of HIV. Faced with this reality, it is important to educate PLWHA and caretakers on the fact that besides taking supplements, preference should be granted to consumption of antioxidants through local diets. The local diet contains beside antioxidants also a wide range of bioactive nutrients capable of improving nutritional and health status of patients.[19]

# INTERACTION BETWEEN ANTIOXIDANTS AND HIV/AIDS

## Endogenous Antioxidants

### *Glutathione*

GSH, an endogenous antioxidant present in all eukaryotic cells is a tripeptide composed of three amino acids: glutamine, cysteine, and glycine. GSH is known for its function in maintaining redox homeostasis.[20] GSH is synthesized or catalyzed by γ-glutamyl cysteine and GSH synthetase. GSH contains selenium and functions in the breakdown of $H_2O_2$. During this process, GSH is oxidized to glutathione disulphide (GSSG) and later on regenerated to $NADPH + H^+$. Glutathione peroxidase (GPX) is also capable of reducing free peroxide fatty acids (LOOH) to GSSG.

Moreover, GSH reacts with other vitamins synergistically to regenerate vitamin E and vitamin C, while vitamin C regenerates vitamin E, which exerts a protective role on β-carotene.

Studies show low levels of GSH and Th1- and Th2-associated cytokines in PLWHA, especially in plasma and bronchoalveolar lavage fluids, due to low levels of GSH de novo synthesis enzymes and high levels of oxidative stress.[21] This process weakens the immune system and is associated with the development of opportunistic infections such as tuberculosis, hepatitis, chronic diseases, etc.

Although the precise mechanism is not clear, evidence exists that administering supplements of liposomal GSH for 13 weeks to HIV-infected individuals resulted in significant increase in Th1 cytokines, IL-1β, IL-12, IFN-γ, and TNF-α. Supplementation with LGSH also resulted in important decrease in the level of free radicals and immunosuppressive cytokines such as IL-10, TGF-β.[22,23] Studies also demonstrate that deficiency in GSH in HIV-infected patients triggers impaired cytokine production, leading to increased susceptibility to opportunistic infections such as *Mycobacterium tuberculosis* infection.[21,24,25] A glycoprotein 120 (gp 120), found on viral envelope can also increase oxidative stress and inhibit redox levels of GSH and GPX associated with increased oxidative stress and cell damage.[26]

### *Superoxide Dismutase*

SOD is considered as one of the key enzymes capable of destroying the superoxide radical. In the presence of oxygen, SOD converts superoxide radical in to hydrogen peroxide, which is later on converted to water and oxygen by CAT, in synergy with GSH $2O_2^{*} + 2H^+ \rightarrow O_2 + H_2O_2$. There exist several forms of SOD in animal cells. The mitochondrial SOD contains manganese (Mn-SOD), the cytosolic SOD contains copper and zinc (Cu–Zn-SOD) and the extracellular SOD. These minerals are therefore able to influence the antioxidative status of the body.[27] It is well known that many HIV gene products play a role in the expression of viral gene. One of these is the regulatory transactivator of transcription (Tat). Tat secreted from HIV infected cells, is capable of entering uninfected cells and activating nuclear factor—kappa binding NF-kB, which is responsible for the upregulation of various cytokines and adhesion molecules (vascular cell adhesion molecule-1 (VCAM-1) and intercellular adhesion molecule (ICAM)) in human endothelial cells and astrocytes.[28] Although the mechanism of Tat-mediated NF-kB activation is not very clear, evidence about the interaction of the Tat protein and SOD exists.[29] Previous and recent studies explored the role of ROS generation in HIV-Tat–induced NF-kB activation by using cell permeable protein—capable of delivering protein into another cell—PEP-1 SOD. Results of these studies show that PEP-1 SOD could significantly inhibit HIV-1 Tat–induced NF-kB activation suppress HIV-1 Tat–induced upregulation of ICAM-1/VCAM-1 expression and enhanced monocyte adhesion to astrocytes in cells.[28,30] This is an indication that SOD can play a regulatory role in HIV-1 Tat–induced NF-kB activation.

TABLE 9.1 Endogenous Antioxidants and Functions

| Antioxidant | Function |
|---|---|
| Glutathione (GSH) | Antioxidant, immune enhancing effect. Interferes hydrogen peroxide breakdown |
| Superoxide Dismutase (SOD) | Antioxidant converts superoxide radical to hydrogen peroxide |
| Catalase (CAT) | Antioxidant, converts hydrogen peroxide to oxygen and water |

Another HIV gene product involved in oxidative stress in HIV patients is glycoprotein 120 (gp 120), found on the surface of the HIV envelope. The gp 120 is required for viral entry, replication, and later on induces oxidative stress.[31] The gp 120 is known to interfere with the level of SOD.

### *Catalase*

Like GSH, CAT is an endogenous antioxidant found in almost all organisms. CAT is an enzyme containing iron. CAT is found mostly in peroxisomes and is capable of detoxifying hydrogen peroxide by catalyzing a reaction between two hydrogen peroxide molecules producing water and oxygen. Like SOD, studies show low CAT level in HAART naïve patients (patient who are not taking ART), as a result rapid depletion of these endogenous enzymes, due to high levels of ROS. Deficiency in CAT may be due to increase utilization of these enzymes, caused by increased oxidative stress.[32] Besides HIV, CAT, such as GSH and SOD, has beneficial effect in the healing of various diseases mediated by ROS[33] (Table 9.1).

## Exogenous Antioxidants

### *Vitamin A*

Vitamin A or retinol is a fat soluble vitamin found mostly in animal products. Sources of vitamin A in the diet of a majority of Africans include red palm oil, carrots, tomatoes, vitamin A-rich fruits (mango, papaya), squash, sweet potatoes, liver, tuna fish; butter, milk, egg yolk. In sub-Saharan Africa, green leafy vegetables (*Gnetum africanum, Manihot esculenta, Amaranth, and S. retroflexum)* are rich sources of vitamin A. In addition, red palm oil is highly available and cheap, and readily used in the preparation of most sauces and dishes such as cowpea cake, and locally known in some African countries as *koki* or *moi moi*. For example, one of the main ingredients in the preparation of *G. africanum* or leaves of *M. esculenta, or Corchorus olitorius* (jute mallow leaves) in many sub-Saharan African countries is red palm oil or palm kernel juice. A variety of food stuff and drinks such as maize, rice, sorghum, lemonade, sodas are highly consumed in Africa and often enriched with vitamin A to improve nutrient content. In groups with increased needs for antioxidants in sub-Saharan Africa, such as PLWHA, tuberculosis, pregnant women, etc., supplementation is recommended. In PLWHA vitamin A, like most micronutrients, is deficient. Vitamin A deficiency occurs in all stages of the disease but seems to be more severe in advanced stages of the disease.[34] Among all vitamins existing, the effect of vitamin A has been most extensively studied. Vitamin A is known to play an important role in cell immunity and oxidative stress. Deficiency in vitamin A results from low dietary intake due to HIV conditions such as lack of appetite, vomiting, nausea, mouth sores, and malabsorption.[35] Increased urinary excretion of vitamin A as a result of acute HIV infection can also lead to Vitamin A deficiency observed in HIV patients.[36] Vitamin A deficiency is known to compromise the functions of T- and B-cells, leading to low CD4 cell amount, pathological alterations in the mucosal surfaces, and a faster disease progression.[17,37] Evidence exists that vitamin A increase function of natural killer cells in vitro and supplementation of vitamin A is capable of modifying the phagocytic cell response by stimulating phagocytosis and cell-mediated cytotoxicity.[9]

Previous and even recent studies show that vitamin A supplementation could not reduce MTCT but could provide benefits for the infant postpartum.[14,17,38] Recent studies show that supplementation of antenatal vitamin A. reduces maternal night blindness, anemia, and maternal infections in HIV positive women.[39].

### *β-Carotene*

Carotenoid are made of eight isoprene units and divided into the two groups—the oxygen-free carotenoids (β-carotene and lycopene) and oxycarotenoids (xanthophylls). Carotenoids confers to fruits and vegetables their yellow, orange, or red color e.g., in tomatoes, oranges, peppers, egg yolk, carrots. Carotenoids are also abundantly found in African fruits and vegetables such as green plants, e.g., spinach, huckleberry, amaranth leaves, etc., always masked by chlorophyll. Food such as sweet potatoes, squash, mostly eaten with tomato sauces, stewed beans, or stewed vegetables are very rich sources of β-carotene. In sub-Saharan Africa, β-carotene–rich food with tubers are yam, cocoyams, plantain, green banana, potatoes, or cereals such as maize, millet, sorghum, known as *couscous* or *fufu* including carrot sauce, stewed spinach, kale, pigweed, cassava leaves, okra sauce, jute mallow leaves, etc. It is also worth saying that most African food is seasoned with hot chili pepper, which are rich sources of β-carotene as well. Of all the approximately 700 carotenoids known, only a few can act as provitamin A.[8] This is β-carotene. β-carotene is capable of inactivating and quenching ROS. Both previous and recent studies showed low concentrations of β-carotene PLWHA. However, the low β-carotene levels in HIV patients could be linked to elevated C-reactive protein, reflecting primarily more active infection.[40] Supplementing β-carotene in HIV patients receiving HAART may be beneficial (improving CD4 cell counts and viral load) to HIV patients, considering its antioxidative effect, in the presence of HAART.[13,34] Scientific results are controversary: A study in Tanzania suggested that supplementation of β-carotene in HIV-infected women leads to an increase risk of subclinical mastitis.[41] Subclinical mastitis is a strong risk factor for MTCT of HIV, through increase in breast milk shedding. In other words, supplementation of β-carotene would increase breast milk concentration of retinol and therefore is not recommended for HIV positive women due to increased risk of MTCT.[42] Also, Leitz and coworkers observed that contrary to the classical supplementation of β-carotene, the intake of 12 g/d of red palm oil by pregnant women during the third trimester, could improve β-carotene, milk concentration of provitamin A.[43]

### Vitamin C

Vitamin C also known as ascorbic acid is a water soluble vitamin found mostly in fruits and vegetables. The rich sources of vitamin C in African food are mostly fresh fruits and vegetables (oranges, lime, lemon, tangerine, guava grapefruits, mango, kiwano, lettuce, amaranth, spinach, pepper varieties, cabbage, tomato, potatoes, etc.). However, due to its high sensitivity to heat and light, in addition to the long preparation methods in Africa—partly for hygienic purposes— most of the water soluble vitamins including vitamin C is lost before consumption. The most available sources of vitamin C in the African diet are in fresh fruits. Also, a variety of fruit and vegetable juices consumed in Africa are fortified with vitamin C and therefore constitutes a rich source of this nutrient. Vitamin C, like most other water soluble vitamins, is very sensitive to atmospheric conditions such as heat and light and already mentioned, thus caution should be taken when preparing vitamin C-rich food. For example, vegetables should be washed before slicing, and not overcooked to reduce vitamin C loss. Other factors affecting vitamin C content of food is duration and period of storage, duration of transport from farm to market and harvesting season The longer the storage and transport time, the more the loss in vitamin C.[44]

Vitamin C is the first line of antioxidants defense in the plasma, capable of protecting the body against superoxide, hydrogen peroxide, and hydroxyl radicals. Vitamin C does this by capturing peroxide radical from the hydrophilic phase before starting the lipid peroxidation. In this way, vitamin C and vitamin E prevent the peroxidation of the biomembranes. Besides being capable of regenerating vitamin E, vitamin C also acts as an antioxidant.

Increased use of vitamin C coupled with low dietary intake can lead to vitamin C deficiency during an HIV infection.[34,45] Studies suggest that high doses of vitamin C in HIV can mildly improve clinical parameters such as CD4 cell count and viral load. Thus enhance immune function and protect blood cells from ROS.[9,45] Another study in Tanzania shows that a combination of vitamin C and vitamin E could protect red blood cells from ROS.[46]

### Vitamin E

There exist eight chemical forms of vitamin E (α-, β-, γ-, δ-Tocopherols and α-, β-, γ-, δ-Tocotrienols) in nature. Of these, α-Tocopherol is the most important that can meet human needs vitamin E has received considerable attention among antioxidants over the past years. Common sources of Vitamin E in the diet of a sub-Saharan Africans are nuts or seeds commonly roasted and eaten as snacks or used as sauce thickeners e.g., peanut sauce, melon seed sauce (melon seeds are commonly known as *egusi* or *ibara* in other African cultures). Seeds from *Ricinodendron heudelotii* plant known as *njansa*, or *kishongo*, *wama*, or *munguella* depending on region, is used to season or thicken soup or stew and is also a rich source of vitamin E, as well as avocado (commonly eaten with boiled cassava or any other boiled tuber) and *Dacryodes edulis* (commonly called African plum). Also, oils typically used in the African cookery (melon seed oil, soy oil, cotton seed oil) enrich African food with vitamin E). As an antioxidant, it is capable of reducing free radical damage, preventing lipid oxidation and inflammatory processes. Studies indicate low plasma vitamin E levels in HIV-infected persons. Vitamin E deficiency is the result of increased utilization of vitamin E in quenching free radicals and poor intake in the diet, impairment of recycling vitamin E through vitamin C, diarrhea, etc.[40,45] Vitamin E deficiency has been observed in HIV patients and is related to disrupting the cell-mediated response and humoral immunity as well as reducing T-cell immunity and natural killer cells, phagocytic response and consequently rapid disease progression. Vitamin E can exert a positive effect on the immune system through the reduction of oxidative stress, improving CD4 cell counts and reducing viral loads.[46] It is also important to underline that vitamin E can provide health benefits not only to HIV-infected persons but also to AIDS patients.[14] Both previous and recent studies show that some HAART regiment, while improving the health status of HIV/AIDS patients also increases oxidative stress. Thus association of vitamin E supplements in the right doses or intake with the local diet will improve antioxidative status and effectiveness of HAART.[4,34,45]

### *Selenium*

Selenium is an essential micronutrient, which plays an important role, for example, in antioxidant defense, cell-mediated immunity, brain, and thyroid function. The concentration in food, respectively, in the diet varies by geographical area. Soils with low selenium concentration, coupled with low dietary intakes may lead to selenium deficiency. Sources of selenium in the African food are stews cooked with seafood such as shrimps and fish. In most African countries, shrimps are dried and used abundantly in the preparation of green leafy vegetable, especially in poor settings as a replacement for meat or fish. Oily fish (tuna, mackerel), stewed with tomato or peanut cream or mashed pumpkin seeds are good sources of selenium in African and snacks of peanuts, cashew nuts and sesame. Studies show low selenium concentrations in HIV-infected patients, as the disease progresses. Researchers hypothesize that selenium becomes depleted in HIV patients because the body uses selenoproteins to suppress viral replication or the virus uses to selenium to create its own selenoenzymes.[47,48] Randomized controlled trials show that selenium supplementation may provide modest benefits for HIV-infected adults. However, how effective the supplement is depends on the form of selenium used. Supplementing selenium in breast-feeding mothers observed that inorganic sodium selenite was somewhat less readily absorbed than selenomethionine. They insisted on the fact that inorganic selenium supplementation will less likely be incorporated in individuals with high plasma selenium concentrations. This is due to the fact that with high concentrations, selenoproteins such as GPX are maximally expressed, thus preventing integration of selenite supplement. It was also mentioned in the above publication that contrary to selenite, good sources of selenomethionine such as selenium-enriched yeast or high-protein food produced on soils that are not deficient in selenium could produce greater increases in tissue selenium concentrations.[48,49] It has also been observed that in populations where selenium concentrations are low at population levels, it is even more severe in HIV patients. Thus, it is necessary to associate selenium supplementation to ART, to attain adequate selenium concentration. An important strategy to reduce selenium deficiency in sub-Saharan Africa would be to increase the intake of selenium-rich food through local diets. However, this strategy can only be effective in areas where local food is rich in selenium. Studies in Malawi showed that such a strategy will hardly be successful since the most common source of selenium there is maize, which is eaten in large quantities and has less than 7 μg Se/100g (normal range).[49] In the same line to improve selenium intakes, it is necessary to educate general populations especially people with a compromised immune system on the use of selenium-containing crop fertilizer, sources on selenium in local diets. Further, organic selenium supplements could also be provided to women during antenatal consultations or immunization visits.

### *Zinc*

Zinc is part of the more than 200 metalloenzymes intervening in the synthesis and breakdown of carbohydrates, lipids, proteins, DNA, etc., in the human body. Important sources of zinc in the diet include nuts, oysters, meat and meat products, calf liver, and wheat products. In African countries close to the Gulf of Guinea, seafood, rich in zinc such as crabs stewed with tomato or *C. olitorius* (jute mallow leaves) are highly consumed. In African countries, beef, chicken, and pork that are rich sources of zinc and are often used in combination with peanut butter, pumpkin seed or vegetables for soups, stews and sauces. Studies indicate a relationship between zinc and impaired immune function, zinc is involved in function of neutrophils, macrophages, natural killer cells, and lymphocytes. Across the years, low zinc concentrations have been observed in HIV patients. Zinc concentration in HIV patients depends on nutritional status (low zinc intake in the diet) and stage of the disease.[50] A recent study carried out on HIV patients with immunological discordance showed that zinc supplementation significantly increased CD4 cell counts from 176 before supplementation to 250 cells/mm$^3$ after supplementation. However, caution should be taken with zinc supplementation since rare cases of minor rash and/or itches, which disappeared after supplement discontinuation, has been reported in some studies[5,48] (Table 9.2).

### *Polyphenols*

In addition to the exogenous antioxidants there exists a group of nutritious elements known as secondary plant substances. They possess a variety of biochemical and pharmacological functions especially in protecting the body

TABLE 9.2 Exogenous Antioxidants and Functions

| Antioxidants | Required Daily Allowance (RDA)mg/d[a] | Source in Diet | Function |
|---|---|---|---|
| Vitamin A | 0.9–1.1 mg [b]RÄ/d | Carrots, palm oil, meat, milk, egg yolk | • important scavenger of single oxygen atoms |
| Vitamin C | 75–150 | Oranges, lemon, lime, pineapple, spinach, squash, guava, passion fruit | • important water soluble antioxidant<br>• regeneration of vitamin E<br>• inhibits the formation of nitrosamine<br>• stabilizes of superoxide-/hydroxyl radical |
| Vitamin E | 15–30 | Soy, groundnut, cotton oil, enriched margarine, nuts, | • Important antioxidant in fat soluble milieu<br>• Stabilizes of superoxide-/hydroxyl radical |
| Selenium | 30–100 μg/d | Fish, meat (muscle) cereals, egg, milk products | • important part of selenoproteins and enzymes<br>• antioxidant<br>• important in the regeneration of vitamin E |
| Zinc | 15 | Wheat products, liver, chicken, cocoa, milk, egg white | • important part of metalloenzymes<br>• immune defensive and modulatory function<br>• involved in the synthesis and degradation of carbohydrates, lipids, proteins, alcohol, etc. |
| β Carotenoids | 2–4 | Carrots, tomatoes, lettuce, spinach, papaya, grape fruits | • capable of quenching free radicals and ROS<br>• capable of influencing cell differentiation |

*ROS*, reactive oxygen species.
[a]*RDA for healthy adults with oxidative stress.*
[b]*RÄ (retinol equivalence).*
*RDA from Leitzmann C, Müller C, Michel P, Brehme U, Triebel T, Hahn A, Laube H.* Ernährung in Prävention und Therapie. Ein Lehrbuch, 3 Auflage. *Stuttgart: Hippokrates Verlag; 2009. p. 123.*

against inflammatory biomarkers and oxidative disorders. A wide variety of African plants (green tea, cocoa), fruits (e.g., oranges, lemon, grapes) mostly consumed as a snack or used to produce fruit juice, vegetables and legumes, e.g., dry beans, peas, soybean (used in the preparation of soup, stews and sauces) are sources of polyphenols. Soybean, for example, is used to produce soydrinks, yoghurt, and milk and is especially in Africa added to groundnut to make sauces or cake. Red wine, fruit drinks, and spices, e.g., celery, parsley, onion, leeks, and garlic are rich sources of anthocyanins, used in the preparation of a majority of dishes. In cocoa for example the main phenolic compounds found are flavonoids and tannins. Flavonoids are capable of reducing free radicals by quenching, upregulating, or protecting antioxidant defense and chelating radical intermediate compounds. Previous and recent research show that polyphenol supplementation in HIV patients could affect lymphocyte proliferation and apoptosis.[9,15,51] Studies in vivo and in vitro show that flavonoids found in dark chocolate can increase antioxidant capacity of the organism and prevent the oxidation of low-density lipoprotein in HIV patients. Also plant extracts from *Clerodendrum volubile* and *Gasteria bicolor* and *Pittosporum viridiflorum*[52] have been shown—due to its polyphenolic content—to exert antioxidative effects, by breaking the free radical chain through donation of a hydrogen atom and scavenging $H_2O_2$. Another recent study indicate that *Apium graveolens*, (Apiaceae) a food spice commonly called celery could exhibit antioxidant properties by significantly decreasing levels of superoxide ion and reducing oxidative stress in rats. Since PLWHA often have impaired antioxidant system, these polyphenol-rich plants could be used to improve their antioxidant status[53] (Table 9.3).

## INTERACTION BETWEEN ANTIOXIDANTS AND HAART

Besides viral protein gp120 and Tat that negatively influences the concentration of endogenous antioxidants (GSH, GPX, SOD, and CAT), another factor that seems to influence oxidative stress and inflammation in HIV/AIDS patients is HAART. HIV patients are assigned to HAART when CD4 cell counts decrease below 500 cell/mm$^3$ and/or at opportunistic infections onset.[2] The introduction of HAART in Africa has greatly reduced the rate of morbidity and mortality of PLWHA. HAART has also helped to improve immunity function, health outcomes, and quality of life in PLWHA. Increasing evidence exists on the fact that HAART is a cause of metabolic disorders, including fat redistribution in patients. This increase in body fat leads to an increase in inflammatory markers such as IL-6, tumor necrosis factor-alpha, C-reactive protein, fibrinogen, etc., which may interfere with antioxidant concentration and further increases oxidative stress in PLWHA in sub-Saharan Africa[4,54]). The results of the study by Kaio and

TABLE 9.3 Secondary Plant Substances and Functions

| Secondary Plant Substances | Sources in Diet | Function |
|---|---|---|
| Polyphenols | Green beans, egg plant, red wine, onion, cabbage, | Antioxidant, anticancerous, antimicrobial, antithrombotic, immunomodulation, inhibit inflammation, reduce blood glucose level, and regulate blood pressure |
| Phytat | Cereals, nuts, bean varieties | Antioxidant, anticancerous, immunomodulation, reduces cholesterol and blood glucose level, |
| Sulfides | Onion, leek, garlic, cabbage | Antioxidant, anticancerous, antimicrobial, antithrombotic, immunomodulation, inhibit inflammations, regulate digestion, and regulate blood pressure |
| Saponin | Soya bean, green beans, peas, lentils | Anticancerous, antimicrobial, immunomodulation, reduces blood-cholesterol |
| Carotenoids | Oranges, tomatoes, pepper varieties, carrots, egg yolk, spinach | Antioxidant, anticancerous effects, immunomodulation |

coworkers showed a decrease of 4.12 μmol/L ($P = .03$) in mean alpha-tocopherol concentrations in patients receiving HAART combination of (nucleoside reverse transcriptase inhibitors (NRTIs) plus other classes (fusion inhibitors, integrase inhibitors, entry inhibitors, protease inhibitor (PI), etc.) when compared to those receiving NRTIs plus non-reverse transcriptase inhibitors (NNRTIs). These lower concentrations of alpha-tocopherol in the group using NRTIs combined with the other classes of antiretroviral drugs (most of them new drugs) may be related to indications of salvage therapy, intolerance or poor treatment compliance, previous use of inadequate regimens, and primary resistance.[34] Results from previous and recent research show that PI together with early start of alpha-tocopherol supplementation may reduce the therapy-induced risk of arteriosclerosis.[4,54,55] Evidence exists that prolonged use of PI may improve antioxidant status, found higher antioxidants levels in subjects using combination therapy with PI and nucleoside analogues compared to those receiving nucleoside analog monotherapy). In vitro, cells previously treated with PIs and later on exposed to naringin—grapefruit-derived flavonoids, commonly found in African countries—showed that the later could significantly reduce lipid peroxidation, SOD activities and also increase GSH and ATP levels in the cells. The results of this finding suggest that naringin as nutritional supplement could prevent impairment of pancreatic β-cell function in patients treated with HIV-1 PIs.[56] The combination of antiretroviral drug therapy, a healthy diet, and a low-dose multivitamin/mineral supplement may be especially effective in the treatment of HIV-1–infected patients.

## ROLE OF NUTRITIONAL EDUCATION IN HIV/AIDS—CARE

Aside health challenges, one of the problems faced by Africans is analphabetism. According to UNESCO Institute of statistics, 85% of adults worldwide are illiterates, with sub-Saharan Africa and South and West Asia having the lowest rates. Of this, 63% are women. It is also well known that 90% of HIV-infected women live in Africa.[1] In Africa, mostly the women are responsible for the nutritional welfare of the household and caretakers. Meanwhile the importance of nutritional support for PLWHA is well established. Nutritional support is needed to improve nutritional status during symptom-free period and to prevent further health deterioration during the symptomatic phase. It has also been shown that any HIV care intervention without nutritional support is likely to fail and compromise the effects of HAART. Considering the fact that HIV affects access to availability and metabolism of food and also coupled with the food insecurity in most African countries, it is intimating to include nutritional education in the care of PLWHA. This will help patients make good use of the little available food[6,57]). To make sure knowledge and good practices acquired by PLWHA is transformed to good dietary practices, support groups are required for effective follow-up. Besides knowledge about drug-food interactions, relationship between diet and disease, dietary recommendations for PLWHA, use of fruits and vegetables as antioxidant-rich food, and use of legumes as complementary source of protein provided to PLWHA, education on HIV/AIDS itself, intake of HAART and managing side effects, hygiene, coping with stigma, and discrimination and good lifestyle practices is required.[58] Proper education of PLWHA by well trained health personnel or dieticians will also solve the problem of nutritional supplement overdo-sage, which ends up acting as prooxidants instead of antioxidant.[9,18]

The management and care of PLWHA in Africa, especially sub-Saharan Africa does not depend only on the implementation of one approach but a sustainable strategy will make use of a combination of HAART, an adequate nutrition (local diet and/or nutritional supplement) and education on the various articulations of the disease.

# CHALLENGES TO EFFECTIVE INTAKE OF ANTIOXIDANTS IN PEOPLE LIVING WITH HIV/AIDS

## Stigmatization

Since the beginning of the HIV epidemic, PLWHA especially in African countries have been experiencing stigma and discrimination. HIV stigma reflects the negative social identity ascribed to PLWHA by HIV negative people. Stigma prevents those who know their HIV status from sharing their diagnosis and seeking appropriate and timely treatment and care for themselves. Consequences include withdrawal, self-rejection, isolation, and avoidance from support groups intended to help PLWHA take better care of their condition. Also, patients avoid adopting healthy eating habits for fear of being suspected for HIV by relatives or friends. They will prefer to continue taking great quantities of alcohol or cigarettes when they go out with friends. It is well known that a healthy nutrition and lifestyle will help improve nutritional and health status for HIV patients, however, to not to arouse suspicion of their HIV status, because of their change of behavior, prefer to continue in their old eating habits and lifestyle, knowing well that it will worsen their health condition. Also, using avoidance as a coping strategy against stigma, PLWHA isolate themselves, even from family members who could provide the required resources for them to feed well and ameliorate their living conditions.[59]

## Food Insecurity and Poverty

In African countries and especially in sub-Saharan Africa, populations are often faced with drought, political instability, war, poor health, and malnutrition, which are some of the causes of food insecurity. Food insecurity is defined as the situation in which all individuals and communities at all times do not have physical, social, and economic access to sufficient, safe, and nutritious food that meets their energy and nutrient requirements for an active and health life.[6,19,60] Most PLWHA in Africa usually suffer from food and nutrition insecurity. This phenomenon is usually a result of loss of productive labor, income, savings, and poverty, in general, leading to limited food availability and inappropriate nutritional care and support. In case of food security, PLWHA would have adequate nutrition including antioxidant-rich local diet or supplementation with vital nutrient, preventing weight loss, enhancing the efficacy of HAART and reducing oxidative stress.[19,60] Studies ascertain the positive effects of nutritional supplement on the immune system of PLWHA, however, a majority of PLWHA still cannot afford these, due to the high cost. In African countries with unfavorable weather conditions, local food that will provide enough antioxidants (fruits, vegetables, legumes etc.) and other essential nutrients is very costly, making it practically difficult for PLWHA to have an optimum intake of antioxidant with the diet.

## Lack of Knowledge

Insufficient knowledge on the proper use of the available food in some African communities is one of the important causes of inadequate intake of antioxidants, leading to increase oxidative stress to an already compromised immune system (see section Role of Nutritional Education in HIV/AIDS—Care).

## Culture

Other factors challenging effective intake of antioxidants are strong cultural influences concerning food choices and preparation methods, especially in sub-Saharan Africa. Preparation of some African traditional meals require several hours of cooking, altering nutrient content of most vegetables. Such practices would lead to destruction of most if not all water soluble vitamins including antioxidants such as vitamin C. Intake of certain food items in some cultural settings are linked to beliefs and taboos. Excluding such food in the diet may cause nutritional deficiencies. In sub-Saharan Africa, it a cultural issue to have large families. The larger the families, the more resources are required to meet nutritional needs. PLWHA in such families may have difficulties meeting antioxidant intake since the available meal will be shared with the rest of the family.[7]

Another challenge to effective intake of antioxidants in an African context includes the lack of sufficiently trained personnel in the area of nutrition and health greatly limits effective communication of important information on adequate nutrition in general and antioxidant intake for PLWHA in particular.

## SUMMARY POINTS

- Previous and recent studies have identified an increase in free radicals in PLWHA worldwide and especially in sub-Saharan Africa, leading to increased oxidative stress and decreased in antioxidants.
- HAART are known not only to improve health status of PLWHA but also to increase oxidative stress.
- Optimal nutrition with local diet or nutritional supplementation of antioxidants in African people infected with HIV/AIDS will go a long way to improve immune system.
- Nutritional education of African populations in general and PLWHA specifically will help to transfer knowledge on proper feeding habits, cooking methods, and drug–food interaction.
- There are challenges faced by PLWHA in Africa, which affect proper implementation of acquired knowledge.

## References


1. UNAIDS. Global AIDS Update 2016.
2. WHO/UNAIDS/UNICEF. *Global update on HIV treatment*. 2013.
3. Bolhaar MG, Karstaedt AS. A high incidence of lactic acidosis and symptomatic hyperlactatemia in women receiving highly active antiretroviral therapy in Soweto, South Africa. *Clin Infect Dis* 2007;**45**(2):254–60.
4. Ngondi JL, Oben J, Musoro DF, Etame HL, Mbaya D. The effects of different combination therapies on oxidative stress markers in HIV infected patients in Cameroon. *AIDS Res Ther* 2006;**3**:19.
5. Lowe NM, Medina MW, Stammers AL, et al. The relationship between zinc intake and serum/plasma zinc concentration in adults: a systematic review and dose-response metaanalysis by the EURRECA Network. *Br J Nutr* 2012;**108**:1962–71.
6. Bukusuba J, Kikafunda JK, Whitehead RG. Nutritional knowledge, attitudes, and practices of women living with HIV in Eastern Uganda. *J Health Popul Nutr* 2010;**28**(2):182–8.
7. Shils ME, Moshe S, Catharine RA, Benjamin C, Cousins RJ. *Modern nutrition in health and disease*. 10th ed. Lippincott Williams & Wilkins,; 2006. p. 110.
8. Leitzmann C, Müller C, Michel P, Brehme U, Triebel T, Hahn A, Laube H. *Ernährung in Prävention und Therapie. Ein Lehrbuch, 3 Auflage*. Stuttgart: Hippokrates Verlag; 2009. p. 123.
9. Amir B, Ghobadi S. Studies on oxidants and antioxidants with a brief glance at their relevance to the immune system. *Life Sci* 2016;**1**(146):163–73.
10. Kedzierska K, Crowe SM. Cytokines and HIV-1: interactions and clinical implications. *Antivir Chem Chemother* 2001;**12**:133–50.
11. Schreck R, Rieber P, Baeuerle PA. Reactive oxygen intermediates as apparently widely used messengers in the activation of NF Kappa-B transcriptional factor and HIV-1. *EMBO J* 1991;**10**:2247–58.
12. Mellor AL, Chandler P, Lee GK, Johnson T, Keskin DB, Lee J, Munn DH. Indoleamine 2,3-dioxygenase, immunosuppression and pregnancy. *J Reprod Immunol* 2002;**57**:143–50.
13. Malik S, Saha R, Seth P. Involvement of extracellular signal-regulated kinase (ERK1/2)-p53-p21 axis in mediating neural stem/progenitor cell cycle arrest in co-morbid HIV-drug abuse exposure. *J Neuroimmune Pharmacol* 2014;**9**:340–53.
14. Müller F, Svardal AM, Nordoy I, Berge RK, Aukrust P, Frøland SS. Virological and immunological effects of antioxidant treatment in patients with HIV infection. *Eur J Clin Invest* 2000;**30**(10):905–14.
15. Dominguez-Avila JA, Alvarez-Parrilla E, de la Rosa-Carrillo LA, Martinez-Martinez A, Gonzalez-Aguilar GA, Gomez-Garcia C, Robles-Sanchez M. Effect of fruit and vegetable intake on oxidative stress and dyslipidemia markers in human and animal models. In: Rasooli I, editor. *Phytochemicals – bioactivities and impact on health*. Intech; 2011.
16. Siegfried N, Irlam JH, Visser ME. Micronutrient supplementation in pregnant women with HIV infection. *Cochrane Database Syst Rev* 2012;**3**:CD009755.
17. Mehta S, Fawzi W. Effects of vitamins, including vitamin A, on HIV/AIDS patients. *Vitam Horm* 2007;**75**:355–83.
18. Birringer M, Ristow M. Efficacy and risks of supplementation with antioxidants. *Ernahrungs Umschau* 2012;**59**:10–4.
19. Lunney KM, Jenkins AL, Tavengwa NV, Majo F, Chidhanguro D, Iliff P, Strickland GT, Piwoz E, Iannotti L, Humphery JH. HIV positive poor women may stop breastfeeding early to protect their infants from HIV infection although available replacement diets are grossly adequate. *J Nutr* 2008;**138**(2):351–7.
20. Lushchak VI. Glutathione homeostasis and functions: potential targets for medical interventions. *J Amino Acids* 2012:736837.
21. Morris D, Ly J, Chi PT, Daliva J, Nguyen T, Soofer C, Chen YC, Lagman M, Venketaraman V. Glutathione synthesis is compromised in erythrocytes from individuals with HIV. *Front Pharmacol* 2014;**5**:73.
22. Guerra C, Morris D, Sipin A, Kung S, Franklin M. Glutathione and adaptive immune responses against *Mycobacterium tuberculosis* infection in healthy and HIV infected individuals. *PLoS One* 2011;**6**(12):e28378.
23. Ly J, Lagman M, Saing T, Kaur Singh M, Tudela EV, Morris D, Anderson J, Daliva J, Ochoa C, Patel N, Pearce D, Venketaraman V. Liposomal glutathione supplementation restores TH1 cytokine response to *Mycobacterium tuberculosis* infection in HIV-infected individuals. *J Interferon Cytokine Res* 2015;**35**:11.
24. Venketaraman V, Morris D, Donohou C, Sipin A, Kung S. Role of cytokines and chemokines in HIV infection, HIV and AIDS. In: Dumais N, editor. *Updates on biology, immunology, epidemiology and treatment strategies*. 2011.

25. Vera TE, Singh MK, Minette L, Judy L, Nishita P, Cesar O, Vishwanath V. Cytokine levels in plasma samples of individuals with HIV infection. *Austin J Clin Immunol* 2014;**1**(1):1003.
26. Samikkannu T, Deepa R, Rao KVK, Atluri VSR, Pimentel E, El-Hage N, Nair MPN. HIV-1 gp120 and morphine induced oxidative stress: role in cell cycle regulation. *Front Microbiol* 2015;**6**:614.
27. Diplock AT. Antioxidant nutrient s and disease prevention: an overview. *Am J Clin Nutr* 1991;**53**:189–93.
28. Song HY, Ryu J, Ju SM, Park LJ, Lee JA, Choi SY, Park J. Extracellular HIV-1 Tat enhances monocyte adhesion by up-regulation of ICAM-1 and VCAM-1 gene expression via ROS-dependent NF-kappaB activation in astrocytes. *Exp Mol Med* 2007;**39**:27–37.
29. Sonia FC, Marecki JC, Harper KP, Bose SK, Nelson SK, Mccord JM. Tat protein of human immunodeficiency virus type1 represses expression of manganese superoxide dismutase in HeLa cells. *Proc Natl Acad Sci USA* 1993Vol 90:7632–6.
30. Eum WS, Kim DW, Hwang IK, Yoo KY, Kang TC, Jang SH, Choi HS, Choi SH, Kim YH, Kim SY, Kwon HY, Kang JH, Kwon OS, Cho SW, Lee KS, Park J, Won MH, Choi SY. *In vivo* protein transduction: biologically active intact pep-1-superoxide dismutase fusion protein efficiently protects against ischemic insult. *Free Radic Biol Med* 2004;**37**:1656–69.
31. Guo L, Xing Y, Pan R, Jiang M, Gong Z, Lin L. Curcumin protects microglia and primary rat cortical neurons against HIV- 1 gp120-mediated inflammation and apoptosis. *PLoS One* 2013;**8**:e70565.
32. Lizette GD, Hernández RG, Ávila JP. Oxidative stress associated to disease progression and toxicity during antiretroviral therapy in human immunodeficiency virus infection. *J Virol Microbiol* 2013:15. ID 279685.
33. Imlay JA. Pathways of oxidative damage. *Annu Rev Microbiol* 2003;**57**:395–418.
34. Kaio IDJ, Rondó P,HC, Luzia LA, Souza JM, Firmino AV, Sigrid SS. Vitamin E concentrations in adults with HIV/AIDS on highly active antiretroviral therapy. *Nutrients* 2014;**6**:3641–52.
35. Byron E, Gillespie S, Hamazakaza P. Local perceptions of HIV risk and prevention in Southern Zambia. *Int Food Policy Res Inst* 2006;**2**:339–44.
36. Stephensen CB, Alvarez JO, Kohatsu J, Hordmeier R, Kennedy JI, Gammon RB. Vitamin A excreted in urine during acute infection. *Am J Clin Nutr* 1994;**60**:388–92.
37. Ross AC. Vitamin A status : relationship to immunity and antibody response. *Proc Soc Exp Biol. Med* 1992;**200**:303–20.
38. McHenry MS, Edith A, Rachel CV. Vitamin A supplementation for the reduction of the risk of mother-to-child transmission of HIV. *Expert Rev Anti Infect Ther* 2015;**13**(7).
39. McCauley ME, van den Broek N, Dou L, Othman M. Vitamin A supplementation during pregnancy for maternal and newborn outcomes. *Cochrane Database Syst Rev* 2015;**27**(10):CD008666.
40. Baetens JM, Mclelland RS, Richardson BA, et al. Relationship between markers of HIV-1 disease progression and serum β - carotene concentrations in kenyan women. *Int J STD AIDS* 2007;**18**:202–6.
41. Arsenault JE, Said A, Manji KP, Wafaie WF, Eduardo V. Vitamin supplementation increases risk of subclinical mastitis in HIV-infected women. *J Nutr* 2010;**140**:1788–92.
42. Webb AL, Said A, Jeremy F, Clare M, Hannia C, Wafaie WF, Eduardo V. Effect of vitamin supplementation on breast milk concentrations of retinol, carotenoids, and tocopherols in HIV-infected Tanzanian women. *Eur J Clin Nutr* 2009;**63**(3):332–9.
43. Lietz G, Generose M, Henry JC, Andrew MT. Xanthophyll and hydrocarbon carotenoid patterns differ in plasma and breast milk of women supplemented with red palm oil during pregnancy and lactation. *J Nutr* 2006;**136**(7):1821–7.
44. Leppälä JM, Virtamo J, Fogelholm R, Albanes D, Taylor PR, Heinonen OP. Vitamin E and beta carotene supplementation in high risk for stroke: a subgroup analysis of the alpha-tocopherol, beta-carotene cancer prevention study. *Arch Neurol* 2000;**57**:1503.
45. Allard JP, Aghdassi E, Chau J, Tam C, Kovacs CM, Salit IE, Walmsley SL. Effects of vitamin E and C supplementation on oxidative stress and viral load in HIV-infected subjects. *AIDS* 1998;**12**:1653–959.
46. Fawzi WW, Msamanga GI, Kupka R, Spiegelman D, Villamor E, Wei R, Hunter D. Multivitamin supplementation improves hematologic status in HIV-infected women and their children in Tanzania. *Am J Clin Nutr* 2007;**85**:1335–43.
47. Hurwitz BE, Klaus JR, Llabre MM, Gonzalez A, Lawrence PJ, Maher KJ, Greeson JM, Baum MK, Shor-Posner G, Skyler JS. Suppression of human immunodeficiency virus type 1 viral load with selenium supplementation: a randomized controlled trial. *Arch Intern Med* 2007;**167**:148–54.
48. Campa A, Baum MK. Micronutrients and HIV infection. *HIV Ther* 2010;**4**:437–69.
49. Chilimba AD, Young SD, Black CR, Rogerson KB, Ander EL, Watts MJ, Lammel J, Broadley MR. Maize grain and soil surveys reveal suboptimal dietary selenium intake is widespread in Malawi. *Sci Rep* 2011;**1**:72.
50. Kupka R, Fawzi W. Zinc nutrition and HIV infection. *Nutr Rev* 2002;**60**:69–79.
51. Petrilli AA, Souza SJ, Teixeira AM, Pontilho PM, Souza JM, Luzia LA, Rondó PH. Effect of chocolate and Yerba mate phenolic compounds on inflammatory and oxidative biomarkers in HIV/AIDS individuals. *Nutrients* 2016;**8**:132.
52. Otang WM, Grierson SD, Ndip NR. Phytochemical studies and antioxidant activity of two South African medicinal plants traditionally used for the management of opportunistic fungal infections in HIV/AIDS patients. *Complement Altern Med* 2012;**12**:43.
53. Wanida S, Pennapa C, Supita T, Nutjanat C, Tulaporn W. Effects of *Apium graveolens* extract on the oxidative stress in the liver of adjuvant-induced arthritic rats. *Prev Nutr Food Sci* 2016 Jun;**21**(2):79–84.
54. Piche ME, Lemieux S, Weisnagel SJ, Corneau L, Nadeau A, Bergeron J. Relation of high-sensitivity C-reactive protein, interleukin-6, tumor necrosis factor-alpha, and fibrinogen to abdominal adipose tissue, blood pressure, and cholesterol and triglyceride levels in healthy post-menopausal women. *Am J Cardiol* 2005;**96**:92–7.
55. Muntcanu A, Ricciarelli R, Zing JM. HIV protease inhibitors-induced atherosclerotic prevention by alpha-tocopherol. *IUBMB* 2004;**56**:629–31.
56. Sanelisiwe N, Ndwandwe DE, Owira PM. Naringin protects against HIV-1 protease inhibitors-induced pancreatic β-cell dysfunction and apoptosis. *Mol Cell Endocrinol* December 5, 2016;**437**:1–10.
57. Bukusuba J, Kikafunda JK, Whitehead RG. Food security status in households of people living with HIV & AIDS (PLWHA) in a Ugandan urban setting. *Br J Nutr* 2007;**98**:211–7.
58. Nkengfack GN, Ndongo JT, Ngogang J, Binting S, Roll S, Tinnemann P, Englert H. Effects of an HIV-care-program on immunological parameters in HIV-positive patients in Yaoundé, Cameroon: a cluster- randomized trial]. *Int J Public Health* June 2014;**59**(3):509–17.
59. Wolitski RJ, Pals SL, Kidder DP, Courtenay-Quirk C, Holtgrave DR. The effects of stigma on health, disclosure of HIV status, and risk behavior of homeless and unstably housed persons living with HIV. *AIDS Behav* 2009;**13**:1222–32.
60. Ndure KS. *Nutrition and HIV: opportunities and challenges*. USAID; 2004.
61. Jasen van Rensburg. *African leafy vegetables in South Africa*. 2007. Available on website http://www.wrc.org.za.

CHAPTER

# 10

# Gene Delivery of Antioxidant Enzymes in HIV-1-Associated Neurocognitive Disorder

*Jean-Pierre Louboutin*[1,2], *David S. Strayer*[1]

[1]Thomas Jefferson University, Philadelphia, PA, United States; [2]University of the West Indies, Kingston, Jamaica

OUTLINE



## Abstract

Human immunodeficiency virus (HIV) encephalopathy covers a range of HIV-1-related brain dysfunctions. HIV-1 infects resident microglia, periventricular macrophages, but neurons themselves are rarely infected by HIV-1. HIV-1 infection of microglia leads to increased production of cytokines and to release of HIV-1 proteins, among which are the envelope glycoprotein gp120 and HIV-1 *trans*-acting protein transactivator of transcription (Tat). Gp120 and Tat, the most likely neurotoxins, induce both oxidative stress in the brain, leading to neuronal apoptosis/death. We examine here the role of oxidative stress in patients with HIV-1-associated neurocognitive disorder (HAND) and in animal models of HAND. We used simian virus 40 (SV40) vectors for gene delivery of antioxidant enzymes, Cu/Zn superoxide dismutase (SOD1) or glutathione peroxidase (GPx1) into the rat caudate-putamen (CP). Intracerebral injection of SV(SOD1) or SV(GPx1)



HIV/AIDS
http://dx.doi.org/10.1016/B978-0-12-809853-0.00010-9

protects neurons from gp120-and Tat-induced apoptosis. Vector administration into the lateral ventricle or cisterna magna protects from intra-CP gp120-induced neurotoxicity comparably to intra-CP vector administration. These models should provide a better understanding of the pathogenesis of HIV-1 in the brain and offer new therapeutic avenues.

**Keywords:** Antioxidant enzymes; Blood-brain barrier; Dementia; Gene therapy; HIV-1; HIV-1-Asssociated neurocognitive disorder; Neuroinflammation; Oxidative stress.

## CLINICAL PRESENTATION OF HIV-1-ASSOCIATED NEUROCOGNITIVE DISORDER

HIV-1 enters the central nervous system (CNS) soon after it enters the body. HIV-1 is largely impervious to highly active antiretroviral therapeutic drugs (HAART) once in the CNS. Survival rates improved dramatically over the past 20 years due to advances in the treatment of HIV-1, but HIV-associated neurocognitive disorders (HANDs) remain highly prevalent and continue to represent a significant public health problem partly because HAART penetrate the CNS poorly. Human immunodeficiency virus (HIV-1) encephalopathy covers a range of HIV-related CNS dysfunction. In early 1990's, the neurologic complications of HIV-1 infection were classified into two types of manifestations: (1) HIV-associated dementia (HAD) with motor, behavioral/psychosocial, or combined features; and (2) minor cognitive motor disorder (MCMD). HAD was considered as the most common cause of dementia in adults under 40[1] and was estimated to affect as many as 30% of patients with advanced AIDS[1] but has become less common since HAART was introduced.[2] This reduction probably reflects better control of HIV in the periphery since antiretroviral drugs penetrate the CNS poorly. Before the introduction of HAART, most neuro-AIDS patients showed subcortical dementia, with predominant basal ganglia involvement, manifesting as psychomotor slowing, behavioral abnormalities, cognitive difficulties, and Parkinsonism.[3] MCMD was described as a less severe presentation that did not meet criteria for HAD.

More recently, the need to update and further structure the diagnostic criteria for HAND has been recognized in light of the changing epidemiology of HIV infection.[4] There are several reasons for this update. First, the applicability of the old criteria appears limited in the present age of HAART. Prior to the advent of HAART, a diagnosis of HAD was associated strongly with high viral loads, low T-cell counts, and opportunistic infections. With HAART limiting viral severity, patients with HIV typically live longer with milder medical symptoms. Secondly, possible neurocognitive impairment due to comorbid conditions with CNS effects (e.g., substance use disorders) was not precisely described in the previous diagnostic scheme. This limitation is particularly important in the era of HAART as those infected with HIV-1 live longer with several CNS risk factors, including substance abuse disorders (e.g., methamphetamine dependence), medical conditions associated with HAART treatment (e.g., hyperlipidemia) and comorbid infectious diseases (e.g., hepatitis C virus).[5]

Thus, the newly redefined criteria allow for three possible research diagnoses: (1) asymptomatic neurocognitive impairment (ANI); 2) HIV-associated mild neurocognitive disorder (MND); and 3) HAD.[4] ANI is now estimated to represent the majority of cases of HAND (i.e., about 50% of diagnosed cases) and 21%– 30% of the asymptomatic HIV-infected individuals. Detecting patients with mild, neurocognitive impairment may help in the effort to preidentify those at risk for more significant cognitive and functional decline, before cognitive deficits contribute to a decline in everyday functioning with serious medical consequences.[5] Formerly referred to as MCMD, MND requires mild-to-moderate neurocognitive impairment in addition to mild everyday functioning impairment.[4] Approximately 30%–50% of persons with a HAND diagnosis experience some degree of functional impairment, and it is estimated that 20% to 40% of HAND diagnoses are of MND, which comprises 5%–20% of the HIV population overall.[5] The most severe form of HAND, HAD is marked by at least moderate-to-severe cognitive impairment along with marked activities of daily living declines that are not fully attributable to comorbidities.[4,5] Furthermore, HAD represents the most severe form of HAND in terms of its functional impact. After the advent of HAART in the late 1990s, estimates appeared to shift downward with approximately 4%–7% of persons with AIDS, with more recent appraisals suggesting that as few as 1%–2% of HIV-positive patients meet criteria for HAD.[5]

If there are currently less cases of HAD, as survival with chronic HIV-1 infection improves, the number of people harboring the virus in their CNS increases, leading to new HIV-1-related neurological manifestations. The prevalence of HAND therefore continues to rise, and less fulminant forms of HAND have become more common than their more severe predecessors.[4,5] HAND remains a significant independent risk factor for AIDS mortality.[2,4–9] Incident cases of HAND are accelerating fastest among drug users, ethnic minorities, and women.[6–9] The number of HIV-infected individuals over 50 years of age is rapidly growing, including patients taking HAART.[9] It has been suggested that in 10 years, 50% of AIDS patients in the United States will be over the age of 50. Moreover, it is becoming clear that the brain is an important reservoir for the virus, and that neurodegenerative and neuroinflammatory changes may continue despite HAART.[8]

# NEUROPATHOGENESIS OF HIV-ASSOCIATED NEUROCOGNITIVE DISORDER

## Role of Human Immunodeficiency Virus-1 Proteins in Neuronal Damage

The principal manifestations of central nervous system in HIV infection result from neuronal injury and loss and from extensive damage to the dendritic and synaptic structures in the absence of neuronal loss. Neurons themselves are rarely infected by HIV-1, and neuronal damage is felt to be mainly indirect. In fact, the pathogenesis of HAND largely reflects the neurotoxicity of HIV-1 proteins.[10] HIV-1 infects resident microglia, periventricular macrophages, and some astrocytes,[11] leading to increased production of cytokines and to release of HIV-1 proteins, the most likely neurotoxins, among which are the envelope (Env) proteins gp120 and gp41 and the nonstructural proteins Nef, Rev, viral protein r (Vpr), and transactivator of transcription (Tat).[9,12]

### **Trans-*Acting Protein Transactivator of Transcription***

The HIV-1 *trans*-acting protein Tat, an essential protein for viral replication, is a key mediator of neurotoxicity. Brain areas that are particularly susceptible to Tat toxicity include the CA3 region and the dentate gyrus of the hippocampus and the striatum. Tat is internalized by neurons primarily through lipoprotein-related protein receptor and by activation of N-methyl-D-aspartate (NMDA) receptor. It also interacts with several cell membrane receptors, including integrins, vascular endothelial growth factor receptor in endothelial cells and possibly CXCR4.[13]

Tat can directly depolarize neuron membranes, independently of $Na^+$ flux and may potentiate glutamate- and NMDA-triggered calcium fluxes and neurotoxicity. It promotes excitotoxic neuron apoptosis by activating endoplasmic reticulum pathways to release intracellular calcium ($[Ca^{2+}]i$). Consequent dysregulation of calcium homeostasis[14] leads to mitochondrial calcium uptake, caspase activation and, finally, neuronal death. Tat also increases levels of lipid peroxidation by generating the reactive oxygen species (ROS) superoxide ($O_2^-$) and hydrogen peroxide ($H_2O_2$).[14] It activates inducible nitric oxide synthase (iNOS) to produce nitric oxide (NO), which binds superoxide anion to form the highly reactive peroxynitrite (ONOO).

Tat neurotoxicity has been reported in cultured cells, but fewer studies have demonstrated its neurotoxic properties in vivo.[15] Despite the evidence that Tat has been detected in the striatum of patients with HIV encephalitis,[16] it is difficult to know the exact levels of Tat generated. There are important differences between the situation in the striatum of patients with AIDS and models where Tat is directly injected into the caudate-putamen (CP).

### *Envelope Glycoprotein Gp120*

The HIV-1 *env* gene codes for a glycoprotein that is cleaved into two major Env glycoproteins, gp120, and gp41. Soluble gp120 can induce apoptosis in a wide variety of cells including lymphocytes, cardiomyocytes, and neurons.[17] Gp120-induced apoptosis has been demonstrated in cortical cell cultures, in rat hippocampal slices and by intracerebral injections in vivo.[18] Gp120 binds neuron cell membrane coreceptors (CCR3, CCR5, and CXCR4) and elicits apoptosis.[19] Soluble gp120 also increases glial cell release of arachidonate, which impairs neuron and astrocyte reuptake of glutamate, leading to prolonged activation of NMDA receptor with consequent disruption of cellular $Ca^{2+}$ homeostasis. This process involves generation of superoxide and peroxide species, with resultant oxidative stress, and leads to neuron cell death after mitochondrial permeabilization, cytochrome *c* release and activation of caspases and endonucleases.[2]

### *Other Human Immunodeficiency Virus-1 Proteins*

The *trans*-membrane protein gp41 is elevated in patients with HAD. In vitro, gp41 can induce neuronal death but requires the presence of astrocytes, suggesting indirect mechanisms, involving iNOS, NO formation, depletion of glutathione, and disruption of mitochondrial function.[20] Other HIV-1 proteins (Vpr, Nef, Rev) are also involved in HAND neuropathogenesis. HIV-1 Vpr plays a role in effective viral replication in the early stages of the infection. Vpr is present as a soluble protein within the blood serum and the CSF of patients infected with HIV-1. Some studies have shown that Vpr can directly induce neuronal apoptosis[21] and can deregulate calcium secretion in neural cells. The nonstructural protein Nef is required for the proper budding of virions from HIV-infected cells. In vitro, Nef can be lethal for astrocytes and neurons and can increase the expression of matrix metalloproteinases (MMPs).[22] Extracellular HIV-1 phosphoprotein Rev has neurotoxic properties, which have been demonstrated in rodents by intracerebroventricular injection.

# ANIMAL MODELS OF HIV-ASSOCIATED NEUROCOGNITIVE DISORDER

There are no perfect models for HAND. Several animal systems have been used to study the pathogenesis of HAND. Many of them are based on other lentiviruses (i,e., simian immunodeficiency virus (SIV) infection of macaques, feline immunodeficiency virus infection of cats, Visna-Maedi virus infection in sheep).[23–25] However, only small percentages of animals develop neurological manifestations in these models. Transgenic expression of gp120 in mice has been studied,[26] but the gp120 in that model is mainly expressed in astrocytes, whereas in humans HIV-1 chiefly infects microglial cells. Other models based on introduction of HIV-infected macrophages into the brains of severe combined immunodeficiency mice have been proposed. Some models of ongoing exposure to Tat have been developed: glial fibrillary acidic protein (GFAP)-driven, doxycycline-inducible Tat transgenic mice have been useful for mechanistic studies of Tat contribution to HAND. However, the reported data concerning neuronal TUNEL positivity are still debated.

We[27–29] and others[30,31] have used model systems in which recombinant gp120, or Tat, proteins are directly injected into the striatum. The neurotoxicity of such recombinant proteins is highly reproducible and can be used as a tool for testing novel therapeutic interventions. Administration of recombinant proteins is useful in understanding the effects of HIV-1 gene products, and so their individual contribution to the pathogenesis of HAND. However, HIV-1 infection of the brain is a chronic process. This is in part the reason why we developed experimental models of chronic HIV-1 neurotoxicity based on recombinant simian virus 40 (rSV40) vector-modified expression of gp120[32] or Tat[27] in the brain.

# OXIDATIVE STRESS IN HIV-ASSOCIATED NEUROCOGNITIVE DISORDER

## Role of Oxidative Stress in Patients With HIV-Associated Neurocognitive Disorder

Oxygen is vital for all living cells whether neuronal or not, but on the other hand it is potentially dangerous in excess. Oxygen has a role in glucose breakdown in mitochondria through oxidative phosphorylation and generates energy currency of cell, i.e., ATP.[33] Under physiologic normal conditions, ROS, which include superoxide ($O_2^-$) , hydrogen peroxide ($H_2O_2$), and hydroxyl radical (OH-) are generated at low levels and play important roles in signaling and metabolic pathways.[34] Oxidative stress arises due to the disturbances of the balance in prooxidant/antioxidant homeostasis that further causes the generation of ROS, which are potentially toxic for neurons. Reactive oxygen and nitrogen species (ROS/RNS) change cellular responses through diverse mechanisms. Oxidative damage and the associated mitochondrial dysfunction may result in energy depletion, accumulation of cytotoxic mediators, and cell death. The brain is more susceptible to ROS for several reasons. Neurons are particularly susceptible to ROS because of their biochemical composition; brain contains high levels of fatty acids, which are particularly susceptible to peroxidation and oxidative modification. Double bonds of unsaturated fatty acids are hot spots for attack by free radicals that initiate cascade to damage neighboring unsaturated fatty acids.[33,34] Membrane lipids can undergo oxidation, producing cytotoxic lipid peroxidation products such as malondialdehyde (MDA) and 4-hydroxynonenal (4-HNE). Glial cells require more oxygen and glucose consumption to generate continuous ATP pool in vivo for normal functioning of the brain as it is one of the busiest organs, making them more susceptible to oxygen overload and thus to free radicals generation.[33] Finally, brain has less antioxidant activity compared to other tissues and has higher levels of iron in some areas.[33]

Organisms respond to oxidative injury by orchestrating a stress response to prevent further damage. An increase in the intracellular levels of antioxidant agents, and at the same time the removal of already damaged components, are both part of the oxidative stress response. ROS levels are controlled by endogenous antioxidants such as superoxide dismutases (SODs), glutathione peroxidase (GPx1), glutathione and catalase. The tripeptide glutathione (γ-L-glutamyl-L-cysteinylglycine, GSH) is the key low-molecular thiol antioxidant involved in the defense of brain cells against oxidative stress. Although GSH is the primary molecule involved in detoxification of ROS in the body, antioxidant enzymes such as GPx1, are also known to play a role in this process.[35] During detoxification of peroxides, the enzyme GPx1 converts GSH to glutathione disulphide (GSSG).

Interaction of ROS with other tissue components produces a variety of other radicals: following activation of iNOS, NO can bind superoxide anion to form the highly reactive peroxynitrite.[36] The latter may attack lipids, proteins, and DNA to enhance oxidant-related injury. Mitochondria are the primary source of ROS involved in many brain tissue injuries (i.e., hypoxia, excitotoxicity). Once generated, mitochondrial ROS influence the release of cytochrome *c* and other apoptotic proteins from the mitochondria into the neuronal cytosol, which leads to apoptosis.[35] A link between oxidative stress and activation of some caspases seems highly probable.

Abnormalities in oxidative metabolism have been reported in many nervous system diseases. These include neurodegenerative diseases (Parkinson's disease, Alzheimer's disease (AD), Huntington's disease, amyotrophic lateral sclerosis and cerebellar degeneration) ,[37] vascular diseases (ischemia–reperfusion)[34] or toxic reactions (chronic alcoholism), as well as aging.[38] A decrease in GSH levels has been connected to physiological processes such as aging and neurological disorders such as AD, epilepsy, and Parkinson's disease.

The highly abundant mitochondria in brain cells are a major site of generation and action of ROS/RNS. Lipid peroxidation is a consistent feature of neurodegenerative diseases and biologically active reactive lipid species, such as HNE, accumulates in brains individuals with Parkinson's disease, AD, and HAND. Mechanisms of protein oxidation are NO-dependent, through generation of ONOO– or S-nitrosylation. These changes have been reported in a broad range of pathologies, including Parkinson's disease, which is associated with both nitrated α-synuclein and S-nitrosated parkin. Likewise, ONOO– dependent modifications of proteins are widespread in brains of individuals with AD. The cross talk between autophagy, oxidative stress, and mitochondrial dysfunction is not well understood in neurodegenerative diseases.

Oxidative stress plays a role in the development of HAND as well.[39] Oxidative stress in HIV-1 dementia has been documented in brain tissue, with increased levels of lipid peroxidation product (i.e., HNE) and presence of oxidized proteins. Serum levels of GSH and GPx1 are decreased in HIV-1 patients while MDA levels are increased.[35] In the late stage of HAND, a diffuse intracellular oxidation was reported in the form of decreased availability of GSH, and augmented lipid oxidation, which triggers a cascade of downstream signaling events.

Membrane-associated oxidative stress correlates with HAND.[9] HNE-positive neurons have been demonstrated in the brains of patients with HAND.[40,41] In the case of HIV-1 infection, Tat and gp120 can elicit such oxidative stress,[9,42] which can induce apoptosis in cultured neurons. Tat and gp120 induce ceramide production in cultured neurons by triggering sphingomyelinase activity via a mechanism that involves induction of oxidative stress by CXCR4 activation.[9] Oxidative stress can play a role in HAND in other ways. Circulating toxins in the CSF, derived from HIV-1-infected cells, may damage mitochondria, leading to release of cytochrome *c* and then to a cascade of events leading to apoptosis.[39–41] HIV-1 gp120 and Tat can cause free radical production, possibly as part of the signal-transduction pathways they activate.[9,42]

It is still unclear whether oxidative stress is the primary initiating event associated with neurodegeneration. Oxidative stress might be involved in at least the propagation of cellular injury that leads to neuron death. Oxidative modifications of macromolecular cell components (lipids, proteins, and nucleic acids) may be an early step in the mechanism of Tat and gp120 neurotoxicity.[15]

## Oxidative Stress in Animal Models of HIV-Associated Neurocognitive Disorder

### *Gp120-Induced Oxidative Stress*

HIV-1 gp120 can cause lipid peroxidation and production of hydroxynonenal esters,[41] which can mediate oxidative stress-induced apoptosis of cultured neurons and can damage neurons and cause cognitive dysfunction in vivo.[43]

We reported that direct injection of recombinant gp120 into the striatum can induce lipid peroxidation attested by the measurement of MDA (Fig. 10.1A) and the production of HNE (Fig. 10.2A–C).[44] After injection of gp120 into the CP, levels of MDA were increased and HNE was expressed in apoptotic cells (Fig. 10.2A). HNE was localized, in endothelial cells (Fig. 10.2B), in neurons (Fig. 10.2C), and in astrocytes (not shown). Intra-CP injection of gp120-induced apoptosis (Fig. 10.3A and B). The number of apoptotic cells was related to the concentration of gp120 injected (Fig. 10.3A) and peaked 1 day after gp120 administration (Fig. 10.3B).

## Oxidative Stress Associated With Transactivator of Transcription

Tat-induced lipid and protein oxidation is well documented.[15] We demonstrated that Tat activates multiple signaling pathways. In one of these, Tat-induced superoxide acts as an intermediate, while the other utilizes peroxide as a signal transducer.[42]

We injected Tat into the rat striatum. We observed elevated MDA levels persisting 1 week after Tat administration (Fig. 10.1B).[27] Furthermore, the number of apoptotic cells peaked 2 days after intra-CP Tat injection (Fig. 10.3C). Thus, Tat can also mediate neurotoxicity through lipid peroxidation. The sustained MDA levels might be due to the long-term neuroinflammation, because it is known that increased production of inflammatory products induced by Tat may cause an excess formation of ROS.[15] Tat-mediated neurotoxicity might also be caused by increased oxidative modifications of proteins.[15] For example, increased protein carbonyl formation, a marker of protein oxidative

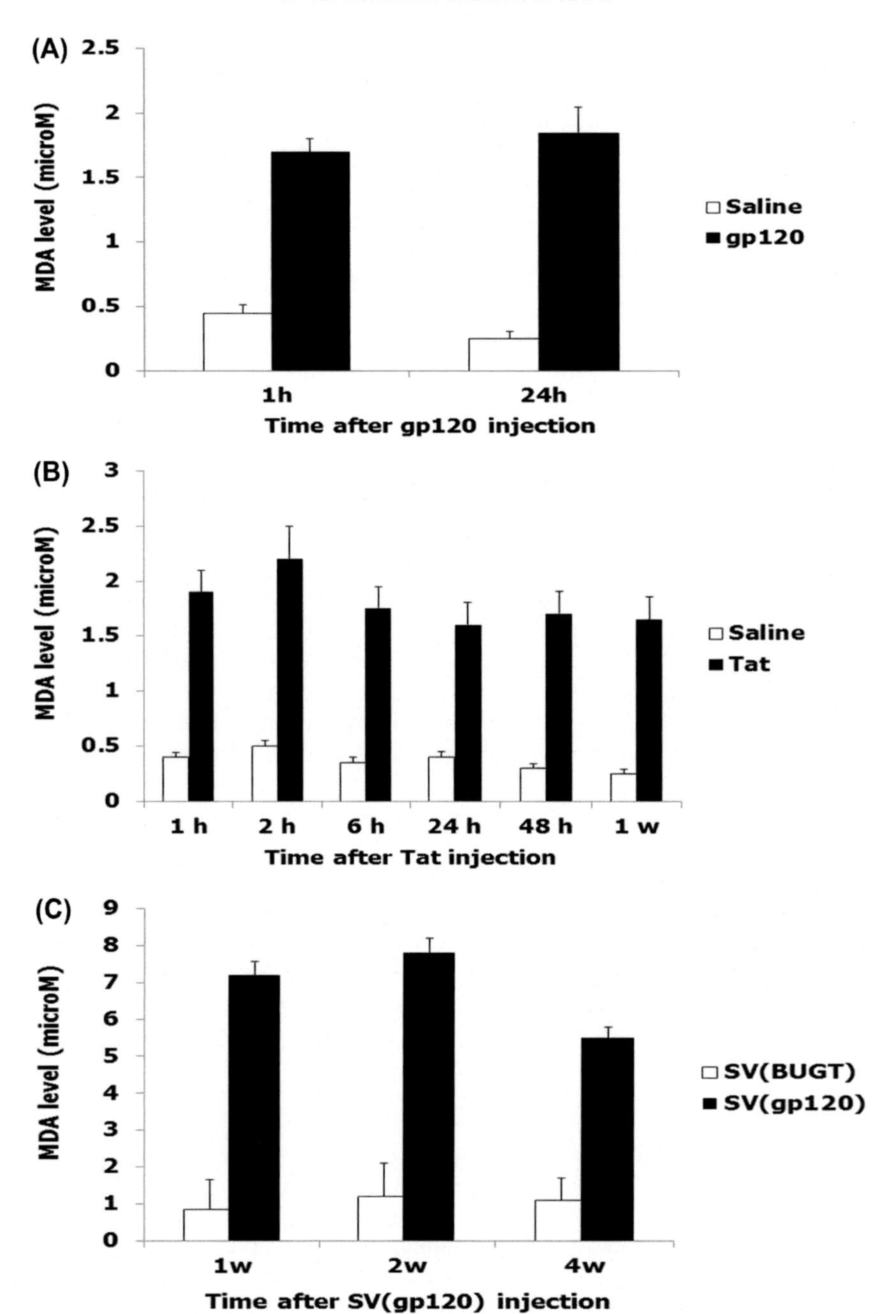


FIGURE 10.1 Increased levels of malondialdehyde (MDA) in experimental models of HIV-1-associated neurocognitive disorder (HAND). (A) Increase in levels of MDA after intra-caudate-putamen (CP) injection of 500ng/microl gp120. (B) Evolution of MDA levels after injection of transactivator of transcription (Tat) in the CP. Note that levels of MDA are still elevated 1 week after the administration of Tat. (C) MDA levels are increased 4 weeks after inoculation of SV(gp120) in a chronic model of HAND. *(C) Modified from Louboutin JP, Agrawal L, Reyes BAS, Van Bockstaele EJ, Strayer DS. A rat model of human immunodeficiency virus 1 encephalopathy using envelope glycoprotein gp120 expression delivered by SV40 vectors.* J Neuropathol Exp Neurol *2009;**68**:456–73.*

damage, occurs early after Tat injection and coincides with the earliest changes in the amount of degenerating striatal neurons.[15]

Tat can decrease levels of GSH available to relieve oxidant stress.[45] Tat can also trigger the expression of iNOS, leading to the overproduction of NO, which can react with superoxide anion to form peroxynitrite, a neurotoxic compound. NO can increase glutamate release from astrocytes, enhancing NMDA excitotoxicity.[35] Tat may induce superoxide and nitrite release in a microglial cell line.[46] An exposure of macrophages and astrocytes to Tat for few minutes in vitro is sufficient for sustained release of cytokines for several hours. Thus, Tat might promote oxidative stress and its consequences (i.e., neuron death) through activation of proinflammatory responses.[15,46]

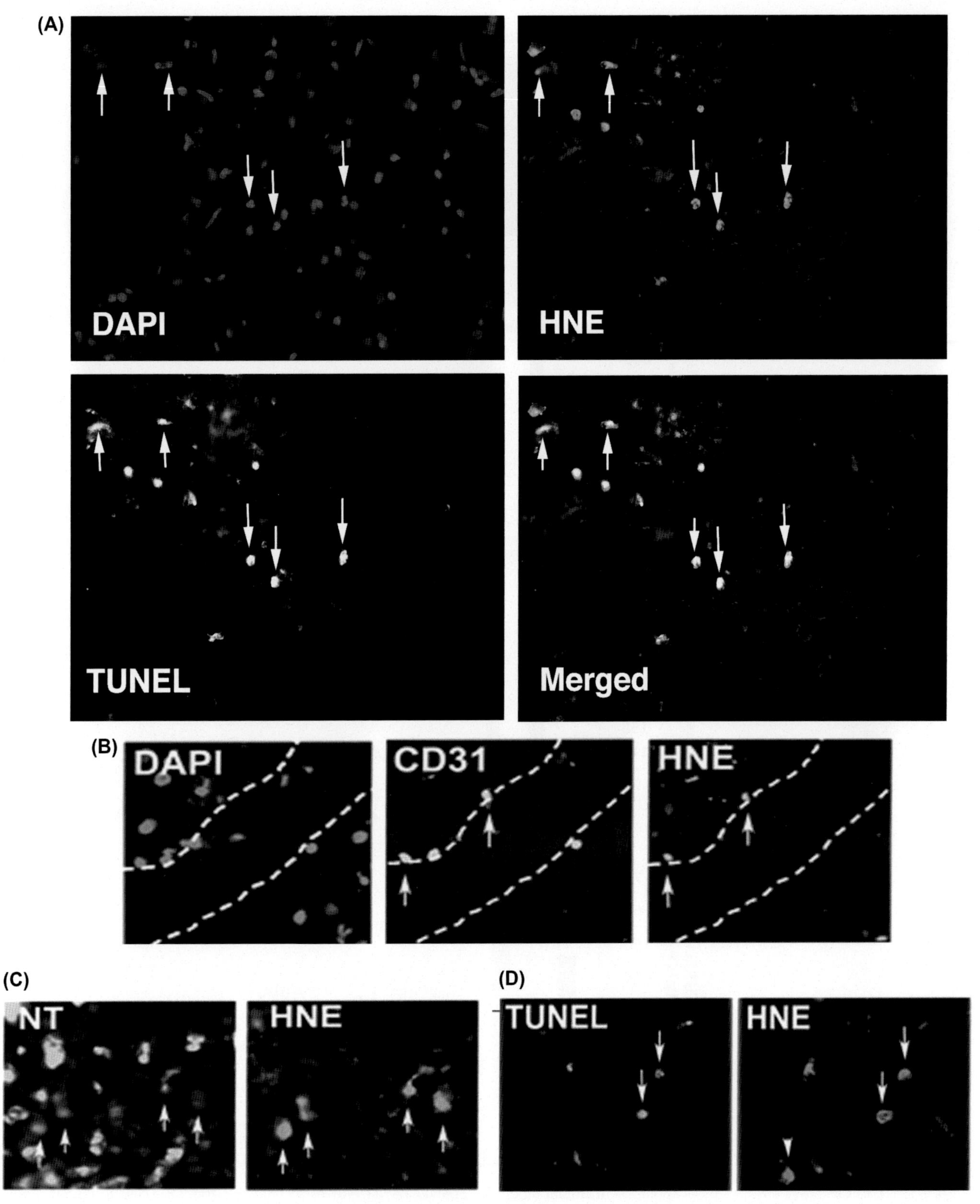


FIGURE 10.2 Expression of 4-hydroxynonenal (HNE) in the caudate-putamen (CP) in experimental models of HAND. (A) HNE, a product of lipid peroxidation was immunocolocalized in apoptotic (TUNEL-positive) cells (*arrows*) 24 h after injection of gp120 into the rat CP. (B) HNE was localized in the vessel walls, colocalizing with endothelial (CD31-positive) cells (*arrows*), 1 hour after injection of 500 ng/microl gp120 into the CP. (C) One hour after injection of 500 ng/microl gp120) into the CP, some neurons (NT-positive cells) were positive for HNE (*arrows*). (D) Some HNE-positive cells in SV(gp120)-injected rats were apoptotic (TUNEL-positive) (*arrows*); some were not (*arrowhead*). *(C) Modified from Louboutin JP, Agrawal L, Reyes BAS, Van Bockstaele EJ, Strayer DS. HIV-1 gp120-induced injury to the blood-brain barrier: role of metalloproteinases 2 and 9 and relationship to oxidative stress.* J Neuropathol Exp Neurol *2010;**69**:801–16. (D) Modified from Louboutin JP, Agrawal L, Reyes BAS, Van Bockstaele EJ, Strayer DS. A rat model of human immunodeficiency virus 1 encephalopathy using envelope glycoprotein gp120 expression delivered by SV40 vectors.* J Neuropathol Exp Neurol *2009;**68**:456–73.*

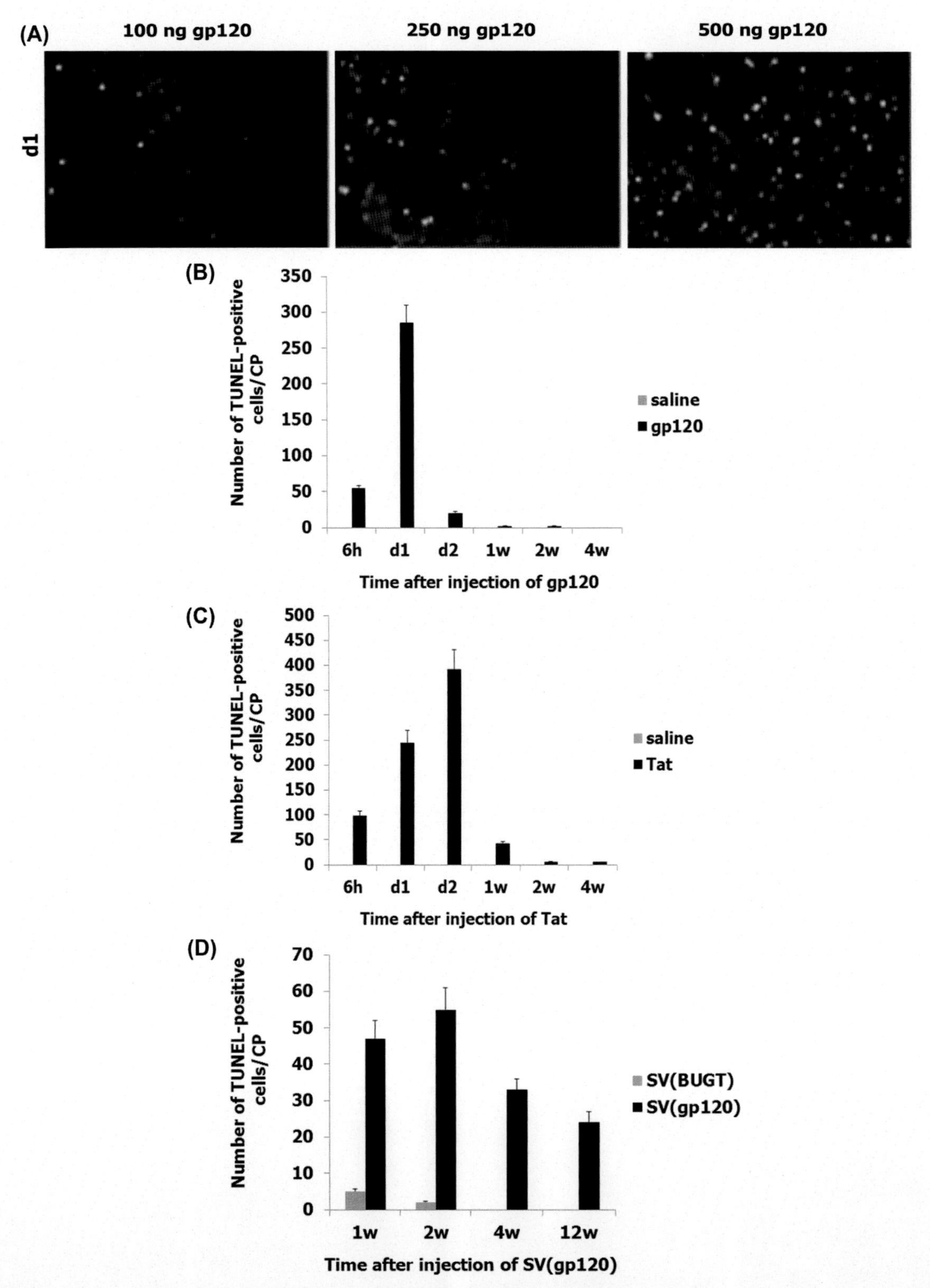


FIGURE 10.3 Oxidative stress induces apoptosis in animal models of HAND. (A) TUNEL assay showing apoptotic cells in the rat caudate-putamen (CP) after injection of different concentrations of gp120. (B) The number of apoptotic (TUNEL-positive) cells peaks 24 h after injection of 500 ng/microl gp120) into the rat CP. (C) The maximum number of TUNEL-positive cells was seen 48 h after administration of 10 ng/microl of Tat. (D) Apoptotic cells were still detected 4 weeks after inoculation of SV(gp120). *(C) Modified from Agrawal L, Louboutin JP, Reyes BAS, Van Bockstaele EJ, Strayer DS. HIV-1 Tat neurotoxicity: a model of acute and chronic exposure, and neuroprotection by gene delivery of antioxidant enzymes.* Neurobiol Dis *2012;**45**:657–70. (D) Modified from Louboutin JP, Agrawal L, Reyes BAS, Van Bockstaele EJ, Strayer DS. A rat model of human immunodeficiency virus 1 encephalopathy using envelope glycoprotein gp120 expression delivered by SV40 vectors.* J Neuropathol Exp Neurol 2009;***68***:*456–73.*

It might be suggested that a direct interaction of Tat with neurons is necessary for inducing early neuron death. However, it is difficult to directly answer the question whether Tat directly induces oxidative stress in neurons or promotes it through activation of proinflammatory responses. It remains plausible that direct interactions of Tat with neurons play the role of a triggering mechanism in the process of the development of oxidative stress and neurodegeneration.[15]

### Oxidative Stress in SV(Gp120) Model

Experimental systems for studying the effects of HIV proteins on the brain have been limited to the acute effects of recombinant proteins in vitro or in vivo, or in chronic situation like SIV-infected monkeys. We described an experimental rodent model of ongoing gp120-induced neurotoxicity in which HIV-1 Env gp120 is expressed in the brain using an SV40-derived gene delivery vector, SV(gp120).[32] We previously demonstrated that SV40-derived vectors deliver long-term transgene expression to brain neurons and microglia, when administered by different routes. rSV40s can transduce a wide range of cell types and deliver genes to cells in Go efficiently, including neurons, to achieve long-term transgene expression in vitro and in vivo. Moreover, they do not elicit immune response.[47] These vectors transduce >95% of cultured human NT2-derived neurons, primary human neurons and microglia[48] without detectable toxicity. When it is inoculated stereotaxically into the rat CP, SV(gp120) caused a lesion in which neuron and other cell apoptosis continue for at least 12weeks. HIV gp120 was expressed throughout this time, and some apoptotic cells are gp120 positive. MDA (Fig. 10.1C) and HNE (Fig. 10.2D) assays showed lipid peroxidation. Similarly, protein oxidation was demonstrated by immunostaining for dinitrophenol in brain cryosections. Thus, in vivo inoculation of SV(gp120) into the rat CP causes ongoing oxidative stress and apoptosis in neurons and may therefore represent a useful animal model for studying the pathogenesis and treatment of HIV-1 Env-related brain damage.

### Oxidative Stress Associated With Viral Protein r

Vpr induces augmented production of ROS related to an increase in the level of oxidized glutathione (GSSG) and a reduction in the overall GSH/GSSG ratio. This event was almost entirely suppressed by treatment with an anti-Vpr antibody or cotreatment with the antioxidant molecule N-acetyl-cysteine (NAC).

## ANTIOXIDANT THERAPEUTIC APPROACHES IN HIV-ASSOCIATED NEUROCOGNITIVE DISORDER

As HIV-1 infection of the brain lasts the lifetime of affected individuals, and as eradication of CNS HIV-1 is currently not possible, control of the damage caused by the virus may represent a useful approach to treatment. This could entail limiting oxidative stress-related neurotoxicity.

### Experimental Data

Antioxidant therapeutic options targeting oxidative stress can be artificially divided as targeting upstream and downstream pathways.

#### *Upstream Antioxidant Therapy*

Upstream preventive treatment is based on prevention of free radical generation, regulation of neuronal protein interaction with redox metals (i.e., Fe) and maintaining normal cellular metabolism.

Some components have been studied in models of HAND or can be useful in this context. Vitamin E can block the neurotoxicity induced by CSF of patients with HIV dementia.[49]

Flavinoids are a group of compounds made by plants that have antioxidant and neuroprotective properties. This class of molecules has weak estrogen-receptor-binding properties and thus, do not have the side effects of estradiol. Diosgenin, a plant-derived estrogen present in yam and fenugreek can prevent neurotoxicity by HIV-1 proteins.[49] Resveratrol, found in grape skins, red wine, and peanuts, and genistein, quercetin, found in soybeans, and polyphenols, have not been tested yet. Curcumin can induce stress response-protective genes, such as heme oxygenase 1 (HO-1), and can offer neuroprotection.[45] Selenium is a key molecule in GPx1 metabolism. Some HIV-infected patients have low levels of selenium. Because selenium supplementation increases GPx1 activity, it might be beneficial in these patients. NAC is a nutritional supplement precursor in the formation of the antioxidant glutathione in the body and its sulfhydryl group confers antioxidant effects and is able to reduce free radicals. NAC injected i.p. into rodents increases glutathione levels in the brain and protects the CNS against the damaging effects of hydroxyl radicals and lipid peroxidation product acrolein.[50] N-acetylcysteine amide (NACA), a modified form of NAC, where the carboxyl group has been replaced by an amide group, is more effective on neurotoxicity because of its ability to permeate the blood-brain barrier (BBB). Treatment of animals injected intravenously with gp120, Tat and methamphetamine METH by NACA significantly rescued the animals from oxidative stress. Further, NACA-treated animals had significantly less BBB permeability as compared to the group treated with gp120 + Tat + METH alone, indicating that NACA can protect the BBB from oxidative stress-induced damage in gp120, Tat, and METH exposed animals.[50]

### *Downstream Antioxidant Therapy*

The therapeutic coverage of postoxidative stress events can be done by downstream antioxidant therapy. Nonsteroidal antiinflammatory drugs limit the infiltration of macrophages and can reduce the inflammatory cascade induced by oxidative stress. CPI-1189, a nitrone-related compound, is supposed to regulate the proinflammatory cytokine cascade of genes in primary glial cells.[33] Minocycline is a tetracycline-derived compound that demonstrated neuroprotective profile in several models of neurodegeneration. It has significant antiinflammatory actions and can easily cross the BBB. In vitro data show that minocycline protected mixed neuronal cultures in an oxidative stress assay and has effective antioxidant properties with radical-scavenging potency similar to that of vitamin E. Furthermore, minocycline treatment suppressed viral load in the brain, decreased the expression of CNS inflammatory markers and reduced the severity of encephalitis in an SIV model of HIV dementia.[51] A chemical moiety that resembles vitamin E in its chemical structure is the female sex hormone estrogen (estradiol) that contains a phenolic free radical-scavenging site and acts as an antioxidant. Estrogen replacement may result in improvement of cognitive function in several neurodegenerative disorders and conversely estrogen deficiency has been considered as a risk factor in some of them. Estradiol can protect against the neurotoxic effects of HIV-1 proteins in human neuronal cultures, probably by protecting the neuronal mitochondria in a receptor-independent manner.[49] However, estradiol has well-known side effects in women (potential risk of developing breast or uterine cancer) and cannot be used in men or children because of feminizing effects. It has been shown that several novel antioxidants (ebselen, diosgenin) can protect in vitro against neurotoxicity induced by CSF from patients with HV dementia.[49]

It is likely that neuroprotective therapies should benefit from multiple and combination approaches targeting different aspects and pathways of the oxidative-stress insult. For example, coupling a potent antioxidant with a compound that modifies downstream signaling pathways (i.e., minocycline) could provide a synergistic neuroprotective effects, at lower doses (and thus with less toxicity) that each molecule could achieve alone. The combination of HAART with an antioxidant compound and a molecule involved in downstream antioxidant therapy could be a promising avenue in the treatment of HAND.[35] However, it should be reminded that one of the challenges in designing antioxidants to protect the CNS against ROS is the crossing of the BBB.

## Clinical Trials

A few antioxidants have been tried in small prospective controlled studies in HAND. However, the findings have all been relatively disappointing so far. Selegiline (L-deprenyl), which might decrease the production of ROS and serve as an antiapoptotic factor, was used in two double-blind controlled studies in the pre-HAART era. The first trial involving patients with MCMD showed improvement in verbal learning and trends for improvement in recall.[52] The second study was a smaller study in patients with MCMD and HIV dementia and showed significant improvement in delayed recall. However, other tests were not improved. A slight improvement was noted in patients treated with OPC-14,117, a lipophilic compound structurally similar to vitamin E that acts as an antioxidant by scavenging superoxide radicals. CPI-1189, a lipophilic antioxidant that scavenges superoxide anion radicals and block the neurotoxicity of gp120 and TNF-alpha, showed no effect on neurocognition in patients with MCMD and HIV dementia.[53]

## Gene Delivery of Antioxidant Enzymes in HIV-Associated Neurocognitive Disorder

### *Introduction*

To deliver potent antioxidant compounds to the brain, we used gene transfer of antioxidant enzymes. Gene transfer of antioxidant enzymes has been studied in numerous models of neurological disorders by using diverse viral vectors.[54] We used rSV40 vectors to deliver SV(SOD1) or SV(GPx1) carrying the antioxidant enzymes Cu/Zn SOD1 or GPx1, respectively, into the rat CP. We demonstrated the safety of SV(SOD1) and SV(GPx1) delivered intra-CP in rats and in Rhesus macaques monkeys. Resulting transgene expression is very durable (Fig. 10.4).[55] We also showed that SV40-based gene delivery of antioxidant enzymes can also be achieved through intravenous injection.[56]

Mitochondria are a major site of production of superoxide in normal cells and probably contribute to increased oxidative stress in numerous diseases. Glutathione is localized in both the cytosol and the mitochondria. Mice overexpressing the cytosolic enzyme $Cu^{2+}Zn^{2+}$-superoxide dismutase develop smaller infarcts than wild-type ones, with a decrease in multiple events associated with mitochondrially mediated apoptosis, including the release of cytochrome *c*.[57] It is thus possible that cytosolic overexpression of antioxidant enzymes delivered by SV40-derived vectors can mitigate the apoptotic events linked to mitochondria.

### *Effects of Gene Delivery of Antioxidant Enzymes on Oxidative Stress, Apoptosis, and Neuronal Loss in Animal Models of HIV-Associated Neurocognitive Disorder*

We showed that prior administration of recombinant SV40 vectors carrying antioxidant enzymes SOD1 or GPx1 protected either from Tat-induced oxidative injury (i.e., lipid peroxidation) caused by intra-CP injection of Tat (Fig. 10.5A), or against SV(gp120)-induced oxidative injury (Fig. 10.5B).

This reduction in oxidative stress due to gene transfer of antioxidant enzymes was associated with a protection against both gp120-and Tat-elicited apoptosis and neuronal loss. Both striatal (either in acute or chronic models of HAND) and dopaminergic neurons were protected against gp120-induced insult.[28,29,32,58,59] Vector administration into the lateral ventricle[29] or cisterna magna,[60] particularly if preceded by intraperitoneal mannitol, protects from intra-CP gp120-induced neurotoxicity comparably to intra-CP vector administration (Fig. 10.6).

Caspases are implicated in neuronal death in neurodegenerative and other CNS diseases, such as HAND. The involvement of caspases in HIV-1 neurotoxicity has been documented in vitro and in vivo. Higher levels of caspase-3 and caspase-6 have been shown in the brains of patients with HAD.[12] Both HIV-1 gp120 and Tat significantly increase caspase-3 activation in striatal neurons in vitro. However, gp120 acts in large part through the activation of caspase(s), while Tat-induced neurotoxicity is also accompanied by activating an alternative pathway involving endonuclease G. Tat can induce both caspases 3/7 and 9 in hippocampal cell cultures. Increased expression of caspase-3 has been shown in neurons following exposure to Tat[61] and to gp120.[62,63] Caspase-3-positive cells were also observed in our model of protracted exposure to gp120, SV(gp120).[32]

We studied the effect of gp120 on different caspases (3, 6, 8, 9) expression. Caspases production increased in the rat CP 6h after gp120 injection into the same structure, peaked by 24h. Caspases colocalized mainly with neurons. There was a relationship with the concentration of gp120 injected. Both initiator (caspases 8 and 9) and effector/executioner (caspases 3 and 6) were increased after gp120 injection. We showed that about 70% of caspase-8- and 9-positive cells were TUNEL-positive while about 60% of caspase-3- and 6-positive cells were TUNEL-positive 1 day after intra-CP injection of gp120.[63,64]

A link between oxidative stress and activation of some caspases seems highly probable. Prior gene delivery of the antioxidant enzymes SOD1 or GPx1 into the CP before injecting gp120 results in reduced levels of gp120-induced caspases, recapitulating the effect of antioxidant enzymes on gp120-induced apoptosis observed by TUNEL. Thus, HIV-1 gp120 increased caspases expression in the CP. Prior antioxidant enzyme treatment mitigated production of these caspases, probably by reducing ROS levels.

### *Gp120-Mediated Abnormalities of the Blood-Brain Barrier Are Mitigated by Gene Transfer of Antioxidant Enzymes*

ROS are important in the pathogenesis of HIV-induced CNS injury and can be induced in brain endothelial cells by HIV-1 gp120 and Tat.[65] Although damage to the BBB has been documented in patients with HIV-related encephalopathy,[66] the exact mechanism by which this injury occurs is still debated.[67] We used animal models of HAND to characterize abnormalities of the BBB in this context. Exposure to gp120, whether acute (by direct intra-CP injection) or chronic [using SV(gp120)], an experimental model of ongoing production of gp120 disrupted the BBB, and led to leakage of vascular contents into the area of gp120 exposure. Gp120 was directly toxic to brain endothelial cells and gp120-mediated BBB abnormalities were related to lesions of brain microvessels.[68] Abnormalities of the BBB may reflect the activity of proteolytic enzymes, particularly MMPs. MMPs are a family of neutral proteases that are grouped according to their protein structures. MMP-2 and MMP-9 target laminin, a major BBB component, and attack the tight junctions between endothelial cells and BBB basal laminae. MMP-2 and MMP-9 were upregulated following intra-CP gp120 injection. Gp120 greatly diminished total CP content of laminin and tight junction proteins. ROS have been reported to activate MMPs. One product of gp120-triggered lipid peroxidation, HNE, was immunolocalized to vascular endothelial cells. Moreover, gene transfer of antioxidant enzymes using recombinant SV(SOD1) and SV(GPx1) protected against gp120-induced BBB abnormalities. BBB injury has also been linked to NMDA, which

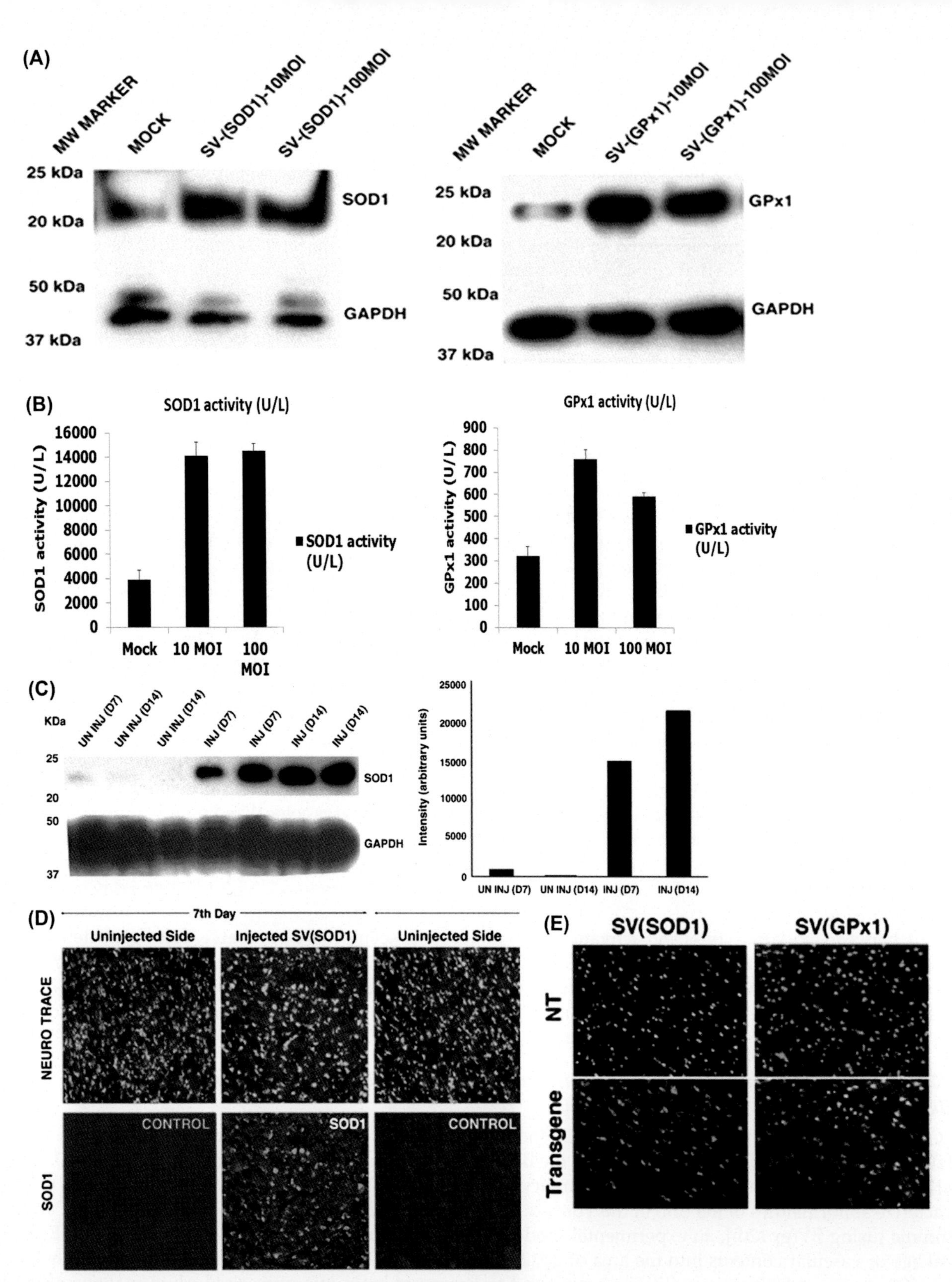


FIGURE 10.4 Transgene expression of antioxidant enzymes. (A) Western blot analysis of superoxide dismutase (SOD1) (left) glutathione peroxidase (GPx1) (right) of differentiated neurons transduced by SV(SOD1) and SV(GPx1) respectively. GAPDH was used as internal loading control. Predominant bands at 22kDa and 24kDa were seen for SOD1 and GPx1, respectively. (B) Kinetic analysis for SOD1 (left) and GPx1 (right). NT-2N cells were transduced with SV(SOD1) and SV(GPx1) on day 0, 3, and 5 with MOI 10, 3, 3, respectively, or MOI 100, 30, 30. Cells were analyzed for SOD1 and GPx1 activity 10days later using kinetic assays. (C) Western blot analysis of the expression of SOD1 after injection of SV(SOD1) into the rat caudate-putamen (CP). Predominant band was seen at 22kDa 7 and 14days postvector inoculation. GAPDH was used as internal loading control (left). Densitometric analysis of the intensity of SOD1 bands, corrected for GAPDH band intensity, shows that expression was slightly increased at day 14 compared to day 7 (right). (D) One month after injection of the vector SV(SOD1) and SV(GPx1) into the caudate nucleus of the rhesus monkey, numerous neurons (NT-positive cells) were expressing the transgene. Morphometric analysis showed that 26.5±2.9% and 26.1±3.1% of NT-positive cells were positive for SOD1 and GPx1 transgenes, respectively. (E) Colocalization between neurotrace and transgene (SOD1 and GPx1) 1month after injection of the vector SV(SOD1) and SV(GPx1) in the CP of the monkey. *(B) Modified from Agrawal L, Louboutin JP, Reyes BAS, Van Bockstaele EJ, Strayer DS. Antioxidant enzyme gene delivery to protect from HIV-1 gp120-induced neuronal apoptosis.* Gene Ther *2006;***13***:1645–56. (E) Modified from Louboutin JP, Chekmasova AA, Marusich E, Chowdhury JR, Strayer DS. Efficient CNS gene delivery by intravenous injection.* Nat Methods *2010;***7***:905–07.*

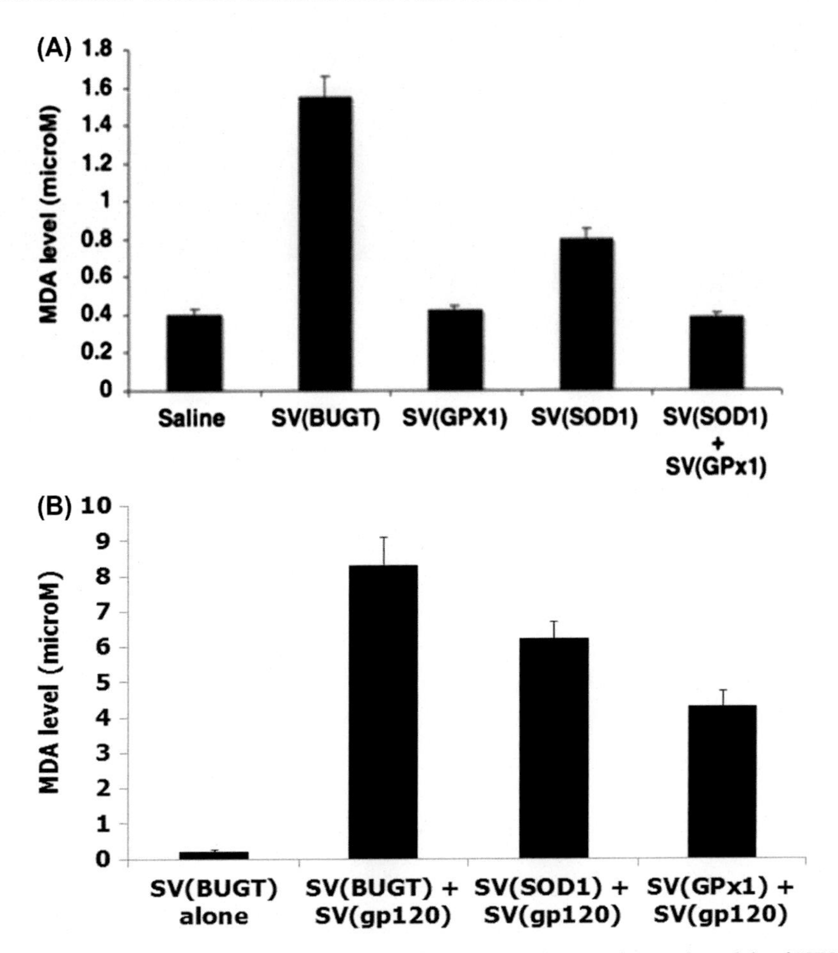


FIGURE 10.5 Gene transfer of antioxidant enzymes reduces oxidative stress in brains of animal models of HIV-associated neurocognitive disorder (HAND). (A) SV(superoxide dismutase (SOD1)), SV(glutathione peroxidase (GPx1)) and a mixture SV(SOD1)/SV(GPx1) were administered into the caudate-putamen (CP) 1 month before 10 ng Tat injection in the same structure. Tat-induced lipid peroxidation, assessed by assay for MDA, was reduced by prior SV40 delivery of SOD1 and antioxidant enzymes GPx1. (B) Prior inoculation of SV(SOD1) or SV(GPx1) protects from SV(gp120)-induced lipid peroxidation. Levels of MDA were assayed in the CP that has been injected with SV(GPX1), or SV(SOD1), or control, SV(BUGT), prior to inoculation of SV(gp120). *(A) Modified from Agrawal L, Louboutin JP, Reyes BAS, Van Bockstaele EJ, Strayer DS. HIV-1 Tat neurotoxicity: a model of acute and chronic exposure, and neuroprotection by gene delivery of antioxidant enzymes.* Neurobiol Dis *2012;***45***:657–70; (B) Modified from Louboutin JP, Agrawal L, Reyes BAS, Van Bockstaele EJ, Strayer DS. A rat model of human immunodeficiency virus 1 encephalopathy using envelope glycoprotein gp120 expression delivered by SV40 vectors.* J Neuropathol Exp Neurol *2009;***68***:456–73.*

upregulates the proform of MMP-9 and increases MMP-9 gelatinase activity. Using the NMDA receptor (NMDAR-1) inhibitor, memantine, we observed partial protection from gp120-induced BBB injury.[68]

MMPs are upregulated in different neurological diseases and models of CNS injury.[69] Various factors, such as ROS, NO, and proteases such as plasmin and stromelysin-1, are involved in MMP activation and upregulation in CNS injury. MMPs have been reported in the cerebrospinal fluid of HIV-1-infected patients as well as in models of HIV-1 encephalopathy.[70]

Relatively little is known about the respective roles of ROS and oxidative stress in the balance between MMPs and their endogenous tissue inhibitors (TIMPs). MMPs and four TIMPs act together to control tightly temporally restricted, focal proteolysis of extracellular matrix (ECM).[69] Once activated, MMPs are subject to inhibition by specific TIMPs that bind MMPs noncovalently.[69] Tissue destruction by MMPs is regulated by TIMPs and TIMPs prevent excessive MMP-related degradation of ECM components. The balance between MMPs and TIMPs is linked to ECM.

We studied the effect of gp120 on TIMP1- and TIMP-2 production. TIMP-1 and TIMP-2 levels increased 6h after gp120 injection into rat CP. TIMP-1 and TIMP-2 colocalized mainly with neurons. By 24h, expression of these protease inhibitors diverged, as TIMP-1 levels remained high but TIMP-2 subsided. Gene delivery of the antioxidant enzymes SOD1 or GPx1 into the CP before injecting gp120 there reduced levels of gp120-induced TIMP-1 and TIMP-2, recapitulating the effect of antioxidant enzymes on gp120-induced MMP-2 and MMP-9. A significant correlation was observed between MMP/TIMP upregulation and BBB leakiness. Thus, HIV-1 gp120 upregulated TIMP-1 and

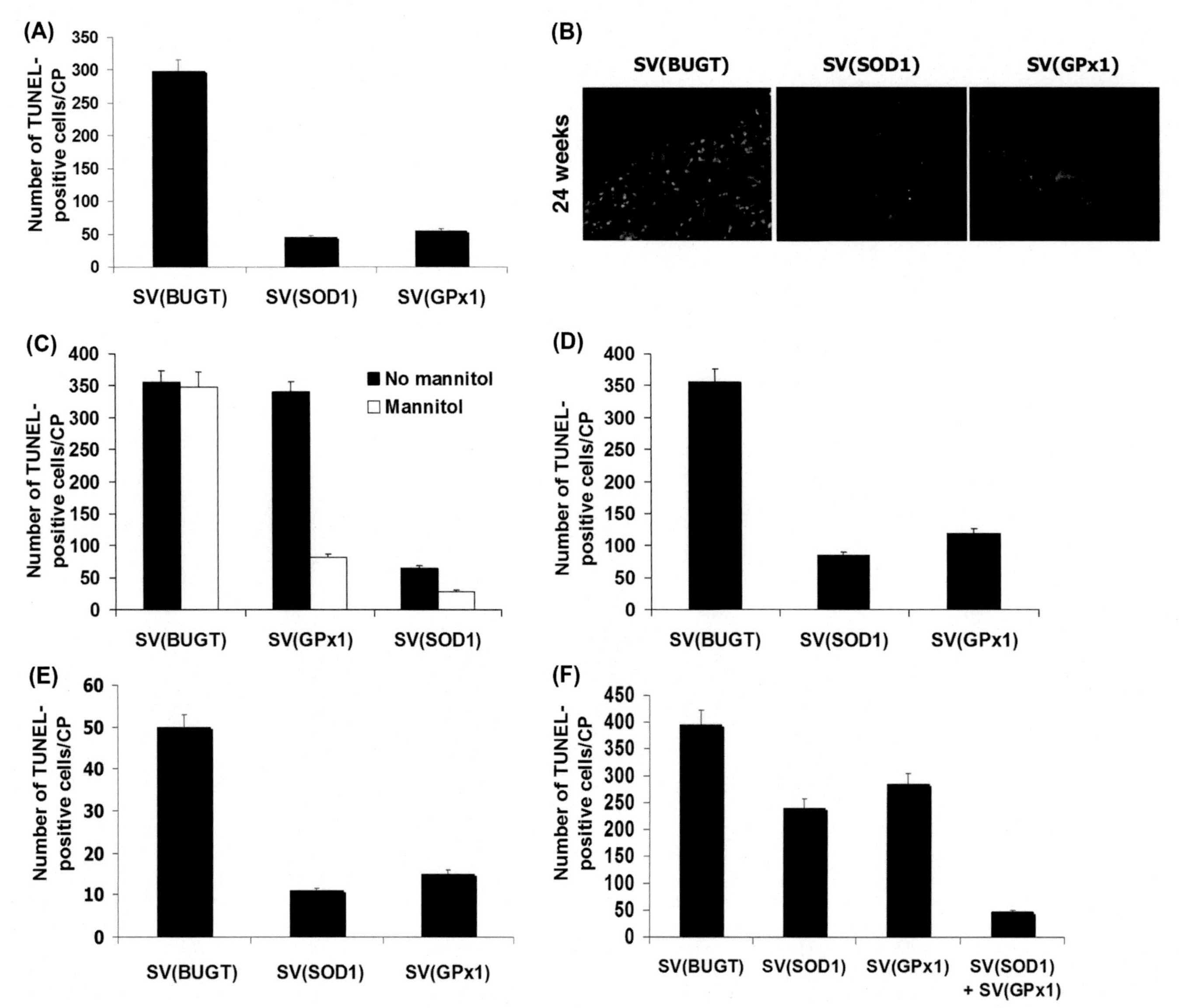


FIGURE 10.6 Gene transfer of antioxidant enzymes reduces the number of apoptotic cells in different animal models of HIV-associated neurocognitive disorder (HAND). (A) SV40-mediated gene delivery of antioxidant enzymes protects against gp120-induced apoptosis. SV(superoxide dismutase (SOD1)) and SV(glutathione peroxidase (GPx1)) were administered into the rat caudate-putamen (CP) 1 month before the inoculation of gp120 into the same structure. Brains were harvested 1 day after gp120 challenge and analyzed for apoptotic cells by TUNEL assay. Significantly less TUNEL-positive neurons were counted in CPs injected with SV(SOD1) and SV(GPx1), compared to CPs injected with a control vector, SV(BUGT). (B) TUNEL assay performed in rats whose CPs was injected with gp120 1 month after inoculation with SV(SOD1), SV(GPx1), or a control vector, SV(BUGT). (C) Intraperitoneal (i.p.) injection of mannitol increases gene delivery and protection efficiency after administration of SV(SOD1) and SV(GPx1) into the lateral ventricle (LV). Mannitol was injected i.p. before the administration of SV(SOD1), SV(GPx1) or a control vector, SV(BUGT), into the LV. Gp120 was injected into the CP 1 month later and TUNEL assay was performed on brain sections harvested 1 day after gp120 inoculation. There were significantly less apoptotic cells when administration of SV(SOD1) and SV(GPx1) was preceded by injection of mannitol. (D) Similarly, mannitol was injected i.p. before the administration of SV(SOD1), SV(GPx1) or a control vector, SV(BUGT), into the cisterna magna. One month later, gp120 was injected into the CP. Significantly less TUNEL-positive less were enumerated in the CPs of rats whose CM has been injected by SV(SOD1) or SV(GPx1), compared to rats administered with SV(BUGT). (E) Gene delivery of SOD1 and GPx1 reduces SV(gp120)-induced apoptosis. SV(SOD1), SV(GPx1) or SV(BUGT), a control vector, were injected into the CP 1 month before the inoculation of SV(gp120) into the same structure. One week thereafter, brains were harvested, and TUNEL-positive cells were enumerated. Significantly less apoptotic cells were seen in the CPs injected with SV(SOD1) or SV(GPx1) compared to the CPs administered with the control vector, SV(BUGT). (F) SV(SOD1) and SV(GPx1) were injected into the CP 1 month before inoculation of Tat in the same structure. Brains were harvested 2 days after Tat challenge and analyzed for apoptotic cells using TUNEL assay. Prior administration of recombinant rSV40 vectors carrying antioxidant enzymes SOD1 or GPx1 protected from Tat-induced apoptosis. *(A) and (B) Modified from Louboutin JP, Agrawal L, Reyes BAS, Van Bockstaele EJ, Strayer DS Protecting neurons from HIV-1 gp120-induced oxidant stress using both localized intracerebral and generalized intraventricular administration of antioxidant enzymes delivered by SV40-derived vectors.* Gene Ther *2007;**14**:1650-61. (D) Modified from Louboutin JP, Reyes BAS, Agrawal L, Van Bockstaele EJ, Strayer DS. Intracisternal rSV40 administration provides effective pan-CNS transgene expression.* Gene Ther *2002;**19**:114–8; (E) Modified from Louboutin JP, Agrawal L, Reyes BAS, Van Bockstaele EJ, Strayer DS. A rat model of human immunodeficiency virus 1 encephalopathy using envelope glycoprotein gp120 expression delivered by SV40 vectors.* J Neuropathol Exp Neurol *2009;**68**:456–73; (F) Modified from Agrawal L, Louboutin JP, Reyes BAS, Van Bockstaele EJ, Strayer DS. HIV-1 Tat neurotoxicity: a model of acute and chronic exposure, and neuroprotection by gene delivery of antioxidant enzymes.* Neurobiol Dis *2012;**45**:657–70.*

TIMP-2 in the CP. Prior antioxidant enzyme treatment mitigated production of these TIMPs, probably by reducing MMP expression. Following prior antioxidant gene delivery, a relationship was also seen between the reduction in Evans Blue extravasation and MMP-9/TIMP-1 decreased production.[71,72]

### *SV40-Mediated Gene Delivery of Antioxidant Enzymes Reduces Gp120-Induced Neuroinflammation*

If neuron loss[29,30,59] and astrogliosis[15] have been described in animals receiving gp120 directly into their brains, a temporal relationship between neuronal degeneration, astrocytic reaction, proinflammatory cytokine production, and microglial proliferation remained to be established. We challenged rat CPs with 100–500 ng HIV-1BaL gp120, with or without prior rSV40-delivered SOD1 or GPx1. CD11b-positive microglia were increased 1 day postchallenge; Iba-1- and ED1-positive cells peaked at 7 days and 14 days, respectively. Astrocyte infiltration was maximal at 7–14 days. MIP-1alpha was produced immediately, mainly by neurons. ED1-and GFAP-positive cells correlated with neuron loss and gp120 dose. Increase in microglia and astrocytes was seen following intra-CP SV(gp120) injection, suggesting that continuing gp120 production increased neuroinflammation. SV(SOD1) or SV(GPx1) significantly reduced MIP-1alpha and limited neuroinflammation following gp120 administration into the CP, as well as microglia and astrocytes proliferation after injection of SV(gp120) in the striatum. Thus, gp120-induced CNS injury, neuron loss, and inflammation may be mitigated by antioxidant gene delivery.[73] Similar results were observed when we injected Tat in the CP instead of gp120.[27] The participation of other chemokines/cytokines in gp120-induced lesions in vivo remains to be established. The modulation of the interaction between these chemokines/cytokines and their ligands needs to be investigated.

## CONCLUSION

- HIV-1-associated neurocognitive disorder is an increasingly common, progressive disease characterized by neuronal loss, and progressively deteriorating CNS function.
- HIV-1 gene products, particularly gp120 and Tat, elicit ROS that lead to oxidant injury, cause neuron apoptosis, and subsequent consequences (e.g., neuroinflammation, abnormalities of the BBB).
- Understanding of, and developing therapies for, HAND requires accessible models of the disease. We have devised experimental approaches to studying the acute and chronic effects of gp120 and Tat on the CNS.
- Gene transfer of antioxidant enzymes by recombinant SV40-derived vectors protects against gp120 and Tat-induced oxidative stress and neuronal apoptosis.
- Gene delivery of antioxidant enzymes opens new avenues for potential therapeutics of HAND.

## References


1. McArthur JC, Hoover DR, Bacellar H, Miller EN, Cohen BA, Becker JT, Graham NM, McArthur JH, Selnes OA, Jacobson LP. Dementia in AIDS patients: incidence and risk factors. Multicenter AIDS Cohort Study. *Neurology* 1993;**43**:2245–52.
2. Major EO, Rausch D, Marra C, Clifford D. HIV-associated dementia. *Science* 2000;**288**:440–2.
3. Koutsilieri E, Sopper S, Scheller C, ter Meulen V, Riederer P. Parkinsonism in HIV dementia. *J Neural Transm* 2002;**109**:767–75.
4. Antinori A, Arendt G, Becker JT, Brew BJ, Byrd DA, Cherner M, Clifford DB, Cinque P, Epstein LG, Gookin K, Gisslen M, Grant I, Heaton RK, Joseph J, Marder K, Marra CM, McArthur JC, Nunn M, Price RW, Pulliam L, Robertson KR, Saktor N, Valcour N, Wojna VE. Updated research nosology for HIV-associated neurocognitive disorders. *Neurology* 2007;**69**:1789–99.
5. Woods SP, Moore DJ, Weber E, Grant I. Cognitive neuropsychology of HIV-associated neurocognitive disorders. *Neuropsychol Rev* 2009;**19**:152–68.
6. McArthur JC, Brew BJ, Nath A. Neurological complications of HIV infection. *Lancet Neurol* 2005;**4**:543–55.
7. Nath A, Sacktor N. Influence of highly active antiretroviral therapy on persistence of HIV in the central nervous system. *Curr Opin Neurol* 2006;**19**:358–61.
8. Ances BM, Ellis RJ. Dementia and neurocognitive disorders due to HIV-1 infection. *Semin Neurol* 2007;**27**:86–92.
9. Mattson MP, Haughey NJ, Nath A. Cell death in HIV dementia. *Cell Death Diff* 2005;**12**:893–904.
10. Rumbaugh JA, Nath A. Developments in HIV neuropathogenesis. *Curr Pharm Des* 2006;**12**:1023–44.
11. Gonzalez-Scarano F, Martin-Garcia J. The neuropathogenesis of AIDS. *Nat Rev Immunol* 2005;**5**:69–81.
12. Kaul M, Garden GA, Lipton SA. Pathways to neuronal injury and apoptosis in HIV-associated dementia. *Nature* 2001;**410**:988–94.
13. Eugenin EA, Osiecli K, Lopez L, Goldstein H, Calderon TM, Berman JW. CCL2/monocyte chemoattractant protein-1 mediates enhanced transmigration of human immunodeficiency virus (HIV)-infected leukocytes across the blood-brain barrier: a potential mechanism of HIV-CNS invasion and neuroAIDS. *J Neurosci* 2006;**26**:1098–106.
14. Haughey NJ, Nath A, Mattson MP. HIV-1 tat through phosphorylation of NMDA receptors potentiates glutamate excitotoxicity. *J Neurochem* 2001;**78**:457–67.
15. Askenov MY, Hasselrot U, Wu G, Nath A, Anderson C, Mactutus CF, Booze RM. Temporal relationship between HIV-1 Tat-induced neuronal degeneration, OX-42 immunoreactivity, reactive astrocytosis, and protein oxidation in the rat striatum. *Brain Res* 2003;**987**:1–9.

16. Hudson L, Liu J, Nath A, Jones M, Raghavan R, Naravan O, Male D, Everall I. Detection of the human immunodeficiency virus regulatory protein tat in CNS tissues. *J Neurovirol* 2000;**6**:144–55.
17. Garden GA, Guo W, Jayadev S, Tun C, Balcaitis S, Choi J, Montine TJ, Moller T, Morrison RS. HIV associated neurodegeneration requires p53 in neurons and microglia. *FASEB J* 2004;**18**:1141–3.
18. Meucci O, Fatatis A, Simen AA, Bushell TJ, Gray PW, Miller RJ. Chemokines regulate hippocampal neuronal signalling and gp120 neurotoxicity. *Proc Natl Acad Sci USA* 1998;**95**:14500–5.
19. Kaul M, Lipton SA. Chemokines and activated macrophages in HIV gp120-induced neuronal apoptosis. *Proc Natl Acad Sci USA* 1999;**96**:8212–6.
20. Adamson DC, Wildemann B, Sasaki MD, Glass JD, McArthur JC, Christov VI, Dawson TM, Dawson VL. Immunologic NO synthase elevation in severe AIDS dementia and induction by HIV-1 gp41. *Science* 1996;**274**:1917–20.
21. Patel CA, Mukhtar M, Pomerantz RJ. HIV-1 Vpr induces apoptosis in human neuronal cells. *J Virol* 2000;**74**:9717–26.
22. Trillo-Pazos G, McFarlane-Abdulla E, Campbell IC, Pilkington GJ, Everall IP. Recombinant nef HIV-IIIB protein is toxic to human neurons in culture. *Brain Res* 2000;**864**:315–26.
23. Lackner AA, Veazey RS. Current concepts in AIDS pathogenesis: insights from the SIV/macaque model. *Annu Rev Med* 2007;**58**:461–76.
24. Thormar H. Maedi-Visna virus and its relationship to human deficiency virus. *AIDS Rev* 2005;**7**:233–45.
25. Hurtrel M, Ganiere JP, Guelfi JF, Chakrabarti L, Maire MA, Gray F, Montagnier L, Hurtrel B. Comparison of early and late feline immunodeficiency virus encephalopathies. *AIDS* 1992;**6**:399–406.
26. Toggas SM, Masliah E, Rockenstein EM, Rall GF, Abraham CR, Mucke L. Central nervous system damage produced by expression of the HIV-1 coat protein gp120 in transgenic mice. *Nature* 1994;**367**:188–93.
27. Agrawal L, Louboutin JP, Reyes BAS, Van Bockstaele EJ, Strayer DS. HIV-1 Tat neurotoxicity: a model of acute and chronic exposure, and neuroprotection by gene delivery of antioxidant enzymes. *Neurobiol Dis* 2012;**45**:657–70.
28. Agrawal L, Louboutin JP, Reyes BAS, Van Bockstaele EJ, Strayer DS. Antioxidant enzyme gene delivery to protect from HIV-1 gp120-induced neuronal apoptosis. *Gene Ther* 2006;**13**:1645–56.
29. Louboutin JP, Agrawal L, Reyes BAS, Van Bockstaele EJ, Strayer DS. Protecting neurons from HIV-1 gp120-induced oxidant stress using both localized intracerebral and generalized intraventricular administration of antioxidant enzymes delivered by SV40-derived vectors. *Gene Ther* 2007;**14**:1650–61.
30. Bansal AK, Mactutus CF, Nath A, Maragos W, Hauser KF, Booze RM. Neurotoxicity of HIV-1 proteins gp120 and Tat in the rat striatum. *Brain Res* 2000;**879**:42–9.
31. Nosheny RL, Bachis A, Acquas E, Mocchetti I. Human immunodeficiency virus type 1 glycoprotein gp120 reduces the levels of brain-derived neurotrophic factor in vivo: potential implication for neuronal cell death. *Eur J Neurosci* 2004;**20**:2857–64.
32. Louboutin JP, Agrawal L, Reyes BAS, Van Bockstaele EJ, Strayer DS. A rat model of human immunodeficiency virus 1 encephalopathy using envelope glycoprotein gp120 expression delivered by SV40 vectors. *J Neuropathol Exp Neurol* 2009;**68**:456–73.
33. Uttara B, Singh AV, Zamboni P, Mahajan RT. Oxidative stress and neurodegenerative diseases: a review of upstream and downstream antioxidant therapeutic options. *Curr Neuropharmacol* 2009;**7**:65–74.
34. Broughton BRS, Reutens DC, Sobey CG. Apoptotic mechanisms after cerebral ischemia. *Stroke* 2009;**40**:e331–9.
35. Steiner J, Haughey N, Li W, Venkatesan A, Anderson C, Reid R, Malpica T, Pocernich C, Butterfield DA, Nath A. Oxidative stress and therapeutic approaches in HIV dementia. *Antioxid Redox Sign* 2006;**8**:2089–100.
36. Bonfoco E, Krainc D, Ankarcrona M, Nicotera P, Lipton SA. Apoptosis and necrosis: two distinct events induced, respectively, by mild and intense insults with N-methyl-D-aspartate or nitric oxide/superoxide in cortical cell cultures. *Proc Natl Acad Sci USA* 1995;**92**:7162–6.
37. Smith MA, Sayre LM, Monnier VM, Perry G. Radical ageing in Alzheimer's disease. *Trends Neurosci* 1995;**18**:172–6.
38. Montoliu C, Valles S, Renau-Piqueras J, Guerri C. Ethanol-induced oxygen radical formation and lipid peroxidation in rat brain: effect of chronic alcohol consumption. *J Neurochem* 1994;**63**:1855–62.
39. Mollace V, Nottet HS, Clayette P, Turco MC, Muscoli C, Salvemini D, Perno CF. Oxidative stress and neuroAIDS: triggers, modulators and novel antioxidants. *Trends Neurosci* 2001;**24**:411–6.
40. Haughey NJ, Cutler RG, Tamara A, McArthur JC, Vargas DL, Pardo CA, Turchan J, Nath A, Mattson MP. Perturbation of sphingolipid metabolism and ceramide production in HIV-dementia. *Ann Neurol* 2004;**5**:257–67.
41. Cutler RG, Haughey NJ, Tamara A, McArthur JC, Nath A, Reid R, Vargas DL, Pardo CA, Mattson MP. Dysregulation of sphingolipids and sterol metabolism by ApoE4 in HIV dementia. *Neurology* 2004;**63**:626–30.
42. Agrawal L, Louboutin JP, Strayer DS. Preventing HIV-1 Tat-induced neuronal apoptosis using antioxidant enzymes: mechanistic and therapeutic implications. *Virology* 2007;**363**:462–72.
43. Bruce-Keller AJ, Li YJ, Lovell MA, Kraemer PJ, Gary DS, Brown RR, Markesbery WR, Mattson MP. 4-Hydroxynonenal, a product of lipid peroxidation, damages cholinergic neurons and impairs visuospatial memory in rats. *J Neuropathol Exp Neurol* 1998;**57**:257–67.
44. Louboutin JP, Agrawal L, Reyes BAS, Van Bockstaele EJ, Strayer DS. HIV-1 gp120-induced injury to the blood-brain barrier: role of metalloproteinases 2 and 9 and relationship to oxidative stress. *J Neuropathol Exp Neurol* 2010;**69**:801–16.
45. Banerjee A, Zhanq X, Manda KR, Banks WA, Ercal N. HIV proteins (gp120 and Tat) and methamphetamine in oxidative stress-induced damage in the brain: potential role of the thiol antioxidant N-acetylcysteine amide. *Free Rad Biol Med* 2010;**48**:1388–98.
46. Bruce-Keller AJ, Barger SW, Moss NI, Pham JT, Keller JN, Nath A. Proinflammatory and pro-oxidant properties of Tat in a microglial cell line: attenuation by 17b-estradiol. *J Neurochem* 2001;**78**:1315–24.
47. McKee HJ, Strayer DS. Immune responses against SIV envelope glycoprotein, using recombinant SV40 as a vaccine delivery vector. *Vaccine* 2002;**20**:3613–25.
48. Cordelier P, Van Bockstaele E, Calarota SA, Strayer DS. Inhibiting AIDS in the central nervous system: gene delivery to protect neurons from HIV. *Mol Ther* 2003;**7**:801–10.
49. Turchan J, Pocernich CB, Gairola C, Chauhan A, Schifitto G, Butterfield DA, Buch S, Naravan O, Sinai A, Geiger G, Berger JR, Elford H, Nath A. Oxidative stress in HIV demented patients and protection ex vivo with novel antioxidants. *Neurology* 2003;**60**:307–14.
50. Pocernich CB, La Fontaine M, Butterfield DA. In-vivo glutathione elevation protects against hydroxyl free radical-induced protein oxidation in rat brain. *Neurochem Int* 2000;**36**:185–91.

51. Zinc MC, Uhrlaub J, DeWitt J, Voelker T, Bullock B, Mankowski J, Tarwater P, Clements J, Barber S. Neuroprotective anti-human immunodeficiency virus activity of minocycline. *JAMA* 2005;**293**:2003–11.
52. Consortium D. A randomized, double-blind, placebo-controlled trial of deprenyl and thioctic acid in human immunodeficiency virus-associated cognitive impairment: dana consortium on the therapy of HIV dementia and related cognitive disorders. *Neurology* 1998;**50**:645–51.
53. Clifford DB, McArthur JC, Schifitto G, Kieburtz K, McDermott MP, Letendre S, Cohen BA, Marder K, Ellis RJ, Marra CM. Neurologic AIDS Research Consortium. A randomized clinical trial of CPI-1189 for HIV-associated cognitive-motor impairment. *Neurology* 2002;**59**:1568–73.
54. Ridet JL, Bensadoun JC, Deglon N, Aebischer P, Zurn AD. Lentivirus-mediated expression of glutathione peroxidase: neuroprotection in murine models of Parkinson's disease. *Neurobiol Dis* 2006;**21**:29–34.
55. Louboutin JP, Marusich E, Fisher-Perkins J, Dufour JP, Bunnell BA, Strayer DS. Gene transfer to the Rhesus monkey brain using SV40-derived vectors is durable and safe. *Gene Ther* 2011;**18**:682–91.
56. Louboutin JP, Chekmasova AA, Marusich E, Chowdhury JR, Strayer DS. Efficient CNS gene delivery by intravenous injection. *Nat Methods* 2010;**7**:905–7.
57. Sims NR, Muyderman H. Mitochondria, oxidative metabolism and cell death in stroke. *Biochim Biophys Acta* 2010;**1802**:80–91.
58. Louboutin JP, Reyes BAS, Agrawal L, Van Bockstaele EJ, Strayer DS. Strategies for CNS-directed gene delivery: in vivo gene transfer to the brain using SV40-derived vectors. *Gene Ther* 2007;**14**:939–49.
59. Louboutin JP, Agrawal L, Reyes BAS, Van Bockstaele EJ, Strayer DS. HIV-1 gp120 neurotoxicity proximally and at a distance from the point of exposure: protection by rSV40 delivery of antioxidant enzyme. *Neurobiol Dis* 2009;**34**:462–76.
60. Louboutin JP, Reyes BAS, Agrawal L, Van Bockstaele EJ, Strayer DS. Intracisternal rSV40 administration provides effective pan-CNS transgene expression. *Gene Ther* 2012;**19**:114–8.
61. Singh IN, Goody RJ, Dean C, Ahmad NM, Lutz SE, Knapp PE, Nath A, Hauser KF. Apoptotic cell death of striatal neurons induced by human immunodeficiency virus-1 Tat and gp120: differential involvement of caspase-3 and endonuclease G. *J Neurovirol* 2004;**10**:141–51.
62. Bachis A, Aden SA, Nosheny RL, Andrews PM, Mocchetti I. Axonal transport of human immunodeficiency virus type 1 envelope protein glycoprotein 120 is found in association with neuronal apoptosis. *J Neurosci* 2006;**26**:6771–80.
63. Louboutin JP, Reyes BAS, Agrawal L, Van Bockstaele EJ, Strayer DS. Gene delivery of antioxidant enzymes inhibits HIV-1 gp120-induced expression of caspases. *Neuroscience* 2012;**214**:68–77.
64. Louboutin JP, Reyes BAS, Agrawal L, Van Bockstaele EJ, Strayer DS. Assessment of apoptosis and neuronal loss in animal models of HIV-1-Associated Neurocognitive Disorder. In: Van Bockstaele EJ, editor. *Transmission electron microscopy methods for understanding the brain. Springer protocols, neuromethods series*, vol. 115. New York: Springer Science + Business Media; 2016. p. 217–43 (Walz W, editor).
65. Price TO, Ercal N, Nakaoke R, Banks WA. HIV-1 viral proteins gp120 and Tat induce oxidative stress in brain endothelial cells. *Brain Res* 2005;**1045**:57–63.
66. Annunziata P. Blood-brain barrier changes during invasion of the central nervous system by HIV-1. Old and new insights into the mechanism. *J Neurol* 2003;**250**:901–6.
67. Kanmogne GD, Schall K, Leibhart J, Knipe B, Gendelman HE, Persidsky Y. HIV-1 gp120 compromises blood-brain barrier integrity and enhance monocyte migration across blood-brain barrier: implication for viral neuropathogenesis. *J Cereb Blood Flow Metab* 2007;**27**:123–34.
68. Louboutin JP, Reyes BAS, Agrawal L, Maxwell CR, Van Bockstaele EJ, Strayer DS. Blood-brain barrier abnormalities caused by exposure to HIV-1 gp120- Protection by gene delivery of antioxidant enzymes. *Neurobiol Dis* 2010;**38**:313–25.
69. Clark IM, Swingler TE, Sampieri CL, Edwards DR. The regulation of matrix metalloproteinases and their inhibitors. Int. *J Biochem Cell Biol* 2008;**40**:1362–78.
70. Conant K, St Hillaire C, Anderson C, Galey D, Wang J, Nath A. Human immunodeficiency virus type 1 Tat and methamphetamine affect the release and activation of matrix-degrading proteinases. *J Neurovirol* 2004;**10**:21–8.
71. Louboutin JP, Reyes BAS, Agrawal L, Van Bockstaele EJ, Strayer DS. HIV-1 gp120 upregulates matrix metalloproteinases and their inhibitors in a rat model of HIV encephalopathy. *Eur J Neurosci* 2011;**34**:2015–23.
72. Louboutin JP. Immunocytochemical assessment of blood-brain barrier structure, function and damage. In: Merighi A, Lossi L, editors. *"Immunocytochemistry and related techniques". Springer Protocols, Neuromethods Series*New York: Springer Science + Business Media; 2015. p. 225–53 (Walz W, editor). Chapter 13.
73. Louboutin JP, Reyes BAS, Agrawal L, Van Bockstaele EJ, Strayer DS. HIV-1 gp120 induced neuroinflammation: relationship to neuron loss and protection by rSV40-delivered antioxidant enzymes. *Exp Neurol* 2010;**221**:231–45.

CHAPTER

# 11

# Genistein and HIV Infection

*Jia Guo*[1,2], *Yuntao Wu*[1]

[1]George Mason University, Fairfax, VA, United States;
[2]Merck Sharp & Dohme Corp., Boston, MA, United States

OUTLINE



## Abstract

Genistein is a tyrosin kinase inhibitor that belongs to the category of naturally occurring isoflavones. HIV infection causes CD4 T cell depletion and immunodeficiency. The virus infects CD4 T cells by binding to the chemokine coreceptor CXCR4/CCR5. This interaction also induces chemotactic signaling that triggers aberrant actin activity necessary for viral infection of blood T cells. In the disease course, HIV infection also causes chronic immune activation and oxidative stress that could be reduced by genistein. In this chapter, we review the effects of genistein on T cell chemotactic signaling and HIV infection of cultured blood CD4 T cells. Genistein has been found to inhibit SDF-1 (stromal cell-derived factor 1)- and HIV-mediated chemotactic signaling and HIV infection of cultured blood T cells and macrophages. These recent studies suggest that genistein could be used in combination with antiretroviral therapy to inhibit HIV infection and reduce HIV-induced oxidative damages.

**Keywords:** Antioxidant; Chemotaxis; Cofilin; Genistein; HIV-1; SDF-1; Tyrosin kinase.

## INTRODUCTION

Genistein is a naturally occurring isoflavones found in a number of plants such as soybeans and *Flemingia vestita*. Similar to many isoflavones, genistein acts as a phytoestrogen in mammals through interaction with estrogen receptors. Genistein is also considered an antioxidant because of its ability to trap singlet oxygen and reduce the damages caused by free radicals in tissues.[1] In addition, genistein is termed as a general tyrosine kinase inhibitor, capable of inhibiting multiple tyrosine-specific protein kinases such as $pp60^{v\text{-}src}$ and $pp110^{gag\text{-}fes}$[2] (Fig. 11.1). Given that tyrosine kinases are frequently involved in tumor growth and metastasis, genistein has been tested for the treatment of cancers such as leukemia[3] and prostate cancer.[4] Dietary genistein has also been shown to inhibit metastasis of human



HIV/AIDS
http://dx.doi.org/10.1016/B978-0-12-809853-0.00011-0

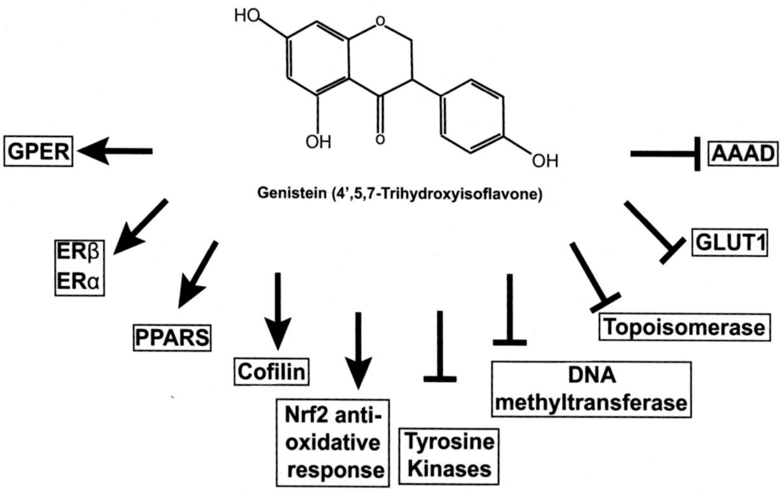

FIGURE 11.1 **Summary of the biological effects of genistein in living cells.**

prostate cancer in mice.[4,5] Mechanistically, genistein inhibits human prostate cancer cell motility through inhibiting promotility signaling, specifically, by inhibiting the activation of focal adhesion kinase (FAK) and the p38 MAPK–HSP27 pathway.[6] Genistein has also been suggested to modulate the cellular distribution of actin-binding proteins in human stromal cells by inducing the perinuclear accumulation of the actin-binding protein formin-2 and profilin.[7]

HIV infection causes CD4 T cell depletion and immunodeficiency. The virus selectively infects CD4 T cells and macrophages by binding to surface CD4 and the chemokine coreceptor CXCR4 or CCR5; this interaction mediates viral fusion and entry. HIV binding to the chemokine coreceptors also triggers the activation of multiple signaling molecules such as Pyk2/FAK, NF-κB, Rac1, PAK1/2, LIMK (LIM domain kinase), WAVE2, and cofilin.[8] In particular, cofilin has been identified as a critical factor required for HIV latent infection of blood resting CD4 T cells.[9–11] Cofilin is an actin-binding protein that binds and depolymerizes filamentous actin (F-actin) and is responsible for the high-turnover rates of actin filaments. In cells, cofilin regulates actin treadmilling, a process in which monomeric actin (G-actin) is preferentially incorporated into F-actin at the (+) end, and then dissociated from the (−) end (Fig. 11.2). During HIV infection of blood CD4 T cells, the virus binding to CXCR4 triggers a transient course of cofilin phosphorylation and dephosphorylation,[9,11] which promotes cortical actin dynamics facilitating viral entry and intracellular migration (Fig. 11.2). Thus, cofilin-regulated actin dynamics are major events targeted by the virus to facilitate infection of blood CD4 T cells.[8,9] Other actin-binding proteins such as Arp2/3[12], filamin-A, and moesin also promote actin activity and facilitate the anchorage of F-actin to membrane proteins, facilitating receptor clustering for viral entry and intracellular migration.[12–15]

The process of HIV-initiated chemotactic signaling involves multiple signaling molecules such as Pyk2, Rac1, PAK, LIMK, cofilin, GTPase Ras, phospholipase C, protein kinase C, Tiam-1, Abl, IRSp53, Wave2, and Arp2/3. It has been known that multiple actin regulators, such as gelsolin, villin, ezrin, cortactin, Rac1, and WASP (Wiskott-Aldrich syndrome protein), require tyrosine phosphorylation for activation. Among these molecules, several including ezrin,[14] Rac1,[11] and WASP[16] have been implicated in HIV infection of CD4 T cells. Given the pleotropic effects of genistein in inhibiting tyrosine kinases, it is expected that genistein would inhibit HIV infection through modulating these actin regulators.

The natural course of HIV infection is also accompanied by chronic immune activation and inflammation.[17] Antiretroviral therapy (ART) can effectively suppress HIV to undetectable level. However, immune function is not fully restored, and major immune activation markers such as IL-6 and soluble CD14/CD163 continue to be high in the presence of ART.[18] The virological cue for this chronic immune activation is HIV-mediated gut mucosal damage that causes microbial translocation and chronic activation of the innate and adaptive immune systems.[19,20] It has also been shown that this chronic immune activation is associated with sustained production of reactive oxygen species.[21] A number of studies have underlined the pathogenic roles of high levels of local and systemic oxidative stress in HIV infection.[21,22] Antioxidant defenses, although present in the bodies of patients, are insufficient to neutralize the damages caused by reactive oxygen species.[21] Antiretroviral drugs do not reduce oxidative stress. On the contrary, ART may increase oxidant injury.[22,23] As such, a current challenge in treating chronic HIV infection is to reduce inflammation and oxidative stress in the immune system. Genistein has attractive antioxidative and anti-HIV properties. In this review, we discuss the possibility to use genistein for managing chronic HIV infection.

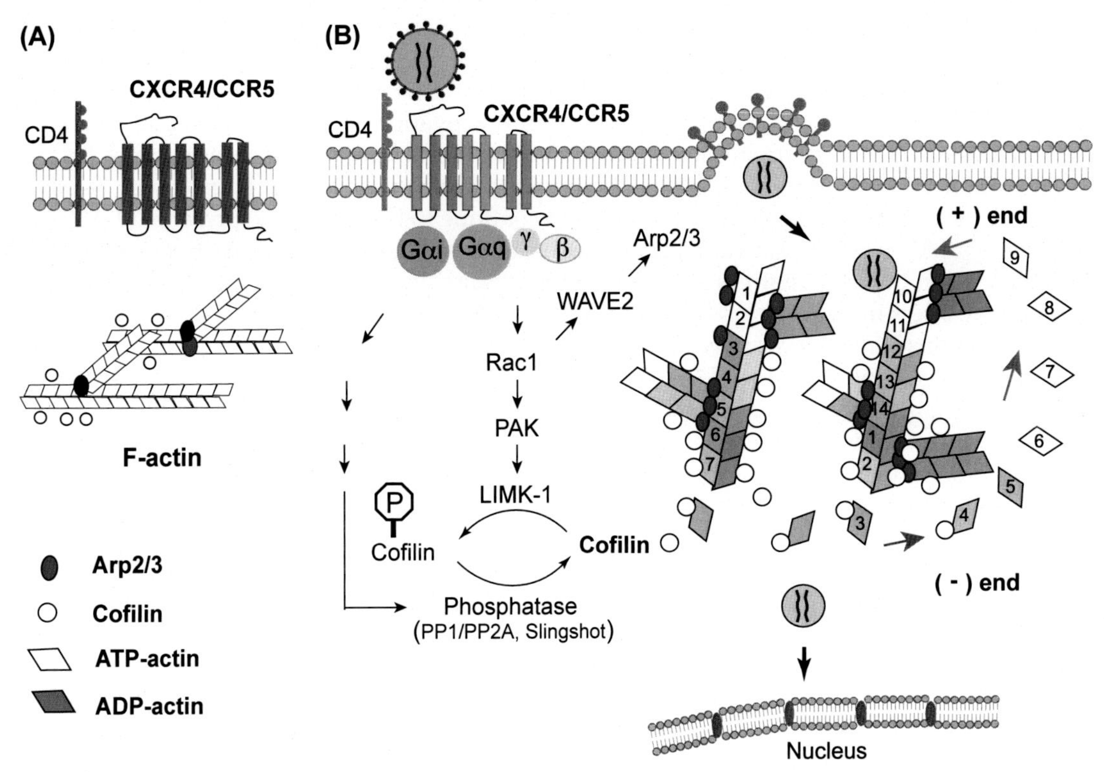


FIGURE 11.2 **Model of HIV-mediated chemotactic signaling in facilitating viral infection.** (A) In the absence of chemotactic signaling, the cortical actin in resting CD4 T cells remains relatively static. (B) Binding of HIV envelope protein gp120 to CXCR4 or CCR5 triggers chemotactic signaling, leading to the activation of the Rac1-PAK1/2-LIMK–cofilin pathway and the WAVE2–Arp2/3 pathway. Arp2/3 and Cofilin activation promotes actin polymerization and depolymerization, facilitating viral entry and postentry nuclear migration. Cofilin-mediated actin treadmilling may directly promote the migration of viral core toward the nucleus.

## BIOLOGICAL EFFECTS OF GENISTEIN IN LIVING CELLS

Genistein (4′,5,7-trihydroxyisoflavone) is a naturally occurring phytoestrogen present in a number of plants such as soybeans, fava beans, lupin, psoralea, and *F. vestita*. Genistein is found to exert a wide variety of biological effects in animal and human cells, including activation of estrogen receptor alpha and beta (,[24] G protein-coupled estrogen receptor 1, peroxisome proliferator-activated receptors,[25] and the Nrf2 antioxidative response.[26] Genistein is also found to inhibit multiple cellular proteins such as tyrosine kinases,[2] topoisomerase,[27] aromatic L-amino acid decarboxylase, glucose transporter 1, and DNA methyltransferase[28] (Fig. 11.1). Dietary genistein has been linked, through epidemiological and animal studies, with a range of potential health benefits such as prevention of breast and prostate cancers. In addition, genistein has been tested for controlling leukemia and the growth of cancers of the prostate, cervix, brain, breast, and colon. Genistein's chief mechanism of action for anticancer is through the inhibition of tyrosine kinases, which are involved in the signal transduction of cell growth and proliferation. Dietary genistein has also been found to inhibit metastasis of human prostate cancer in mice.[4,5] The inhibition of prostate cancer cell motility is through inhibiting promotility signaling, specifically, the activation of FAK and the p38 MAPK–HSP27 pathway.[6,29] Genistein can also modulate the cellular distribution of actin-binding proteins in human stromal cells by inducing the perinuclear accumulation of the actin-binding protein formin-2 and profilin.[7] Recently, genistein has been shown to inhibit the chemotactic migration of blood memory CD4 T cells toward SDF-1 (stromal cell-derived factor 1).[30] Genistein blocks CD4 T cell migration mainly through interfering with the cofilin signaling pathway.[30]

## HIV INFECTION AND THE ROLE OF HIV-MEDIATED CHEMOTACTIC SIGNALING

HIV infection causes severe CD4 T cell depletion, leading to the development of AIDS. The natural course of HIV infection results in multiple T cell dysfunctions and the eventual depletion of helper T cells. It has long been recognized that during the disease course, there is a time-dependent gradual loss of T helper function.[31] In general, the CD4 T cells in the peripheral blood of HIV-infected individuals carry numerous functional abnormalities, such as loss of T cell responses to recall antigens,[31,32] T cell anergy,[33,34] and abnormal T cell homing and migration.[35,36] These defects are also accompanied by chronic immune activation that correlates with an increased T cell turnover and proliferation.[17] The virological cue for this chronic immune activation is HIV-mediated gut mucosal damage that causes microbial translocation and chronic activation of the innate and adaptive immune systems.[19,20] Nevertheless, even with near complete viral suppression from ART, immune functions are not fully restored, and major immune activation markers such as IL-6 and soluble CD14/CD163 continue to be high in the presence of ART.[18] In addition, this HIV-mediated chronic immune activation is also associated with generation of reactive oxygen species.[21] High levels of local and systemic oxidative stress are present in the bodies of HIV-infected patients.[21,22] ART is not effective in reducing oxidative stress in patients.[22]

In recent years, various clinical trials have been conducted for immune reconstitution in patients receiving ART.[18] For example, in two large clinical trials (SILCAAT and ESPRIT), patients were given ART plus IL-2.[37] Several other phase 1 or phase 2 trials have also tested multiple nonspecific immunomodulators or broad antiinflammatory drugs, including sevelemer (anti-lipopolysaccharide (LPS)), statins, aspirin, COX-2 inhibitor, methotrexate, chloroquine/hydroxychloroquine, prebiotics/probiotics, bovine colostrum, rifaximin, aciclovir/valaciclovir, mesalazine (anti-LPS), or interleukin 7. Most of these clinical studies focused on reducing chronic immune activation and inflammation.

In the human immune system, T cell activities are mainly regulated by receptor signaling, such as signaling through the receptors of cytokines and chemokines. In HIV infection, the virus binding to the chemokine coreceptor CXCR4 or CCR5 also aberrantly activates G-protein and certain downstream signaling molecules such as PYK2/FAK2, NF-κB, WAVE2-Arp2/3, and LIMK-cofilin. In particular, among them, cofilin is identified as a critical factor required for HIV latent infection of blood resting CD4 T cells.[9–11] Cofilin is an actin-binding protein that binds and depolymerizes filamentous actin (F-actin) and is responsible for the high turnover rates of actin filaments (Fig. 11.2). The activity of cofilin is mainly regulated by phosphorylation of serine 3 at the N-terminal, which inhibits cofilin binding to actin, whereas cofilin is activated through dephosphorylation of serine 3 by phosphatases such as PP1α/PP2A, slingshot 1L. During HIV infection of cultured blood CD4 T cells, the virus binding to CXCR4 triggers a transient course of cofilin phosphorylation and dephosphorylation through the Rac1-PAK1/2-LIMK–cofilin pathway.[9,11] This HIV-initiated signaling cascade promotes cortical actin dynamics, facilitating viral entry and intracellular migration. In addition to activating the LIMK–cofilin pathway, HIV also activates the WAVE2-Arp2/3 signaling pathway to trigger actin polymerization to facilitate viral nuclear migration[12] (Fig. 11.2).

The HIV requirement for actin activity has also been demonstrated by several studies showing that cofilin is the key regulator activated by chemokines such as CCL19/CCL21 to facilitate the establishment of postintegration latency in R5 and X4 viral infection of resting memory CD4 T cells.[38–41] In addition to CCL19/CCL21, HIV infection results in upregulation of multiple proinflammatory cytokines and chemokines. Certain chemokines, such as CCL2 and CXCL-10, are found to be at high levels in patients' peripheral blood.[42–45] Overexpression of these proinflammatory chemokines likely results in aberrant chemotactic signaling, promoting aberrant T cell migration and oxidative stress.

## GENISTEIN INHIBITS CD4 T CELL CHEMOTAXIS AND CHEMOTACTIC SIGNALING

Genistein inhibits multiple tyrosine kinases that are involved in chemotactic signaling. For examples, multiple actin regulators such as gelsolin,[46] villin,[47] ezrin,[48] cortactin,[49] Rac1,[50] and WASP[51] require tyrosine phosphorylation for activation. It is expected that genistein would inhibit these tyrosine kinases and chemotaxis regulated by them. Recently, we have tested the effects of genistein on CD4 T cell chemotaxis. We used human resting $CD45RO^+$/$CD45RA^-$ memory CD4 T cells purified from peripheral blood. Cultured cells were treated with genistein (1 μg/mL) and tested for their responses to SDF-1-mediated chemotactic induction in an in vitro transwell assay (Fig. 11.3). We observed inhibition of memory T cell migration by genistein. We have also studied the dosage effects of genistein in inhibiting chemotaxis. Interestingly, at lower dosages (2.5–10 μM), genistein inhibits T cell chemotaxis in a dosage-dependent manner. However, at higher concentrations (20 and 40 μM), although genistein inhibits chemotaxis, the inhibition and the drug concentration are linearly correlated.[30] Given that genistein likely targets multiple tyrosine kinases, the nonlinear inhibition may result from different sensitivities of these kinases to genistein.

We have also studied the direct effect of genistein on T cell actin dynamics. We observed that at certain doses (e.g., 3.7 μM), genistein treatment can directly trigger actin polymerization in blood memory CD4 T cells (Fig. 11.4A). When cultured cells were stimulated with SDF-1, genistein did not inhibit SDF-1-mediated actin polymerization,

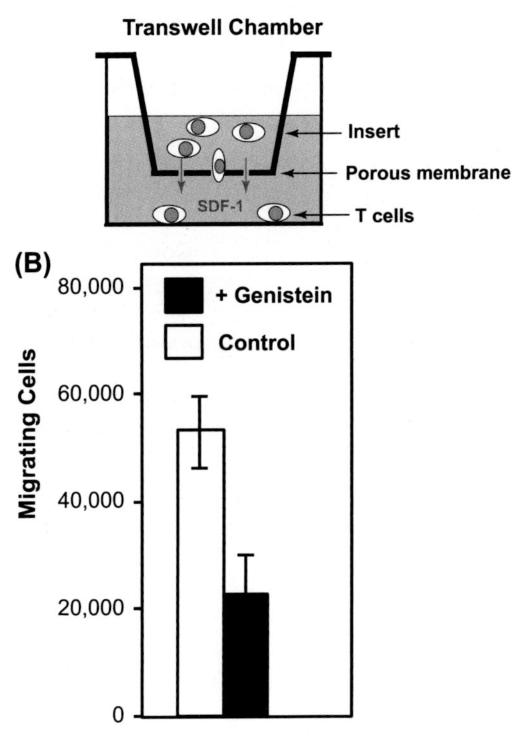


FIGURE 11.3 **Genistein inhibits T cell chemotaxis.** (A) Schematic diagram of transwell assay for quantification of SDF-1 (stromal cell-derived factor 1)-mediated T cell chemotaxis. (B) Resting $CD45RO^+/CD45RA^-$ memory CD4 T cells were treated with genistein (1 μg/mL) or DMSO (1%, control), and then placed in the upper chambers of transwell plates, and assayed for migration toward SDF-1 (12.5 nM) that was placed in lower chambers. Following incubation at 37°C for 2 h, cells migrating into the lower chamber were enumerated.

but can alter the temporal and spatial characteristics of actin polymerization. Specifically, genistein treatment can increase the nuclear accumulation of F-actin in human CD4 T cells (Fig. 11.4C, 60 min). In addition, genistein appears to promote a faster actin depolymerization (Fig. 11.4B, 5 min), which decreases the duration of actin polymerization following SDF-1 stimulation. The mechanisms of how genistein alters chemotactic actin activity are likely complex and not currently understood.

## GENISTEIN INHIBITS HIV INFECTION OF CD4 T CELLS AND MACROPHAGES

Given that CXCR4/CCR5-mediated actin dynamics are required for HIV infection of blood CD4 T cells,[9] genistein may inhibit HIV infection through interfering with HIV-mediated actin dynamics. We have treated resting CD4 T cells with genistein and infected cells with HIV-1. Following infection, cells were cultured for 5 days, and then activated with anti-CD3/CD28 antibodies.[52] We observed inhibition of HIV replication by multiple dosages of genistein (2.5–40 μM)[30] (Fig. 11.5). At lower dosages (below 5 μM), we observed dosage-dependent inhibition of HIV. However, at higher dosages (10 and 40 μM), the inhibition and genistein concentrations were not linearly correlated, although genistein inhibited HIV-1 replication at all dosages tested.

We have also tested genistein inhibition of HIV infection of peripheral blood monocyte-derived macrophages cultured in vitro. Cells were treated with 37 μM (or 10 μg/mL)[53] genistein and infected with the HIV stain THRO.c/2626.[54] We observed inhibition of HIV by genistein (Fig. 11.5B), similar to a previously report by Stantchev and coauthors.[53] Stantchev et al.[53] also reported that 5–10 μg/mL (or 18.5–37 μM) genistein inhibited HIV infection of macrophages; genistein blocked viral infection at the step of viral entry and early postentry.

We have also studied the molecular mechanisms of how genistein inhibit HIV infection. Stepwise mapping of HIV replication cycle demonstrated that genistein does not inhibit HIV entry. However, it inhibits viral DNA synthesis and, to a less extent, viral nuclear entry in resting CD4 T cells.[30] In conclusion, genistein mainly inhibits the slow accumulation of viral DNA in resting CD4 T cells following viral entry.[30]

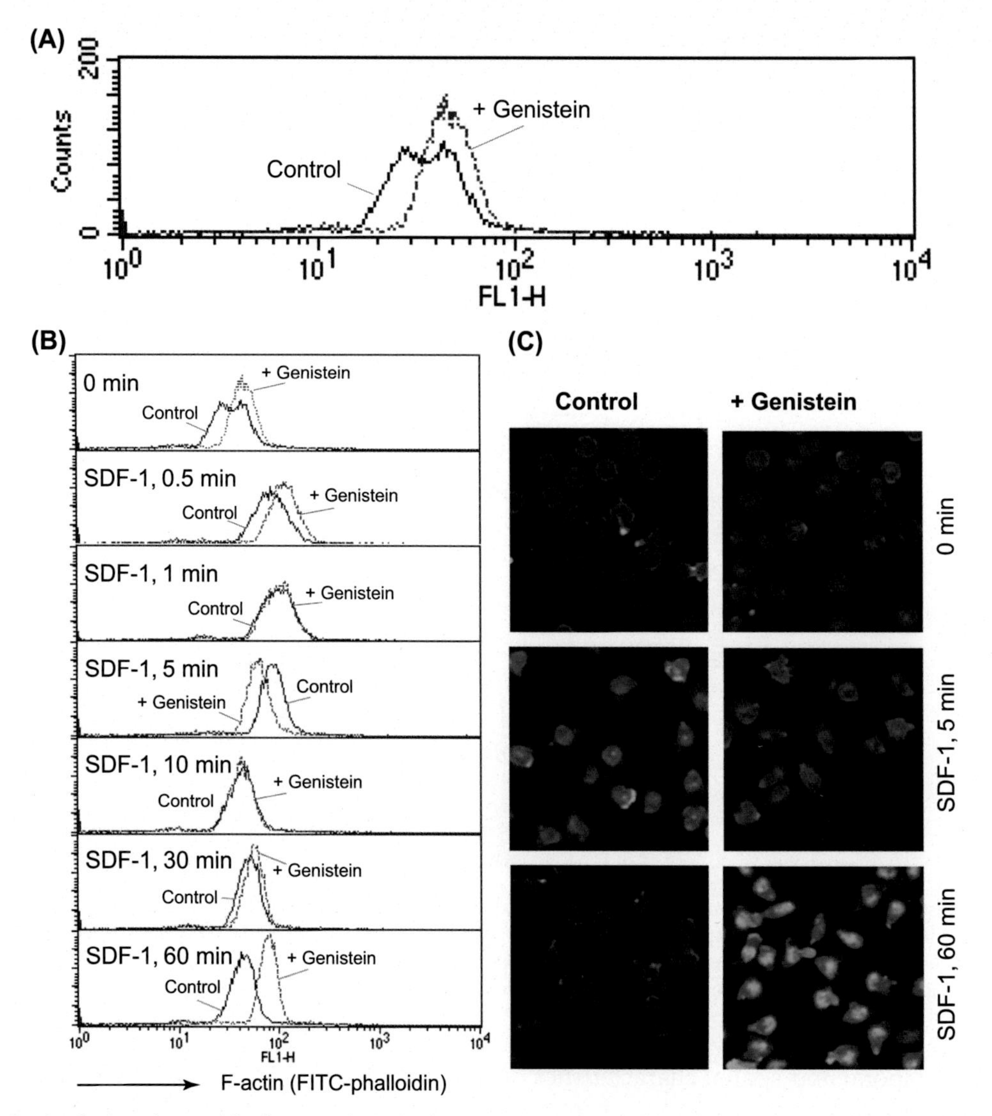


FIGURE 11.4 **Genistein interferes with chemotactic actin dynamics in CD4 T cells.** (A) Resting memory CD4 T cells were treated with genistein (1 μg/mL) or mock-treated for 1 h at 37°C, and then fixed, stained with fluorescein isothiocyanate (FITC)-phalloidin for F-actin, and then analyzed with flow cytometer. (B) Cells were also stimulated with stromal cell-derived factor 1 (SDF-1) (12.5 nM) for various times, from 1 min to 60 min, and then fixed, stained with FITC-phalloidin for F-actin, and then analyzed with flow cytometer. (C) Confocal images of cells stained in (B), showing nuclear accumulation of F-actin.

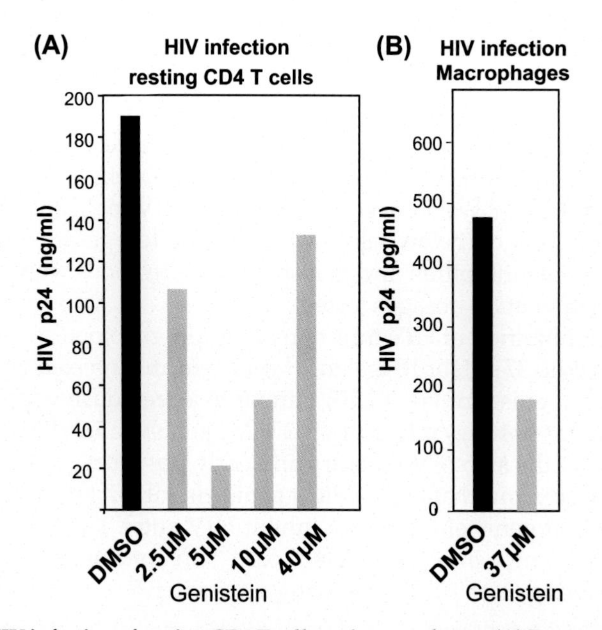


FIGURE 11.5 **Genistein inhibits HIV infection of resting CD4 T cells and macrophages.** (A) Resting CD4 T cells were treated with genistein, and then infected with HIV-$1_{NL4-3}$. Following infection, cells were washed, cultured for 5 days, and then activated with anti-CD3/CD28 magnetic beads to stimulate viral replication. (B) Genistein inhibits HIV infection of primary macrophages. Human peripheral blood monocyte-derived macrophages were treated with genistein, and then infected with HIV. Following infection, cells were cultured to measure viral replication.

## GENISTEIN INTERFERES WITH HIV-INITIATED CHEMOTACTIC SIGNALING

Given the viral requirement for interaction with actin cytoskeleton in HIV infection of blood T cells, genistein inhibits HIV infection likely through inhibiting HIV-initiated actin dynamics. As described above, genistein has been shown to inhibit metastasis of cancer cells by inhibiting cell signaling and the redistribution of actin-binding proteins such as formin-2 and profilin.[7] We have measured the effects of genistein on HIV-mediated actin dynamics and found that genistein can promote faster actin depolymerization (Figs. 11.6, 1 min), reducing the overall actin activity (Fig. 11.6). This enhanced actin depolymerization is associated with an increased activation of cofilin following HIV infection.[30] The genistein effect on coflin phosphorylation is likely indirect, possibly resulting from inhibition of upstream tyrosine kinases, as cofilin is directly phosphorylated on serine 3.[55,56]

The inhibition of HIV-mediated actin activity and HIV infection by genistein is consistent with the role of actin dynamics in HIV infection of blood resting T cells. Previous studies have also shown that chemokines such as CCL2 and CCL19 increase gp120-induced chemotactic signaling in CD4 T cells, facilitating HIV DNA synthesis and nuclear migration.[38,44] Interestingly, similar interference of the HIV-mediated chemotactic signaling has also been reported in CR2 (cannabinoid receptor 2)-mediated inhibition of HIV infection of CD4 T cells.[57]

## PERSPECTIVE

The success of ART has turned HIV infection from a deadly affliction to a manageable chronic disease that requires life-long treatment. However, ART does not fully restore immune function. Even with near complete viral suppression from ART, major immune activation markers such as IL-6, soluble CD14 and CD163 continue to be high. In addition, HIV-mediated chronic immune activation is also associated with high levels of local and systemic oxidative stress. As such, a major current goal in treating chronic HIV infection is to reduce chronic immune activation and

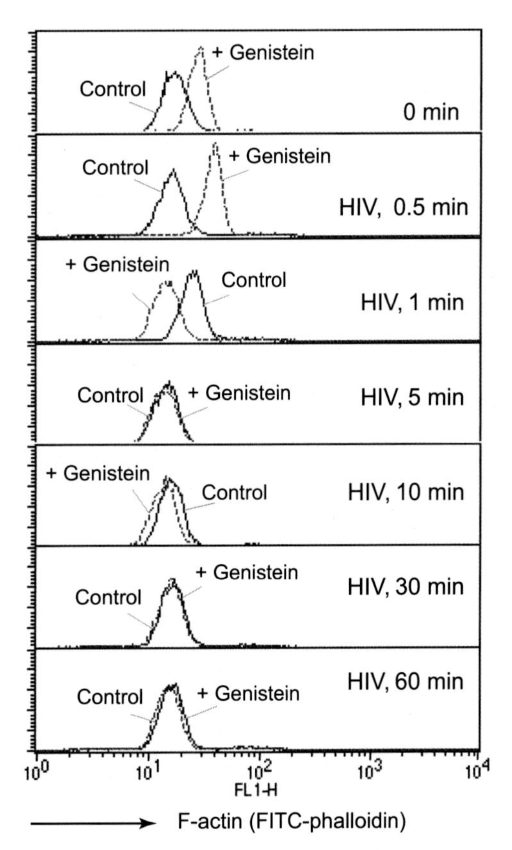


FIGURE 11.6 **Genistein interferes with HIV-initiated chemotactic signaling.** Resting memory CD4 T cells were treated with genistein (1 μg/mL), and then stimulated with HIV-1. Cells were fixed, stained with fluorescein isothiocyanate (FITC)-phalloidin for F-actin, and then analyzed with flow cytometer.

systemic oxidative stress. Genistein has attractive antixoidative and antiinflammatory properties,[58,59] and it can also block HIV infection of blood T cells and macrophages, as shown in our recent studies.[30]

Genistein can be found in a number of plants such as soybeans, and its consumption is associated with a lower incidence of metastatic prostate cancer in Southeast Asians who subsist on a soybean-based diet.[60] In a phase I clinical trial, subjects were given genistein (2–8 mg/kg) orally, and sustained a maximal total plasma genistein concentration between 4.3 and 16.3 μM, with a drug half-life of 15–22 h.[5] No cytotoxicities were observed among subjects.[5] Recently, we also conducted a preliminary animal trial to evaluate the safety of daily administration of genistein. Three rhesus macaques of Chinese origin (*Macaca mulatta*) were chronically infected with SIVmac251 (plasma viral loads between $10^2$ to $10^4$ copies/mL) and treated with a monotherapy of genistein at 10 mg/kg orally for 12 weeks. No adverse effects were observed, and two of the animals had a reduction of viral load to undetectable level.[30] These results suggest that genistein has the potential to be used, in combination with ART, for the long-term management of HIV infection.

## SUMMARY POINTS

- HIV infection causes chronic immune activation and oxidative stress that may be reduced by the use of genistein
- Success HIV infection of human CD4 T cells and macrophages requires CXCR4/CCR5 signaling that promotes actin dynamics necessary for HIV nuclear migration.
- Genestein is a tyrosine kinase inhibitor that can block chemotactic signaling, and reduce HIV-initiated actin activity. Genistein inhibits HIV infection of blood CD4 T cells and macrophages and blocks viral DNA accumulation in T cells.
- Genistein inhibits SDF-1-mediated actin polymerization and T cell chemotaxis.
- Genistein can cause overactivation cofilin during HIV infection, interfering with HIV-mediated chemotactic signaling.

## Acknowledgments

This work was funded in part by Public Health Service grants 1R01MH102144 and 1R03AI110174 from NIAID and NIMH to YW.

## References

1. Han RM, Tian YX, Liu Y, et al. Comparison of flavonoids and isoflavonoids as antioxidants. *J Agric Food Chem* 2009;**57**(9):3780–5.
2. Akiyama T, Ishida J, Nakagawa S, et al. Genistein, a specific inhibitor of tyrosine-specific protein kinases. *J Biol Chem* 1987;**262**(12):5592–5.
3. Raynal NJ, Momparler L, Charbonneau M, Momparler RL. Antileukemic activity of genistein, a major isoflavone present in soy products. *J Nat Prod* 2008;**71**(1):3–7.
4. Lakshman M, Xu L, Ananthanarayanan V, et al. Dietary genistein inhibits metastasis of human prostate cancer in mice. *Cancer Res* 2008;**68**(6):2024–32.
5. Takimoto CH, Glover K, Huang X, et al. Phase I pharmacokinetic and pharmacodynamic analysis of unconjugated soy isoflavones administered to individuals with cancer. *Cancer Epidemiol Biomarkers Prev* 2003;**12**(11 Pt 1):1213–21.
6. Xu L, Bergan RC. Genistein inhibits matrix metalloproteinase type 2 activation and prostate cancer cell invasion by blocking the transforming growth factor beta-mediated activation of mitogen-activated protein kinase-activated protein kinase 2-27-kDa heat shock protein pathway. *Mol Pharmacol* 2006;**70**(3):869–77.
7. Shieh DB, Li RY, Liao JM, Chen GD, Liou YM. Effects of genistein on beta-catenin signaling and subcellular distribution of actin-binding proteins in human umbilical CD105-positive stromal cells. *J Cell Physiol* 2010;**223**(2):423–34.
8. Wu Y, Yoder A. Chemokine coreceptor signaling in HIV-1 infection and pathogenesis. *PLoS Pathog* 2009;**5**(12):e1000520.
9. Yoder A, Yu D, Dong L, et al. HIV envelope-CXCR4 signaling activates cofilin to overcome cortical actin restriction in resting CD4 T cells. *Cell* 2008;**134**(5):782–92.
10. Wu Y, Yoder A, Yu D, et al. Cofilin activation in peripheral CD4 T cells of HIV-1 infected patients: a pilot study. *Retrovirology* 2008;**5**:95.
11. Vorster PJ, Guo J, Yoder A, et al. LIM kinase 1 modulates cortical actin and CXCR4 cycling and is activated by HIV-1 to initiate viral infection. *J Biol Chem* 2011;**286**(14):12554–64.
12. Spear M, Guo J, Turner A, et al. HIV-1 triggers WAVE2 phosphorylation in primary CD4 T cells and macrophages, mediating Arp2/3-dependent nuclear migration. *J Biol Chem* 2014;**289**(10):6949–59.
13. Jimenez-Baranda S, Gomez-Mouton C, Rojas A, et al. Filamin-A regulates actin-dependent clustering of HIV receptors. *Nat Cell Biol* 2007;**9**(7):838–46.
14. Naghavi MH, Valente S, Hatziioannou T, et al. Moesin regulates stable microtubule formation and limits retroviral infection in cultured cells. *Embo J* 2007;**26**(1):41–52.
15. Barrero-Villar M, Cabrero JR, Gordon-Alonso M, et al. Moesin is required for HIV-1-induced CD4-CXCR4 interaction, F-actin redistribution, membrane fusion and viral infection in lymphocytes. *J Cell Sci* 2009;**122**(Pt 1):103–13.

16. Komano J, Miyauchi K, Matsuda Z, Yamamoto N. Inhibiting the Arp2/3 complex limits infection of both intracellular mature vaccinia virus and primate lentiviruses. *Mol Biol Cell* 2004;**15**(12):5197–207.
17. Lempicki RA, Kovacs JA, Baseler MW, et al. Impact of HIV-1 infection and highly active antiretroviral therapy on the kinetics of CD4+ and CD8+ T cell turnover in HIV-infected patients. *Proc Natl Acad Sci USA* 2000;**97**(25):13778–83.
18. Deeks SG, Lewin SR, Havlir DV. The end of AIDS: HIV infection as a chronic disease. *Lancet* 2013;**382**(9903):1525–33.
19. Veazey RS, DeMaria M, Chalifoux LV, et al. Gastrointestinal tract as a major site of CD4+ T cell depletion and viral replication in SIV infection. *Science* 1998;**280**(5362):427–31.
20. Brenchley JM, Price DA, Schacker TW, et al. Microbial translocation is a cause of systemic immune activation in chronic HIV infection. *Nat Med* 2006;**12**(12):1365–71.
21. Coaccioli S, Crapa G, Fantera M, et al. Oxidant/antioxidant status in patients with chronic HIV infection. *Clin Ter* 2010;**161**(1):55–8.
22. Jiang B, Hebert VY, Khandelwal AR, Stokes KY, Dugas TR. HIV-1 antiretrovirals induce oxidant injury and increase intima-media thickness in an atherogenic mouse model. *Toxicol Lett* 2009;**187**(3):164–71.
23. Hulgan T, Morrow J, D'Aquila RT, et al. Oxidant stress is increased during treatment of human immunodeficiency virus infection. *Clin Infect Dis* 2003;**37**(12):1711–7.
24. Patisaul HB, Melby M, Whitten PL, Young LJ. Genistein affects ER beta- but not ER alpha-dependent gene expression in the hypothalamus. *Endocrinology* 2002;**143**(6):2189–97.
25. Wang L, Waltenberger B, Pferschy-Wenzig EM, et al. Natural product agonists of peroxisome proliferator-activated receptor gamma (PPARgamma): a review. *Biochem Pharmacol* 2014;**92**(1):73–89.
26. Zhai X, Lin M, Zhang F, et al. Dietary flavonoid genistein induces Nrf2 and phase II detoxification gene expression via ERKs and PKC pathways and protects against oxidative stress in Caco-2 cells. *Mol Nutr Food Res* 2013;**57**(2):249–59.
27. Markovits J, Linassier C, Fosse P, et al. Inhibitory effects of the tyrosine kinase inhibitor genistein on mammalian DNA topoisomerase II. *Cancer Res* 1989;**49**(18):5111–7.
28. Fang M, Chen D, Yang CS. Dietary polyphenols may affect DNA methylation. *J Nutr* 2007;**137**(1 Suppl.):223S–8S.
29. Huang X, Chen S, Xu L, et al. Genistein inhibits p38 map kinase activation, matrix metalloproteinase type 2, and cell invasion in human prostate epithelial cells. *Cancer Res* 2005;**65**(8):3470–8.
30. Guo J, Xu X, Rasheed TK, et al. Genistein interferes with SDF-1- and HIV-mediated actin dynamics and inhibits HIV infection of resting CD4 T cells. *Retrovirology* 2013;**10**:62.
31. Clerici M, Stocks NI, Zajac RA, et al. Detection of three distinct patterns of T helper cell dysfunction in asymptomatic, human immunodeficiency virus-seropositive patients. Independence of CD4+ cell numbers and clinical staging. *J Clin Invest* 1989;**84**(6):1892–9.
32. Schweneker M, Favre D, Martin JN, Deeks SG, McCune JM. HIV-induced changes in T cell signaling pathways. *J Immunol* 2008;**180**(10):6490–500.
33. Gurley RJ, Ikeuchi K, Byrn RA, Anderson K, Groopman JE. CD4+ lymphocyte function with early human immunodeficiency virus infection. *Proc Natl Acad Sci USA* 1989;**86**(6):1993–7.
34. Deeks SG, Tracy R, Douek DC. Systemic effects of inflammation on health during chronic HIV infection. *Immunity* 2013;**39**(4):633–45.
35. Poggi A, Carosio R, Fenoglio D, et al. Migration of V delta 1 and V delta 2 T cells in response to CXCR3 and CXCR4 ligands in healthy donors and HIV-1-infected patients: competition by HIV-1 Tat. *Blood* 2004;**103**(6):2205–13.
36. Mavigner M, Cazabat M, Dubois M, et al. Altered CD4+ T cell homing to the gut impairs mucosal immune reconstitution in treated HIV-infected individuals. *J Clin Invest* 2012;**122**(1):62–9.
37. Group I-ES, Committee SS, Abrams D, et al. Interleukin-2 therapy in patients with HIV infection. *N Engl J Med* 2009;**361**(16):1548–59.
38. Cameron PU, Saleh S, Sallmann G, et al. Establishment of HIV-1 latency in resting CD4+ T cells depends on chemokine-induced changes in the actin cytoskeleton. *Proc Natl Acad Sci USA* 2010;**107**(39):16934–9.
39. Saleh S, Solomon A, Wightman F, Xhilaga M, Cameron PU, Lewin SR. CCR7 ligands CCL19 and CCL21 increase permissiveness of resting memory CD4+ T cells to HIV-1 infection: a novel model of HIV-1 latency. *Blood* 2007;**110**(13):4161–4.
40. Evans VA, Khoury G, Saleh S, Cameron PU, Lewin SR. HIV persistence: chemokines and their signalling pathways. *Cytokine Growth Factor Rev* 2012;**23**(4–5):151–7.
41. Jones KL, Smyth RP, Pereira CF, et al. Early events of HIV-1 infection: can signaling be the next therapeutic target? *J Neuroimmune Pharmacol* 2011;**6**(2):269–83.
42. Ploquin MJ, Madec Y, Casrouge A, et al. Elevated basal pre-infection CXCL10 in plasma and in the small intestine after infection are associated with more rapid HIV/SIV disease onset. *PLoS Pathog* 2016;**12**(8):e1005774.
43. Ramirez LA, Arango TA, Thompson E, Naji M, Tebas P, Boyer JD. High IP-10 levels decrease T cell function in HIV-1-infected individuals on ART. *J Leukoc Biol* 2014;**96**(6):1055–63.
44. Campbell GR, Spector SA. CCL2 increases X4-tropic HIV-1 entry into resting CD4+ T cells. *J Biol Chem* 2008;**283**(45):30745–53.
45. Ansari AW, Bhatnagar N, Dittrich-Breiholz O, Kracht M, Schmidt RE, Heiken H. Host chemokine (C-C motif) ligand-2 (CCL2) is differentially regulated in HIV type 1 (HIV-1)-infected individuals. *Int Immunol* 2006;**18**(10):1443–51.
46. De Corte V, Gettemans J, Vandekerckhove J. Phosphatidylinositol 4,5-bisphosphate specifically stimulates PP60(c-src) catalyzed phosphorylation of gelsolin and related actin-binding proteins. *FEBS Lett* 1997;**401**(2–3):191–6.
47. Zhai L, Zhao P, Panebra A, Guerrerio AL, Khurana S. Tyrosine phosphorylation of villin regulates the organization of the actin cytoskeleton. *J Biol Chem* 2001;**276**(39):36163–7.
48. Bretscher A. Regulation of cortical structure by the ezrin-radixin-moesin protein family. *Curr Opin Cell Biol* 1999;**11**(1):109–16.
49. Huang C, Liu J, Haudenschild CC, Zhan X. The role of tyrosine phosphorylation of cortactin in the locomotion of endothelial cells. *J Biol Chem* 1998;**273**(40):25770–6.
50. Chang F, Lemmon C, Lietha D, Eck M, Romer L. Tyrosine phosphorylation of Rac1: a role in regulation of cell spreading. *PLoS One* 2011;**6**(12):e28587.
51. Cory GO, Garg R, Cramer R, Ridley AJ. Phosphorylation of tyrosine 291 enhances the ability of WASp to stimulate actin polymerization and filopodium formation. Wiskott-Aldrich Syndrome protein. *J Biol Chem* 2002;**277**(47):45115–21.
52. Wu Y, Marsh JW. Selective transcription and modulation of resting T cell activity by preintegrated HIV DNA. *Science* 2001;**293**(5534):1503–6.

53. Stantchev TS, Markovic I, Telford WG, Clouse KA, Broder CC. The tyrosine kinase inhibitor genistein blocks HIV-1 infection in primary human macrophages. *Virus Res* 2007;**123**(2):178–89.
54. Keele BF, Giorgi EE, Salazar-Gonzalez JF, et al. Identification and characterization of transmitted and early founder virus envelopes in primary HIV-1 infection. *Proc Natl Acad Sci U S a* 2008;**105**(21):7552–7.
55. Yang N, Higuchi O, Ohashi K, et al. Cofilin phosphorylation by LIM-kinase 1 and its role in Rac-mediated actin reorganization. *Nature* 1998;**393**(6687):809–12.
56. Arber S, Barbayannis FA, Hanser H, et al. Regulation of actin dynamics through phosphorylation of cofilin by LIM-kinase. *Nature* 1998;**393**(6687):805–9.
57. Costantino CM, Gupta A, Yewdall AW, Dale BM, Devi LA, Chen BK. Cannabinoid receptor 2-mediated attenuation of CXCR4-tropic HIV infection in primary CD4+ T cells. *PLoS One* 2012;**7**(3):e33961.
58. Verdrengh M, Jonsson IM, Holmdahl R, Tarkowski A. Genistein as an anti-inflammatory agent. *Inflamm Res* 2003;**52**(8):341–6.
59. Trieu VN, Dong Y, Zheng Y, Uckun FM. In vivo antioxidant activity of genistein in a murine model of singlet oxygen-induced cerebral stroke. *Radiat Res* 1999;**152**(5):508–16.
60. Severson RK, Nomura AM, Grove JS, Stemmermann GN. A prospective study of demographics, diet, and prostate cancer among men of Japanese ancestry in Hawaii. *Cancer Res* 1989;**49**(7):1857–60.

CHAPTER

# 12

# Opportunistic Infections in HIV Individuals and Enhanced Immunity by Glutathione*

*Alexis Alejandre, Leslie Gonzalez, Parveen Hussain, Judy Ly, Anand Muthiah, Tommy Saing, Anddre Valdivia, Vishwanath Venketaraman*

**Western University of Health Sciences, Pomona, CA, United States**

OUTLINE



## Abstract

The incidence of human immunodeficiency virus/acquired immunodeficiency syndrome (HIV/AIDS) has dramatically increased since it was first reported in 1981 and now stands as one of the most prevalent disease in the world. The consequences of HIV infection lead to a compromised immune system and increase susceptibility to opportunistic infections such as the one caused by *Mycobacterium tuberculosis*. This results in many complications that often become problematic during the course of treatment and finding supplements that can deal with both the viral and bacterial infection become scarce. Glutathione (GSH), a natural antioxidant, has been identified to have beneficial effects on individuals infected with HIV by alleviating excessive oxidative stress and enhancing the immunity. This is in part by the chemical properties of GSH as well as the direct effect it has on cells of the immune system. Overall, GSH's beneficial effects should be considered as an added supplement to the current treatment for HIV/AIDS.

**Keywords:** Antimicrobial; Antioxidant; Glutathione; HIV; Immunity; *Mycobacterium tuberculosis*; N-acetylcysteine.

* **All authors contributed equally.**



http://dx.doi.org/10.1016/B978-0-12-809853-0.00012-2

## List of Abbreviations

**GSH** Glutathione
**HIV** Human immunodeficiency virus
***M. tb*** *Mycobacterium Tuberculosis* Opportunistic infections in HIV positive individuals and immune enhancement by glutathione
**ROI** Reactive oxidative intermediates

# HUMAN IMMUNODEFICIENCY VIRUS GLOBAL INCIDENCE

According to the World Health Organization, approximately 37 million individuals worldwide are currently living with human immunodeficiency virus (HIV) and acquired immunodeficiency syndrome (AIDS).[1] HIV/AIDS was first reported in 1981 and has quickly become a global health crisis claiming more than 34 million lives thus far.[1,2] The region most affected includes sub-Saharan Africa, which accounts for almost 70% of new HIV infections globally.[1] In the United States, more than 1.2 million people are living with HIV and about 12.8% are unaware of their infection.[3] Overall, the number of people living with HIV has increased and the average number of new infections has remained fairly stable, but there are still groups especially at risk, including men who have sex with men, injection drug users, and people living in middle to low-income countries.[3] There is currently no cure or vaccine for HIV/AIDS. HIV is composed of an enveloped RNA virus containing a capsid, two single positive sense RNA strands, reverse transcriptase, integrase, and two transfer RNAs.[4] Specific glycoproteins, gp120 and gp41, surround the envelope and are primarily responsible for fusion and binding to the host cell during infection, respectively.[5] HIV targets CD4+ T lymphocytes (CD4+ T cells) and macrophages and replicates using its own reverse transcriptase that forms a DNA intermediate (Fig. 12.1). The nascent DNA intermediate then integrates itself into the host's genome, while enabling the virus to remain within the host. The virus weakens the host defense mechanism against infections by infecting CD4+ T cells and monocytes/macrophages through binding of CD4 receptors and chemokine coreceptor (CXCR4 or CCR5).[6] HIV can be transmitted through exchange of infected body fluids including blood, breast milk, semen, and vaginal secretions.[1] Depending on the individual and treatment interventions, it may take 2–15 years for AIDS to develop, which is characterized by the inability of the immune system to fight infections.[1] This chapter will discuss the effects of HIV infection on redox imbalance and immune dysfunctions along with the increase in susceptibility to opportunistic infections. Furthermore, it will discuss how glutathione (GSH) can alleviate redox imbalance and enhance host immunity opportunistic infections in HIV positive individuals and immune enhancement by GSH.

# HUMAN IMMUNODEFICIENCY VIRUS INDUCES CHANGES IN INNATE AND ADAPTIVE IMMUNITY

HIV-infected cells trigger immune responses that initially begin as an innate response and will further modulate adaptive immune responses.[7] Pathogen-associated molecular patterns specifically viral products are recognized by pathogen recognition receptors (PRRs) on the innate immune cells such as macrophages and dendritic cells (DCs), initiating a cascade of immune responses against both viral replication and activity.[7] Proinflammatory cytokines, type I and III interferon (IFN), and chemokines produced in response to HIV infection are responsible for recruitment and activation of natural killer cells (NK), macrophages, and DCs.[7] HIV infection that is recognized by PRR

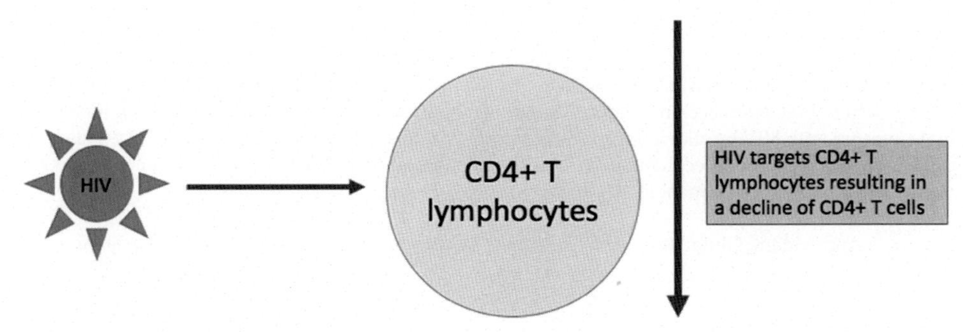


FIGURE 12.1 **Human immunodeficiency virus (HIV) targets CD4+ T lymphocytes.** Chronic HIV infection results in the decline of CD4+ T lymphocytes further compromising the immune and antioxidant response.

recruiting innate immune cells such macrophages and dendritic will lead to an activation of NK cells.[7] NK cells mediate antiviral and cytotoxic activities by releasing granzymes and perforin.[7]

CD4+ T cells play an important role in adaptive immune responses against viral and bacterial infections (Fig. 12.2). In an HIV-1 positive individual compromised CD4+ T cells makes them increasingly susceptible to the opportunistic infections such as those caused by *Mycobacterium tuberculosis* (*M. tb*), human papillomavirus, and cytomegalovirus.[8–10]

Exacerbated production of proinflammatory cytokines such as TNF-$\alpha$, IL-1, and IL-6 can lead to systemic oxidative stress resulting in apoptosis and decline of CD4+ T cells.[9,11] HIV disrupts the balance of proinflammatory cytokines versus antiinflammatory cytokines in a host, which further weakens and compromises the immune system.[7,9] TNF-$\alpha$, IL-1, and IL-6 can also promote HIV replication and transcription (Fig. 12.2). Furthermore, elevation in the levels of IL-10 during HIV infection causes immune suppression, enhancing the risks for susceptibility to opportunistic infections.[9]

Clinical stages of HIV infection can be divided into three phases.[8] The first phase appears after a couple of weeks following infection and that includes infection of large numbers of CD4+ T cells accompanied by acute symptoms. The second phase of HIV infection includes a consistent balance in the viral load and infected CD4+ T cells. Latent phase is defined as the second phase of HIV infection, which is characterized by high turnover of CD4+ T cell numbers and increase in viral load. During the last phase of HIV infection there is a large decline in the number of CD4+ T cells. The decline in the CD4+ T cell numbers during the late phase is characterized by high rates of CD4+ T cell apoptosis. Therefore, an exacerbated inflammatory response involving systemic increase in the levels of proinflammatory cytokines and reactive oxygen species (ROI) can result in a decline in the number of CD4+ T cells, which can further disrupt the overall immune responses, resulting in increased susceptibility to infections.[7,12]

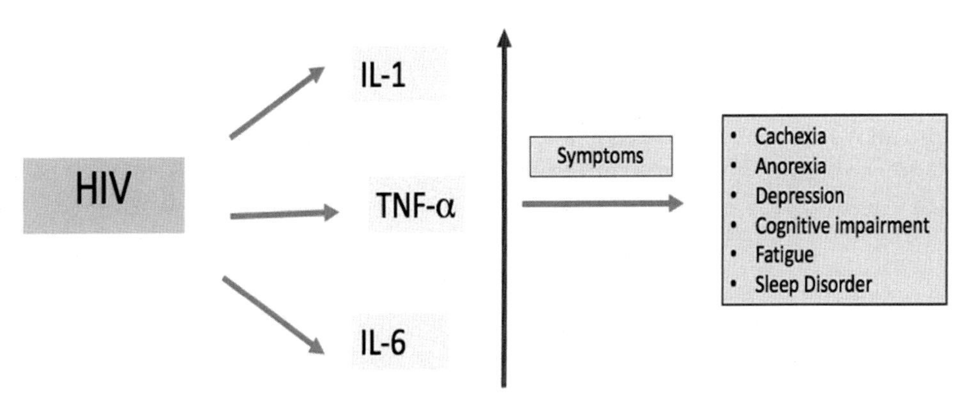


FIGURE 12.2 **Overproduction of inflammatory cytokines in human immunodeficiency virus (HIV)-infected individuals.** Chronic HIV causes an overproduction of inflammatory cytokines such as interleukins-1 (IL-1), TNF-$\alpha$, and IL-6. These result in symptoms such as cachexia, anorexia, and depression, which are unique to both chronic HIV and AIDS individuals.

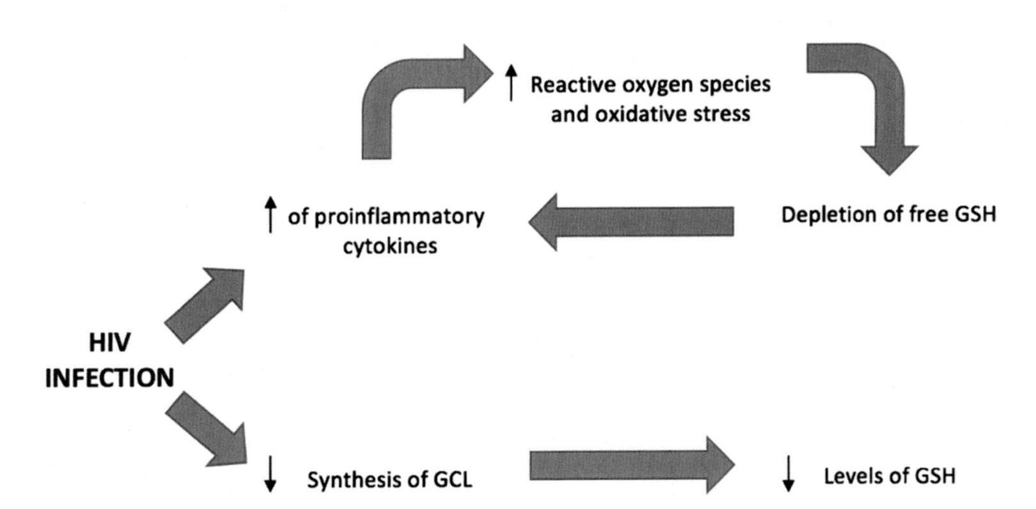


FIGURE 12.3 **Diagram depicting the various reasons for the decrease of glutathione levels in human immunodeficiency virus + individuals.**

## HUMAN IMMUNO VIRUS SUSCEPTIBILITY TO PULMONARY TUBERCULOSIS

*M. tb* is one of the leading causes of death in HIV-infected individuals.[9] According to the World Health Organization approximately a third of the world population is latently infected with *M. tb* (LTBI), with an estimated 9 million individuals with active tuberculosis (TB).[9] It is estimated that approximately 2 million individuals die each year from TB infection.[9] In particular, 14.4% of these individuals have HIV and *M. tb* coinfection.[9] Approximately 50–60% of these individuals with HIV and *M. tb* coinfection are from sub-Saharan Africa.[9,13] HIV and *M. tb* coinfection continues to burden health-care systems in developing countries in both African and Asian subcontinents.[13]

## IMMUNE RESPONSES IN HUMAN IMMUNODEFICIENCY VIRUS–*MYCOBACTERIUM TUBERCULOSIS* COINFECTED INDIVIDUALS

*M. tb* infection begins when an individual inhales infectious aerosol droplets containing *M. tb*.[14] The inhaled bacilli are phagocytized by alveolar macrophages, which are believed to provide the first line of defense against *M. tb* infection. TNF-$\alpha$ released by *M. tb*-infected macrophages is responsible for formation and maintenance of granuloma, a critical immune response required to restrict and localize *M. tb* infection in the lungs thereby preventing systemic dissemination of *M. tb* infection to other parts of the body.[14] Studies have shown that the T-helper 1 ($T_H1$) subset of CD4+ T cell immunity plays an important role in augmenting the effector functions of macrophages to combating *M. tb* infection.[15] It is believed that 90% of the healthy individuals mount effective immune responses against *M. tb* infection in the lungs, causing the bacteria to become dormant inside the granuloma and this condition is referred to as LTBI.[16]

Chronic stages of HIV infection is usually accompanied by progressive decline in the number of CD4+ T cells, which leads to disruption in the macrophage effector functions and weakened granulomatous responses against *M. tb* causing active TB.[7,11,17] In HIV–TB coinfected individuals, *M. tb* can also systemically disseminate to other parts of the body to cause extrapulmonary TB.[8]

Studies have also shown that patients with active TB have an elevated expression of the coreceptors CCR5 and CXCR4, which further promotes HIV invasion in CD4+ T cells and macrophages cell populations in the host.[18] HIV infection also result in depletion of $T_H17$ cell line, a subset of CD4+ T cells and regulatory (Treg) cells further causing imbalance and impaired immune responses against *M. tb* infection.[1,13]

Coinfection with *M. tb* is therefore primarily responsible for the increased risk of morbidity and mortality among HIV-1 individuals worldwide.[1] The risks for reactivation of LTBI in an immunocompetent individual's lifetime is between 5% and 10%, however, for HIV-1 individuals, the risk of reactivation is 5%–15% annually.[1]

## HUMAN IMMUNO VIRUS SUSCEPTIBILITY TO EXTRAPULMONARY TUBERCULOSIS

HIV-*M. tb* coinfected individuals are also at high risk for developing extrapulmonary TB.[14,19] Extrapulmonary TB is characterized by dissemination of *M. tb* that occurs from the lungs to other parts of the body such as in vascular regions like the skeletal system, lymph nodes, gastrointestinal (GI) organs, heart, and brain. Compromised innate and adaptive immune responses in the lungs can facilitate dissemination of *M. tb* from its primary site of infection to other parts of the body via bloodstream or lymphatic system.[1,14,15,19–21]

Miliary TB, a fatal and lethal form of the disease common in patients with AIDS, is caused by widespread dissemination and invasion of *M. tb* in vital organs throughout the entire body, including but not limited to the lungs, liver, spleen, and brain. Miliary TB is characterized by large amounts of TB bacilli with tiny tubercles that are millet seed in size on gross examination.[15]

Another dangerous form of extrapulmonary TB is spinal TB also known as Pott's disease. The rich vasculature and abundant nutrients found in the spine allows mycobacterial survival, which results in the disease. Immunosuppression during advanced stages of AIDS can cause primary *M. tb* infection in the lungs to disseminate to the bones, leading to osteomyelitis and arthritis, the third common cause of extrapulmonary TB.[22,23] The disease is life-threatening as it can cause spinal compression, deformity, neurological impairment, and paraplegia.[22,24] Although the incidence and prevalence of spinal TB is not exactly known worldwide, 10% of extrapulmonary TB cases involve skeletal TB and of these skeletal TB cases, 50% are accounted, with the highest number of cases occurring in Africa.[25]

Peripheral lymphadenitis, which affects mainly the cervical lymph node, is the most common form of extrapulmonary TB infection seen in HIV-1 patients in South Africa and India,[19,26] and it is also common in Asian and Pacific

Islanders women and children.[27] About 20–40% of extrapulmonary TB cases occur in the lymph nodes. The symptoms include painless swelling and multiple and matted forms of lymph nodes. The pathology includes irregular shapes and sizes of the lymph nodes[19] and distinct histomorphological patterns are observed.[28]

GI TB is the sixth common occurrence among extrapulmonary TB.[15] About 15–20% of patients with active pulmonary TB experience GI TB[23] and the risks are greater in HIV and AIDS patients due to low levels of CD4+ T cells.[29] GI TB, which includes areas such as the abdomen and the peritoneum and organs such as the liver, spleen and pancreas, commonly occurs in the ileocecal area, ileum, and colon.[29] *M. tb* is able to travel into the GI tract from infected lymph nodes, hematogenous spread, primary lung TB, or ingestion of infected sputum.[15] Some of the clinical manifestations of GI TB include weight loss, dysphagia, abdominal pain, and diarrhea.[30] In a study that examines hepatic TB, of the 164 patients that were analyzed, 17.4% of HIV-1 patients were infected with hepatic TB when compared to 4.3% non-HIV patients.[29] Splenomegaly and pancytopenia are other diseases contributed by *M. tb*.[31]

In Africa and other endemic regions where *M. tb* is prevalent, TB pericarditis is very common and there are 80,000–160,000 new cases reported annually. In one of the largest study conducted in South Africa, more than 50% of the individuals infected with TB pericarditis were individuals diagnosed with HIV-1.[32] *M. tb* can travel to the heart in three different ways; by the mediastinal, peritracheal, and peribronchial lymph nodes via hematogenous spread compared to immunocompetent individuals where the mycobacteria travel via the lymph nodes.[32] Depending on the four different stages of TB pericarditis, symptoms range from chest pain, pericardial effusion to cardiac compression.[32] Diagnostic and treatment in HIV-infected individuals can be challenging due to immunological suppression.

Although TB meningitis (TBM) is a rare occurrence it manifests by the inflammation of the meninges. This severe form of extrapulmonary TB has a high rate of morbidity and mortality, especially in sub-Saharan Africa and Asia[33] as well as in endemic TB regions because of the delayed and difficult diagnosis and treatment of it. HIV and AIDS individuals who are immunocompromised are at greater risk of developing TBM.

Extrapulmonary TB diagnosis and treatment can be challenging for clinicians. The risk of extrapulmonary TB in HIV/AIDS individuals is greater than in immunocompetent individuals. Proactive measures and clinical follow-ups are vital to decrease the morbidity and mortality, especially in regions where TB is endemic.

## GLUTATHIONE TRANSPORT AND THE SYNTHESIS PATHWAY

Composed of glutamate, glycine and cysteine, GSH is a natural antioxidant that serves as a key regulator of redox balance and maintenance of cellular homeostasis.[34,35] GSH is naturally found in the reduced (rGSH) and oxidized glutathione (GSSG) forms, with the reduced form comprising the majority of total GSH.[36–38] The reduced form is the active form that provides GSH with its antioxidant properties and the deficiency has been linked to have major implications in many diseases including HIV and bacterial infections.[36–40] Maintaining adequate levels of GSH then becomes crucial, and understanding the causes for GSH deficiency in various disease conditions can shed light on how to effectively restore redox homeostasis to achieve beneficial effects.

Intracellular transport of GSH requires breakdown of GSH into its individual components before cellular uptake[41] (Fig. 12.4). Although, natural transport requires the breakdown of GSH, a synthetic version of rGSH that is encapsulated within a liposome (LGSH) can be transported as a complete molecule and provide GSH to the cell in its complete form,[42] (Fig. 12.4). This ability to be transported as a whole molecule comes from the encapsulating liposome, which allows phagocytosis of the entire liposome and its contents.[43]

The major components for the synthesis of GSH include its three core amino acids, which can be obtained through dietary sources. However, cysteine becomes the rate limiting step within the synthesis of GSH due to the lack of abundance of this amino acid in a typical diet.[34,35] Cysteine is a crucial component since it provides GSH with a thiol group that acts as the source of the antioxidant properties of GSH. This can be overcome by supplementing with N-acetylcysteine (NAC), which after undergoing hydrolysis provides cysteine for de novo synthesis. Synthesis occurs in two steps in which glutamate and cysteine are covalently linked by glutamate-cysteine ligase (GCL) to form $\gamma$-glutamylcysteine, which is followed by the addition of glycine by glutathione synthase (GSS),[44–46] (Fig. 12.5).

After the synthesis of GSH, the reduced form (rGSH) is available to undergo oxidation as a mechanism for neutralizing reactive oxygen intermediates (ROI). Two molecules of rGSH are required for the production of GSSG via the reactive thiol group. Following the production of GSSG, the reactive thiol group becomes inactive and is unavailable for further use, however, the cell is able to replenish rGSH via glutathione reductase (GSR) (Fig. 12.6). This balanced system is the principal mechanism for eliminating ROI from the body and thus can be overwhelmed if ROI are produced in excess or the availability of rGSH becomes compromised.

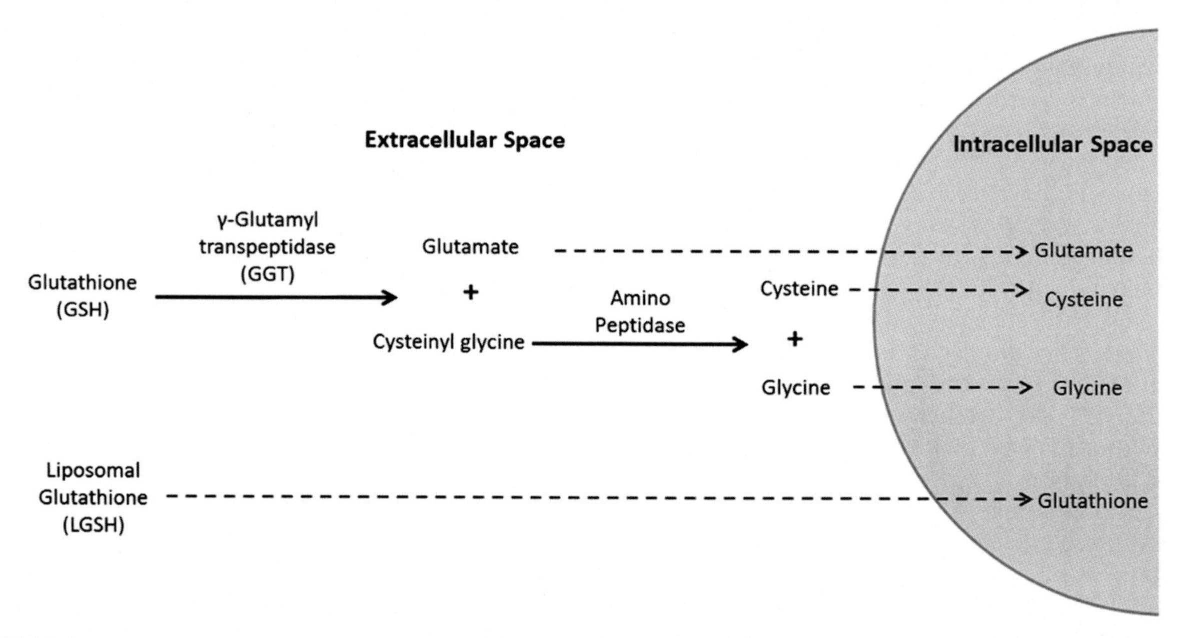


FIGURE 12.4 **Extracellular transport of glutathione (GSH) and lipsome (GSH). GSH needs to be broken down into its individual components in order for the cell to intake the building blocks.** However, LGSH can be transported as a whole molecule due to its encapsulation in the liposome.

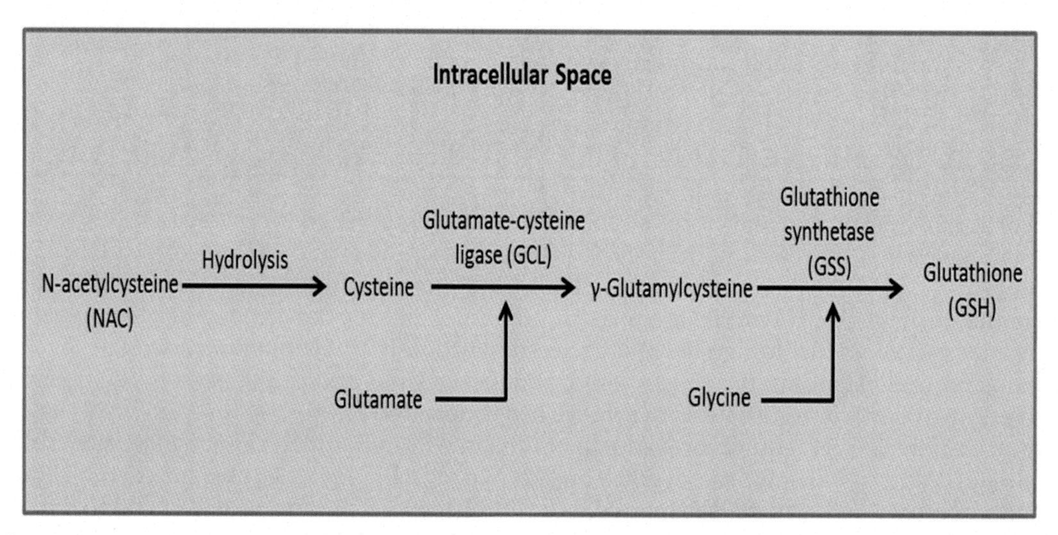


FIGURE 12.5 **Metabolism of N-acetylcysteine and biosynthesis of glutathione (GSH).** Cysteine is the rate limiting step in the synthesis of GSH and the amino acid containing the thiol group that provides GSH with its antioxidant properties.

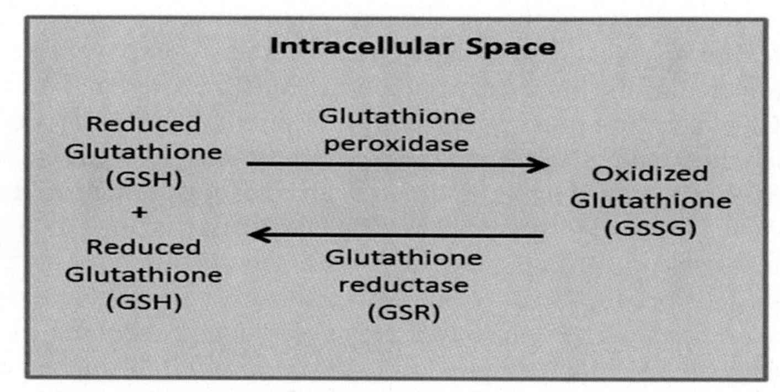


FIGURE 12.6 **Oxidation and reduction of glutathione (GSH)** The redox balance is maintained within the cell and favors the rGSH. The antioxidant property of GSH requires two molecules of GSH to neutralize ROI, resulting in the formation of GSSG.

## GLUTATHIONE AND ITS ANTIMYCOBACTERIAL PROPERTIES

GSH has been shown to have antimycobacterial properties and theories have explored the antioxidant effects it has in redox homeostasis as well as the nature of its chemical structure. GSH as an antioxidant serves to neutralize reactive intermediates and maintains redox balance within the cell. The maintenance of this balance becomes crucial when analyzing its effects on the growth and survival of *M. tb* since macrophages utilize the production of ROIs and reactive nitrogen intermediates to successfully eliminate intracellular bacterial infections. *M. tb* is a successful intracellular pathogen that is able to evade the intracellular effector mechanisms by interrupting the fusion of the lysosome with the phagosome.[47] Via this interruption the microbe is able to survive inside the host cell (macrophage) and remain persistent for several decades.[48]

Studies have focused on this pathway and the role GSH has in the control of mycobacterial growth within macrophages. Early studies have determined that during activation of macrophages with IFN-γ and lipopolysaccharides (LPS), there is an increase transport of L-arginine.[49] L-arginine is an important amino acid since it is a source of nitrogen that is utilized by the macrophage to produce nitric oxide (NO) via the induced form of inducible nitric oxide synthase (iNOS). NO is used by macrophages to eliminate bacterial infections, however, NO being a short-lived gas rapidly gets detoxified to nitrate and nitrite, both of which lack antimicrobial properties. Intracellular GSH can form a complex with NO known as nitrosoglutathione (GSNO), a stable form of NO and is utilized by the macrophage as a carrier molecule to deliver NO as an antimicrobial agent. Given the natural instability of this complex, experiments were designed to test the full effects GSH and GSNO have on the growth of mycobacteria. For example, studies demonstrated that under normal conditions, activation of macrophages via IFN-γ and LPS was able to successfully reduce the survival of intracellular mycobacteria.[50] However, if the macrophage was infected with a mycobacterium and is unable to transport dipeptides such as the ones derived from GSH and GSNO then this resulted in survival and proliferation of the microbe. These results indicate that the inability to transport dipeptides provides the microbe with resistance against the direct effects of GSH and GSNO, which were later explored and found that direct exposure to GSH and GSNO had a bacteriostatic and bactericidal effect, respectively, on mycobacteria.[51]

The same studies not only tested the direct effects of GSH and GSNO but also the effects of intracellular levels of GSH and GSNO on mycobacterial growth.[51] The results demonstrated that although IFN-γ and LPS treatment can activate other antimicrobial mechanisms on macrophages, NO did not play a major role in the control of intracellular growth of mycobacteria but rather it was primarily controlled by GSH and GSNO.[51] Similarly, other studies focused on the direct effects GSH and GSNO have on bacterial redox homeostasis.[36] Contrasting mycobacteria to eukaryotic cells, the maintenance of redox balance is primarily mediated by mycothiol.[52] These studies demonstrated that interrupting redox balance via exposure to GSH and GSNO led to growth inhibition and decreased survival of mycobacteria respectively. In addition, macrophages that were infected with a mycobacterial mutant that is unable to express γ-glutamyl transpeptidase exhibited hypersurvival within the macrophage. The results demonstrated that the mutant was able to successfully avoid elimination by the macrophages because of their inability to process GSNO. Furthermore, GSH is thought to be an evolutionary precursor to penicillium and cephalosporium and mycobacteria might have a component in their cell wall that gives them sensitivity to GSH.[53]

To summarize, GSH and GSNO both have been demonstrated to have antimicrobial effects. GSH has static effects on the growth of *M. tb* by causing reductive stress whereas GSNO has cidal effects on *M. tb* by releasing NO. These findings demonstrate that GSH can potentially be used as an adjunctive therapy or mycobacterial infections.

## GLUTATHIONE AND INNATE/ADAPTIVE IMMUNITY

The effects of GSH are not only restricted to its antimycobacterial properties but it has also been observed to have enhancement effects in both innate and adaptive immunity. During infection the innate immune response is the first line of defense against pathogens, which is later tailored and specialized to deal with pathogens in a more specific manner via adaptive immunity. The interaction of the various cells of the immune system is an intricate process involving complex set of signals and responses that ultimately lead to the control of infection, however, there are certain pathogens that are able to evade this defense and maintain chronic infection in the host. Two pathogens that effectively evade host defense are *M. tb*, infecting cells responsible for innate immunity, and HIV, which infects cells responsible for adaptive immunity.

*M. tb* primarily infects macrophages and is able to evade digestion by the macrophage, resulting in intracellular survival of the pathogen. NK cells contribute to innate immune responses against intracellular *M. tb* infection by mounting various effector mechanisms.[54] It has been demonstrated that the cytolytic function of NK cells can be consistently enhanced by treatment with NAC and stimulatory cytokines (IL-2 and IL-12), which correlated with an improved control of *M. tb* growth. In addition, NK cells were also found to have an increase in the production of proinflammatory cytokines (IL-1β, TNF-α, and IFN-γ) during infection and were regulated by the presence of NAC and stimulatory cytokines. Overall these studies demonstrate that GSH modulates NK cell activity to inhibit the intracellular growth of *M. tb*.

Subsequent studies showed that individuals infected with HIV have reduced levels of GSH in NK cells, which correlated with a poor ability to control intracellular *M. tb* infection.[55] The activity of NK cells is primarily modulated by the activation of cell surface receptors (NKp30, NKp44, NKp46, and NKG2D) and cytotoxic ligands (FasL and CD40L), which were all found to be regulated by intracellular levels of GSH.[54,55] These results suggest that low levels of GSH in HIV+ individuals not only increased susceptibility to opportunistic infections such as *M. tb* but also compromise the activity of NK cells, which form part of the innate immune system to control infections.

Studies have consistently demonstrated that individuals infected with HIV have a depletion of GSH levels and an increase in the production of free radicals.[37,55] Diminished levels of GSH in individuals with HIV infection is due to the decrease in the production of the catalytic subunit of the enzyme glutamate cysteine ligase (GCLC) and the depletion of cysteine due to the rapid incorporation of cysteine during viral replication[37,56,57] (Fig. 12.3). The reduction of this catalytic subunit was found to be primarily reduced in macrophages, which are an important component of the innate immune system.[37,40] These findings demonstrate that macrophages have a compromised ability to maintain homeostatic levels of GSH. In addition, it was noted that when macrophages from HIV+ individuals were infected with *M. tb*, there was a notable decrease in the levels of rGSH, which correlated with an increase in malondialdehyde (MDA) levels, an indicator of increased oxidative damage. Interestingly, rGSH and MDA levels were restored to baseline when macrophages were treated with LGSH. Overall, infection of HIV leads to compromised macrophage function, a depletion of rGSH, and an increase in MDA levels which all were to be improved by the sole supplementation of GSH. These findings further support the immune enhancement effects of GSH.

Just as GSH was observed to enhance innate immunity, other studies demonstrated that GSH also enhances adaptive immunity.[38,39,56] GSH was observed to enhance the function of DCs and T cells when infected with *M. tb*.[38] In addition, when infected with *M. tb*, DCs produced increased levels of IL-10, an immunosuppressive cytokine that can polarize CD4+ T cells to differentiation to $T_H2$ subset.[38] The levels of IL-10 were decreased by treatment with GSH. At the same time, levels of IL-12, a $T_H1$ polarizing cytokine were noted to increase with GSH treatment. After infection and treatment with GSH, DCs were also observed to have an increase in expression of CD80, CD86, and HLA-DR, which were correlated with an enhanced ability to control *M. tb* growth. Similarly, GSH supplementation was observed to improve T cell function by the downregulation of transforming growth factor-β (TGF-β) and IL-10.[38] Proliferation of T cells was also found to be improved by DCs that were treated with GSH.[38]

Enhancing the levels of GSH in T cells has been shown to improve the intracellular control of *M. tb* infection inside monocytes and this correlated with increased production of IFN-γ.[39] Importantly, the levels of GSH are compromised in T cells isolated from individuals with HIV infection.[39]

Considering these studies, we begin to understand the immune enhancement effects of GSH. These effects are observed in cells of the innate immune system (NK cells and macrophages) along with responses from the adaptive immune system (DCs and T cells) (Fig. 12.7). Together, both aspects of the immune system collaborate to control infections and GSH begins to emerge as a molecule that can shift the immune response to control *M. tb* infection in HIV+ individuals. Opportunistic infections in HIV positive individuals and immune enhancement by GSH.

## REDOX IMBALANCE AND ITS SIGNIFICANCE IN INFECTIONS OF THE CENTRAL NERVOUS SYSTEM

The beneficial effects observed in GSH are not only confined to enhancement of the immune system, but they are beginning to emerge as a possible intervention to treat cases of meningitis in HIV+ individuals. Due to the increase in susceptibility to opportunistic infections in HIV+ individuals they are at a higher risk of acquiring infections of the central nervous system (CNS). Together with a decline of CD4+ T cell numbers outside the CNS and in conjunction with widespread infection of macrophages, pathogens infiltrate the CNS and effectively infect microglia.[25] Once infected, the immune cells in the CNS are depleted of GSH due to the increase in oxidative stress.[26] Similarly, recent studies have demonstrated that just as levels of GSH are depleted in cells of the immune system, a similar decrease is observed in brain tissue samples of HIV+ individuals.[58] This decrease was also correlated with a decrease in the production of GCLC, GSS, and GSR, which are essential for the synthesis of GSH.[58] The same studies further demonstrated that MDA levels were elevated in brain tissue samples of HIV+ individuals. With elevated levels of oxidative

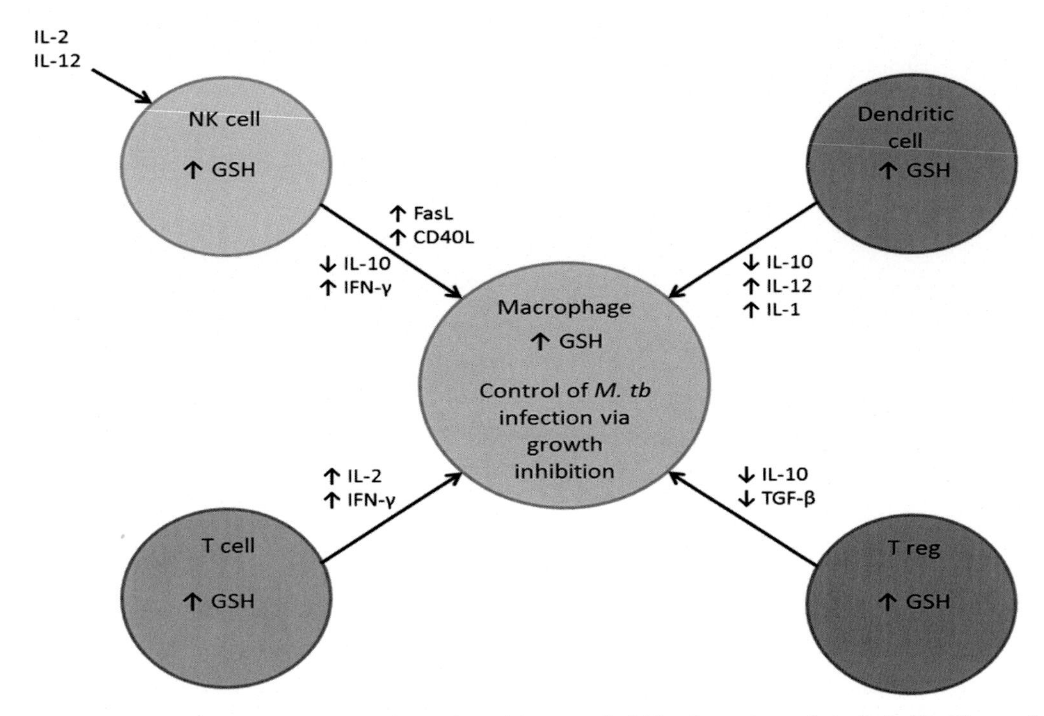


FIGURE 12.7 **Model describing the effects of glutathione (GSH) in control of *Mycobacterium tuberculosis* (*M. tb*) growth in individuals infected with HIV.** The increase of GSH in T cells, T regs, natural killer (NK) cells, dendritic cells, and macrophages lead to a network of interactions via cytokines and cell surface receptors that work together to control the growth of *M. tb*. The production of IL-2 and IL-12 from other cells can be seen how it activates NK cells and lead to its interaction with macrophages.

stress and with reduction of enzymes involved in the synthesis and restoration of GSH, redox balance comes up as part of the pathophysiology of this disease and an important target for therapeutic intervention. In addition, the antimycobacterial properties of GSH can be beneficial to the treatment of infections of the CNS.

## SUPPLEMENTING LGSH IN INDIVIDUALS INFECTED WITH HIV

In a study by Ly et al., 2015, the effects of LGSH on restoring cytokine response was tested in healthy and HIV-infected individuals in a double blind clinical trial.[59] Previous findings show that those infected with HIV had lower levels of GSH.[17,39] Following this previous report, the clinical trial focused on the effects of LGSH supplementation in replenishing the levels of GSH and inducing cytokine balance. The study participants included 10 healthy individuals and 15 HIV positive patients. On the first visit with blood draw, patients were given either an LGSH supplement or a placebo supplement. The participants were instructed to take one and half teaspoon of formula, twice a day, once in the morning and evening. This amounts to about 1260 mg of rGSH. Seven weeks worth of supplement was given to the patients. During their second visit, participants received the second half of the supplement and instructed to continue taking the supplement until the completion of 13 weeks. Patients then returned for their third and final visit, where the second blood draw occurred.

Using the proper techniques, the blood samples were processed and components were isolated. The plasma component was used to measure cytokine levels. The measurement of cytokine levels in both healthy and HIV groups were measured and recorded. This is referred to as the baseline measurement. As seen in Table 12.1, baseline levels of cytokines such as TGF-β, IL-6, and IL-10 were significantly increased in HIV individuals compared to healthy. Baseline levels of cytokines IL-2, IFN-γ, IL-1β, IL-17F, and TNF-α were significantly decreased in HIV individuals compared to healthy. To reaffirm previous findings, this study also measured baseline GSH levels and found that HIV patients had a 12-fold decrease of GSH compared to healthy individuals. This study also examined ROI levels in terms of MDA and CellROX staining of CD4, CD8, and CD14 T cells. The baseline ROI levels are seen in Table 12.2. In all baseline categories, HIV patients have significantly increased levels of ROI compared to healthy individuals. After 13 weeks of taking either an LGSH supplement or a placebo, the participant's blood samples were analyzed

TABLE 12.1 Summary of Cytokine Production on Individuals Infected With Human Immunodeficiency Virus

| Cytokines | Baseline HIV Levels |
|---|---|
| TGF-β | 2-fold ↑ increase |
| GSH | 12-fold ↓ decrease |
| IL-6 | 16-fold ↑ increase |
| IL-2 | 2-fold ↓ decrease |
| IFN-γ | 3-fold ↓ decrease |
| IL-1β | 12-fold ↓ decrease |
| IL-17F | 3-fold ↓ decrease |
| TNF-α | 11-fold ↓ decrease |
| IL-10 | 6-fold ↑ increase |

Measurements are compared to a healthy group not infected with HIV. *GSH*, glutathione; *IFN*, interferon; *IL*, interleukins; *TGF*, transforming growth factor.

TABLE 12.2 Summary of Free Radical Production on Individuals Infected With Human Immunodeficiency Virus

| Free Radicals | Baseline HIV Levels |
|---|---|
| MDA | 9-fold ↑ increase |
| CD14 | 5-fold ↑ increase |
| CD4 | 2-fold ↑ increase |
| CD8 | 8-fold ↑ increase |

Measurements are compared to a healthy group not infected with HIV. *MDA*, malondialdehyde.

TABLE 12.3 Summary of Cytokine Production on Individuals Infected With Human Immunodeficiency Virus After 3 Months of Taking Liposomal Glutathione

| Cytokines | 13 week HIV Levels |
|---|---|
| TGF-β | 3-fold ↓ decrease |
| GSH | 2-fold ↑ increase |
| IL-6 | 2-fold ↓ decrease |
| IL-2 | 3-fold ↑ increase |
| IFN-γ | 2-fold ↑ increase |
| IL-1β | 10-fold ↑ increase |
| IL-17F | No change |
| TNF-α | 2.5-fold ↑ increase |
| IL-10 | 6-fold ↓ decrease |

Measurements are compared to baseline values of infected individuals with HIV.

for the same components of cytokines and ROI. None of the placebo show a significant difference compared to the baseline measurement. However, for HIV patients that received the LGSH supplement, measurements were compared to their respective baseline levels. In cytokines TGF-β, IL-6, and IL-10, there was a significant decrease when comparing the 13-week HIV with supplement with their baseline. In cytokines IL-2, IFN-γ, IL-1β, and TNF-α there was a significant increase in HIV patients after 13 weeks of LGSH supplementation compared to their respective baseline measurements. These levels are also shown in Table 12.3. All measurements relating to ROI also significantly

TABLE 12.4 Summary of Free Radical Production on Individuals Infected With Human Immunodeficiency Virus After 3 Months of Taking Liposomal Glutathione

| Free Radicals | 13 week HIV Levels |
|---|---|
| MDA | 2-fold ↓ decrease |
| CD14 | 6-fold ↓ decrease |
| CD4 | 4-fold ↓ decrease |
| CD8 | 8-fold ↓ decrease |

Measurements are compared to baseline values of infected individuals with HIV. *MDA*, malondialdehyde.

decreased in the HIV with 13-week LGSH supplementation compared to their baseline levels. More specific details can be seen in Table 12.4.

Furthermore, to show the importance of these cytokine levels in opportunistic infections within immunocompromised individuals such as HIV patients, the blood samples were infected in vitro with H37Rv, a laboratory strain of *M. tb*. Both baseline and 13-week blood samples were infected with H37Rv. Comparing the baseline HIV intracellular growth to the 13-week post-LGSH supplementation intracellular growth, there was a significant decrease of survival of H37Rv. In comparison to the placebo group, the LGSH group showed a twofold decrease of intracellular survival of H37Rv.

Overall, this study showed that deficiency in GSH within those with HIV infection leads to an impairment in the $T_H1$ cytokine production. As for other cytokines, increased levels of IL-6 limits T cell growth and induce oxidative stress and systemic inflammation. With both decreased levels of $T_H1$ cytokines and increased levels of TGF-β and IL-6, there will be an increased susceptibility to opportunistic infections such as *M. tb*. With proper LGSH supplementation, immune response imbalance can be restored, therefore aiding HIV patients in managing opportunistic infections.

## CONCLUSION

In conclusion, chronic HIV infection can lead to compromised levels of CD4+ T cells leading to a dysfunctional immune response.[60] Overproduction of proinflammatory cytokines can contribute to the decline of CD4+ cells, including innate immune effector cells such as NK, macrophages, and DCs.[7] The decline of CD4+ T cells outside of the CNS can also increase oxidative stress in microglia and brain parenchyma.[60] The weakened innate and adaptive immune response will further increase host susceptibility to opportunistic infections such as pulmonary and extrapulmonary TB infection.[8] Chronic HIV infection results in the disruption of antioxidant balance and increases oxidative stress in the immune system.[60] GSH is a tripeptide antioxidant with antimicrobial properties, reduces oxidative stress, and protects cells from free radical damage.[12] In HIV individuals, there is a decline in rGSH and GSH synthase enzyme levels, which also disrupts antioxidant homeostatic balance.[12] GSH supplemented with antiviral treatment, works to replenish rGSH levels in HIV individuals and can decrease the risk of opportunistic infections.[7] GSH supplementation continues to show promising effects in slowing down the progression of HIV infection while reducing the risk of opportunistic infections for HIV individuals.[12]

## SUMMARY POINTS

- HIV is an enveloped positive RNA Lentivirus that targets both CD4+ T cells and macrophages through the binding of coreceptors and glycoproteins on the surface of these cells.
- Overproduction of innate immune responses targeting HIV-infected cells leads to the decline in CD4+ T cells furthermore disrupting the adaptive immune response.
- CD4+ T cell apoptosis and HIV replication is increased by the overproduction of proinflammatory cytokines and increased ROI which both activate NF-kB transcription factor.
- HIV and TB coinfection results in the disruption of granuloma formation in the lungs and accelerating both HIV and TB coinfection.

- Extrapulmonary TB results from systemic dissemination of TB outside the lungs and can invade vascularized regions of the body, such as skeletal systemic, lymph codes, GI organs, heart, and brain.
- CD4+ T cell decline can result in oxidative stress outside the CNS, however, can still affect CNS cells such as microglia and brain parenchyma.
- GSH is an antioxidant with antimicrobial properties that maintains homeostatic redox balance and enhances both innate and adaptive immunity.
- Chronic HIV infection can disrupt antioxidant homeostatic balance, such as GSH redox cycle responsible for minimizing oxidative stress in cells.
- Levels of rGSH in HIV individuals are reduced and GSS enzymes further resulting in an increase in oxidative stress, causing oxidative damage to cells.
- Supplemented liposomal rGSH with HIV antiviral treatment replenishes rGSH and improves immune response against opportunistic infections such as TB.

## References


1. HIV/AIDS. (n.d.). Retrieved from: http://www.who.int/mediacentre/factsheets/fs360/en/.
2. From the Centers for Disease Control, Prevention. Pneumocystis pneumonia–Los Angeles, 1981. *JAMA* 1996;**276**(13):1020–2. http://dx.doi.org/10.1001/jama.276.13.1020.
3. U.S. Statistics. (n.d.). Retrieved from: https://www.aids.gov/hiv-aids-basics/hiv-aids-101/statistics/index.html.
4. Turner BG, Summers MF. Structural biology of HIV. *J Mol Biol* 1999;**285**(1):1–32.
5. Engelman A, Cherepanov P. The structural biology of HIV-1: mechanistic and therapeutic insights. *Nat Rev Microbiol Nat Rev Micro* 2012;**10**(4):279–90.
6. Moore JP, Trkola A, Dragic T. Co-receptors for HIV-1 entry. *Curr Opin Immunol* 1997;**9**(4):551–62.
7. Altfeld M, Gale Jr M. Innate immunity against HIV-1 infection. *Nat Immunol* 2015;**16**(6):554–62.
8. Ahmed A, Rakshit S, Vyakarnam A. HIV-TB co-infection: mechanisms that drive reactivation of *Mycobacterium tuberculosis* in HIV infection. *Oral Dis* 2016;**22**(Suppl. 1):53–60.
9. Breen EC. Pro- and anti-inflammatory cytokines in human immunodeficiency virus infection and acquired immunodeficiency syndrome. *Pharmacol Ther* 2002;**95**(3):295–304.
10. Luetkemeyer AF. Current issues in the diagnosis and management of tuberculosis and HIV coinfection in the United States. *Top HIV Med* 2010;**18**(4):143–8.
11. Baxter AE, Kaufmann DE. Tumor-necrosis factor is a master of T cell exhaustion. *Nat Immunol* 2016;**17**(5):476–8.
12. Staal FJ, Roederer M, Israelski DM, Bubp J, Mole LA, McShane D, et al. Intracellular glutathione levels in T cell subsets decrease in HIV-infected individuals. *AIDS Res Hum Retroviruses* 1992;**8**(2):305–11.
13. Juffermans NP, Speelman P, Verbon A, Veenstra J, Jie C, van Deventer SJ, van Der Poll T. Patients with active tuberculosis have increased expression of HIV co-receptors CXCR4 and CCR5 on CD4(+) T cells. *Clin Infect Dis* 2001;**32**(4):650–2.
14. Allen M, Bailey C, Cahatol I, Dodge L, Yim J, Kassissa C, Venketaraman V. Mechanisms of control of *Mycobacterium tuberculosis* by NK cells: role of glutathione. *Front Immunol* 2015;**6**:508.
15. Sharma SK, Mohan A, Sharma A. Challenges in the diagnosis & treatment of miliary tuberculosis. *Indian J Med Res* 2012;**135**(5):703–30.
16. Ahmad S. Pathogenesis, immunology, and diagnosis of latent *Mycobacterium tuberculosis* infection. *Clin Dev Immunol* 2011;**2011**:17. Article ID 814943.
17. Morris D, Guerra C, Donohue C, Oh H, Khurasany M, Venketaraman V. Unveiling the mechanisms for decreased glutathione in individuals with HIV infection. *Clin Dev Immunol* 2012;**2012**:734125.
18. Bruchfeld J, Correia-Neves M, Kallenius G. Tuberculosis and HIV coinfection. *Cold Spring Harb Perspect Med* 2015;**5**(7):a017871.
19. Swaminathan S, Narendran G. Extra-pulmonary tuberculosis. *Med Update* 2005:651–6.
20. Pawlowski A, Jansson M, Sköld M, Rottenberg ME, Källenius G. Tuberculosis and HIV Co-Infection. *PLoS Pathog* 2012;**8**(2):e1002464.
21. Wu JQ, Dwyer DE, Dyer WB, Yang YH, Wang B, Saksena NK. Genome-wide analysis of primary CD4+ and CD8+ T cell transcriptomes shows evidence for a network of enriched pathways associated with HIV disease. *Retrovirology* 2011;**8**:18.
22. Pigrau-Serrallach C, Rodriguez-Pardo D. Bone and joint tuberculosis. *Eur Spine J* 2013;**22**(Suppl. 4):556–66.
23. Marschall J, Evison JM, Droz S, Studer UC, Zimmerli S. Disseminated tuberculosis following total knee arthroplasty in an HIV patient. *Infection* 2008;**36**(3):274–8.
24. Turgut M. Spinal tuberculosis (Pott's disease): its clinical presentation, surgical management, and outcome. A survey study on 694 patients. *Neurosurg Rev* 2001;**24**(1):8–13.
25. Garg RK, Somvanshi DS. Spinal tuberculosis: a review. *J Spinal Cord Med* 2011;**34**(5):440–54.
26. Reddy DL, Venter WDF, Pather S. Patterns of lymph node pathology; fine needle aspiration biopsy as an evaluation tool for lymphadenopathy: a retrospective descriptive study conducted at the largest hospital in Africa. *PLoS One* 2015;**10**(6):e0130148.
27. Center for Disease, Control (CDC). *Tuberculosis*. [Fact Sheet] 2014http://www.cdc.gov/tb/publications/factsheets/statistics/tbtrends.htm.
28. Mahe E, Ross C, Sur M. Lymphoproliferative lesions in the setting of HIV infection: a five-year retrospective case series and review. *Pathol Res Int* 2011;**2011**:618760.
29. Hickey AJ, Gounder L, Moosa M-YS, Drain PK. A systematic review of hepatic tuberculosis with considerations in human immunodeficiency virus co-infection. *BMC Infect Dis* 2015;**15**:209.
30. Bhaijee F, Subramony C, Tang S-J, Pepper DJ. Human immunodeficiency virus-associated gastrointestinal disease: common endoscopic biopsy diagnoses. *Pathol Res Int* 2011;**2011**:247923.

31. Chandni R, Rajan G, Udayabhaskaran V. Extra pulmonary tuberculosis presenting as fever with massive splenomegaly and pancytopenia. *IDCases* 2016;**4**:20–2.
32. Ntsekhe M, Mayosi BM. Tuberculous pericarditis with and without HIV. *Heart Fail Rev* 2013;**18**(3):367–73.
33. Thinyane KH, Motsemme KM, Cooper VJL. Clinical presentation, aetiology, and outcomes of meningitis in a setting of high HIV and tb prevalence. *J Trop Med* 2015;**2015**:423161.
34. Wu G, Fang YZ, Yang S. Glutathione metabolism and its implications for health. *J Nutr* 2004;**134**:489–92.
35. Foreman HJ, Zhang H, Rinna A. Glutathione: overview of its protective roles, measurement, and biosynthesis. *Mol Aspects Med* 2009;**30**:1–12.
36. Dayaram YK, Talaue MT, Connell ND, Venketaraman V. Characterization of a glutathione metabolic mutant of *Mycobacterium tuberculosis* and its resistance to glutathione and nitrosoglutathione. *J Bacteriol* 2006;**188**(4):1364–72.
37. Morris D, Guerra C, Khurasany M, Guilford F, Saviola B, Huang Y, Venketaraman V. Glutathione supplementation improves macrophage functions in HIV. *J Interferon Cytokine Res* 2013;**33**(5):270–9.
38. Morris D, Gonzalez B, Khurasany M, Kassissa C, Luong J, Kasko S, Venketaraman V. Characterization of dendritic cell and regulatory T cell functions against *Mycobacterium tuberculosis* infection. *Biomed Res Int* 2013;**2013**:402827.
39. Guerra C, Morris D, Sipin A, Kung S, Franklin M, Gray D, Venketaraman V. Glutathione and adaptive immune responses against *Mycobacterium tuberculosis* infection in healthy and HIV infected individuals. *PLoS One* 2011;**6**(12):e28378.
40. Morris D, Khurasany M, Nguyen T, Kim J, Guilford F, Mehta R, Gray D, Saviola B, Venketaraman V. Glutathione and infection. *BBA Gen Subjects* 2013;**1830**:3329–49.
41. Lushchak V. Glutathione homeostasis and functions: potential targets for medical interventions. *J Amino Acids* 2012;**2012**:26. Article ID 736837.
42. Suntres Z. Liposomal antioxidants for protection against oxidant-induced Damage. *J Toxicol* 2011;**2011**:16. Article ID 152474.
43. Bozzuto G, Molinari A. Liposomes as nanomedical devices. *Int J Nanomed* 2015;**10**:975–99.
44. Anderson ME, Meister A. Transport and direct utilization of Gamma-glutamylcyst(e)ine for glutathione synthesis. *Proc Natl Acad Sci* 1983;**80**:707–11.
45. Meister A, Anderson ME. Glutathione. *Annu Rev Biochem* 1983;**52**:711–60.
46. Meister A. Selective modification of glutathione metabolism. *Science* 1983;**220**(4596):472–7.
47. Goren MB, ArcyHart PD, Young MR, Arm-strong JA. Prevention of phagosome lysosome fusion in cultured macrophages by sulfatides of *Mycobacterium tuberculosis*. *Proc Natl Acad Sci USA* 1976;**73**:2510–4.
48. de Boer AS, Borgdorff MW, Vynnycky E, Sebek MM, van Soolingen D. Exogenous reinfection as a cause of recurrent tuberculosis in a low-incidence area. *Int J Tuberc Lung Dis* 2003;**7**:145–52.
49. Venketaraman V, Talaue MT, Dayaram YK, Peteroy-Kelly MA, Bu W, Connell ND. Nitric oxide regulation of L-arginine uptake in murine and human macrophages. *Tuberculosis (Edinb)* 2003;**83**(5):311–8.
50. Venketaraman V, Dayaram YK, Amin AG, Ngo R, Green RM, Talaue MT, Connell ND. Role of glutathione in macrophage control of mycobacteria. *Infect Immun* 2003;**71**(4):1864–71.
51. Venketaraman V, Dayaram YK, Talaue MT, Connell ND. Glutathione and nitrosoglutathione in macrophage defense against *Mycobacterium tuberculosis*. *Infect Immun* 2005;**73**(3):1886–9.
52. Penninckx MJ, Elskens MT. Metabolism and functions of glutathione in micro-organisms. *Adv Microb Physiol* 1993;**34**:239–301.
53. Spallholz JE. Glutathione: is it an evolutionary vestige of the penicillins? *Med Hypotheses* 1987;**23**(3):253–7.
54. Millman AC, Salman M, Dayaram YK, Connell ND, Venketaraman V. Natural killer cells, glutathione, cytokines, and innate immunity against *Mycobacterium tuberculosis*. *J Interferon Cytokine Res* 2008;**28**(3):153–65.
55. Guerra C, Johal K, Morris D, Moreno S, Alvarado O, Gray D, Venketaraman V. Control of *Mycobacterium tuberculosis* growth by activated natural killer cells. *Clin Exp Immunol* 2012;**168**(1):142–52.
56. Morris D, Ly J, Chi PT, Daliva J, Nguyen T, Soofer C, Chen YC, Lagman M, Venketaraman V. Glutathione synthesis is compromised in erythrocytes from individuals with HIV. *Front Pharmacol* 2014;**5**:73.
57. Papi A, Contoli M, Gasparini P, Bristot L, Edwards MR, Chicca M, Leis M, Ciaccia A, Caramori G, Johnston SL, Pinamonti S. Role of xanthine oxidase activation and reduced glutathione depletion in rhinovirus induction of inflammation in respiratory epithelial cells. *J Biol Chem* 2008;**283**:28595–606.
58. Saing T, Lagman M, Castrillon J, Gutierrez E, Guilford FT, Venketaraman V. Analysis of glutathione levels in the brain tissue samples from HIV-1 positive individuals and subject with Alzheimer's disease and its implication in the pathophysiology of the disease process. *BBA Clin* 2016;**29**(6):38–44.
59. Ly J, Lagman M, Saing T, Singh MK, Tudela EV, Morris D, Anderson J, Daliva J, Ochoa C, Patel N, Pearce D. Liposomal glutathione supplementation restores $T_H1$ cytokine response to *Mycobacterium tuberculosis* infection in HIV-infected individuals. *J Interferon Cytokine Res* 2015;**35**(11):875–88.
60. Romero-Alvira D, Roche E. The keys of oxidative stress in acquired immune deficiency syndrome apoptosis. *Med Hypotheses* 1998;**51**(2):169–73.

CHAPTER

# 13

# *Plectranthus barbatus*; Antioxidant, and Other Inhibitory Responses Against HIV/AIDS

*Petrina Kapewangolo*[1], *Debra Meyer*[2]

[1]University of Namibia, Windhoek, Namibia; [2]University of Johannesburg, Auckland Park, South Africa

OUTLINE



## Abstract

*Plectranthus barbatus*, a member of the Lamiaceae family, exhibits strong antioxidant activity, which appears to contribute to the reduction of viral load and other symptoms in human immunodeficiency virus (HIV)-1 infected individuals using this plant as treatment in a traditional medicine setting. *P. barbatus* extract has been shown to inhibit HIV-1 protease enzyme activity and whole virus replication in vitro, better than three other *Plectranthus* species (*P. ciliatus*, *P. ecklonii*, and *P. neochilus*), explaining its preferential use in the informal traditional medicine management of HIV. How the extract was prepared (the solvent and drying of leaves or lack thereof) influenced its behavior in a routinely used free radical scavenging assay (2,2-diphenyl-1-picrylhydrazyl radical). Other members of the Lamiaceae family also demonstrated promising activity against HIV/AIDS in vitro either through antioxidant or antiinflammatory responses or as immune modulators since HIV stimulates reactive oxygen species production directly or through its effects on inflammatory cytokines.

**Keywords:** Antioxidant; HIV/AIDS; Lamiaceae; Oxidative stress; *Plectranthus barbatus*; Reactive oxidant species; ROS.

## List of Abbreviations

**AIDS** Acquired immunodeficiency syndrome
**ARV** Antiretroviral
**ART** Antiretroviral therapy
**CD4+** Cluster determinant 4
**DPPH** 2,2-diphenyl-1-picrylhydrazyl radical
**Env** Envelope
**Gag** Group-specific antigen
**HAART** Highly active antiretroviral therapy


http://dx.doi.org/10.1016/B978-0-12-809853-0.00013-4

**HIV** Human immunodeficiency virus
**IL** Interleukin
**LTR** Long terminal repeats
**NFκβ** Nuclear factor kappa light chain enhancer of activated B cells
**OS** Oxidative stress
**P** Plectranthus
**PBMCs** Peripheral blood mononuclear cells
**PMA** Phorbol 2-myristate 13-acetate
**Pol** polymerase
**RA** Rheumatoid arthritis
**ROS** Reactive oxygen species
**SOD** Superoxide dismutase
**TNF-α** Tumor necrosis factor alpha

# INTRODUCTION—THE RELATIONSHIP BETWEEN OXIDATIVE STRESS, HIV/AIDS, AND PLANT-BASED ANTIOXIDANTS

Oxidative stress (OS) is explained as the imbalance between oxidants, also called free radicals, and antioxidants. There are many types of free radicals, but those derived from oxygen and collectively known as reactive oxygen species (ROS) are of great importance in many disease states. Free radicals (atoms or groups of atoms that have one or more unpaired electrons) are produced during normal cellular processes like the processing of nutrients but can also be produced by external factors such as infection, exposure to radiation, and smoking. If generated in excess, free radicals can cause oxidative damage to cell membranes and macromolecules such as DNA (through hydroxyl reactions with the deoxyribose sugar or the nucleotide base) and proteins (fragmentation of amino acids or protein–protein cross-linkages leading to loss of function). The most important intracellular source of ROS is the mitochondria.[1] The bulk of oxygen consumed by mammals is utilized for ATP production in this organelle, but a small percentage is completely reduced to produce superoxide anion radical, which is an oxygen-containing free radical involved in many disease states.[1] Examples of other free radicals include hydroxyl radical, hydrogen peroxide, oxygen singlet, hypochlorite, nitric oxide radical, and peroxynitrite radicals.

The accumulation of oxidants (such as ROS) has been implicated in cellular senescence and aging[2,3] and is associated with neurodegenerative[4] and chronic diseases (for example, different types of cancers, nonalcoholic liver disease, and cardiovascular disease) and also diseases associated with chronic inflammation such as rheumatoid arthritis and infectious diseases such as human immunodeficiency virus (HIV)/acquired immunodeficiency syndrome (AIDS). In various disease states, the increased production of ROS leads to tissue damage and other complications.[4,5] Animal studies[6] demonstrated shorter life expectancies in the presence of increased OS.

To minimize the OS caused by elevated levels of free radicals, oxidants are removed by antioxidants (electron donors that neutralize free radicals). Examples of antioxidants that are produced or are active during normal cellular processes are glutathione, ubiquinol, superoxide dismutase (SOD), and uric acid. Other antioxidants such as selenium, vitamins E (α-tocopherol), C (ascorbic acid), and B-carotene are found through the diet.

During the increased immune cell activity or chronic inflammation of HIV infection, high levels of apoptosis of uninfected cluster determinant 4 ($CD4^+$) T-cells occur through direct viral action or its induction of OS (schematically represented in Fig. 13.1). During the apoptotic process there is a rise in ROS, which places the host under OS. The excessive production of ROS[7,8] during HIV infection, is well documented. Elevated levels of ROS eventually lead to a deficiency in the antioxidant capacity of the infected individual, mostly due to the utilization of antioxidant molecules for the protection of cells against ROS-induced damage. OS is involved in many aspects of HIV pathogenesis including viral replication, the host's inflammatory responses, decreased immune cell proliferation, loss of immune function, cellular apoptosis, chronic weight loss, and increased sensitivity to drug toxicity.[9]

HIV reportedly induces the generation of ROS through the regulatory protein Tat and the envelope glycoprotein gp120.[10] Furthermore, these viral proteins (gp120 and Tat) negatively influence the concentrations of the antioxidant enzymes glutathione reductase and glutathione peroxidase.[10] Free radicals activate nuclear factor κβ (NF-κβ), a cellular protein that triggers viral replication (Fig. 13.2). OS therefore supports HIV replication. The activation of HIV-1 is also believed to be linked to OS leading to the development of AIDS.[9,11,12] HIV replicates in a highly oxidized environment[9] and biomarkers of OS have been documented in HIV infected and in AIDS patients.[13]

Micronutrient supplementation, including antioxidants, is suggested to be beneficial as a cost-effective strategy for improving oxidative and nutritional status in HIV infection. Antioxidants such as vitamin E and C have been shown to reduce OS and viral load in HIV-infected individuals.[8]

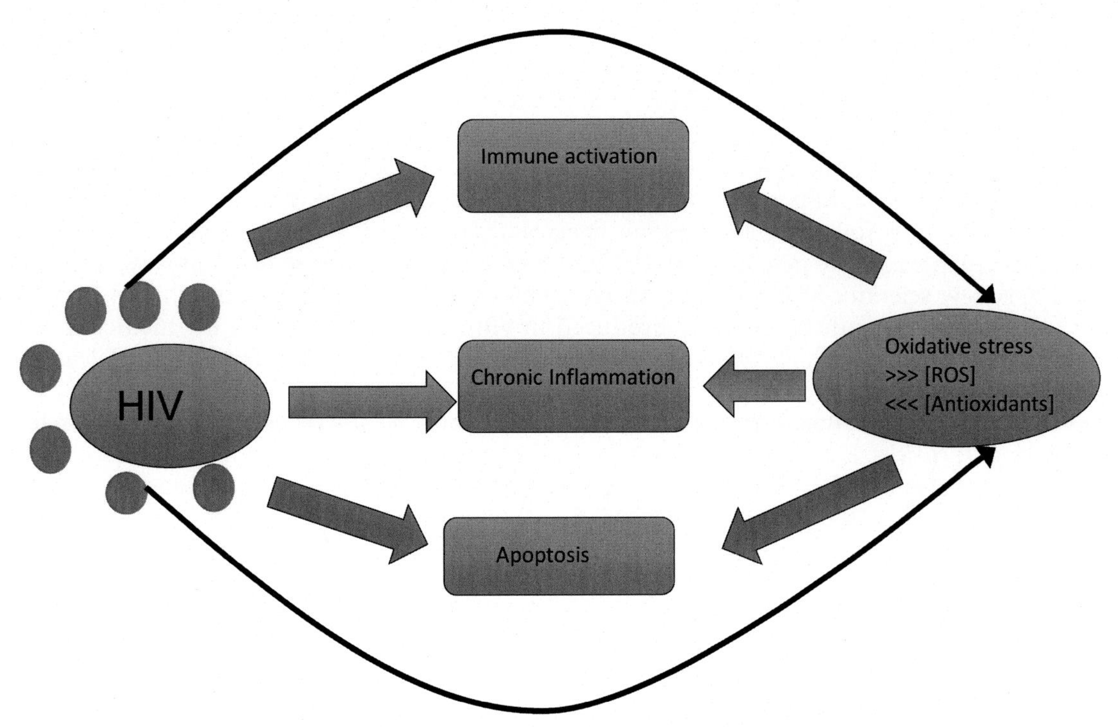


FIGURE 13.1 **Implication of human immunodeficiency virus (HIV) infection and oxidative stress in cellular reactions.** Reports suggest that the mode of action for triggering these cellular events during HIV/AIDS is by the virus, through its proteins such as gp120 (represented here by the smaller circles) initiating oxidative stress (excessive free radical/reactive oxygen species—ROS production). In initiating the excessive production of ROS, HIV infection also causes the depletion of natural antioxidant resources.

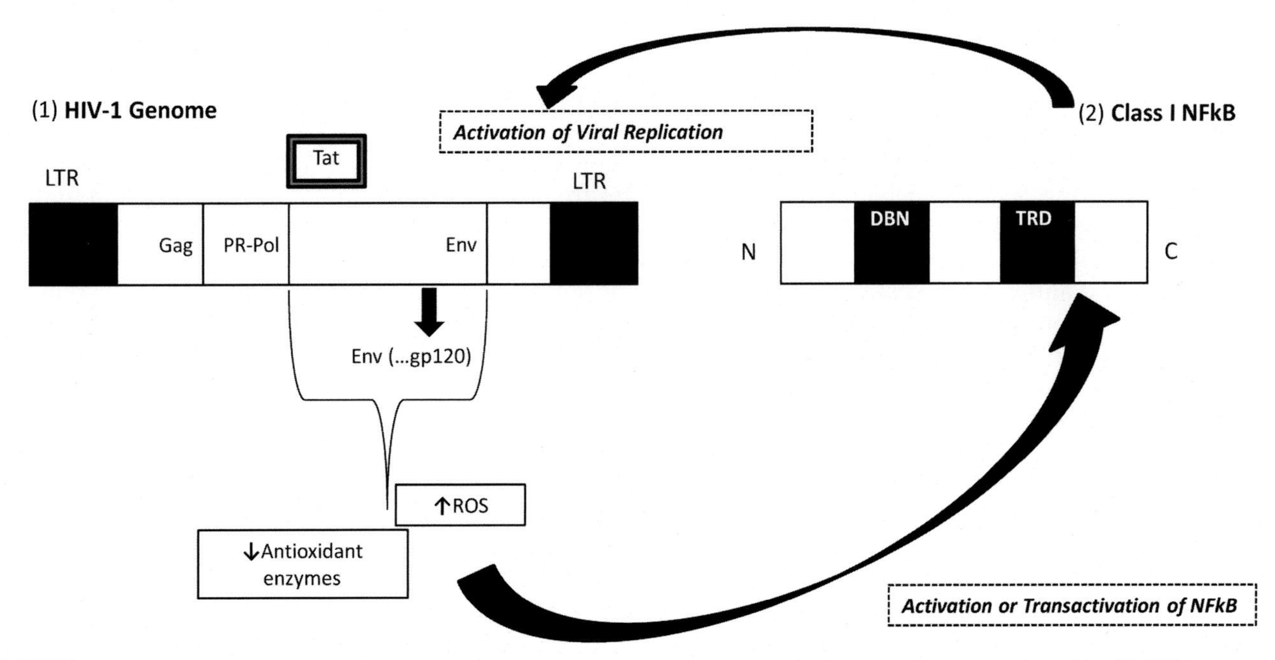


FIGURE 13.2 **Effect of human immunodeficiency virus (HIV) gene products on oxidative stress.** Representation of the (1) HIV-1 gene products (group-specific antigen—Gag, envelope—Env, regulatory proteins—tat, rev, and long terminal repeats—LTR) and (2) NFκβ and the interrelated responses between the protein and the virus during oxidative stress. HIV proteins (gp120 and tat) stimulate the production of ROS, which activates NFκβ (through the production of TNF-α). The latter in turn triggers viral replication. The black square closer to the N terminus of NFκβ (3) represents the DNA binding domain (DBN) of the protein, while the one closer to the C-terminus represents the transrepression domain (TRD). What is represented here is class 1 of NFκβ, if it were class II, the TRD would be replaced by a transactivation domain (TAD). Transactivation of NFκβ plays a role in HIV-1 replication.

Plants are also known to exhibit strong antioxidant activity[14] and natural product preparations may play a crucial role in reducing virus-induced OS in HIV-1 infected individuals. Numerous plant extracts with anti-HIV properties have been reported to possess antioxidant activity,[15] and some of these plants (*Hypoxis hemerocallidea*[16] and *Sutherlandia frutescens*[17]) are in use as African herbal medicines in the management of HIV/AIDS.

Lamiaceae, a plant family commonly known as mints, has 980 species distributed in Southern Africa alone.[18] One of the most significant, prolific, and most used Lamiaceae genus in Southern Africa is *Plectranthus*.[19] A literature search identified the following *Plectranthus* species; *P. barbatus*, *P. comosus*, and *P. ciliatus* as some of the common medicinal plants used by persons with HIV/AIDS.[20–22] These plants are used mostly based on anecdotal evidence, and very few scientific studies on the mechanisms involved in extracts from these plants being active against HIV/AIDS can be found. We recently produced in vitro scientific proof for the medical potency of two Lamiaceae family members; *P. barbatus* (direct activity against HIV and as antioxidant[23]) and *Ocimum labiatum* (as antioxidant[24]).

Here-in evidence is presented that the in vitro anti-HIV activity of *P. barbatus* is better in comparison to three other *Plectranthus* species (*P. ciliatus*, *P. ecklonii*, and *P. neochilus*). In addition, a brief review is provided of what has been demonstrated in terms of antioxidant and other protective behavior by *Plectranthus* species and other members of the Lamiaceae family, especially in the context of HIV/AIDS.

## THE LAMIACEAE FAMILY, ANTI–HIV, AND ANTIOXIDANT RESPONSES

Plants belonging to the Lamiaceae (also called Labiatae) family are used traditionally to treat viral and microbial ailments.[25] Surveys carried out in Uganda, Tanzania, and Namibia on the use of medicinal plants for HIV/AIDS treatment revealed that Lamiaceae plants, especially those from the genera *Ocimum*, *Plectranthus*, and *Leonotis*, were used traditionally in managing HIV/AIDS and opportunistic infections associated with the disease. Other Lamiaceae species namely the genus *Salvia*, *Stachys*, and *Tetradenia* are also commonly used to alleviate symptoms in HIV-infected individuals.[20,22,26,27] Leaves from these plants are administered orally by traditional medicine practitioners as decoctions or infusions for the treatment of HIV/AIDS, tuberculosis, oral candidiasis, herpes simplex, etc.[22,27]

Plants from the Lamiaceae family (Order: Lamiales) are considered to be good sources of antioxidants (neutralizers of ROS) due to the presence of high concentrations of phenolic compounds, terpenes, and essential oils. Antioxidants from plants come in various forms; as ascorbic acid and tocopherols, polyphenolic compounds, or terpenoids. Terpenes are one of natures most varied structural compounds and display a variety of biological and pharmacological activities including strong antioxidant properties. Due to their antioxidant behavior, terpenes have been shown to provide relevant protection under OS conditions.

Lamiaceae plants contain various terpenoid (also called isoprenoid) compounds. Terpenes are a large diverse class of naturally occurring organic chemicals derived from five-carbon isoprene units. Many of the active ingredients isolated from medicinal plants are terpenoid in nature. Plant isolates containing terpenoids have been found to suppress NF-κB signaling (this protein complex has been linked to the pathogenesis of inflammatory diseases, cancer, viral infection, and autoimmune diseases).[28] *Leonotis leonurus* reportedly contain diterpenoids, which are identified as the active ingredients of that plant. Our group isolated bioactive terpenoids with strong antioxidant and antiinflammatory behavior from *O. labiatum*.[24] Diterpenoids are the more common secondary metabolites in *Plectranthus* species.[29]

*Mentha longifolia* and *Tetradenia riparia* plants, also members of the Lamiaceae family, contain numerous monoterpenoids, which have also been identified as the active ingredients in those plants.[25]

Given the promising health benefits of the Lamiaceae plant family, it would be worthwhile to explore plants from this family for possible drug leads. *In vitro* evidence already supports the vast anecdotal evidence that products from these plants alleviate HIV/AIDS symptoms through the antioxidant process and by modulating the immune system.

The *Plectranthus* (spurflowers) genus is a well-known herbal remedy in Africa, China, parts of South America, and India where these plants play a huge role in the traditional medicine regimen for the treatment of a wide range of conditions. Different parts of *Plectranthus* species (stems, leaves, roots, and tubers) are used to treat various ailments.[19] In Africa, *P. barbatus* is used as treatment for malaria and for treating symptoms associated with HIV/AIDS and associated conditions (e.g., oral candidiasis, herpes simplex, herpes zoster, and skin rashes).[20,23] The aromatic-smelling *P. barbatus* contains a number of bioactive secondary metabolites and essential oils in leaves and roots. The main components of essential oils are mono- and diterpenes for which ample evidence of antioxidant behavior exists. Diterpenes isolated from *P. barbatus* have demonstrated numerous bioactivities (Table 13.1).

TABLE 13.1 Bioactive Diterpenoids Isolated From *Plectranthus barbatus*

| *Plectranthus* Species | Isolated Diterpenoids | Bioactivity |
|---|---|---|
| *P. barbatus* | Barbatusin[40] | Anticancer activity[40] |
| | Barbatusol[40] | Antihypertensive effect[40] |
| | Coleolic acid[40] | Antitumor activity[41] |
| | Coleon C[40] | Antiproliferative effect[40] |
| | (16S)-Coleon E[42] | Antiacetylcholinesterase activity[43] |
| | Dehydroabietane[44] | Antiaspergillus activity[45] |
| | Demethylcryptojaponol (11-hydroxysugiol)[40] | Antibacterial activity[46] |
| | Ferruginol[40] | Gastroprotective activity[47] |
| | Forskoditerpenoside A,B, C,D,E[40] | Antihistamine effect[48] |
| | Forskolin[49] | Anticonvulsant activity[50] |
| | Manoyl oxide[49] | Antimicrobial activity[51] |
| | Plectrin[42] | Antifeedant activity[40] |
| | (16R)-Plectrinone A[42] | Antisecretory acid/ulcer effect[52] |
| | Plectranthone J[42] | Anticariogenic activity[53] |
| | Sugiol[40] | Antiinflammatory activity[54] |
| | Taxodione[44] | Antifungal[55] |
| | 1-Acetylforskolin[40] | Antihypertensive effect[40] |
| | 1,9-Dideoxyforskolin[40] | Demonstrates several cAMP-related effects[40] |
| | 5,6-didehydro-7-hydroxy-taxodone[44] | Antiprotozoal activity[44] |
| | 7-Deacetylforskolin[40] | Antihypertensive effect[40] |
| | 9-Deoxyforskolin[40] | Activity on cAMP generating systems[40] |
| | 13-Epi-9-deoxycoleonol[40] | Antihypertensive activity[56] |
| | 14-Deoxycoleon U[40] | Antifungal[55] |
| | 20-Deoxocarnosol[40] | Antioxidant activity[57] |

A number of compounds have been isolated from *P. barbatus* and the most common group of compounds isolated from the plant with reported bioactivity is diterpenoids.

*P. barbatus* has about eight synonyms including *Coleus forskohlii* where the most important active ingredient from the roots of the plant is forskolin (a terpene). *C. forskohlii* is traditionally used in Ayurvedic medicine for various health issues from serious (cardiovascular and central nervous system problems) to those less so (e.g., gastrointestinal ailments).

# SCIENTIFIC EVIDENCE FOR IN VITRO EFFICACY OF *PLECTRANTHUS BARBATUS*

As discussed previously, HIV replicates in a highly oxidized environment where antioxidants could potentially decrease the pathogen's effect. *P. barbatus* is implicated as active against HIV based on anecdotal and recent in vitro evidence (see sections Introduction—The Relationship Between Oxidative Stress, HIV/AIDS, and Plant-Based Antioxidants and The Lamiaceae Family, Anti–HIV, and Antioxidant Responses). Demonstrating substantial natural antioxidant activity in extracts of *P. barbatus* defends its successful anecdotal use in a traditional medicine setting. Three recent publications[1,23,30] demonstrated the antioxidant activity of *P. barbatus*. The ability of *P. barbatus* extract to scavenge 2,2-diphenyl-1-picrylhydrazyl radical, a stable free radical potentially reactive with all compounds capable of donating a hydrogen atom, is shown in Table 13.2. The $IC_{50}$ of the extract prepared by Kapewangolo et al.[23] was

TABLE 13.2 Free Radical Scavenging Activity of *Plectranthus barbatus* Extracts

| Extract/Solvent Of Leaves | $IC_{50}$ (Antioxidant Activity-DPPH) μg/ml | References |
|---|---|---|
| Ethanol[a] | 15.87 ± 0.3 | Kapewangolo et al.[23] |
| Aqueous[b] | 35.87 ± 0.27 | Maioli et al.[1] |
| Ethanol[c] | 75.71 ± 10.57 | Silva et al.[30] |
| Aqueous[c] | 318.83 ± 21.02 | Silva et al.[30] |
| Acetone[c] | 198.14 ± 5.45 | Silva et al.[30] |

The in vitro antioxidant activity of *P. barbatus* extracts was measured by the ability of the samples to scavenge the 2,2-diphenyl-1-picrylhydrazyl (DPPH) radical. $IC_{50}$ = 50% inhibitory concentration.
[a]*Fresh leaves.*
[b]*Leaves dried in an oven with air renewal and circulation at 37°C to complete dehydration.*
[c]*Leaves dried in an oven or solar dryer.*

twice lower than that obtained by Maioli et al.[1] utilizing a similar assay. The former authors also utilized vitamin C as positive control with an $IC_{50}$ of 1.1 ± 0.02 μg/mL. Data presented indicate that the type of solvent used for extract preparation (aqueous[1] and ethanol[23]) determined the potency (presented as $IC_{50}$ values) in terms of free radical scavenging ability of *P. barbatus* extract. Whether leaves were dried prior to extraction or not, also influenced the strength of the prepared antioxidant.

For the purposes of this review, *P. barbatus* extract was compared to that of three other members of the *Plectranthus* species for direct in vitro antiviral activity. The former extract demonstrated a stronger ability to inhibit HIV-1 protease (Table 13.3). Two *Plectranthus* species could decrease HIV replication in an in vitro model of HIV-1 infection (see Fig. 13.3). Extracts that demonstrated moderate anti-HIV-1 activity were tested for their ability to suppress HIV-1 expression in the chronically HIV-1 infected U1 cell line. Due to the extreme cytotoxicity demonstrated by *P. ciliatus* extract in TZM-bl and peripheral blood mononuclear cells (Table 13.3), it was excluded from the HIV-1 expression study. *P. barbatus* significantly reduced HIV-1 replication in U1 cells (Fig. 13.3).[23] *P. neochilus* also reduced HIV-1 replication in U1 cells, however, the difference was not significant ($P > .05$) when compared to phorbol 2-myristate 13-acetate (PMA)-stimulated control cells. U1 cells express minimal HIV-1 p24 protein under normal growth conditions. The levels of HIV-1 p24 core protein increased remarkably in the presence of PMA (stimulant), and a reduction in p24 secretion is indicative of replication/activation inhibition. Viability of PMA-treated U1 cells in the presence/absence of extracts was again tested after the supernatant was collected for p24 determination, and the data obtained confirmed that the ability of the extracts to reduce HIV-1 expression in U1 cells was not due to cytotoxicity because viability of the cells for all treatments was more than 80% (Fig. 13.3). These data provide empirical evidence for the anecdotal use of *P. barbatus* rather than *P. neochilus/ciliatus* in HIV management.

## HIV/AIDS, HIGHLY ACTIVE ANTIRETROVIRAL THERAPY, AND HERBAL REMEDIES

HIV replication triggers OS through ROS accumulation (section Introduction—The Relationship Between Oxidative Stress, HIV/AIDS, and Plant-Based Antioxidants). A variety of enzymatic (SOD, catalase, glutathione peroxidase, etc.) and nonenzymatic (carotenoids, tocopherols, ascorbate, bioflavonoids, bilirubin, uric acid, etc.) antioxidants present in human blood become insufficient for controlling or decreasing the effects of HIV[13] during untreated (and also treated) infection. Highly active antiretroviral therapy (HAART) reportedly, only partially, restores the antioxidant capacity by suppressing HIV.[29] In some cases, it seems HAART causes more harm than good because the production of free radical species in HIV-1 infected individuals receiving antiretroviral treatment (ART) including HAART has been reported to be higher than that of infected individuals not on treatment or normal and healthy subjects.[31]

Concurrent herb-antiretroviral (ARV) usage is becoming common in different areas of the world including Africa, North America, and Europe.[21] There are various reasons for why herbal remedies are often used with HIV drugs and one of the reasons is to apparently counteract some of the negative side effects of ARV.[21] The inability of the current HIV regimen to completely eradicate the virus is another reason for concurrent herb-ARV use by patients[21] and the antioxidant support that herbal remedies may offer, may be another.

TABLE 13.3 In vitro Cytotoxic Activity and Inhibition of HIV-1 Enzymes by Four *Plectranthus* Extracts

| | | | | $CC_{50}$ (μg/mL) ± SD[a] | | $IC_{50}$ (μg/mL) ± SD[a] | |
|---|---|---|---|---|---|---|---|
| **Plant extract** | **Voucher specimen number** | **Initial weight of fresh plant leaves (g)** | **Yield[b] (%, w/w)** | **TZM-bl** | **PBMCs** | **HIV-1 RT** | **HIV-1 PR** |
| *Plectranthus barbatus* | 117198 | 74.8 | 28.6 | 50.4 ± 2.7 | 84.7 ± 2.2 | >100 | 62 ± 0.2 |
| *Plectranthus ciliatus* | 117199 | 86.7 | 17.3 | 9.7 ± 0.2 | 4.2 ± 0.0 | >100 | >100 |
| *Plectranthus ecklonii* | 117695 | 85.7 | 26.1 | 33.7 ± 1.0 | 8.0 ± 0.1 | >100 | >100 |
| *Plectranthus neochilus* | 117696 | 282.9 | 7.4 | 33.0 ± 0.8 | 16.8 ± 0.2 | >100 | >100 |
| Auranofin[c] | – | – | – | <10 μM | <10 μM | – | – |
| Doxorubicin[d] | – | – | – | – | – | <25 | – |
| Acetyl pepstatin[e] | – | – | – | – | – | – | <0.3 |

The yield of the *Plectranthus* extracts obtained from fresh plant leaves is shown as well as the in vitro activity of the extracts against 2 cell types; TZM-bl (adherent cell line) and PBMCs (suspension cells). Anti-HIV activity of the extracts was tested against HIV-1 reverse transcriptase (RT) and protease (PR).

[a] *Each value represents the mean ± Standard Deviation (SD) computed from repeats of 3–6 independent experiments.*

[b] *Yield represents the percentage recovery of dried extract per weight as compared to the original fresh plant material.*

[c] *Positive control for cytotoxicity.*

[d] *Positive control for HIV-1 reverse transcriptase (RT).*

[e] *Known inhibitor of HIV-1 protease (PR).*

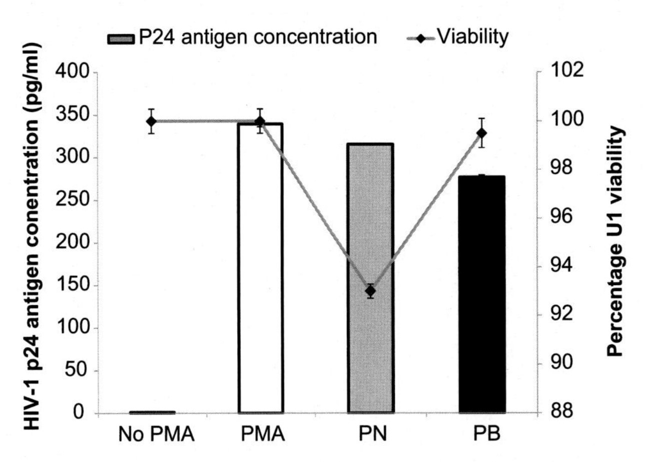


FIGURE 13.3 **Inhibition of human immunodeficiency virus (HIV)-1 replication by *Plectranthus neochilus* and *Plectranthus barbatus* in U1 cells.** The bars shows the mean ± SD of six experimental repeats, analyzing the effect of *Plectranthus* extracts at noncytotoxic concentrations on HIV-1 expression in chronically infected U1 cells. Cells were incubated with noncytotoxic concentrations of *PN, P. neochilus* (12.5 µg/mL) and *PB, P. barbatus* (25 µg/mL) for 72 h in the presence of phorbol 2-myristate 13-acetate (PMA) (2 ng/mL). Controls included unstimulated (no PMA) and stimulated/PMA-treated (PMA) cells. HIV-1 expression was monitored by measuring HIV-1 p24 antigen in culture supernatant. Both extracts decreased HIV-1 inhibition in U1 cells. HIV-1 reduction in U1 cells by *P. neochilus* ($CC_{50}$ = 24.6 ± 0.3 µg/mL) and *P. barbatus* ($CC_{50}$ > 100 µg/mL) was not affected by cytotoxicity. Viability of PMA-induced U1 cells in the presence of extracts was more than 50% after 72 h.

## DIAGNOSIS OF AND MONITORING HUMAN IMMUNODEFICIENCY VIRUS–INDUCED OXIDATIVE STRESS

As mentioned previously (section HIV/AIDS, Highly Active Antiretroviral Therapy, and Herbal Remedies) HIV proteins, chronic use of HAART, and metabolic complications caused by the virus leads to OS and inflammation measured by the increased biomarkers of both. Conventional biochemical analysis is primarily used to detect the presence of OS markers but recently, metabolomics methodology (different types of spectroscopy utilized to detect disease markers in fluids or tissue) has been successfully applied, e.g., the difference between glutathione levels in sera from HIV-infected versus uninfected individuals was demonstrated with Raman spectroscopy and confirmed by conventional biochemical methodology.[32] Biomarkers serve to diagnose OS but can also be used to monitor the success of antioxidants.

Total antioxidant capacity can be used as a novel, early biochemical marker of OS in HIV-1 infected patients that may result in reduced tissue damage by free radicals and help to monitor and optimize antioxidant therapy in such patients.[13]

A cohort study showing an independent association of OS with death in HIV-infected patients demonstrated that OS constituted an additional predictor of mortality, independent of established HIV-associated predictors such as CD4 cell count, viral load, and inflammation.[3]

## OXIDANTS AND ANTIOXIDANTS AS IMMUNOMODULATORS/STIMULATORS

HIV stimulates the production of ROS through its proteins or through its effects on cellular systems like the immune system. The immune system is a collection of cells and proteins (cytokines) influenced by infection, foreign particles, etc. Cytokines (indicators of immune system function) are minute signaling protein molecules that are involved in intercellular communication. Some cytokines have prooxidant effects, while others are proinflammatory (interleukin/IL-1 and tumor necrosis factor/TNF-α) or central mediators in major inflammatory diseases. IL-1 and TNF-α, need to be downregulated/suppressed, while dual function cytokines (e.g., IL-6) could require inhibitory or stimulatory responses depending on their function at the time. The ability to regulate the production of certain cytokines can be of therapeutic importance. Interferon is an interesting cytokine because it is generally produced during early stages of any viral infection, but in the case of HIV infection this cytokine is apparently undetectable and is only detected during the later stages when HIV progresses to AIDS.[33] HIV activates macrophages through TNF-α release, and activated polymorphonuclear leukocytes, also contribute to the generation and accumulation of ROS.

Accumulation of free radicals in animal tissues is a major cause of cell damage or death and is considered instrumental to various cancers and other diseases.[4] ROS in low concentrations act as significant cell signaling molecules and regulate the biological conditions of cytokines, hormones, and growth factors. High levels of free radicals, however, overcome the normal cellular antioxidant defenses and end up being cytotoxic to the biological system.[34]

ROS have also been reported to be involved in the activation of NF-κβ by proinflammatory cytokines (Fig. 13.2) such as TNF-α.[34] Given the importance of activated NF-κβ in inflammatory disease progression, suppression of this protein directly or through inhibition of ROS or proinflammatory cytokines preferably by antioxidants remains therapeutically important because of the ability of the latter to combat pathogenic chain reactions initiated by free radicals. ROS can also act as modulators of the immune system during the course of HIV infection.[35] Herbal medicines as immune modulators may offer novel approaches in the treatment of a variety of diseases.

Plants investigated for HIV or antioxidant activity have been screened for immune-stimulatory/modulatory activity as well. The ability of a single natural product to produce more than one health-related quality makes it possible for these products/compounds to offer many health benefits in a single dose when developed for clinical use. Plant-based products can act as immune stimulants or immune suppressants. Being potential sources of antiinflammatory agents, natural products are being investigated for potent immune-modulatory capacity. The Lamiaceae family demonstrates antioxidant and antiinflammatory activity as exemplified by the data collected using *Thymus vulgaris, Ocimum canum, Ocimum adscendens*, and *Leucas linifolia*, respectively. Our work on *O. labiatum* and *P. barbatus* supports these data.

Inflammation is a way in which an organism responds to harmful stimuli and is one of the processes the immune system uses to modulate itself. Inflammation has been reported to be important in the onset of various diseases such as cancer, diabetes, and neurodegenerative diseases.[36] When exposed to harmful substances, cells produce nitric oxide (NO) as part of an inflammatory response.[37] This is experimentally quantified by the Griess reagent, which is a chemical that detects the presence of organic nitrites.[38] Plant-derived flavonoids have been reported to be potential antiinflammatory agents, and more research is being conducted to determine their beneficial effects against various diseases.

A phenolic compound, phloroglucinol alpha pyrone arzanol, which has an ability to reduce HIV replication as well as possess antiinflammatory properties by inhibiting the production of proinflammatory cytokines in primary monocytes qualifying it as a plant-derived antiinflammatory and antiviral chemotype worthy of further investigation.[39]

Our group demonstrated that *O. labiatum* extract had promising antiinflammatory and antioxidant properties, while a terpenoid isolated from the plant showed strong antiinflammatory activity in vitro.[24] These data support the development of naturally derived antiinflammatory treatments.

## CONCLUSION

Natural products, especially those belonging to the Lamiaceae family, have great potential as sources of antiviral, antiinflammatory, and immune enhancing agents. Scientific validation of many of the members of this family has yet to be done but in the case of *Plectranthus* species, a number of studies have been under taken and are referenced here. *P. barbatus* is one of the Lamiaceae plants extensively studied and diterpenoids isolated from this plant have demonstrated numerous bioactivities, both in vitro and in vivo. The scientific evidence is in support of the traditional success of *P. barbatus* as an herbal remedy used in alleviating various symptoms of both communicable (bacterial, viral) and noncommunicable (allergies, cancer, hypertension, etc.) ailments. Especially promising is the growing amount of evidence supporting the traditional medicine use of *P. barbatus* against HIV/AIDS symptoms. The potential of *P. barbatus* and other promising Lamiaceae plants demands further validation and development as possible natural supplements or drugs. These are plants already popularly used in Africa, mostly in poverty stricken areas and largely contribute toward reducing disease burdens for most of communities.

## SUMMARY POINTS

- *P. barbatus* extract is active against HIV/AIDS through a direct in vitro effect on the virus or as antioxidant.
- Many Lamiaceae plants demonstrate antioxidant and anti-HIV activities.
- The induction of OS explains HIV's ability to trigger numerous cellular events.
- HIV stimulates ROS production directly or through its effects on inflammatory cytokines
- Plants can serve as natural antioxidants but also as immune modulators and antiinflammation agents.

# References


1. Maioli MA, Alves LC, Campanini AL, Lima MC, Dorta DJ, Groppo M, Cavalheiro AJ, Curti C, Mingatto FE. Iron chelating-mediated antioxidant activity of *Plectranthus barbatus* extract on mitochondria. *Food Chem* 2010;**122**:203–8.
2. Salminen A, Ojala J, Kaarniranta K, Kauppinen A. Mitochondrial dysfunction and oxidative stress activate inflammasomes: impact on the aging process and age-related diseases. *Cell Mol Life Sci* 2012;**69**:2999–3013.
3. Masiá M, Padilla S, Fernández M, Rodríguez C, Moreno A, Oteo JA, Antela A, Moreno S, del Amo J, Gutierrez F. Oxidative stress predicts all-cause mortality in HIV-infected patients. *PLoS One* 2016;**11**:e0153456.
4. Valko M, Rhodes CJ, Moncol J, Izakovic M, Mazur M. Free radicals, metals and antioxidants in oxidative stress-induced cancer. *Chem Biol Interact* 2006;**160**:1–40.
5. Mirshafiey A, Mohsenzadegan M. The role of reactive oxygen species in immunopathogenesis of *Rheumatoid arthritis*. *Iran J Allergy Asthma Immunol* 2008;**7**:195–202.
6. Sinha JK, Ghosh S, Swain U, Giridharan NV, Raghunath M. Increased macromolecular damage due to oxidative stress in the neocortex and hippocampus of WNIN/Ob, a novel rat model of premature aging. *Neuroscience* 2014;**269**:256–64.
7. Allard JP, Aghdassi E, Chau J, Salit I, Walmsley S. Oxidative stress and plasma antioxidant micronutrients in humans with HIV infection. *Am J Clin Nutr* 1998;**67**:143–7.
8. Allard JP, Aghdassi E, Chau J, Tam C, Kovacs CM, Salit IE, Walmsley SL. Effects of vitamin E and C supplementation on oxidative stress and viral load in HIV-infected subjects. *AIDS* 1998;**12**:1653–9.
9. Gil del Valle L, Hernández RG, Ávila JP. Oxidative stress associated to disease progression and toxicity during antiretroviral therapy in human immunodeficiency virus infection. *J Virol Microbiol* 2013. http://dx.doi.org/10.5171/2013.279685.
10. Price TO, Ercal N, Nakaoke R, Banks WA. HIV-1 viral proteins gp120 and Tat induce oxidative stress in brain endothelial cells. *Brain Res* 2005;**1045**:57–63.
11. Gil L, Mart G, González I, Tarinas A, Álvarez A, Giuliani A, Molina R, Tapanas R, Perez J, Sonia O. Contribution to characterization of oxidative stress in HIV/AIDS patients. *Pharmacol Res* 2003;**47**:217–24.
12. Legrand-Poels S, Vaira D, Pincemail J, Vorst A, Piette J. Activation of human immunodeficiency virus type 1 by oxidative stress. *AIDS Res Hum Retroviruses* 1990;**6**:1389–97.
13. Suresh DR, Annam V, Pratibha K, Prasad BVM. Total antioxidant capacity–a novel early bio-chemical marker of oxidative stress in HIV infected individuals. *J Biomed Sci* 2009;**16**:61.
14. Gupta VK, Sharma SK. Plants as natural antioxidants. *Nat Prod Radiance* 2006;**5**:326–34.
15. Chen K, Plumb GW, Bennett RN, Bao Y. Antioxidant activities of extracts from five anti-viral medicinal plants. *J Ethnopharmacol* 2005;**96**:201–5.
16. Nair VDP, Dairam A, Agbonon A, Arnason JT, Foster BC, Kanfer I. Investigation of the antioxidant activity of African potato (*Hypoxis hemerocallidea*). *J Agric Food Chem* 2007;**55**:1707–11.
17. Katerere DR, Eloff JN. Antibacterial and antioxidant activity of *Sutherlandia frutescens* (Fabaceae), a reputed anti-HIV/AIDS phytomedicine. *Phyther Res* 2005;**19**:779–81.
18. Klopper RR, Chatelain C, Bänninger V, Habashi C, Steyn HM, De Wet BC, Arnold TH, Gautlier L, Smith GF, Spichiger R. *Checklist of the flowering plants of Sub-Saharan Africa. An index of accepted names and synonyms*. Southern African Botanical Diversity Network Report. Southern African Botanical Diversity Network; 2006.
19. Rice LJ, Brits GJ, Potgieter CJ, Van Staden J. Plectranthus: a plant for the future? *South Afr J Bot* 2011;**77**:947–59.
20. Kisangau DP, Lyaruu HVM, Hosea KM, Joseph CC. Use of traditional medicines in the management of HIV/AIDS opportunistic infections in Tanzania: a case in the Bukoba rural district. *J Ethnobiol Ethnomed* 2007;**8**:1–8.
21. Nagata JM, Jew AR, Kimeu JM, Salmen CR, Bukusi EA, Cohen CR. Medical pluralism on Mfangano Island: use of medicinal plants among persons living with HIV/AIDS in Suba District, Kenya. *J Ethnopharmacol* 2011;**135**:501–9.
22. Semenya SS, Potgieter MJ, Erasmus LJC. Ethnobotanical survey of medicinal plants used by Bapedi traditional healers to manage HIV/AIDS in the Limpopo Province, South Africa. *J Med Plants Res* 2013;**7**:434–41.
23. Kapewangolo P, Hussein AA, Meyer D. Inhibition of HIV-1 enzymes, antioxidant and anti-inflammatory activities of *Plectranthus barbatus*. *J Ethnopharmacol* 2013;**149**:184–90.
24. Kapewangolo P, Omolo JJ, Bruwer R, Fonteh P, Meyer D. Antioxidant and anti-inflammatory activity of *Ocimum labiatum* extract and isolated labdane diterpenoid. *J Inflamm* 2015;**12**:1–13.
25. Van Wyk B-E, Van Oudtshoorn B, Gericke N. *Medicinal plants of South Africa*. Pretoria: Briza Publications; 2009.
26. Lamorde M, Tabuti JRS, Obua C, Kukunda-byobona C, Lanyero H, Byakika-kibwika P, Bbosa GS, Lubega A, Ogwal-okeng J, Ryan M, Waako PJ, Merry C. Medicinal plants used by traditional medicine practitioners for the treatment of HIV/AIDS and related conditions in Uganda. *J Ethnopharmacol* 2010;**130**:43–53.
27. Yamasaki K, Nakano M, Kawahata T, Mori H, Otake T, Ueba N, Oishi I, Inami R, Yamane M, Nakamura M, Murata H, Nakanishi T. Anti-HIV-1 activity of herbs in Labiatae. *Biol Pharm Bull* 1998;**21**:829–33.
28. Salminen A, Lehtonen M, Suuronen T, Kaarniranta K, Huuskonen J. Terpenoids: natural inhibitors of NF-kappaB signaling with anti-inflammatory and anticancer potential. *Cell Mol Life Sci* 2008;**65**:2979–99.
29. de Martino M, Chiarelli F, Moriondo M, Torello M, Azzari C, Galli L. Restored antioxidant capacity parallels the immunologic and virologic improvement in children with perinatal human immunodeficiency virus infection receiving highly active antiretroviral therapy. *Clin Immunol* 2001;**100**:82–6.
30. Silva C, Mendes M, Almeida V, Michels R, Sakanaka L, Tonin L. Quality of the leaves of *Plectranthus barbatus* Andr. (Lamiaceae) dried in solar dryer and oven. *Botucatu* 2016;**18**. http://dx.doi.org/10.1590/1983–084X/15_021.
31. Sharma B. Oxidative stress in HIV patients receiving antiretroviral therapy. *Curr HIV Res* 2014;**12**:13–21.
32. Sitole L, Steffens F, Meyer D. Raman spectroscopy-based metabonomics of HIV-infected sera detects amino acid and glutathione changes. *Curr Metab* 2015;**3**:65–75.
33. Francis ML, Meltzer MS, Howard GE. Interferons in the persistence, pathogenesis, and treatment of HIV infection. *AIDS Res Hum Retroviruses* 1992;**8**:199–207.

34. Fang FC. Antimicrobial reactive oxygen and nitrogen species: concepts and controversies. *Nat Rev Microbiol* 2004;**2**:820–33.
35. Salmen S, Berrueta L. Immune modulators of HIV infection: the role of reactive oxygen species. *J Clin Cell Immunol* 2012;**03**:1–9.
36. García-Lafuente A, Guillamón E, Villares A, Rostagno MA, Martínez JA. Flavonoids as anti-inflammatory agents: implications in cancer and cardiovascular disease. *Inflamm Res* 2009;**58**:537–52.
37. Luiking YC, Engelen PKJ, Deutz NEP. Regulation of nitric oxide production in health and disease. *Curr Opin Clin Nutr Metab Care* 2011;**13**:97–104.
38. Choi CY, Kim JY, Kim YS, Chung YC, Seo JK, Jeong HG. Aqueous extract isolated from *Platycodon grandiflorum* elicits the release of nitric oxide and tumor necrosis factor-alpha from murine macrophages. *Int Immunopharmacol* 2001;**1**:1141–51.
39. Appendino G, Ottino M, Marquez N, Bianchi F, Giana A, Ballero M, Sterner O, Fiebich BL, Munoz E. Arzanol, an anti-inflammatory and anti-HIV-1 phloroglucinol alpha-Pyrone from *Helichrysum italicum* ssp. microphyllum. *J Nat Prod* 2007;**70**:608–12.
40. Paul M, Radha A, Kumar DS. On the high value medicinal plant, *Coleus forskohlii* Briq. *Hygeia J Drugs Med* 2013;**5**:69–78.
41. Huang D, Yang Y, Ai L, Lu Y, Wu H. Studies on the chemical constituents of *Coleus forskohlii* transplanted in Tongcheng and their antitumor activity. *J Chin Med Mater* 2011;**34**:375–8.
42. Abdel-Mogib M, Albar HA, Batterjee SM. Chemistry of the genus *Plectranthus*. *Molecules* 2002;**7**:271–301.
43. Falé PL, Borges C, Madeira PJA, Ascensão L, Araújo MEM, Florêncio MH, Serralheiro. Rosmarinic acid, scutellarein 4′-methyl ether 7-O-glucuronide and (16S)-coleon E are the main compounds responsible for the antiacetylcholinesterase and antioxidant activity in herbal tea of *Plectranthus barbatus* ("falso boldo"). *Food Chem* 2009;**114**:798–805.
44. Mothana RA, Al-Said MS, Al-Musayeib NM, El Gamal AA, Al-Massarani SM, Al-Rehaily AJ, Abdulkader M, Maes L. In vitro antiprotozoal activity of abietane diterpenoids isolated from *Plectranthus barbatus* Andr. *Int J Mol Sci* 2014;**15**:8360–71.
45. Gonzalez MA, Perez-Guaita D, Correa-Royero J, Zapata B, Agudelo L, Mesa-Arango A, Betancur-Galvis L. Synthesis and biological evaluation of dehydroabietic acid derivatives. *Eur J Med Chem* 2010;**45**:811–6.
46. Yang Z, Kitano Y, Chiba K, Shibata N, Kurokawa H, Doi Y, Arakawa Y, Tada M. Synthesis of variously oxidized abietane diterpenes and their antibacterial activities against MRSA and VRE. *Bioorg Med Chem* 2001;**9**:347–56.
47. Rodriguez JA, Theoduloz C, Yanez T, Becerra J, Schmeda-Hirschmann G. Gastroprotective and ulcer healing effect of ferruginol in mice and rats: assessment of its mechanism of action using in vitro models. *Life Sci* 2006;**78**:2503–9.
48. Shan Y, Wang X, Zhou X, Kong L, Niwa M. Two minor diterpene glycosides and an eudesman sesquiterpene from *Coleus forskohlii*. *Chem Pharm Bull* 2007;**55**:376–81.
49. Valdes LJ, Mislankar SG, Paul AG. *Coleus barbatus* (*C. forskohlii*) (Lamiaceae) and the potential new drug forskolin (Coleonol). *Econ Bot* 1987;**41**:474–83.
50. Soto-Blanco B, Borges Fernandes LC, Campos Cmara C. Anticonvulsant activity of extracts of *Plectranthus barbatus* leaves in mice. *Evid Bence Based Compl Altern Med* 2012. http://dx.doi.org/10.1155/2012/860153.
51. Demetzos C, Loukis A, Spiliotis V, Zoakis N, Stratigakis N, Katerinopoulos H. Composition and antimicrobial activity of the essential oil of *Cistus creticus* L. *J Essent Oil Res* 1995;**7**:407–10.
52. Schultza C, Bossolania MP, Torresb LMB, Lima-Landmana MTR, Lapaa AJ, Souccara C. Inhibition of the gastric H+,K+-ATPase by plectrinone A, a diterpenoid isolated from *Plectranthus barbatus* Andrews. *J Ethnopharmacol* 2007;**111**:1–7.
53. Figueiredo NL, Falé PL, Madeira PJA, Florêncio MH, Ascensão L, Serralheiro MLM, Lino ARL. Phytochemical analysis of *Plectranthus* sp. extracts and application in inhibition of dental bacteria, *Streptococcus sobrinus* and *Streptococcus mutans*. *Eur J Med Plants* 2014;**4**:794–809.
54. Chao KP, Hua KF, Hsu HY, Su YC, Chang ST. Anti-inflammatory activity of sugiol, a diterpene isolated from *Calocedrus formosana* bark. *Planta Med* 2005;**71**:300–5.
55. Kusumoto N, Ashitani T, Murayama T, Ogiyama K, Takahashi K. Antifungal abietane-type diterpenes from the cones of taxodium distichum rich. *J Chem Ecol* 2010;**36**:1381–6.
56. Tandon JS, Roy R, Balachandran S, Vishwakarma RA. Epi-deoxycoleonol, a new antihypertensive labdane diterpenoid from *Coleus forskohlii*. *Bioorg Med Chem Lett* 1992;**2**:249–54.
57. Escuder B, Torres R, Lissi E, Labbé C, Faini F. Antioxidant capacity of abietanes from *Sphacele salviae*. *Nat Prod Lett* 2002;**16**:277–81.

CHAPTER

# 14

# Methyl Gallate as an Antioxidant and Anti-HIV Agent

*Tzi B. Ng*[1], *Jack H. Wong*[1], *Chit Tam*[1], *Fang Liu*[2], *Chi F. Cheung*[1], *Charlene C. W. Ng*[3], *Ryan Tse*[4,5,6], *Tak F. Tse*[4,5,6], *Helen Chan*[4,5,6]

[1]The Chinese University of Hong Kong, Shatin, Hong Kong, China; [2]Nankai University, Tianjin, China; [3]King's College London, London, United Kingdom; [4]Vita Green Health Products (HK) Ltd, Tai Po, Hong Kong, China; [5]Hong Kong Institute of Medical Research, Central, Hong Kong, China; [6]Genning Partners Company Limited, Causeway Bay, Hong Kong, China

OUTLINE



*HIV/AIDS*
http://dx.doi.org/10.1016/B978-0-12-809853-0.00014-6

## Abstract

Methyl gallate is an antioxidant found in a great variety of plant species including some edible beans and also mushrooms. It protects a diversity of cells including liver cells, kidney cells, heart cells, neuronal cells, and fat cells from oxidative stress induced by chemicals such as hydrogen peroxide. It upregulates antioxidant enzymes. It inhibits three key enzymes crucial to the HIV life cycle including reverse transcriptase, protease, and integrase. Methyl gallate inhibits HIV-1 replication in pseudovirus-infected TZM-BL cells.

**Keywords:** Antioxidant; HIV integrase; HIV protease; HIV reverse transcriptase; Methyl gallate.

# INTRODUCTION

Methyl gallate (methyl 3,4,5-trihydroxybenzoate) has been isolated from the following plants:*Bauhinia racemosa*,[1] *Bergenia ligulata*,[2] *Byrsonima bucidaefolia*,[3] *Byrsonima crassa*,[4] *Calliandra haematocephala*,[5] *Camellia sinensis*,[6] *Canavalia gladiata*,[7] *Dimocarpus longan*,[8] *Drosera burmannii*,[9] *Euphorbia supine*,[10] *Glochidion hypoleucum*,[11] *Glycine max*,[7] *Klainedox gabonensis*,[12] *Loranthus micranthus*,[13] *Mangifera indica*,[14] *Mangifera pajang*,[15] *Paeonia rockii*,[16],*Phragmanthera austroarabica*,[17] *Plicosepalus curviflorus*,[18] *Spondias pinnata*,[19] and *Tachigalia paniculata*.[20].

Methyl gallate is a molecule, which displays a myriad of biological activities encompassing antioxidant, antiviral, antimicrobial, and anticancer activities.[21–25]

The intent of the present article is to review the antioxidant and anti-HIV activities of methyl gallate. Its other aforementioned activities are not covered.

# ANTIOXIDANT ACTIVITY IN TRIACYLGLYCEROLS OF KILKA FISH OIL AND ITS OIL-IN-WATER EMULSION

Gallic acid displayed higher anti-DPPH radical potency (IC50 = 29.5 μM) and lower hydrophobicity (log $P = -0.28$) than methyl gallate (IC50 = 38.0 μM, log $P = -0.23$) and α-tocopherol (IC50 = 105.3 μM, log $P = .70$). Methyl gallate exhibited higher antiperoxide activity in the bulk Kilka fish oil system than gallic acid and α-tocopherol, and higher activity than gallic acid although lower activity than α-tocopherol in the oil-in-water emulsion system stabilized by soy protein isolate at 55°C. Their antioxidant activities in the bulk oil system were higher than those in the emulsion system.[26]

# PEROXYNITRITE RADICAL AND NITRITE RADICAL SCAVENGING ACTIVITIES

Among the polyphenolics isolated from *Caesalpinia pulcherrima* pods, methyl gallate exhibited the highest potency in scavenging peroxynitrite radicals.[27]

Methyl gallate, gallic acid, and ellagic acid were main phenolics present in nonfermented/*Bacillus subtilis*–fermented red sword beans. Fermented red sword beans displayed more potent DPPH radical scavenging and ferric-reducing antioxidant activities than other beans. Nonfermented/fermented red sword beans had higher nitrite scavenging activity and hyaluronidase inhibitory activity (an index of antiinflammatory activity) than the corresponding soybean counterparts.[7]

# UPREGULATION OF MRNA ENCODING ANTIOXIDANT ENZYMES

miR-17-3p is a microRNA, which controls the cellular redox status and inhibits transcription of mRNAs involved in the synthesis of antioxidant enzymes. Methyl-3-O-methyl gallate induced in peripheral blood mononuclear cells and EVC-304cells a downregulation of miR-17-3p levels and an upregulation of the levels of mRNA encoding antioxidant enzymes.[28]

# PROTECTIVE EFFECT ON FAT CELLS AGAINST OXIDATIVE STRESS

Methyl gallate suppresses the expression of terminal adipogenic transcription factors. It downregulates intracellular reactive oxygen species and upregulates HO-1, Nrf2, and PRDX3 thereby shielding fat cells from oxidative stress.[29]

## PROTECTIVE EFFECT ON NAF-INDUCED OXIDATIVE STRESS IN ERYTHROCYTES

Treatment of male rats with NaF (600 ppm in drinking water for 1 week) brought about an approximately twofold rise in erythrocyte lipid peroxidation, a two- to threefold reduction of activities of catalase and superoxide dismutase. A decline of GSH level was also observed. Prior administration of methyl-3-O-methyl gallate (10 mg/kg and 20 mg/kg) or vitamin C (10 mg/kg) to the animals mitigated NaF-induced oxidative stress.[30]

## PROTECTIVE EFFECT ON ERYTHROCYTES AGAINST PEROXYL RADICALS

Methyl, ethyl, and propyl gallates synergized with water-soluble tocopherol Trolox in producing a protective effect on red blood cells against peroxyl radicals[31].

## PROTECTIVE ACTION AGAINST TERT-BUTYLHYDROPEROXIDE-INDUCED OXIDATIVE STRESS IN LIVER CELLS

Methyl gallate undermined production of reactive oxygen species and augmented concentrations of total glutathione in hepatoma HepG2 cells in response to oxidative stress caused by tert-butylhydroperoxide.[32]

## PROTECTIVE ACTION AGAINST ALCOHOL-INDUCED OXIDATIVE STRESS IN LIVER

Following administration of (−)-Epigallocatechin 3-O-(3-O-methyl gallate) from Chinese oolong tea (100 mg/kg body weight/day), serum activities of alanine and aspartate aminotransferases and hepatic malondialdehyde level were lowered, whereas hepatic activities of glutathione peroxidase and superoxide dismutase were upregulated in mice with alcohol-induced hepatic damage. This suggested that the methyl gallate exerted a protective action against alcohol-induced oxidative stress.[33]

## ATTENUATION OF HEPATIC INJURY INDUCED BY IRON OVERLOAD

Orally administered methyl gallate normalized the concentrations of hepatic antioxidants, serum markers, and intracellular reactive oxygen species. Lipid peroxidation and protein oxidation were inhibited. Free iron was chelated and iron bound to ferritin decreased, thereby removing excess iron, which brought about liver damage.[19]

## INHIBITION OF INTRACELLULAR REACTIVE OXYGEN SPECIES PRODUCTION STIMULATED BY HYDROGEN PEROXIDE IN HEK-293 HUMAN EMBRYONIC KIDNEY CELLS

Polyphenolics from the methanolic extract of *Glochidion hypoleucum* leaves, including methyl gallate, gallic acid, apigenin-8-C-β-D-glucopyranoside, luteolin-6-C-β-D-glucopyranoside, and luteolin-8-C-β-D-glucopyranoside, scavenged picrylhydrazyl radicals (IC50 values = $2.46 \pm 0.05$ to $40.0 \pm 0.3$ μg/mL). The methanolic extract inhibited intracellular reactive oxygen species production stimulated by hydrogen peroxide in HEK-293 human embryonic kidney cells.[34]

## PROTECTION OF CULTURED MADIN–DARBY CANINE KIDNEY CELLS AGAINST HYDROGEN PEROXIDE–INDUCED OXIDATIVE STRESS

Methyl gallate diminished lipid peroxidation, inhibited decline of intracellular glutathione, lowered 8-oxoguanine, which is an indicator of oxidative DNA damage, and suppressed formation of intracellular reactive oxygen species in Madin–Darby canine kidney epithelial cells exposed to hydrogen peroxide[35].

## ANTIOXIDANT ACTION IN SH-SY5Y NEURONAL CELLS

Methyl-3-O-methyl gallate from *Peltiphyllum peltatum* leaves, at concentrations up to 1 mM, demonstrated an antioxidant action in SH-SY5Y neuronal cells and cell-free assay models and was devoid of prooxidant activity.[36]

## ANTIOXIDANT ACTION IN PC12 NEURONAL CELLS

Methyl gallate forestalls depolarization of mitochondria, caspase-9 activation, and DNA degradation in neuronal PC12 cells exposed to hydrogen peroxide.[37] Methyl gallate at 50 uM enhanced cell viability of hydrogen peroxide–stressed PC12 cells, repressed hydrogen peroxide–induced as well as cobalt chloride–induced reactive oxygen species formation, and augmented nitric oxide levels in unstressed and also hydrogen peroxide–stressed cells.[37]

## PROTECTIVE EFFECT ON NAF-INDUCED OXIDATIVE STRESS IN BRAIN

Methyl-3-O-methyl gallate also reduced the brain concentration of thiobarbituric acid reactive substances and increased the brain activities of catalase and superoxide dismutase in rat brains, which were altered due to NaF-induced oxidative stress.[38]

In balb/c mice with poststroke depression, methyl-3-O-methyl gallate and propyl-3-O-methyl gallate (at 25 and 50 mg/kg) alleviated oxidative stress and mitigated symptoms of depression, reinstated normal behavior and antioxidant defense. Superoxide dismutase, TBARS, and GSH all correlated well with behavioral parameters[39].

## PREVENTION OF HYDROGEN PEROXIDE–INDUCED OXIDATIVE STRESS IN HUMAN UMBILICAL VEIN ENDOTHELIAL CELLS

Methyl gallate exerted free radical scavenging, lipid peroxidation inhibiting, and reactive oxygen species inhibiting effects on human umbilical vein endothelial cells exposed to hydrogen peroxide. The expression levels of type 1 sigma receptor, regulator of chromatin condensation 1, and phosphate carrier protein were enhanced[40].

## CARDIOPROTECTIVE EFFECT ON HYDROGEN PEROXIDE–EXPOSED NEONATAL RAT CARDIAC MYOCYTES

Methyl gallate maintained the viability of cardiac myocytes of newborn rats, which were exposed to hydrogen peroxide by suppressing formation of intracellular reactive oxygen species, maintaining the potential of mitochondrial membranes, enhancing the level of endogenous reduced glutathione, and inhibiting apoptosis and DNA fragmentation.[41]

## PREVENTION OF UVB-INDUCED SKIN PHOTOAGING IN HAIRLESS MICE

*Galla chinensis* extracts containing methyl gallate and gallic acid as the chief constituents prevented UVB irradiation–induced skin photoaging in SKH: HR-1 hairless mice by downregulating the levels of intracellular reactive oxygen species, interleukin-6 and matrix metalloproteinase-1, in skin fibroblasts and keratinocytes, and elevating generation of procollagen type I,elastin, and transforming growth factor-β1. Thickness of the skin and formation of wrinkles were decreased. On the other hand, the skin became more elastic.[42]

## INHIBITORY ACTIVITY ON HIV-1 REVERSE TRANSCRIPTASE, PROTEASE, AND INTEGRASE

Methyl gallate from the edible mushroom *Pholiota adiposa* exhibited HIV-1 reverse transcriptase, integrase, and protease inhibitory activities (IC50 value = 80.1 μM and 228.5 μM, and 17.1% inhibition at 10 mM concentration, respectively). Methyl gallate inhibited HIV-1 replication in pseudovirus-infected TZM-BL cells (IC50 value = 11.9 μM).[43]

Methyl gallate at a concentration of 1.36 mM scavenged 71.4% of superoxide anions and 85.6% of DPPH radicals and inhibited erythrocyte hemolysis by 82.4% at 1.36 mM concentration. Hydrogen peroxide stimulation not only activated cellular oxidative stress responses but also expedited HIV-1 long terminal repeat promotion in TZM-BL cells, which was lowered by methyl gallate from 18% to about 2%. This indicated an association between the antioxidant and anti-HIV activities of methyl gallate. Nuclear transcription factor kappa B (NF-κB) signal pathways play a pivotal part in oxidative stress responses. There is κB target sequence in HIV promoter long terminal repeat, which is important for HIV replication and gene expression. Methyl gallate undermined NF-κB signal pathway induced by murine splenocytes through impeding nuclear import of NF-κB (p65) and cytosolic breakdown of NF-κB inhibitor. Methyl gallate inhibit HIV-1 replication through a number of target sites.[43]

## DISCUSSION

The methyl ester of gallic acid, methyl gallate (methyl 3,4,5-trihydroxybenzoate), is found in many plants including grapes. The foregoing account discloses that it has potential therapeutic applications in view of its multifarious biological activities on a diversity of cell types. It does not have significant adverse effects. Its anti-HIV actions may be related to its antioxidative activity. Its protective action against chemical-induced oxidative stress may spare the various organs from damage.

Tables about plant sources of methyl gallate and methyl gallate–derived compounds (Table 14.1), antioxidative activities of methyl gallate in cellular (Table 14.2) and noncellular (Table 14.3) systems, anti-HIV-1 enzyme and anti-HIV-1 replication activities of methyl gallate (Table 14.4), and other activities of methyl gallate including antiinflammatory, antimicrobial, antitumor, and inhibition of melanin synthesis (Table 14.5) summarizing the various aspects of methyl gallate are presented at the end of this article. Enzymes catalyzing formation of methyl gallate are present in mushrooms.[46,47]

TABLE 14.1 Plant Sources of Methyl Gallate and Methyl Gallate–Derived Compounds

| Plant Species | References |
|---|---|
| *Bauhinia racemosa* | 1 |
| *Bergenia ligulata* | 2 |
| *Byrsonima bucidaefolia* | 3 |
| *Byrsonima crassa* | 4 |
| *Calliandra haematocephala* | 5 |
| *Camellia sinensis* | 6 |
| *Canavalia gladiata* | 7 |
| *Dimocarpus longan* | 8 |
| *Drosera burmannii* | 9 |
| *Euphorbia supine* | 10 |
| *Glochidion hypoleucum* | 11 |
| *Glycine max* | 7 |
| *Klainedoxa gabonensis* | 12 |
| *Loranthus micranthus* | 13 |
| *Mangifera indica* | 14 |
| *Mangifera pajang* | 15 |
| *Paeonia rockii* | 16 |
| *Phragmanthera austroarabica* | 17 |
| *Plicosepalus curviflorus* | 18 |
| *Spondias pinnata* | 19 |
| *Tachigalia paniculata* | 20 |

TABLE 14.2 Antioxidative Activities of Methyl Gallate in Cellular Systems

| | References |
|---|---|
| Antioxidative activity | 43 |
| Peroxynitrite radical scavenging activity | 29 |
| Nitrite radical scavenging activity | 7 |
| Downregulation of miR-17-3p levels and upregulation of levels of mRNA encoding antioxidant enzymes | 28 |
| Protective effect on fat cells against oxidative stress | 29 |
| Protective effect on oxidative stress in erythrocytes and brain induced by NaF | 30,38,39 |
| Protective effect on erythrocytes against peroxyl radicals | 31 |
| Protective action against tert-butylhydroperoxide-induced oxidative stress in liver cells | 32 |
| Protective action against alcohol-induced oxidative stress in liver | 33 |
| Attenuation of hepatic injury induced by iron overload | 19 |
| Inhibition of intracellular reactive oxygen species production stimulated by hydrogen peroxide in HEK-293 human embryonic kidney cells | 34 |
| Protection of cultured Madin–Darby canine kidney cells against hydrogen peroxide–induced oxidative stress | 35 |
| Antioxidant action in SH-SY5Y neuronal cells | 36 |
| Antioxidant action in PC12 neuronal cells | 27,37 |
| Prevention of hydrogen peroxide–induced oxidative stress in human umbilical vein endothelial cells | 40 |
| Cardioprotective effect on neonatal rat cardiac myocytes exposed to hydrogen peroxide | 41 |
| Prevention of UVB irradiation–induced skin photoaging in hairless mice | 42 |

TABLE 14.3 Antioxidative Activities of Methyl Gallate in Noncellular Systems

| | References |
|---|---|
| DPPH radical scavenging activity | 26 |
| Antioxidative activity in oil-in-water emulsion system of the Kilka fish oil | 26 |
| Antioxidative activity in bulk phase system of the Kilka fish oil | 26 |

TABLE 14.4 Anti-HIV-1 Enzyme and Anti-HIV-1 Replication Activities of Methyl Gallate[43]

| Inhibitory Activity | |
|---|---|
| HIV-1 reverse transcriptase inhibitory activity | IC50 = 80.1 μM |
| HIV-1 integrase inhibitory activity | IC50 = 28.5 μM |
| HIV-1 protease inhibitory activity | 17.1% inhibition at 10 mM |
| HIV-1 replication inhibitory activity in pseudovirus-infected TZM-BL cells | IC50 = 11.9 μM |

TABLE 14.5 Other Activities of Methyl Gallate

| | References |
|---|---|
| Antiinflammatory | 24 |
| Antimicrobial | 22 |
| Antitumor | 44 |
| Inhibiting melanin synthesis | 45 |

## SUMMARY POINTS

- Methyl gallate is an antioxidant found in a great variety of plant species including some edible beans and also mushrooms.
- It protects a diversity of cells including liver cells, kidney cells, erythrocytes, heart cells, neuronal cells, and fat cells from chemical-induced oxidative stress.
- It prevents UVB-induced skin photoaging in hairless mice.
- It inhibits three key enzymes crucial to the HIV life cycle including reverse transcriptase, protease, and integrase.
- It inhibits HIV-1 replication in pseudovirus-infected TZM-BL cells.
- It exhibits other activities such as antiinflammatory, antimicrobial, antitumor, and inhibition of melanin synthesis.

## Acknowledgments

The award of NSFC grants number 81201270 and number 81471927 is gratefully acknowledged.

## References

1. Rashed K, Butnariu M. Antimicrobial and antioxidant activities of *Bauhinia racemosa* Lam. and chemical content. *Iran J Pharm Res* 2014;**13**(3):1073–80.
2. Sadat A, Uddin G, Alam M, Ahmad A, Siddiqui BS. Structure activity relationship of bergenin, p-hydroxybenzoyl bergenin, 11-O-galloylbergenin as potent antioxidant and urease inhibitor isolated from *Bergenia ligulata*. *Nat Prod Res* 2015;**29**(24):2291–4.
3. Castillo-Avila GM, García-Sosa K, Peña-Rodríguez LM. Antioxidants from the leaf extract of *Byrsonima bucidaefolia*. *Nat Prod Commun* 2009;**4**(1):83–6.
4. Bonacorsi C, Raddi MS, da Fonseca LM, Sannomiya M, Vilegas W. Effect of *Byrsonima crassa* and phenolic constituents on *Helicobacter pylori*-induced neutrophils oxidative burst. *Int J Mol Sci* 2012;**13**(1):133–41.
5. Moharram FA, Marzouk MS, Ibrahim MT, Mabry TJ. Antioxidant galloylated flavonol glycosides from *Calliandra haematocephala*. *Nat Prod Res* 2006;**20**(10):927–34.
6. Kawase M, Wang R, Shiomi T, Saijo R, Yagi K. Antioxidative activity of (-)-epigallocatechin-3-(3′-O-methyl) gallate isolated from fresh tea leaf and preliminary results on its biological activity. *Biosci Biotechnol Biochem* 2000;**64**(10):2218–20.
7. Han SS, Hur SJ, Lee SK. A comparison of antioxidative and anti-inflammatory activities of sword beans and soybeans fermented with *Bacillus subtilis*. *Food Funct* 2015;**6**(8):2736–48.
8. Sudjaroen Y, Hull WE, Erben G, Würtele G, Changbumrung S, Ulrich CM, Owen RW. Isolation and characterization of ellagitannins as the major polyphenolic components of Longan (*Dimocarpus longan* Lour) seeds. *Phytochemistry* 2012;**77**:226–37.
9. Ghate NB, Chaudhuri D, Das A, Panja S, Mandal N. An antioxidant extract of the insectivorous plant *Drosera burmannii* Vahl. alleviates iron-induced oxidative stress and hepatic injury in mice. *PLoS One* 2015;**10**(5):e0128221.
10. Nugroho A, Rhim TJ, Choi MY, Choi JS, Kim YC, Kim MS, Park HJ. Simultaneous analysis and peroxynitrite-scavenging activity of galloylated flavonoid glycosides and ellagic acid in *Euphorbia supina*. *Arch Pharmacal Res* 2014;**37**(7):890–8.
11. Anantachoke N, Kitphati W, Mangmool S, Bunyapraphatsara N. Polyphenolic compounds and antioxidant activities of the leaves of *Glochidion hypoleucum*. Natural Product Communications 2015;**10**(3):479–82.
12. Wansi JD, Chiozem DD, Tcho AT, Toze FA, Devkota KP, Ndjakou BL, Wandji J, Sewald N. Antimicrobial and antioxidant effects of phenolic constituents from *Klainedoxa gabonensis*. *Pharm Biol* 2010;**48**(10):1124–9.
13. Agbo MO, Lai D, Okoye FB, Osadebe PO, Proksch P. Antioxidative polyphenols from Nigerian mistletoe *Loranthus micranthus* (Linn.) parasitizing on *Hevea brasiliensis*. *Fitoterapia* 2013;**86**:78–83.
14. Maisuthisakul P, Gordon MH. Characterization and storage stability of the extract of Thai mango (*Mangifera indica* Linn. Cultivar Chok-Anan) seed kernels. *J Food Sci Technol* 2014;**51**(8):1453–62.
15. Ahmad S, Sukari MA, Ismail N, Ismail IS, Abdul AB, Abu Bakar MF, Kifli N, Ee GC. Phytochemicals from *Mangifera pajang* Kosterm and their biological activities. *BMC Compl Altern Med* 2015;**15**:83.
16. Picerno P, Mencherini T, Sansone F, Del Gaudio P, Granata I, Porta A, Aquino RP. Screening of a polar extract of *Paeonia rockii*: composition and antioxidant and antifungal activities. *J Ethnopharmacol* 2011;**138**(3):705–12.
17. Badr JM. Chemical constituents of *Phragmanthera austroarabica* A. G. Mill and J. A. Nyberg with potent antioxidant activity. *Pharmacognosy Res* 2014;**7**(4):335–40.
18. Badr JM, Ibrahim SR, Abou-Hussein DR. Plicosepalin, A, a new antioxidant catechin-gallic acid derivative of inositol from the mistletoe *Plicosepalus curviflorus*. *Z für Naturforsch C* 2016;**71**(11–12):375–80.
19. Chaudhuri D, Ghate NB, Singh SS, Mandal N. Methyl gallate isolated from *Spondias pinnata* exhibits anticancer activity against human glioblastoma by induction of apoptosis and sustained extracellular signal-regulated kinase 1/2 activation. *Pharmacognosy Mag* 2015;**11**(42):269–76.
20. Cioffi G, D'Auria M, Braca A, Mendez J, Castillo A, Morelli I, De Simone F, De Tommasi N. Antioxidant and free-radical scavenging activity of constituents of the leaves of *Tachigalia paniculata*. *J Nat Prod* 2002;**65**(11):1526–9.
21. Lee JK. Anti-Depressant Like Effect of methyl gallate isolated from *Acer barbinerve* in mice. *Korean J Physiol Pharmacol* 2013;**17**(5):41–46.
22. Choi JG, Mun SH, Chahar HS, Bharaj P. Methyl gallate from *Galla rhois* successfully controls clinical isolates of *Salmonella* infection in both in vitro and in vivo systems. *PLoS One* 2014;**9**(7):e102697.
23. Acharyya S, Sarkar P, Saha DR, Patra A, Ramamurthy T, Bag PK. Intracellular and membrane-damaging activities of methyl gallate isolated from *Terminalia chebula* against multidrug-resistant *Shigella* spp. *J Med Microbiol* 2015;**64**(8):901–9.

24. Correa LB, Pádua TA, Seito LN, Costa TE, Silva MA, Candéa AL, Rosas EC, Henriques MG. Anti-inflammatory effect of methyl gallate on experimental arthritis: inhibition of neutrophil recruitment, Production of inflammatory mediators, and activation of macrophages. *J Nat Prod* 2016;**79**(6):1554–66.
25. Kim H, Lee G, Sohn SH, Lee C, Kwak JW, Bae H. Immunotherapy with methyl gallate, an inhibitor of Treg cell migration, enhances the anti-cancer effect of cisplatin therapy. *Korean J Physiol Pharmacol* 2016;**620**(3):261–8.
26. Asnaashari M, Farhoosh R, Sharif A. Antioxidant activity of gallic acid and methyl gallate in triacylglycerols of Kilka fish oil and its oil-in-water emulsion. *Food Chem* 2014;**159**:439–44.
27. a Crispo JA, Ansell DR, Piche M, Eibl JK, Khaper N, Ross GM, Tai TC. Protective effects of polyphenolic compounds on oxidative stress-induced cytotoxicity in PC12 cells. *Can J Physiol Pharmacol* 2010;**488**(4):429–38.
b Hsu FL, Huang WJ, Wu TH, Lee MH, Chen LC, Lu HJ, Hou WC, Lin MH. Evaluation of antioxidant and free radical scavenging capacities of polyphenolics from pods of *Caesalpinia pulcherrima*. *Int J Mol Sci* 2012;**13**(5):6073–88.
28. Curti V, Capelli E, Boschi F, Nabavi SF, Bongiorno AI, Habtemariam S, Nabavi SM, Daglia M. Modulation of human miR-17-3p expression by methyl 3-O-methyl gallate as explanation of its *in vivo* protective activities. *Mol Nutr Food Res* 2014;**58**(9):1776–84.
29. Rahman N, Jeon M, Kim YS. Methyl gallate, a potent antioxidant inhibits mouse and human adipocyte differentiation and oxidative stress in adipocytes through impairment of mitotic clonal expansion. *Biofactors* 2016;**42**(6):716–26.
30. Nabavi SM, Habtemariam S, Nabavi SF, Moghaddam AH, Latifi AM. Prophylactic effects of methyl-3-O-methyl gallate against sodium fluoride-induced oxidative stress in erythrocytes *in vivo*. *J Pharm Pharmacol* 2013a;**65**(6):868–73.
31. Wu J, Sugiyama H, Zeng LH, Mickle D, Wu TW. Evidence of Trolox and some gallates as synergistic protectors of erythrocytes against peroxyl radicals. *Biochem Cell Biol* 1998;**76**(4):661–4.
32. Oidovsambuu S, Kim CY, Kang K, Dulamjav B, Jigjidsuren T, Nho CW. Protective effect of *Paeonia anomala* extracts and constituents against tert-butylhydroperoxide-induced oxidative stress in HepG2 cells. *Planta Med* 2013;**79**(2):116–22.
33. Zhang X, Wu Z, Weng P. Antioxidant and hepatoprotective effect of (-)-epigallocatechin 3-O-(3-O-methyl) gallate (EGCG3″Me) from Chinese oolong tea. *J Agric Food Chem* 2014;**62**(41):10046–54.
34. Anantachoke N, Kitphati W, Mangmool S, Bunyapraphatsara N. Polyphenolic compounds and antioxidant activities of the leaves of *Glochidion hypoleucum*. *Nat Prod Commun* 2015;**10**(3):479–82.
35. Hsieh TJ, Liu TZ, Chia YC, Chern CL, Lu FJ, Chuang MC, Mau SY, Chen SH, Syu H, Chen CH. Protective effect of methyl gallate from *Toona sinensis* (Meliaceae) against hydrogen peroxide-induced oxidative stress and DNA damage in MDCK cells. *Food Chem Toxicol* 2004;**42**(5):843–50.
36. Habtemariam S. Methyl-3-O-methyl gallate and gallic acid from the leaves of *Peltiphyllum peltatum*: isolation and comparative antioxidant, prooxidant, and cytotoxic effects in neuronal cells. *J Med Food* 2011;**14**(11):1412–2148.
37. Crispo JA, Piché M, Ansell DR, Eibl JK, Tai IT, Kumar A, Ross GM, Tai TC. Protective effects of methyl gallate on H2O2-induced apoptosis in PC12 cells. *Biochem Biophys Res Commun* 2014;**393**(4):773–8.
38. Nabavi SF, Nabavi SM, Habtemariam S, Moghaddam AH, Sureda A, Mirzaei M. Neuroprotective effects of methyl-3-O-methyl gallate against sodium fluoride-induced oxidative stress in the brain of rats. *Cell Mol Neurobiol* 2013;**33**(2):261–7.
39. Nabavi SF, Habtemariam S, Di Lorenzo A, Sureda A, Khanjani S, Nabavi SM, Daglia M. Post-stroke depression modulation and in vivo anti-oxidant activity of gallic acid and its synthetic derivatives in a murine model system. *Nutrients* 2016;**8**(5). pii:E248.
40. Whang WK, Park HS, Ham IH, Oh M, Namkoong H, Kim HK, Hwang DW, Hur SY, Kim TE, Park YG, Kim JR, Kim JW. Methyl gallate and chemicals structurally related to methyl gallate protect human umbilical vein endothelial cells from oxidative stress. *Exp Mol Med* 2005;**37**(4):343–52.
41. Khurana S, Hollingsworth A, Piche M, Venkataraman K, Kumar A, Ross GM, Tai TC. Antiapoptotic actions of methyl gallate on neonatal rat cardiac myocytes exposed to $H_2O_2$. *Oxidative Med Cell Longevity* 2014:657512.
42. Sun ZW, Hwang E, Lee HJ, Lee TY, Song HG, Park SY, Shin HS, Lee DG, Yi TH. Effects of *Galla chinensis* extracts on UVB-irradiated MMP-1 production in hairless mice. *J Nat Med* 2015;**69**(1):22–34.
43. Wang CR, Zhou R, Ng TB, Wong JH, Qiao WT, Liu F. First report on isolation of methyl gallate with antioxidant, anti-HIV-1 and HIV-1 enzyme inhibitory activities from a mushroom (*Pholiota adiposa*). *Environ Toxicol Pharmacol* 2014;**37**(2):626–37.
44. Lee H, Lee H, Kwon Y, Lee JH, Kim J, Shin MK, Kim SH, Bae H. Methyl gallate exhibits potent antitumor activities by inhibiting tumor infiltration of CD4+CD25+ regulatory T cells. *J Immunol* 2010;**185**(11):6698–705.
45. Kim IW, Jeong HS, Kim JK, Lee JK, Kim HR, Yun HY, Baek KJ, Kwon NS, Park KC, Kim DS. Methyl gallate from *Acer barbinerve* decreases melanin synthesis in Mel-Ab cells. *Pharmazie* 2015;**70**(1):55–9.
46. Kirita M, Tanaka Y, Tagashira M, Kanda T, Maeda-Yamamoto M. Purification and characterization of a novel O-methyltransferase from *Flammulina velutipes*. *Biosci Biotechnol Biochem* 2014;**78**(5):806–11.
47. Kirita M, Tanaka Y, Tagashira M, Kanda T, Maeda-Yamamoto M. Cloning and characterization of a novel O-methyltransferase from *Flammulina velutipes* that catalyzes methylation of pyrocatechol and pyrogallol structures in polyphenols. *Biosci Biotechnol Biochem* 2015;**79**(7):1111–8.

CHAPTER

# 15

# Taurine and Oxidative Stress in HIV

*Roberto C. Burini*[1], *Maria D. Borges-Santos*[1], *Fernando Moreto*[1], *Yong Ming-Yu*[2]

[1]Sao Paulo State University, Botucatu, Brazil; [2]Shriners Burns Hospital – Massachusetts General Hospital, Harvard Medical School, Boston, MA, United States

OUTLINE



## Abstract

Taurine (Tau) (2-aminoethanesulfonic acid) is an amino acid, lacking the carboxyl group and messenger ribonucleic acid. As a major intracellular free amino acid, Tau regulates the osmolality and cell membranes stabilization. Moreover, due to its ability to generate conjugates with bile acids, xenobiotics, retinoic acid, and chloramine, Tau is involved in a diverse array of physiological functions, including detoxification, osmoregulation, membrane stabilization, calcium modulation, neurotransmitter agonist, antioxidation, and immunomodulation. As antioxidant Tau scavenges the phagocyte microbicidal agent HOCl to form the more stable and less toxic taurine chloramine (Tau-Cl) therefore acting as cytoprotectant, in the attenuation of apoptosis. As antiinflammatory agent Tau-Cl suppresses superoxide anion and decreases both NO and proinflammatory cytokines secretion by the activated phagocytes, body Tau comes from diet or from its endogenous biosynthesis from methionine (Met) and cysteine (Cys). Tau occurs naturally in food, especially in seafood and meat. The main organs involved in Tau metabolism are the gut, liver, and kidneys. Usually, plasma Tau levels decrease in response to surgical injury, trauma, sepsis, and cancer. In our data with HIV+ patients the low plasma Tau followed the other thiol-antioxidant pattern. Cys supplementation and methionine loading in HIV+ resulted in higher production of glutathione (GSH) and Tau than non-HIV+ controls. Tau normalization was obtained by methionine loading and was found associatively to similar transmethylation and remethylation of Met and lower transsulfuration of homocysteine compared with controls. Thus, the increased flux of Cys into GSH and Tau pathways seems to be a host strategy to strengthen the cellular antioxidant capacity against the HIV progression.

**Keywords:** Antiinflammatory actions; Antioxidant actions; Taurine in diseases; Taurine in HIV+; Taurine metabolism.

## List of Abbreviations

**ATP** adenosine triphosphate
**AUC** area under the curve


http://dx.doi.org/10.1016/B978-0-12-809853-0.00015-8

**CBS** cystathionine B-synthase
**CD4+** lymphocyte CD4+
**CD8+** lymphocyte CD8+
**CDO** cysteine dioxygenase
**CNS** central nervous system
**$CO_2$** carbon dioxide
**CSD** cysteine sulfinic acid decarboxylase
**Cys** cysteine
**GABA** gamma-aminobutyric acid
**GG** glutathione genesis
**Gln** glutamine
**GSH** glutathione reduced form
**GSSG** oxidized glutathione
**Hcy** homocysteine
**HHcy** hyperhomocysteinemia
**HIV+** human virus
**$H_2O_2$** hydroperoxyde
**HOCl** hypochlorous acid
**HPLC** high-performance liquid chromatography
**IKK** IkB kinase
**iNOS** inducible nitric oxide synthase
**IL-6** interleukin 6
**IL-8** interleukin 8
**LPS** lipopolysaccharide
**Met** methionine
**MetLo** methionine load
**MPO** myeloperoxidase
**mRNA** messenger ribonucleic acid
**NAC** N-acetyl cysteine
**NF-kB** nuclear factor kappa B
**NO** nitric oxide
**PMN** polymorphonuclear leukocyte
**RM** remethylation
**ROS** reactive oxygen species
**SAMe** S-adenosylmethionine
**Ser** serine
**Tau** taurine
**Tau-Cl** taurine chloramine
**TG** taurine genesis
**TM** transmethylation
**TNF** Tumor necrosis factor
**tRNA** transport ribonucleic acid
**TS** transsulfuration
**UD** usual diet
**yIFN** gamma interferon

# INTRODUCTION

## Importance of Taurine

Taurine (Tau) is an aminosulfonic acid (2-aminoethanesulfonic acid) present in mammalian tissues in millimolar concentrations. The chemical structure of Tau reveals that it lacks the carboxyl group typical of other amino acids but does contain a sulfonate group. Tau was first identified and isolated from the bile of the ox (*Bos taurus*), from which it derives its name.[1–3]

Many studies have revealed that Tau is involved in various physiological processes.[4–6] In fact, as a major intracellular free amino acid, Tau acts as an organic osmolyte involved in cell volume regulation, modulates calcium transport, and plays a role in the modulation of intracellular free calcium concentration.[7] Therefore, Tau regulates the osmolality and stabilizes cell membranes, and the ability of Tau to generate Tau conjugated with bile acids, xenobiotics, retinoic acid, and Tau-Cls are also examples of protective functions.[8] Overall, Tau is involved in a diverse array of biological and physiological functions, including detoxification, osmoregulation, membrane stabilization, calcium modulation, antioxidation, and immunomodulation.[3] However, Tau has thus far not been found as a component of a protein or nucleic acid and its precise biochemical mechanism(s) is unclear. Although not incorporated into proteins, Tau is considered to be an essential amino acid for felines and a conditionally indispensable amino acid for humans and nonhuman primates.[5]

The first known function of Tau in the body was for bile acid conjugation. Traditionally, bile acids are known for their ability to act as solubilizing agents in the gut, aiding in the absorption of dietary lipids through the formation of mixed micelles. More recently evidence has been collected to prove that bile acids also act as endocrine-signaling molecules that activate nuclear and membrane receptors to control integrative metabolism and energy balance.[9–12]

Perhaps the most enigmatic question regarding Tau is whether it is a neurotransmitter. Tau is one of the most abundant-free amino acids in the central nervous system (CNS) and plays multiple roles, including osmoregulation, neuroprotection, and neuromodulation.[13] The structural resemblance between gamma-aminobutyric acid (GABA) and the similar distributions of these amino acids and their synthesizing enzymes in various regions of the brain, added to the evidence that Tau, when applied to CNS neurons, exerts an inhibitory effect on their firing rate has all contributed to the view that Tau is indeed a neurotransmitter. In fact, Tau can interact with $GABA_B$ receptors to activate a metabotropic pathway and has been known to act as an agonist at receptors of the inhibitory GABA and glycinergic neurotransmitter systems.[14]

Moreover, the effect of Tau has a similar magnitude to that produced by insulin. Insulin produces a catabolic effect by reducing food intake and increasing energy expenditure. It is well known that insulin is a potent anorexigenic hormone, and insulin receptors are widely distributed throughout the CNS.[15]

Analysis of hypothalamic protein expression suggested that Tau exhibits an anorexigenic action in the hypothalamus and enhances the effect of insulin on the control of food intake. It has been reported that a Tau-supplemented diet increases hypothalamic concentrations of Tau,[16] and the level of satiety was significantly greater after a Tau-rich fish meal as compared with a beef or chicken meal.[17] Therefore, it is suggested that Tau exhibited an anorexigenic action in the hypothalamus and enhanced the effect of insulin on the control of food intake.[3] Nevertheless, neither the intracellular link nor a Tau-specific receptor has yet to be identified at the molecular level. Hence, besides being a potent neuronal protective agent, this ubiquitous amino acid is presently considered only to be a neurotransmitter candidate.[2]

## Taurine as Antioxidant

Early events in inflammation include migration of leukocytes to the site of injury. These inflammatory cells produce high levels of hypochlorous acid (HOCl), a major neutrophil microbicidal agent, which is produced by the myeloperoxidase (MPO)-catalyzed oxidation of $Cl^-$ by $H_2O_2$ during the respiratory burst (Fig. 15.1).

Neutrophils and monocytes contain high levels of MPO, which, along with $H_2O_2$, catalyze the formation of the potent oxidant, HOCl. Tau reacts with HOCl to produce the less reactive and long-lived oxidant Tau-Cl. From this, Tau has been shown to be tissue-protective in many models of oxidant-induced injury.[18] Tau is the single most abundant amino acid in leukocytes with an intracellular concentration of 20–50 mM.[19] Stimulated neutrophils and macrophages release large quantities of Tau that are rapidly chlorinated by the reaction with hypochlorous acid, generating Tau-Cl.[8]

The protective role of Tau in mitochondrial morphology and function has been demonstrated in a variety of organs and cells, including the brain,[20] heart,[21] kidney,[22] liver,[23] retina,[24] pancreatic islets,[25] and vascular cells.[26]

Tau, by virtue of its antioxidant activity, has been shown to play a crucial role as a cytoprotectant and in the attenuation of apoptosis. It was found that Tau deficiency reduces the expression of the respiratory chain components required for normal translation of mitochondrial-encoded proteins. It has been proposed that the dysfunctional respiratory chain accumulates electron donors, thereby diverting electrons from the respiratory chain to oxygen, and forming superoxide anion in the process. Tau seems to serve as a regulator of mitochondrial protein synthesis; thereby, enhancing the electron transport chain activity and protecting the mitochondria against excessive superoxide generation. Increasing Tau levels restores respiratory chain activity and increases the synthesis of adenosine triphosphate at the expense of superoxide anion production.[27] Tau indirectly acts as an antioxidant by providing sufficient pH buffering in the mitochondrial matrix. These facts suggest that Tau could maintain mitochondrial function.[3]

Moreover, molecular studies on the function of Tau have provided evidences that Tau can be a constituent of biologic macromolecules. Specifically, two novel Tau-containing modified uridines have been found in human mitochondrial transport ribonucleic acids, Tau-containing modified uridines (5-taurinomethyluridine and 5-taurinomethyl-2-thiouridine).[18] An absence of Tau modified mitochondrial uridine was found in the cells from the mitochondrial diseases, what suggest that Tau is essential for maintaining the normal function of mitochondria.[28]

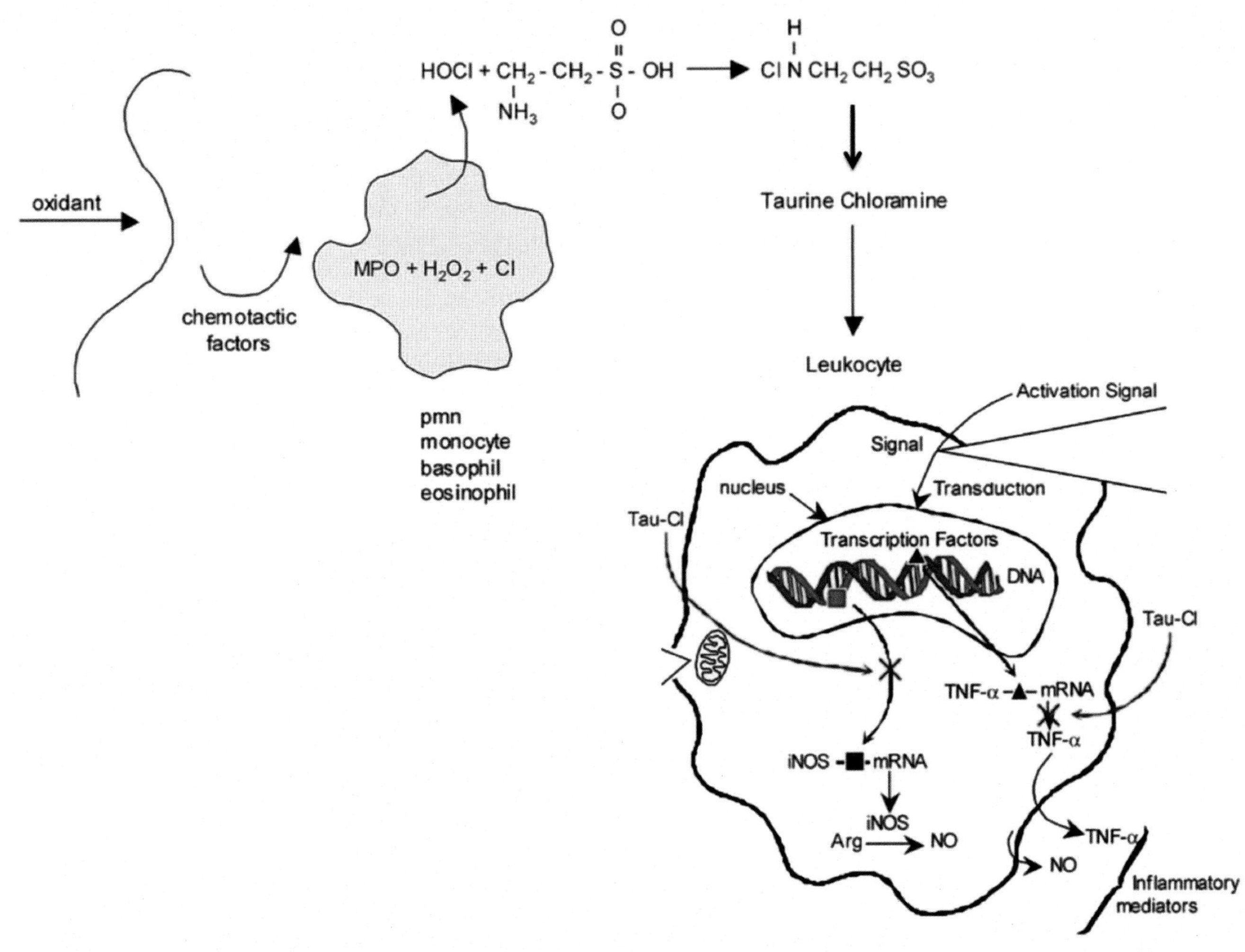


FIGURE 15.1 Formation of taurine chloramine (Tau-Cl) during inflammation and mechanism(s) utilized by Tau-Cl to inhibit production of inflammatory mediators by immune-responsive cells. *GG*, glutathione synthesis; *iNOS*, inducible nitric oxide synthase; *MPO*, myeloperoxidase; *RM*, remethylation; *TG*, taurine synthesis; *TM*, transmethylation; *TNF*, tumor necrosis factor; *TS*, transsulfuration. *Adapted in part from Schuller-Levis GB, Park E. Taurine and its chloramine: modulators of immunity.* Neurochem Res 2004;**29**:117–126; *Jong CJ, Azuma J, Schaffer S. Mechanism underlying the antioxidant activity of taurine: prevention of mitochondrial oxidant production.* Amino Acids 2012;**42**:2223–32.

## Taurine as Antiinflammatory

Tau-Cl is considered as a physiological modulator of the inflammatory response.[29] Data have shown that Tau-Cl can be actively transported into leukocytes and can downregulate the production of inflammatory mediators[18] (Fig. 15.1).

In fact, Tau-Cl inhibits the synthesis and production of proinflammatory substances, ROS, and reactive nitrogen species in inflammatory cells.[30–33]

In activated cell lysates, the presence or absence of Tau-Cl, showed that NO synthase (inducible nitric oxide synthase (iNOS)) protein was absent from cells activated with lipopolysaccharide and gamma interferon–Q. Tau-Cl significantly inhibited iNOS messenger ribonucleic acid (mRNA) by inhibiting the transcription of the iNOS gene.[34] Additionally, Tau-Cl delayed the peak expression of tumor necrosis factor (TNF)-alpha mRNA indicating that although TNF-alpha mRNA is present, the translation of this message is impaired by the presence of Tau-Cl.[34] Thus, Tau-Cl decreases both NO and TNF secretion by the activated macrophages in a manner that involves changes at the transcriptional levels of iNOS and translational levels of TNF expression, as well as by inhibiting iNOS itself (Fig. 15.1).

Studies have demonstrated that Tau-Cl suppresses superoxide anion, interleukin 6 (IL-6), and interleukin 8 (IL-8) production in activated human peripheral blood polymorphonuclear leukocytes,[35] and that Tau-Cl inhibited the production of proinflammatory cytokines IL-6 and IL-8 by fibroblast-like synoviocytes isolated from rheumatoid arthritis patients.[36] The antiinflammatory effect of Tau is attributed to the inhibition of nuclear factor kappa B (NF-kB) activation.[37] In fact, Tau-Cl diminishes the activity of NF-kB and to a lesser extent that of AP-1 transcription factor.

Barua et al.[38] have demonstrated that Tau-Cl depresses the NF-kB migration into the nucleus and caused a more sustained presence of IkB in the cytoplasm. Furtheron, Kanayama et al.[37] reported a Tau-Cl-induced inhibition of NF-kB activation by the oxidation of IkB-alpha. In additional experiments, Tau-Cl did not directly inhibit IkB kinase (IKK) activity suggesting that Tau-Cl exerts its effects at some level upstream of IKK in the signaling pathway. Hence, the mechanism of action of Tau-Cl has shown that it inhibits the activation of NF-kB, a potent signal transducer for inflammatory cytokines, by oxidation of IKK at methionine (Met).[18] (Fig. 15.1). Presently, the involvement of heme oxygenase-1 has also been reported in the antiinflammatory activity of Tau-Cl.[39]

It is assumed that Tau-Cl produced by immune cells such as macrophages and leukocytes plays an immunomodulatory role at the site of inflammation by suppressing not only the production of reactive oxygen species (ROS) but also the proinflammatory cytokines.[29] Since Tau-Cl is formed at the site of inflammation, neutrophil cell death and neutrophil-induced death at the inflammatory site would likely be apoptotic. Apoptotic cell death, in contrast to necrotic cell death, is a physiologic advantage in that cells are cleared by phagocytosis lessening tissue damage. Hence, Tau-Cl may promote apoptotic cell death and thereby decrease the detrimental effects of inflammation.[18] Thus, Tau-Cl, a stable oxidant, can be produced at the site of inflammation and downregulate proinflammatory cytokine production leading to a significant reduction in the immune response. Tau may provide a useful prophylactic approach to prevent resulting tissue damage. Taurolidine, a derivative of Tau, is commonly used in Europe as an adjunctive therapy for various infections as well as for tumor therapy.[18]

## ENDOGENOUS TAURINE

Tau is one of the most abundant amino acids in the body, making up around 0.1% of our body weight. In mammalian tissues, Tau is ubiquitous and is the most abundant-free amino acid in the brain and spinal cord, leukocytes, heart and muscle cells, the retina, and indeed almost every tissue throughout the body.[2,18]

Tau is endogenously synthesized from Met and cysteine (Cys). The first step in the metabolism of Met is its conversion to the intermediate, S-adenosylmethionine (SAMe). In transmethylation (TM) reactions, SAMe serves primarily as the universal methyl donor to a variety of acceptors including nucleic acids, proteins, phospholipids, and biologic amines.[40]

SAMe occupies a central position in the metabolism of all cells as a precursor molecule to three main pathways: methylation, transsulfuration (TS), and aminopropylation. In the TS pathway, homocysteine (Hcy) condenses with serine to form cystathionine, an irreversible reaction catalyzed by cystathionine B-synthase (CBS) in the presence of vitamin B-6 as a cofactor. Cystathionine is hydrolyzed by β-cystathionase (EC 4.4.1.1) to form Cys and α-ketobutyrate. Cys reacts with glutamate and glycine through two consecutive reactions to form the tripeptide glutathione (GSH).[41,42]

The key regulatory enzyme for Tau metabolism is cysteine dioxygenase (CDO, EC 1.13.11.20) that catalyzes the conversion of L-Cys to Cys sulfinate and via cysteine sulfinic acid decarboxylase (CSD) the oxidation of hypotaurine(2-aminoethane sulfinate) to Tau as the final step.[2,43] CSD, the rate-limiting enzyme for Tau biosynthesis, whereas CDO has a critical role in determining the flux of Cys between Cys catabolism/Tau synthesis and GSH[18] (Fig. 15.3).

Key enzymes for Tau biosynthesis have been cloned. CDS has been cloned and sequenced in the mouse, rat, and human. Human CSD has been registered in the GenBank (access on number: AF116548). Another key enzyme for Cys metabolism, CDO, has also been cloned from rat liver.[18] CSD mRNA is expressed in kidney and liver and was not detected in lymphoid tissues and lung. These data suggest that lymphoid tissue may not synthesize Tau directly and may rely on transport of Tau.[18]

The levels of CDO activity changed by dietary protein level, in addition to Cys availability, are key factors in determining the flux of Cys between Cys catabolism/Tau synthesis and GSH synthesis.[43] Excess sulfur amino acids or proteins increase CDO activity and CDO protein.

The main organs involved in Tau metabolism are the gut, liver, and kidneys. The gut regulates Tau uptake from the diet by a specific Tau transporter.[44] The liver is involved in endogenous Tau production and in formation of bile acids containing Tau. The kidneys are the main sites of excretion of Tau.[45]

Tau synthesis occurs in the liver and brain from Met or Cys via cysteic acid or hypotaurine. In the heart and kidney Tau synthesis occurs via cysteamine. Tau is actively synthesized also in the white adipose tissue.[46]

The gut release of Tau depends on intraluminal turnover as part of the enterohepatic cycle because the gut lacks the appropriate enzymes for Tau production.[47] Normal human duodenum has a very high intramucosal concentration of Tau (90 times higher than plasma level) suggesting a gut storage function for Tau. Probably, the gut releases Tau during the fasted state as results from uptake of deconjugated Tau containing bile acids.[1] In the human adult,

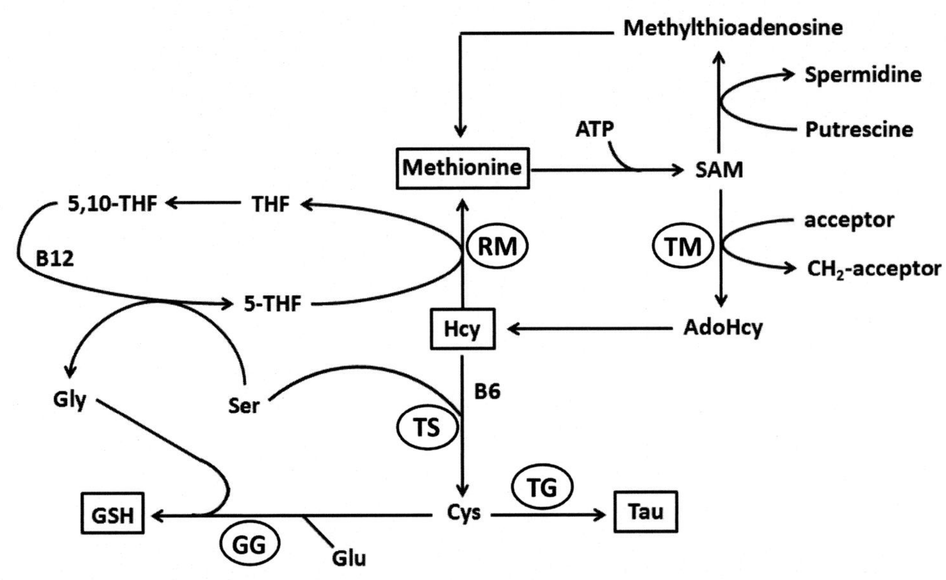


FIGURE 15.2 The methionine cycle and derived synthesis of homocysteine, glutathione, and taurine. *Cys*, cysteine; *GG*, glutathione synthesis; *GSH*, glutathione; *Hcy*, homocysteine; *RM*, remethylation; *SAMe*, S-adenosylmethionine; *Ser*, serine; *Tau*, taurine; *TG*, taurine synthesis; *TM*, transmethylation; *TS*, transsulfuration. *Adapted from Burini RC, Lamonica VC, Moreto F, Yu YM, Henry MACA. Thiol metabolic changes induced by oxidative stress and possible role of B-Vitamins supplements in esophageal cancer Patients HIV. In: Wilber A, editors.* Glutathione: dietary sources, role in cellular functions and therapeutic effects. *New York: Nova Biomedical; 2015. p. 101–26 (Chapter 5).*

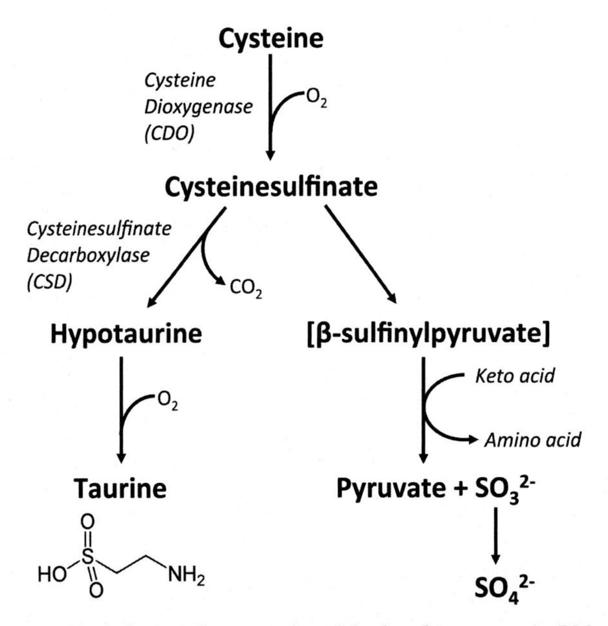


FIGURE 15.3 The main steps in the conversion of L-cysteine to taurine. *GG*, glutathione genesis; *RM*, remethylation; *TG*, taurine genesis; *TM*, transmethylation; *TS*, transsulfuration, $*P < .05$ (HIV≠Co). *Adapted from Ripps H., Shen W. Review: taurine: a "very essential" amino acid.* Mol Vis *2012;**18**:2673–86.*

about one-fourth of bile acids are conjugated with Tau and a small fraction of Tau is also converted to isethionate by either bacterial or tissue enzymes and may be converted in part to sulfate, $CO_2$, water, and ammonia, the last being converted to urea.[48]

Tau concentrations in whole blood are 3.5 higher than in plasma.[1] Total body Tau is regulated by the kidney. Tau is a major urinary amino acid in humans because the capacity of renal uptake is low.[48] Daily Tau losses in urine are diet-dependent but generally range from 65 to 250 mg (0.5–2.0 mmoL). To maintain adequate level of Tau in the tissues,

Tau is tightly regulated by excretion and reabsorption by the kidney.[49] The Tau transporter in proximal tubule brush-border membranes appears to be the primary target for adaptive regulation by dietary availability of Tau.

## EXOGENOUS SOURCES AND HUMAN REQUIREMENTS

Because it is one of the few amino acids not used in protein synthesis, Tau is often referred to as a "nonessential" amino acid, or more generously as a "conditionally essential" amino acid, for humans.[2]

Tau occurs naturally in food, especially in seafood (fish and shellfish) and meat. The mean daily intake from omnivore diets was determined to be around 58 mg.[50] Because the level of CSD is low in humans, the Tau has been added to infant formula as well as to parenteral solutions. On the other hand, some foods or drinks boosters, eye drops, and eardrops, contain a considerable amount of Tau.[51] In fact, Tau-containing health drinks, such as Red Bull[52] usually containing about 1 g of Tau, are marketed worldwide for the treatment of various conditions, for improvement of athletic performance, and for general well-being.[53]

Tau is a relatively nontoxic substance and a normal constituent of the human diet. Animal studies have not indicated toxicity due to Tau. Tau has already been used intravenously in humans in doses of up to 5 g[54] and 2–6 g/day orally for a period of 6 months in children with fatty liver[55] without any toxic side effect. The therapeutic effects of Tau on several diseases[56–59] have been reported.

In physiologic fluid, extracellular Tau concentrations range from 50 to 100 mM after Tau supplementation.[60] Pharmacokinetics of Tau in healthy male volunteers following orally administration (4g Tau) in the fasting state showed that the oral Tau was absorbed from the gastrointestinal tract 1–2.5 h following administration, and then eliminated from plasma by first-order kinetics. The maximum plasma Tau concentration measured at $1.5 \pm 0.6$ h after administration was $86.1 \pm 19.0$ mg/L ($0.69 \pm 0.15$ mmol) and plasma elimination half-life $1.0 \pm 0.3$ h. Plasma Tau returned to endogenous concentrations after 6–8 h of study.[50]

## TAURINE IN CHRONIC DISEASES

The overproduction of reactive oxygen species and the body's inability to stem the accumulation of highly reactive free radicals has been implicated in cardiovascular disease, diabetes-induced renal injury, inflammatory disease, light-induced lipid peroxidation in photoreceptors, reperfusion injury,[61] and several of the major disorders of the CNS.[62] Despite this diversity of pathophysiology in so varied a group of seemingly unrelated disorders, there is a growing consensus that oxidative stress is linked to mitochondrial dysfunction,[62,63] and that the beneficial effects of Tau[64] are a result of its antioxidant properties,[33,65] as well as its ability to improve mitochondrial function by stabilizing the electron transport chain and inhibiting the generation of reactive oxygen species.[27,66]

Human studies showed that the Tau content of the body is lower in subjects with obesity[67] and diabetes.[68] A global survey showed that 24-h urinary Tau excretion was inversely associated with BMI, blood pressure, and plasma cholesterol in humans.[3]

Epidemiological studies revealed that people who consume seafood, a good source of Tau, have a lower risk of developing metabolic diseases such as obesity, diabetes, dyslipidemia, and hypertension.[69,70] Also, it is described that Tau supplementation increases plasma Tau levels, reduces plasma levels of inflammatory and oxidative markers, and increases plasma adiponectin levels in humans.[67] In addition, Tau-Cl, an endogenous product derived from activated neutrophils, has been reported to suppress obesity-induced oxidative stress and inflammation in adipocytes.[3]

## TAURINE IN CRITICAL ILLNESS

Although reports of decreased plasma levels of Tau in trauma, sepsis, and critical illness are available, very little is known about the relationships among changes in plasma Tau, other amino acid levels, and metabolic variables. A large series of plasma amino acid profiles were obtained in 250 trauma patients with sepsis who were undergoing total parenteral nutrition.[71] The results, which characterized the relationships between plasma Tau and other amino acid levels in sepsis, provide evidence that the more severe decreases in plasma Tau correlate with the worsening of metabolic and cardiorespiratory patterns.[18] In patients with esophagus cancer, plasma Tau levels correlated with their positive evolution and survival.[72] Thus, Tau plasma levels decrease in response to surgical injury, trauma, sepsis, and cancer, which suggests an increased metabolic need.[73,74]

## PLASMA THIOL-ANTIOXIDANT DEFENSES IN HIV+ PATIENTS

### Methodological Design

We had conducted experiments involving HIV+ patients in a randomized controlled supplementation design already described elsewhere.[75–77]

The HIV+ group had been diagnosed clinically and by the laboratory viral load and CD4+ and CD8+ lymphocyte counts. All patients had been under highly active antiretroviral treatment for at least 1 year. The healthy control group consisted of adults who were negative for HIV and clinically healthy. After a baseline assessment the two groups were randomly assigned to different dietary supplements (N-acetyl cysteine (NAC) 1 g/d or glutamine (Gln) 20 g/d) with their usual diet (UD) as the baseline and washout in a crossover design. All dietary supplements were administered throughout the consecutive 7-d period preceded and followed by fasting blood sampling. In all dietary regimen, the subjects were submitted, immediately after the overnight-fast blood drawn, to an oral MetLo (0.1 g Met/kg body wt) with blood samples taking 2 and 4h after the MetLo. The area under the curve (AUC) was calculated for sulfur amino acids and GSH assayed by the high-performance liquid chromatography (LC10AD Shimadzu, Japan).[75]

The GSH and free amino acids were used as metabolic markers either isolated or combined to designate thiol-antioxidant capacity (GSH:oxidized glutathione (GSSG) ratio),[78] remethylation (RM) of Hcy (Met:Hcy ratio),[79] or trans-sulfuration of Met(Cys/Met ratio) or of Hcy (Cys:Hcy + Ser ratio) all in situations of either UD or NAC-supplemented diet under overnight fasting.[80] Additionally, the plasma Tau/Cys ratio was defined arbitrarily as "taurine genesis" (TG) and GSH/Cys as "glutathione genesis" (GG).[77,81]

### Data Results and Discussion

HIV+ group presented lower levels of amino acids (Met, Hcy, and Cys), similar RM (Met:Hcy ratio), and lower transsulfuration (Cys:Hcy + Ser ratio) in association with lower GSH concentrations and lower antioxidant capacity (GSH:GSSG ratio) than controls (Co).[80]

Muller et al.[82] found lower Met concentrations in HIV+ than control individuals as we did. The Met metabolism-derived polyamines (putrescine, spermine, and spermidine Fig. 15.2) present in all living cells have been implicated in the replication of some viruses, and elevated levels of these polyamines have been found in lymphocytes of patients infected with HIV-1. In our study the HIV+ plasma levels of Met were comparable to the non-HIV+ group only in the presence of NAC supplementation.[80] In healthy humans it has been described a spared effect of dietary Cys to the Met oxidation.[83]

When compared to the UD the presence of NAC resulted in increased Hcy 3 × higher in HIV+ than the healthy control group.[80] The conservation of Hcy seems crucial (at least beneficial) to the HIV+ milieu due to its prooxidant properties and both chronic and acute Met-induced homocysteinemia are associated with higher status of oxidative stress.[84]

The Hcy levels in blood are a sensitive indicator of folate, and vitamin B12 deficiencies and plasma Hcy can be lowered with B-vitamin supplementation.[85] In our studies, MetLo led to increased Met and Hcy approaching the control values. One way to HIV+ reduce the RM is by limiting folate levels. However, presently TM was statistically similar in both groups. Therefore, it seems that the clearance of Hcy, in this case was normal by the RM pathway even considering the lower plasma folate state found along with normal vitamin B12, in these antiretroviral therapy (ART)-treated patients.[75,77]

The prevalence of hyperhomocysteinemia (HHcy) in HIV+ patients has been reported to be between 12.3% and 35% [113]. In our study the prevalence of HHcy (>10 μmol/L) was 65% in Co and 50% in human virus (HIV+). HHcy in HIV+ patients has been correlated with B-vitamin deficiencies, ARTs, and HIV+ comorbidities [107]. Hence, in this study the NAC supplementation normalized not only Met but also Hcy and GSH in association of normal RM and reduced TS. In spite of normalizing GSH, the antioxidant capacity (GSH/GSSG) continued lower in HIV+, compared to Co.[80]

The conclusion in these series of experiments was that NAC supplementation along with reduced TS pathway spared Hcy (which increased significantly) and serine. The existing serine was then diverted from the Cys to the glycine pathway, which was consumed with the spared Cys (from NAC) to normalize GSH levels. On the other hand, the increasing and normalization of GSH, induced by Gln supplementation, was probably achieved by generating glycine (from glutamic acid) and sparing serine to form Cys (from Hcy), with all three (glutamic acid, glycine, and Cys) together generating GSH.[76,86] Thus, the NAC and Gln, through different mechanisms, were able to supply substrates to increase GSH levels. However, normalized GSH was unable to restore Cys or glutamic acid concentrations to the normal control level, and neither of the two supplements had significant effects on the GSSG/GSH ratio.[75]

Regularly, the endogenous sources of Cys are half derived from Met (through the TS pathway) and half derived from GSH breakdown.[80] In the fasting state, Cys is maintained mostly by the GSH turnover,[87] whereas after methionine loading, most Cys might originate from the TS pathway[43] (through Hcy, Ser, cystathionine, Cys) by de novo synthesis.

At baseline the HIV+ patients presented TS lower than controls, and after MetLo this difference was reduced loosing its statistical significance. Cys is not a precursor for Met because of the irreversibility of the cystathionine synthase reaction.[88] In this sense, the MetLo test is a useful in vivo test to assay the TS reactions and alterations in plasma distribution of redox species.[89]

### *Glutathione and Taurine Formations in HIV*

Under conditions of plenty, Met is directed toward Cys synthesis via the TS pathway for use in synthesis of proteins, GSH, and other cellular functions or directed toward catabolism such as Tau. Conversely, it is expected that when Met levels are low, flux through the TS pathway is downregulated to conserve the amino acid backbone in the Met cycle.[90] Hence, cellular methylation and antioxidant metabolism are linked by the TS pathway, which converts the Met cycle intermediate, Hcy, to Cys the limiting reagent in GSH synthesis, a major intracellular hydrosoluble antioxidant. Cys is also the precursor of Tau, an important phagocyte's antioxidant agent.

Regarding the GSH in HIV+, its level was lower than controls at baseline and after MetLo, but the participation of Cys in the formation of GSH (GSH/Cys + Ser ratio) was always higher in HIV+ patients than controls. As expected, the participation of Cys in GSH was increased in MetLo by its induced TS.[43,75] Therefore, it was clear the priority of Cys for GSH formation is the HIV+ condition. Moreover, part of the circulating Cys might be originated from GSH breakdown, which was highly increased in HIV+ on its oxidized form (GSSG).[75]

Under their UD all HIV+ amino acids and GSH were lower than controls, at baseline as well as after methionine load (MetLo). However, in this UD situation, HIV+ patients showed baseline values similar to controls for TM and RM, associated with a 40% more intense reduction in TS along with 56% higher TG and 15% higher GG, than controls (Fig. 15.4).

These baseline patterns persisted even after MetLo (AUC) for TM, RM, and GG but attenuating the reducted TS to 20.2%. Additionally, MetLo reversed the found baseline TG pattern by a 55% reduction in relation to controls[77] (Fig. 15.5).

Regarding the TS pathway, the HIV+ patients presented a Cys/Met ratio always lower than controls, 20.5% in UD, aggravated by MetLo to 50.6% lower, and indifferently to NAC either alone (42%) or associated with MetLo (31.8%). Hence, the higher MetLo-induced decreasing of Met–Cys conversion occurs in the absence of NAC, probably due to the Met-sparing effect of dietary Cys.[83]

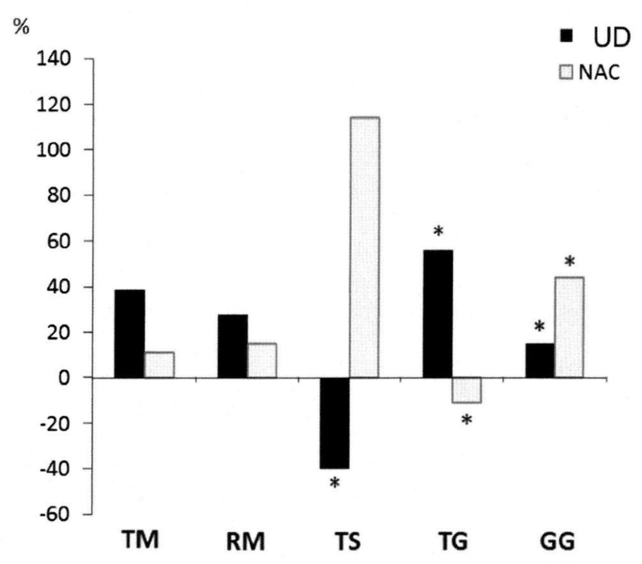


FIGURE 15.4 HIV+ Differences (%) related to control values (as 100%) at baseline in both dietary conditions, usual diet (UD) and NAC-supplemented (NAC). *GG*, glutathione genesis; *NAC*, N-acetyl cysteine; *RM*, remethylation; *TG*, taurine genesis; *TM*, transmethylation; *TS*, transsulfuration; *UD*, usual diet; *$P < .05$ (HIV ≠ Co). *Adapted from Burini RC, Moreto F, Yu YM. HIV-positive patients respond to dietary supplementation with cysteine or glutamine. In: Watson RR, editor.* Food, nutrition and lifestyle in health of HIV infected people. *Tucson: Elsevier; 2015. p. 245–64.*

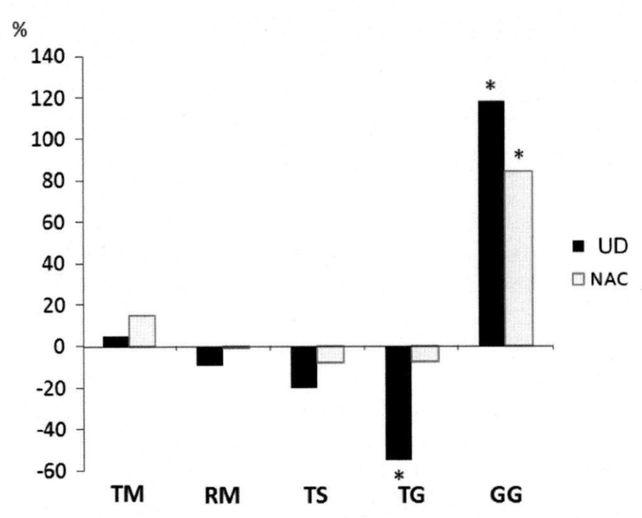


FIGURE 15.5 HIV+ Differences (%) related to control values (as 100%) after MetLo in both dietary conditions, usual diet (UD) and NAC-supplemented (NAC). *GG*, glutathione genesis; *NAC*, N-acetyl cysteine; *RM*, remethylation; *TG*, taurine genesis; *TM*, transmethylation; *TS*, trans-sulfuration; $*P < .05$ (HIV ≠ Co). *Adapted from Burini RC, Moreto F, Yu YM. HIV-positive patients respond to dietary supplementation with cysteine or glutamine. In: Watson RR, editor.* Food, nutrition and lifestyle in health of HIV infected people. *Tucson: Elsevier; 2015. p. 245–64.*

In the presence of NAC, the HIV+ group increased significantly TS and decreased GG, and consequently, the HIV+ patients became similar to controls in TM, RM, and also for TS (Fig. 15.4 and 15.5).

However, even with NAC, HIV patients continued with higher TG and GG, compared to controls. Hence, NAC seemed to boost the antioxidant defenses of HIV+ patients by increasing GG and TG through increasing TS pathway. The MetLo given to NAC-supplemented patients led to even higher TS along with higher GG but with lower TG. Hence, NAC + MetLo resulted in similar responses for controls and patients regarding the variables TM, RM, TS, and TG. Only GG remained 84% higher in HIV+ compared to controls[77] (Fig. 15.4 and 15.5).

Finally, the Tau/<Cys ratio (TG) higher in HIV+ patients was corrected by MetLo rather than by NAC (Figs. 15.4 and 15.5). This occurs probably through Met cycle affecting the TS pathway. Under normal conditions, coordinate regulation of methylation and antioxidant metabolism is achieved by the allosteric activation of CBS by SAMe. Under conditions of Met restriction, CBS protein levels are diminished 10-fold. This led to uncover a mechanism by which a reduction in the methyl donor, SAMe, leads to destabilization of CBS and, in turn, to a reduction in antioxidant capacity. This decrease in CBS levels correlates with reduced GSH that is, in turn, associated with increased vulnerability to oxidative stress. Diminished CBS levels are associated with reduced cell viability. Hence, the TS pathway represents the metabolic link between antioxidant and methylation metabolism. Under pathological conditions where SAMe levels are diminished, CBS, and therefore GSH and Tau levels, will be reduced. SAMe rather than Met signals a change in cellular Met availability and regulates CBS levels.[91,92]

Overall, HIV+ group presented similar TM and RM to controls irrespective to the presence of NAC and/or MetLo. However, under these sulfur supplementation there was a trend to keep Hcy close to the control values. Moreover, the lower than control TS found in HIV+ patients was corrected by NAC supplementation, either in absence or presence of MetLo. Associatively to that, HIV+ presented higher than controls TG and GG, irrespective to the presence of NAC. However, the higher than controls TG was corrected by MetLo in both diets. Therefore, only the higher than control GS was not corrected by any of the studied interventions. In conclusion, the failure of restoring normal Cys by MetLo, in addition to NAC, in HIV+ patients seemed related to increased flux of Cys into GSH and Tau pathways to strengthen the cell antioxidant capacity against the HIV progression.[77]

## SUMMARY POINTS

- Tau is a noncarboxylic amino acid lacking mRNA but considered as conditionally indispensable amino acid for humans;
- as major intracellular osmolyte, Tau stabilizes cell membranes and in CNS acts as neurotransmitter agonist presenting anorexigenic actions;

- as detoxificant agent, Tau generates conjugates with bile acids, retinoic acid, xenobiotics, and the phagocyte microbicidal hypochlorous acid (HOCl);
- by scavenging HOCl to form a more stable and less oxidant compound (Tau-Cl), Tau attenuates apoptosis and plays a role as a mitochondrial, and therefore cytoprotectant;
- as antiinflammatory agent Tau-Cl suppresses superoxide anion and decreases both NO and proinflammatory cytokines (TNF, IL-6 and IL-8) secretion by the activated phagocytes;
- Tau-Cl plays an immunomodulatory role at the site of inflammation by suppressing not only the production of reactive oxygen species (ROS) but also the proinflammatory cytokines;
- body Tau comes from diet or from its endogenous biosynthesis from Met and Cys;
- dietary protein and Cys availability are key factors in determining the flux of Cys between Cys catabolism/Tau synthesis and GSH synthesis;
- usually, plasma Tau levels decrease in response to critical illness and also in some chronic inflammatory diseases;
- HIV+ patients have lower plasma levels of Tau, GSH, and all other sulfur-related amino acids than healthy controls;
- these thiol-antioxidant-depleted patients had higher production of GSH and Tau than non-HIV+ controls, when under Cys supplementation (NAC) and methionine loading;
- Tau normalization was obtained after methionine loading of HIV+ patients, associatively to similar TM and RM of Met and lower TS of Hcy, compared to controls;
- the increased flux of Cys into GSH and Tau pathways can be interpreted as a host strategy to strengthen the cellular antioxidant capacity against the HIV progression.

## Acknowledgments

Special thanks to the Brazilian Research Funding FAPESP(MDBS PhD fellowship and RCB project-financial support) and CNPq (RCB researcher fellowship).

## References

1. Van Stijn MFM, Vermeulen MAR, Siroen MPC, Wong LN, Van den Tol MP, Ligthart-Melis GC, Houdijk APJ, Van Leeuwen PAM. Human taurine metabolism: fluxes and fractional extraction rates of the gut, liver, and kidneys. *Metabolism* 2012;**61**(7):1036–40. http://dx.doi.org/10.1016/j.metabol.2011.12.005.
2. Ripps H, Shen W. Review: taurine: a "very essential" amino acid. *Mol Vis* 2012;**18**:2673–86.
3. Murakami S. Role of taurine in the pathogenesis of obesity. *Mol Nutr Food Res* 2015;**59**:1353–63. http://dx.doi.org/10.1002/mnfr.201500067.
4. Huxtable RJ. Physiological actions of taurine. *Physiol Rev* 1992;**72**:101–63.
5. Sturman JA. Taurine in development. *Physiol Rev* 1993;**73**:119–47.
6. Schuller-Levis G, Mehta PD, Rudelli R, Sturman JA. Immunologic consequence of taurine deficiency in cats. *J Leukoc Biol* 1990;**47**:321–33.
7. Voss JW, Pedersen SF, Christensen ST, Lambert IH. Regulation of the expression and subcellular localization of the taurine transporter TauT in mouse NIH3T3 fibroblasts. *Eur J Biochem* 2004;**271**:4646–58.
8. Weiss SJ, Klein R, Slivka A, Wei M. Chlorination of taurine by human neutrophils. Evidence for hypochlorous acid generation. *J Clin Invest* 1982;**70**:598–607.
9. Hylemon PB, Zhou H, Pandak WM, Ren S, Gil G, Dent P. Bile acids as regulatory molecules. *J Lipid Res* 2009;**50**:1509–20.
10. Sharma R, Long A, Gilmer JF. Advances in bile acid medicinal chemistry. *Curr Med Chem* 2011;**18**:4029–52.
11. Lefebvre P, Cariou B, Lien F, Kuipers F, Staels B. Role of bile acids and bile acid receptors in metabolic regulation. *Physiol Rev* 2009;**89**:147–91.
12. Duboc H, Tache Y, Hofmann AF. the bile acid TGR5 membrane receptor: from basic research to clinical application. *Dig Liver Dis* 2014;**46**:302–12.
13. Kumari N, Prentice H, Wu JY. Taurine and its neuroprotective role. *Adv Exp Med Biol* 2013;**775**:19–27.
14. Albrecht J, Schousboe A. Taurine interaction with neurotransmitter receptors in the CNS: an update. *Neurochem Res* 2005;**30**:1615–21.
15. Pliquett RU, Führer D, Falk S, Zysset S, von Cramon DY, Stumvoll M. The effects of insulin on the central nervous system - focus on appetite regulation. *Horm Metab Res* July 2006;**38**(7):442–6.
16. Murakami T, Furuse M. The impact of taurine- and beta-alanine-supplemented diets on behavioral and neurochemical parameters in mice: antidepressant versus anxiolytic-like effects. *Amino Acids* 2010;**39**:427–34.
17. Uhe AM, Collier GR, O'Dea K. A comparison of the effects of beef, chicken and fish protein on satiety and amino acid profiles in lean male subjects. *J Nutr* 1992;**122**:467–72.
18. Schuller-Levis GB, Park E. Taurine: new implications for an old amino acid. *FEMS Microbiol Lett* 2003;**226**:195–202.
19. Fukuda K, Hirai Y, Yoshida H, Hakajima T, Usii T. Free-amino acid content of lymphocytes and granulocytes compared. *Clin Chem* 1982;**28**:175–61.
20. Sun M, Gu Y, Zhao Y, Xu C. Protective functions of taurine against experimental stroke through depressing mitochondria-mediated cell death in rats. *Amino Acids* 2011;**40**:1419–29.
21. Chang L, Xu J, Yu F, Zhao J, Tang X, Tang C. Taurine protected myocardial mitochondria injury induced by hyperhomocys- teinemia in rats. *Amino Acids* 2004;**27**:37–48.

22. Roy A, Manna P, Sil PC. Prophylactic role of taurine on arsenic mediated oxidative renal dysfunction via MAPKs/NF-κB and mitochondria dependent pathways. *Free Radic Res* 2009;**43**:995–1007.
23. Das J, Ghosh J, Manna P, Sil PC. Protective role of taurine against arsenic-induced mitochondria-dependent hepatic apoptosis via the inhibition of PKCδ-JNK pathway. *PLoS One* 2010;**5**:e12602.
24. Chen K, Zhang Q, Wang J, Liu F, Mi M, Xu H, Chen F, Zeng K. Taurine protects transformed rat retinal ganglion cells from hypoxia- induced apoptosis by preventing mitochondrial dysfunction. *Brain Res* 2009;**1279**:131–8.
25. YY1 L, Lee HJ, Lee SS, Koh JS, Jin CJ, Park SH, Yi KH, Park KS, Lee HK. Taurine supplementation restored the changes in pancreatic islet mitochondria in the fetal protein-malnourished rat. *Br J Nutr* 2011;**6**:1198–206.
26. Chang L, Xu JX, Zhao J, Pang YZ, Tang CS, Qi YF. Taurine antagonized oxidative stress injury induced by homocysteine in rat vascular smooth muscle cells. *Acta Pharmacol Sin* 2004;**25**:341–6.
27. Jong CJ, Azuma J, Schaffer S. Mechanism underlying the antioxidant activity of taurine: prevention of mitochondrial oxidant production. *Amino Acids* 2012;**42**:2223–32.
28. Schaffer SW, Jong CJ, Ito T, Azuma J. Role of taurine in the pathologies of MELAS and MERRF. *Amino Acids* 2014;**46**:47–56.
29. Schuller-Levis GB, Park E. Taurine and its chloramine: modulators of immunity. *Neurochem Res* 2004;**29**:117–26.
30. Park E, Quinn MR, Wright CE, Schuller-Levis G. Taurine chloramine inhibits the synthesis of nitric oxide and the release of tumor necrosis factor in activated RAW 264.7 cells. *J Leukoc Biol* 1993;**54**:119–24.
31. Marcinkiewicz J, Grabowska A, Bereta J, Stelmaszynska T. Taurine chloramine, a product of activated neutrophils, inhibits in vitro the generation of nitric oxide and other macrophage inflammatory mediators. *J Leukoc Biol* 1995;**58**:667–74.
32. Marcinkiewicz J, Grabowska A, Bereta J, Bryniarski K, Nowak B. Taurine chloramine down-regulates the generation of murine neutrophil inflammatory mediators. *Immunopharmacology* 1998;**40**:27–38.
33. Marcinkiewicz J, Kontny E. Taurine and inflammatory diseases. *Amino Acids* 2014;**46**:7–20.
34. Park E, Schuller-Levis G, Quinn MR. Taurine chloramine inhibits production of nitric oxide and TNF-K in activated RAW 264.7 cells by mechanisms that involve transcriptional and translational events. *J Immunol* 1995;**154**:4778–84.
35. Park E, Alberti J, Quinn MR, Schuller-Levis G. Taurine chloramine inhibits production of superoxide anion, IL-6 and IL-8 in activated human polymorphonuclear leukocytes. *Adv Exp Med Biol* 1998;**42**:177–82.
36. Kontny E, Szczepanska K, Kowalczewski J, Kurowska M, Janicka I, Marcinkiewicz J, Maslinski W. The mechanism of taurine chloramine inhibition of cytokine (interleukin-6, interleukin-8) production by rheumatoid arthritis ¢broblast-like synoviocytes. *Arthritis Rheum* 2000;**43**:2169–77.
37. Kanayama A, Inoue J, Sugita-Konishi Y, Shimizu M, Miyamoto Y. Oxidation of IkBα at methionine 45 is one cause of taurine chloramine-induced inhibition of NF-kB activation. *J Biol Chem* 2002;**277**:24049–56.
38. Barua M, Liu Y, Quinn MR. Taurine chloramine inhibits inducible nitric oxide synthase and TNF-alpha gene expression in activated alveolar macrophages: decreased NF-kB activation and IkB kinase activity. *J Immunol* 2002;**167**:2275–81.
39. Muz B, Kontny E, Marcinkiewicz J, Maśliński W. Heme oxygenase-1 participates in the anti-inflammatory activity of taurine chloramine. *Amino Acids* 2008;**35**:397–402.
40. Mato JM, Alvarez L, Ortiz P, Pajares MA. S-adenosylmethionine synthesis: molecular mechanisms and clinical implications. *Pharmacol Ther* 1997;**73**:265–80.
41. Lu SC. S-adenosylmethionine. *Int J Biochem Cell Biol* 2000;**32**:391–5.
42. Selhub J. Homocysteine metabolism. *Annu Rev Nutr* 1999;**19**:217–46.
43. Stipanuk MH. Sulfur amino acid metabolism: pathways for production and removal of homocysteine and cysteine. *Annu Rev Nutr* 2004;**24**:539–77. http://dx.doi.org/10.1146/annurev.nutr. 24.012003.132418.
44. Shimizu M, Satsu H. Physiological significance of taurine and the taurine transporter in intestinal epithelial cells. *Amino Acids* 2000;**19**:605–14.
45. Hansen SH. Taurine homeostasis and its importance for physiological functions. In: Cynober LA, editor. *Metabolic & therapeutic aspects of amino acids in clinical nutrition*. 2nd ed. USA: Routledge; 2003. p. 739–47.
46. Ide T, Kushiro M, Takahashi Y, Shinohara K, Cha S. mRNA expression of enzymes involved in taurine biosynthesis in rat adipose tissues. *Metabolism* 2002;**51**:1191–7.
47. Garcia RA, Stipanuk MH. The splanchnic organs, liver and kidney have unique roles in the metabolism of sulfur amino acids and their metabolites in rats. *J Nutr* 1992;**122**:1693–701.
48. Chesney RW. Taurine: its biological role and clinical implications. *Adv Pediatr* 1985;**32**:1–42.
49. Han X, Budreau AM, Chesney RW. Cloning and characterization of the promoter region of the rat taurine transporter (Tau T) gene. *Adv Exp Med Biol* 2000;**483**:97–108.
50. Ghandforoush-Sattari M, Mashayekhi S, Krishna CV, Thompson JP, Routledge PA. Pharmacokinetics of oral taurine in healthy volunteers. *J Amino Acids* 2010;**2010**:1–5. http://dx.doi.org/10.4061/2010/346237.
51. Gupta RC, Win T, Bittner S. Taurine analogues; a new class of therapeutics: retrospect and prospects. *Curr Med Chem* 2005;**12**(17):2021–39.
52. Alford C, Cox H, Wescott R. The effects of red bull energy drink on human performance and mood. *Amino Acids* 2001;**21**(2):139–50.
53. Parcell S. Sulfur in human nutrition and applications in medicine. *Altern Med Rev* 2002;**7**:22–44.
54. Milei J, Ferreira R, Llesuy S, Forcada P, Covarrubias J, Boveris A. Reduction of reperfusion injury with preoperative rapid intravenous infusion of taurine during myocardial revascularization. *Am Heart J* 1992;**123**(2):339–45.
55. Obinata K, Maruyama T, Hayashi M, Watanabe T, Nittono H. Effect of taurine on the fatty liver of children with simple obesity. *Adv Exp Med Biol* 1996;**403**:607–13.
56. Fennessy FM, Moneley DS, Wang JH, Kelly CJ, Bouchier-Hayes DJ. Taurine and vitamin C modify monocyte and endothelial dysfunction in young smokers. *Circulation* 2003;**107**(3):410–5.
57. Chauncey KB, Tenner Jr TE, Lombardini JB, Jones BG, Brooks ML, Warner RD, Davis RL, Ragain RM. The effect of taurine supplementation on patients with type 2 diabetes mellitus. *Adv Exp Med Biol* 2003;**526**:91–6.
58. M1 C, Sener G, Sehirli AO, Ekşioğlu-Demiralp E, Ercan F, Sirvanci S, Gedik N, Akpulat S, Tecimer T, Yeğen BC. Taurine protects against methotrexate-induced toxicity and inhibits leukocyte death. *Toxicol Appl Pharmacol* 2005;**209**(1):39–50.

59. Singh RB, Kartikey K, Charu AS, Niaz MA, Schaffer S. Effect of taurine and coenzyme Q10 in patients with acute myocardial infarction. *Adv Exp Med Biol* 2003;**526**:41–8.
60. Cantin AM. Taurine modulation of hypochlorous acid-inducing lung epithelial cell injury in vitro, role of anion transport. *J Clin Invest* 1994;**93**:606–14.
61. Crompton M, Andreeva L. On the involvement of mitochondrial pore in reperfusion injury. *Basic Res Cardiol* 1993;**88**:513–23.
62. Menzie J, Pan C, Prentice H, Wu JY. Taurine and central nervous system disorders. *Amino Acids* January 2014;**46**(1):31–46. http://dx.doi.org/10.1007/s00726-012-1382-z.
63. Perfeito R, Cunha-Oliveira T, Cristina Rego A. Revisiting oxidative stress and mitochondrial dysfunction in the pathogenesis of Parkinson disease - resemblance to the effect of amphetamine drugs of abuse. *Free Radic Biol Med* November 1, 2012;**53**(9):1791–806. http://dx.doi.org/10.1016/j.freeradbiomed.2012.08.569.
64. Yamori Y, Taguchi T, Hamada A, Kunimasa K, Mori H, Mori M. Taurine in health and diseases: consistent evidence from experimental and epidemiological studies. *J Biomed Sci* 2010;**17**(Suppl. 1):S6.
65. Chen G, Nan C, Tian J, Jean-Charles P, Li Y, Weissbach H, Huang XP. Protective effects of taurine against oxidative stress in the heart of MsrA knockout mice. *J Cell Biochem* 2012;**113**:3559–66.
66. Schaffer SW, Azuma J, Mozaffari M. Role of antioxidant activity of taurine in diabetes. *Can J Physiol Pharmacol* 2009;**87**:91–9.
67. Rosa FT, Freitas EC, Deminice R, Jordão AA, Marchini JS. Oxidative stress and inflammation in obesity after taurine supplementation: a double-blind, placebo-controlled study. *Eur J Nutr* 2014;**53**:823–30.
68. Merheb M, Daher RT, Nasrallah M, Sabra R, Ziyadeh FN, Barada K. Taurine intestinal absorption and renal excretion test in diabetic patients: a pilot study. *Diabetes Care* 2007;**30**:2652–4.
69. Yamori Y, Liu L, Ikeda K, Miura A, Mizushima S, Miki T, Nara Y, WHO-Cardiovascular Disease and Alimentary Comprarison (CARDIAC) Study Group. Distribution of twenty-four hour urinary taurine excretion and association with ischemic heart disease mortality in 24 populations of 16 countries: results from the WHO-CARDIAC study. *Hypertens Res* July 2001;**24**(4):453–7.
70. Yamori Y, Taguchi T, Mori H, Mori M. Low cardiovascular risks in the middle aged males and females excreting greater 24-hour urinary taurine and magnesium in 41 WHO-CARDIAC study populations in the world. *J Biomed Sci* 2010;**1**(17 Suppl):S21.
71. Chiarla C, Giovannini I, Siegel JH, Boldrini G, Castagneto M. The relationship between plasma taurine and other amino acid levels in human sepsis. *J Nutr* 2000;**130**:2222–7.
72. Lamonica-Garcia VC, Marin FA, Lerco MM, Moreto F, Henry MA, Burini RC. [Plasma taurine levels in patients with esophagus cancer]. *Arq Gastroenterol* 2008;**45**(3):199–203.
73. Paauw JD, Davis AT. Taurine concentrations in serum of critically injured patients and age- and sex-matched healthy control subjects. *Am J Clin Nutr* 1990;**52**:657–60.
74. Rijssenbeek AL, Melis GC, Oosterling SJ, et al. Taurine and the relevance of supplementation in humans, in health and disease. *Curr Nutr Food Sci* 2006;**2**:381–8.
75. Borges-Santos MD, Moreto F, Pereira PC, Ming-Yu Y, Burini RC. Plasma glutathione of HIV(+) patients responded positively and differently to dietary supplementation with cysteine or glutamine. *Nutrition* 2012;**28**(7–8):753–6. http://dx.doi.org/10.1016/j.nut.2011.10.014.
76. Burini RC, Moreto F, Yu YM. HIV-positive patients respond to dietary supplementation with cysteine or glutamine. In: Watson RR, editor. *Food, nutrition and lifestyle in health of HIV infected people*. Tucson: Elsevier; 2015. p. 245–64.
77. Burini RC, Borges-Santos MD, Moreto F, Yu YM. *Cysteine metabolism in response to daily N-acetylcysteine supplementation and acute methionine loading in HIV+ patients*. 2017. submitted.
78. Fraternale A, Paoletti MF, Casabianca A, Nencioni L, Garaci E, Palamara AT, et al. GSH and analogs in antiviral therapy. *Mol Aspects Med* 2009;**30**(1–2):99–110. http://dx.doi.org/10.1016/j.mam.2008.09.001.
79. Vilaseca MA, Sierra C, Colome C, Artuch R, Valls C, Munoz-Almagro C, et al. Hyperhomocysteinaemia and folate deficiency in human immunodeficiency virus- infected children. *Eur J Clin Invest* 2001;**31**(11):992–8.
80. Burini RC, Moreto F, Borges-Santos MD, Yu YM. Plasma homocysteine and thiol redox states in HIV+ patients. In: McCully KS, editor. *Homocysteine: biosynthesis and health implications*. Nova Science Publishers; 2013. p. 14.
81. Burini RC, Borges-Santos MD, Moreto F, Ming-Yu Y. *The failure of methionine load to restore plasma values of cysteine in HIV+ patients with oral N- acetylcysteine-induced glutathione-normal levels*. Barcelona: Clinical Nutrition; 2012. p. 1. ESPEN 2012.
82. Muller F, Svardal AM, Aukrust P, Berge RK, Ueland PM, Froland SS. Elevated plasma concentration of reduced homocysteine in patients with human immunodeficiency virus infection. *Am J Clin Nutr* 1996;**63**(2):242–8.
83. Fukagawa NK, Yu YM, Young VR. Methionine and cysteine kinetics at different intakes of methionine and cysteine in elderly men and women. *Am J Clin Nutr* 1998;**68**(2):380–8.
84. Tousoulis D, Antoniades C, Marinou K, Vasiliadou C, Bouras G, stefanadi E, Latsios G, Siasos G, Toutouzas K, Stefanadis C. Methionine-loading rapidly impairs endothelial function, by mechanisms independent of endothelin-1: evidence for an association of fasting total homocysteine with plasma endothelin-1 levels. *J Am Coll Nutr* 2008;**27**(3):379–86.
85. Homocysteine Lowering Trialists' Collaboration. Dose-dependent effects of folic acid on blood concentrations of homocysteine: a meta-analysis of the randomized trials. *Am J Clin Nutr* 2005;**82**(4):806–12. 82/4/806 [pii].
86. Burini RC, Borges-Santos MD, Moreto F, Yu YM. Plasma antioxidants and glutamine supplementation in HIV. In: Rajendram R, Preedy VR, Patel VB, editors. *Glutamine in clinical nutrition*. Springer; 2015.
87. Bogden JD, Kemp FW, Han S, Li W, Bruening K, Denny T, et al. Status of selected nutrients and progression of human immunodeficiency virus type 1 infection. *Am J Clin Nutr* 2000;**72**(3):809–15.
88. Rose WC. The nutritive significance of the amino acids. *Physiol Rev* 1938;**18**:27.
89. Di Giuseppe D, Ulivelli M, Bartalani S, Battistini S, Cerase A, Passero S, Summa D, Frosali S, Priora R, Margaritis A. Di Simplicio P-Regulation of redox forms of plasma thiols by albumin in multiple sclerosis after fasting and methionine loading test. *Amino Acids* 2010;**38**(5):1461–71.
90. Martinov MV, Vitvitsky VM, Mosharov EV, Banerjee R, Ataulla-khanov FI. The logic of the hepatic methionine metabolic cycle. *J Theor Biol* 2000;**204**:521–32.

91. Prudova A, Bauman Z, Braun A, Vitvitsky V, Lu SC, Banerjee R. S-adenosylmethionine stabilizes cystathionine β-synthase and modulates redox capacity. *PNAS* 2006;**103**:6489–94. http://dx.doi.org/10.1073/pnas.0509531103.
92. Lieber CS, Packer L. S-Adenosylmethionine: molecular, biological, and clinical aspects—an introduction. *Am J Clin Nutr* 2002;**76**(Suppl.):1148S–50S.
93. Burini RC, Lamonica VC, Moreto F, Yu YM, Henry MACA. Thiol metabolic changes induced by oxidative stress and possible role of B-vitamins supplements in esophageal cancer Patients HIV. In: Wilber A, editor. *Glutathione: dietary sources, role in cellular functions and therapeutic effects*. New York: Nova Biomedical; 2015. p. 101–26 (Chapter 5).

CHAPTER

# 16

# Mg-Supplementation Protects Against Oxidative Stress and Cardiac Dysfunction in Chronic Highly Active Antiretroviral Therapy–Treated Rats

*Jay H. Kramer*[1], *Chris F. Spurney*[2], *Joanna J. Chmielinska*[1], *William B. Weglicki*[1], *Ivan T. Mak*[1]

[1]The George Washington University, Washington, DC, United States;
[2]Children's National Medical Center, Washington, DC, United States

OUTLINE



## Abstract

Highly active antiretroviral therapy (HAART) agents, azidothymidine (AZT), ritonavir (RTV), and efavirenz (EFV), were assessed in vivo for oxidative stress, cardiac injury, and dysfunction in rats, and dietary Mg-supplementation (Mg-Sup sixfold higher) was evaluated for antioxidant benefits. Three weeks AZT or up to 8 weeks of RTV- and EFV treatments led to substantial elevations in neutrophil basal superoxide production, plasma 8-isoprostane, and RBC oxidized glutathione (GSSG) levels. Echocardiography revealed that AZT had minimal impact on left ventricular systolic function but depressed diastolic function. RTV and EFV led to diastolic and systolic dysfunction ≥5 weeks and decreased left ventricular posterior wall thickness (LVPW), suggesting onset of dilated cardiomyopathy; RTV and EFV caused ventricular fibrosis at 8 weeks. Mg-Sup suppressed HAART-induced neutrophil superoxide production, isoprostane, and GSSG elevations, attenuated systolic and diastolic dysfunction, lessened LVPW wall thinning, and reduced fibrosis. Mg-Sup provided protection against oxidative cardiac toxicity associated with different classes of HAART agents.

**Keywords:** Antioxidant protection; Fibrosis; Highly active antiretroviral therapy; In vivo cardiac dysfunction; Inflammation; Magnesium supplementation; Oxidative stress; Rodent models.


http://dx.doi.org/10.1016/B978-0-12-809853-0.00016-X

## List of Abbreviations

**AZT** Azidothymidine (also known as zidovudine)
**CO** Cardiac output
**E/A** Mitral valve E/A ratio
**EFV** Efavirenz
**eNOS** Endothelial nitric oxide synthase
**GSSG** Oxidized glutathione
**HAART** Highly active antiretroviral therapy
**IVSd & s** Interventricular septum diameter in diastole and systole.
**LBNF1** Lewis x Brown Norwegian F1 hybrid rat
**LV %FS** Left ventricular % fractional shortening
**LVEF** Left ventricular ejection fraction
**LVPWd & s** Left ventricular posterior wall in diastole & systole
**NNRTI** Nonnucleoside reverse transcriptase inhibitor
**NRTI** Nucleoside reverse transcriptase inhibitor
**PI** Protease inhibitor
**PMA** Phorbol myristate acetate
**RNS** Reactive nitrogen species
**ROS** Reactive oxygen species
**RTV** Ritonavir
**SOD** Superoxide dismutase

# INTRODUCTION

Cardiovascular disorders including ventricular arrhythmias, coronary vasculopathy, and dilated cardiomyopathy (DCM)[1] are frequently identified in HIV-1 infected patients (upward of 5000 cases per year). In instances of advanced HIV infection, evidence of tissue iron accumulation has been observed,[2] and this prooxidant metal may pose a further risk with respect to disease progression,[2] opportunistic microbial infection,[3] oxidative tissue injury,[2] and survival rate.[3] Although treatments aimed at lessening the impact of HIV infection have allowed the disease to become more manageable, the literature[4–6] does suggests that chronic use of highly active antiretroviral therapy (HAART) drugs may present significant independent risk factors for cardiovascular disease in HIV-infected individuals. Nucleoside reverse transcriptase inhibitors (**NRTIs**), nonnucleoside reverse transcriptase inhibitors (**NNRTIs**), and protease inhibitors (**PIs**) are the major classes of HAARTs used to treat HIV infection. Recent Department of Health and Human Services (DHHS) Guidelines[4] suggested that the preferred HAART agent combination can include tenofovir ( an NRTI), efavirenz (EFV, an NNRTI), and ritonavir (RTV, a universal PI-booster). However, it has been reported that some NRTIs such as azidothymidine (AZT), NNRTIs such as EFV, and PIs, including RTV, can promote endothelial cell (EC) reactive oxygen species (ROS) generation and cellular dysfunction, and these may be contributory factors in development of HAART-induced cardiac dysfunction.

The **NRTI**, AZT (or Zidovudine), was the first FDA-approved agent introduced in the late 1980s showing clinical efficacy against HIV[7] and remains in use as a component of the drug regimens for pregnant women (reduces maternal-fetal HIV-1 transmission). Deleterious side effects of AZT were generally attributed to damaged mitochondrial mtDNA synthesis resulting in mitochondrial myopathy in heart and skeletal muscle of animals and humans.[7,8] Alternatively, AZT treatment in rats was also reported to increase ROS and peroxynitrite production in the heart prior to significant changes in mtDNA content.[9] In addition, we[5] reported enhanced CD11b positive cell staining in cardiac tissue from AZT-treated rats, suggesting recruitment of inflammatory cells into tissue. By contrast, the direct oxidative effects produced by the NRTI, TFV was considered milder[10]; however, there was a higher risk of renal failure and associated Mg-wasting in TFV-treated patients,[11,12] which may lead to a prooxidative hypomagnesemic condition.[13]

**NNRTIs** are noncompetitive inhibitors of reverse transcriptase. While some studies indicated that NNRTI-based HAART regimens caused lower incidences of lipid disturbance and cardiovascular toxicity,[14] others report increased oxidative stress (elevated plasma isoprostanes)[15] and elevated plasma triglycerides and cholesterol in EFV-treated patients.[16] Moreover, an in vitro study indicated that EFV increases ROS generation with reduced ATP production in hepatocytes,[17] and that EFV also increased endothelial superoxide anion production, which can be blocked by mitochondria-targeted antioxidants.[18] These reports[15–18] support the notion that EFV treatment leads to potential oxidative and injurious endothelial effects.

**PIs** were introduced clinically in the mid-1990s and used with other HIV/AIDS medications to diminish HIV-1 replication in the patients.[19] RTV inhibits hepatic cytochrome p450 3A isozymes and is universally used as a booster PI coadministered with other PIs such as lopinavir and atazanavir.[4] In addition to clinical studies showing an association between PI therapy and increased insulin resistance, lipodystrophy, hyperlipidemia, and possibly atherosclerosis,[20,21] experimental findings indicated that several PIs, including RTV, caused oxidative stress within the endothelium and impaired endothelium-dependent relaxation function.[21–25] In vitro studies showed that RTV can induce endothelial dysfunction in isolated coronary arteries[23–25] with heightened superoxide formation[23,24] and directly decrease endothelial nitric oxide synthase (eNOS) expression.[21,25] Increased superoxide formation may limit reactive nitrogen species (RNS such as NO·) availability due to peroxynitrite formation.[23–25] Several PIs have been shown to promoted adhesion molecule expression (ICAM, E-selectin) as well as WBC recruitment, facilitating tissue inflammation.[22] HIV patients receiving PI-based regimens experienced higher peroxide levels, again supporting the concept of therapy-induced oxidative stress.

The aforementioned literature suggests that selected drugs representing different classes of HAART agents may induce, to varying degrees, ROS/RNS formation, endothelial dysfunction and WBC tissue recruitment, leading to subsequent oxidative stress, cardiac inflammation, and impairment of contractile function. Since HAART agents may generate or amplify free radical toxicity, this argues in favor of employing antioxidant supplementation[26–28] during HAART treatment. In one study,[26] supranutritional doses of vitamins C and E substantially attenuated both the losses of skeletal muscle total glutathione (GSH + GSSG), and the fivefold elevation in the GSSG/GSH ratio resulting from AZT treatment in mice. In search for low risk, effective, and economical treatments to combat HAART-induced toxicities, we discovered that Mg supplements may have potential benefits. Since a large subpopulation of HIV-1 infected patients also experience Mg deficiency[12] as a potential consequence of poor dietary intake/uptake, nephrotoxicity, and associated Mg-wasting caused by specific drug therapy, use of Mg supplements was considered a reasonable adjunct therapy worth investigating.

This chapter reviews findings related to AZT, RTV, and EFV toxicity in rodent models and the antioxidant benefits afforded by dietary Mg-supplementation (Mg-Sup). The dosages used for the three agents as well as the dietary Mg-supplement reflected the much higher drug metabolic rates in rodents. The FDA recommends that the human equivalent dosage for rodents is 6.2–11-fold higher to reach clinically comparable circulating drug levels.[29] The described animal experiments were guided by the principles for the care and use of laboratory animals as recommended by the US Department of Health and Human Services and were approved by The George Washington University Institutional Animal Care and Use Committee.

## Azidothymidine-Mediated Oxidative Stress In Vitro

The concept that AZT therapy may amplify free radical toxicity was supported by observations of Szabados et al.[9] who stated that "short-term treatment of rats with AZT increased ROS [reactive oxygen species] and peroxynitrite production in [the] heart...[and] activates ROS in mitochondria that triggers a sequence of events unfavorable for the cell; these ROS-mediated processes could be important factors in the cardiomyopathy in AZT-treated AIDS patients." AZT can also biotransform into several metabolites,[30] including 3′-amino-3′-deoxythymidine and AZT-monophosphate (AZT-MP), which also appear to be cytotoxic. We observed that acute AZT treatment displayed moderate, but significant prooxidative activity and that metabolites of AZT possess greater prooxidant potential.[31] Lipid peroxidation (LPO as TBARS (thiobarbituric acid reactive substance)-formation) in isolated rat hepatic microsomal membranes induced by an iron-catalyzed, superoxide-driven free radical system was dose-dependently enhanced by treatment (7–50 μM) with either AZT or AZT-MP. AZT and AZT-MP treatment of cultured aortic bovine ECs also had a prooxidative impact.[31] AZT alone lowered EC survival by 18% after 24h, but survival was worsened significantly (36% lower) when iron (15 μM $FeSO_4$) was present; iron alone had minimal effect. Moreover, AZT-MP dose-dependently enhanced free radical–induced losses of total glutathione and further reduced EC survival after 24h. Some of the earliest observations that high Mg levels (2 x normal) may afford antioxidant protection against HAART-mediated toxicity came from findings that the prooxidative activities of both AZT and AZT-MP were almost completely blocked by 2mM $MgSO_4$ in isolated membrane and EC models.[31]

## Chronic Azidothymidine Exposure In Vivo, *Iron Status*, and Altered Myocardial Tolerance to Postischemic Stress

In pilot studies,[32] AZT treatment of Sprague Dawley rats for 6 weeks at a relatively low dose (0.5 mg/mL, drinking water [30–80 mg/kg/day]) caused oxidative stress-induced pathology. Oxidative stress in vivo was assessed by changes in RBC glutathione and plasma-conjugated dienes levels. AZT alone caused a 15% decrease in total RBC glutathione, a 28% increase (nonsignificant due to small sample size) in plasma-conjugated diene levels, and a 17% increase (nonsignificant) in plasma lactate dehydrogenase activity (tissue injury marker) versus the untreated Mg-normal group. Collectively, these findings suggested that prolonged AZT exposure, even at a relatively low dose, could induce mild to moderate oxidative stress in vivo. Interestingly, while cardiac tissue iron levels were largely unaffected during this 6-week AZT treatment regimen, hepatic iron content increased 95%. These findings are consistent with that of Pollack and Weaver[33] who demonstrated that AZT treatment (oral, 1–2mg/mL) of normal mice for up to 4weeks caused a 50% elevation in nonheme liver iron content and a 20% increase in peritoneal macrophage iron. AZT may interfere with heme synthesis, causing upregulation of transferrin receptor synthesis and a slowing of receptor endocytosis, which may lead to increased cellular iron uptake. AZT was also shown to induce lactic acidosis in a human neonate, and this is likely to mobilize iron from storage proteins.[34]

Clinical features of AZT-DCM after several years of treatment included congestive heart failure, left ventricle dilatation, and reduced left ventricular ejection fractions (LVEFs). The used HAART was associated with a significantly higher incidence of myocardial infarction in HIV-infected individuals.[35] Furthermore, ischemic heart disease was found to be an increasing cause of non-AIDS-related death in HAART-treated patients. Using the 6-week AZT treatment protocol (0.5mg/mL, oral),[32] it was determined if AZT treatment of rats predisposed their hearts to subsequent postischemic stress. Isolated-perfused hearts from these 6-week animals showed greater susceptibility to injury resulting from global ischemia/reperfusion (I/R) stress. Functional recovery of the cardiac pressure–volume work index during reperfusion following 25min ischemia was 28.3% lower in hearts from AZT-treated rats (42.8% recovery) compared to untreated (59.7%). Since iron-mediated oxidative stress is a major component of the reperfusion injury mechanism,[36] we assessed postischemic myocardial iron release. I/R-stress led to a substantial elevation in total effluent iron content from hearts of untreated rats, and chronic AZT treatment further increased ($P < .05$) effluent iron by more than 80%. In association, postischemic-conjugated diene production was much more exaggerated in the AZT-treatment group, with a 2.8-fold increase in total postischemic levels versus untreated control. These studies suggested that prolonged AZT treatment in vivo, even at a relatively low dose, could further promote cardiac dysfunction and oxidative injury in hearts subjected to postischemic stress.

## Impact of Short-Term Azidothymidine Treatment on Cardiac Function In Vivo

A limited study was conducted to assess exposure time- and the dose-dependent impact of AZT treatment on rat heart function using noninvasive echocardiography.[13] Sprague Dawley rats received 0.5 or 1.5mg/mL AZT daily in their drinking water for 3weeks, and echocardiography was performed at 1.5 and 3weeks (Table 16.1). AZT treatment for 1.5weeks at either dose was largely ineffective, and 3weeks of treatment failed to induce substantial changes in left ventricular systolic function (LVEF, and % fractional shortening [LV % FS]), cardiac output (CO), aortic pressure maximum (not shown), or heart rate (not shown). However, at 3weeks, there were dose-dependent decreases in the dimensions of some ventricular structures (interventricular septum diameter in diastole and systole [IVSd & s]; and left ventricular posterior wall thickness in systole [LVPWs]), and this might be consistent with early signs of cardiomyopathy. Although left ventricular systolic function was not substantially affected after only 3weeks of treatment, mitral valve E/A ratio (early/late [atrial=A] ventricular filling velocity) was dose-dependently reduced by AZT, suggesting a trend toward early development of diastolic dysfunction.

## Short-Term Azidothymidine Treatment-Induced Oxidative Stress In Vivo and Mg-Supplementation was Protective

It was initially determined if exposure of Mg-normal Sprague Dawley rats to AZT for 3weeks at the higher dose (1.0mg/mL, oral) elicited significant oxidative toxicity, and if this would be affected by a sixfold higher dietary Mg-supplement.[5] AZT-treated rats tolerated high-dietary Mg-Sup very well during the 3-week drug exposure period. AZT did cause a significant ($P < .05$) decrease (20% lower) in the rate of body weight gain, implicating the systemic toxicity of AZT. However, concurrent dietary Mg-Sup (sixfold higher MgO in food) elevated plasma Mg levels 33% versus control, while having no effect on daily AZT consumption (+Mg vs. – Mg: 98 vs. 95mg/kg/day), and it completely prevented the AZT-mediated reduction in the rate of body weight gain.

AZT treatment for 3weeks aggravated inflammatory cell free radical production in rats, as determined by the ability of blood neutrophil isolates (by gradient centrifugation) to generate superoxide anions.[5] As shown in Fig. 16.1 A, AZT treatment of Mg-normal rats led to a significant ($P < .05$) fivefold elevation in neutrophil basal activity of superoxide anion production (determined by superoxide dismutase (SOD)-inhibitable cytochrome *c* reduction), compared to those isolated from untreated control. Unlike neutrophils obtained from untreated controls, cell isolates from AZT-treated rats were not sensitive to phorbol myristate acetate (PMA) stimulation (not shown), indicating impaired responsive function. These findings suggested that AZT itself could activate neutrophils, and this might be a significant source of endogenous oxidative stress during AZT treatment. AZT treatment also led to a substantial enhancement of WBC infiltration into rat cardiac tissue after 3weeks. AZT caused inflammatory lesions in atria as well as ventricular tissue,

TABLE 16.1 Rat Heart Echocardiography Parameters During 3 Weeks of Azidothymidine (AZT) Exposure

| 3 Weeks Exposure | Control | 0.5 AZT (% ↓ vs. Ctl) | 1.5 AZT (% ↓ vs. Ctl) |
|---|---|---|---|
| LVEF | 0.89 | 1.1 | 1.1 |
| LV %FS | 52.64 | 5 | 3.3 |
| E/A | 1.36 | 8.2 | 22 |
| CO, L/min | 0.195 | 2.6 | 5.1 |
| IVSd, mm | 2.158 | 12.4 | 30.5 |
| IVSs, mm | 3.728 | 15.6 | 23.5 |
| LVPWd, mm | 3.47 | 6.5 | 0.3 |
| LVPWs, mm | 2.563 | 17 | 22.9 |

A decrease (↓) in percentage values for AZT versus control are means of N = 2. Sprague–Dawley male rats received 0.5 or 1.5 mg/mL AZT daily in their drinking water for 3 weeks. Echocardiography was performed with a GE VingMed System Five with a 10 MHz probe (GE Healthcare, Canada). *CO*, cardiac output; *E/A*, mitral valve E/A ratio; *IVSd*, interventricular septum diameter in diastole; *IVSs*, interventricular septum diameter in systole; *LV %FS*, left ventricular % fractional shortening; *LVEF*, left ventricular ejection fraction; *LVPWd*, LV posterior wall in diastole; *LVPWs*, LV posterior wall in systole.

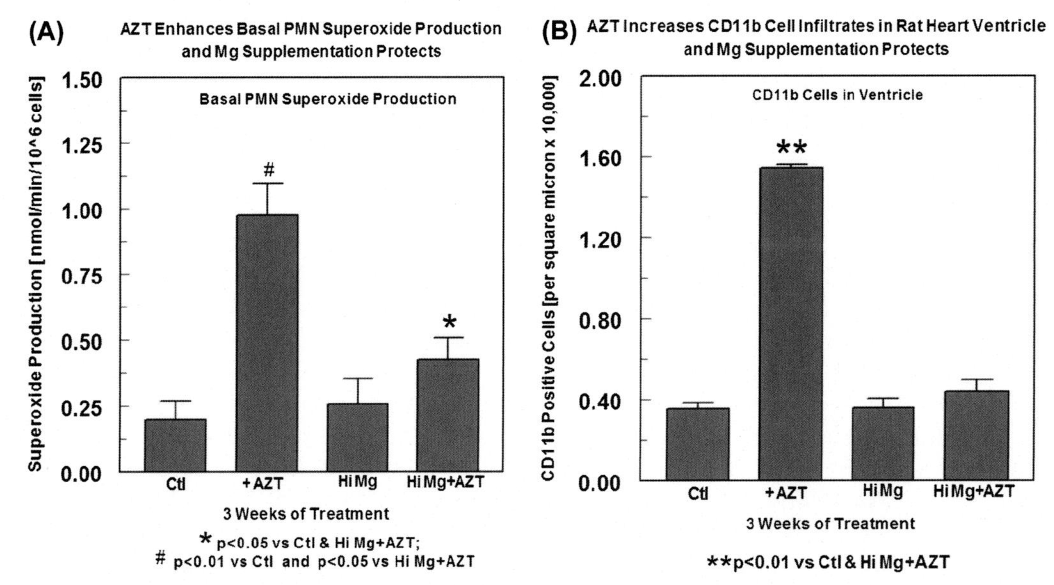


FIGURE 16.1 Effect of azidothymidine (AZT) (1 mg/mL, oral) on (A) neutrophil (polymorphonuclear leukocyte (PMN)) basal free radical production (superoxide anion) in male Sprague Dawley rats after 3 weeks and protective effects of dietary Mg-supplementation (sixfold above normal [Hi Mg], 0.6% MgO/kg food). Basal radical production was determined by SOD-inhibitable cytochrome *c* reduction. AZT-induced changes in (B) numerical density of CD11b positive cell infiltrate in ventricular tissue from Mg-normal (Ctl ± AZT) and Mg-supplemented (Hi Mg ± AZT) rats after 3 weeks. CD11b positive cells in the frozen tissue sections (5–10 μm) were visualized by indirect, immunohistochemical staining using mouse antirat CD11b antibody (Chemical International, Temecula, CA) and Vectastain Elite ABC kit immunoperoxidase system (Vector Laboratories Inc., Burlingame, CA); Samples were examined with an Olympus BX60 microscope. Data are derived from 7 to 8 different fields for each rats. Means ± SE of five rats/group for each endpoint. *Graphic representation of the AZT results modified from Mak IT, Chmielinska JJ, Kramer JH, Weglicki WB. AZT-induced cardiovascular toxicity—attenuation by Mg-supplementation.* Cardiovasc Toxicol. *2009;9:78–85.*

identified as CD11b positive cells (monocyte/macrophage surface marker).[5] These findings were reminiscent of those observed in 3-week AZT-treated mice, in which AZT (0.7 mg/mL, oral) caused mild to moderate cardiac inflammatory lesions as well as fibrosis, especially in the atrium.[37] Using quantitative morphometric analysis, it was estimated that the numerical density of CD11b positive cell infiltrates in ventricular tissue increased 4.36-fold in AZT-treated rats versus control (Fig. 16.1B), and a similar effect was seen in atrial tissue (not shown). AZT-treated animals, fed the Mg-supplemented diet, experienced a nearly 60% reduction in blood neutrophil basal superoxide generation (Fig. 16.1A), and a 72% lowering of WBC infiltration into cardiac tissue (Fig. 16.1B) compared to AZT-treated Mg-normal animals. High Mg alone (in the absence of AZT) had no effect on the levels of positive CD11b cells in either tissue.

A significant increase (2.41-fold) was found in plasma 8-isoprostane levels (Fig. 16.2A), a quantitative index of systemic oxidative stress, during 3 weeks of AZT treatment of Mg-normal rats; $F_2$-like isoprostanes, derived from nonenzymatic peroxidation of polyunsaturated fatty acids,[5] were assessed using an 8-isoprostane enzyme

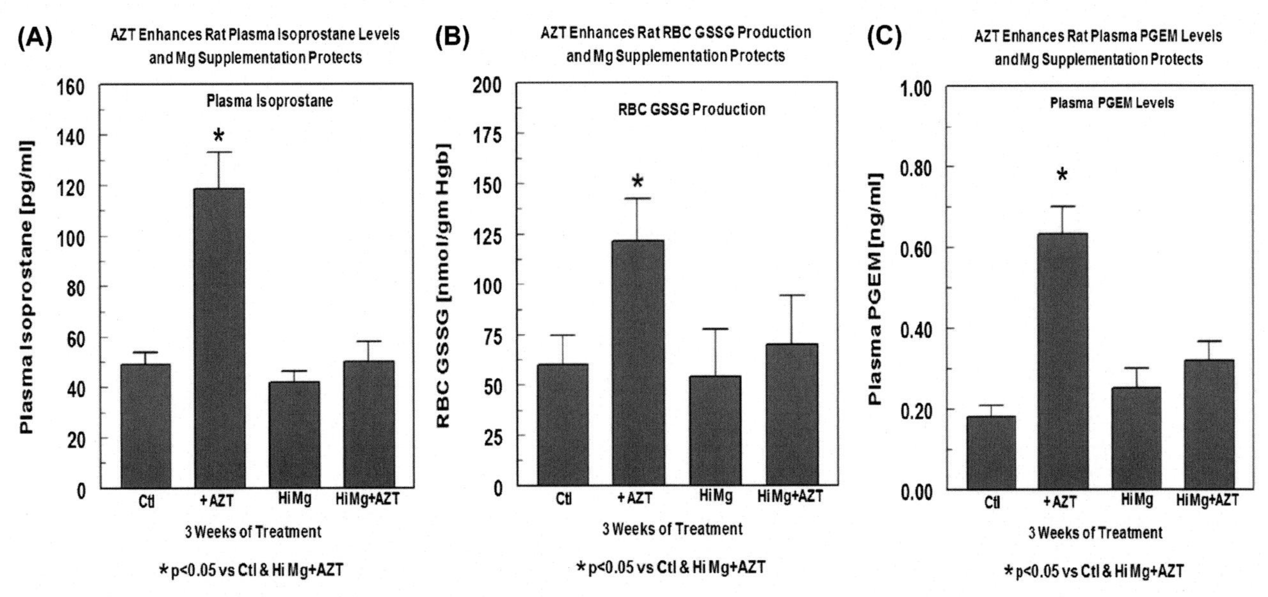


FIGURE 16.2 Effect of azidothymidine (AZT) (1 mg/mL, oral) on circulating (A) 8-isoprostane, (B) RBC glutathione disulfide (GSSG) levels, and (C) plasma PGEM levels in male Sprague Dawley rats after 3 weeks and protective effects of dietary Mg-supplementation (sixfold above normal, Hi Mg). Both plasma 8-isoprostane and PGEM (PGE2-metabolites) were determined by EIA immunoassay kits (Cayman Chemicals). GSSG was assessed by the enzymatic 'recycling' method. Other conditions are described in Fig. 16.1 legend. Means ± SE of 5 rats/group for each endpoint.

immunoassay. A twofold elevation in RBC glutathione disulfide (GSSG, Fig. 16.2B) was also observed during 3 weeks of AZT treatment,[5] further suggesting increased oxidative stress in vivo. GSSG normally constitutes less than 1% (0.8± 0.2%) of total glutathione in RBCs from Mg-normal control rats; however, AZT administration caused GSSG to rise ($P < .01$) to 2.2 ± 0.4% of RBC total glutathione. It is worth noting that the total glutathione, on a blood volume basis, deceased significantly by 25% and is consistent with increased hemolytic degradation of RBCs during AZT treatment. AZT-treated animals placed on the Mg-supplemented diet displayed nearly complete attenuation of the drug-mediated elevations in plasma isoprostane (Fig. 16.2A), and RBC GSSG (Fig. 16.2B) levels, again suggesting the substantial antioxidant protection afforded by increased dietary Mg. Previously, it was found that AZT treatment for 3 weeks in a murine model provoked a three- to fourfold increase in circulating cyclooxygenase-1 (COX-1)-derived vasoactive prostanoid products such as $TXA_2$, $PGI_2$, or $PGE_2$.[37] Increased prostaglandin generation, especially $TXA_2$ and $PGE_2$, may be associated with human vascular diseases such as atherosclerosis and hypertension.[37] Blood pressure was not measured of in our AZT-treated murine study; however, others[38] reported that prolonged exposure of rats to AZT (1 mg/mL, oral) caused a sustained higher systolic blood pressure and greater heart weight. In keeping with the earlier mouse study,[37] this current investigation demonstrated that AZT treatment of Mg-normal rats for 3 weeks led to a 3.52-fold higher ($P < .05$) plasma level of $PGE_2$-derived metabolites (PGEM) (Fig. 16.2C). AZT-induced elevations in vasoactive mediators may also indicate the presence of generalized endothelium activation during drug treatment. The report by Sutliff et al.[39] indicating that AZT-treated mice displayed increased superoxide production from the endothelium are consistent with this notion. When fed a sixfold higher Mg diet, the enhanced plasma PGEM level induced by 3 weeks of AZT exposure in rats was attenuated by 50% (Fig. 16.2C, $P < .05$).

Collectively, all of the above findings suggest that AZT treatment in rodent models can lead to dose- and exposure-time dependent systemic oxidative stress, culminating in development of tissue inflammation, pathology, and cardiac abnormalities. Multiple studies have reported that AZT treatment in rodents can cause elevated ROS production in various tissues, such as skeletal muscle, liver, endothelium, and the heart.[5,9,26,39] The mitochondrion is a primary subcellular target of AZT toxicity[26,40] that may contribute to the bulk of elevated ROS generation (superoxide anions and hydrogen peroxide) due to AZT-mediated partial inhibition of the respiratory electron transport chain.[40] Current data also demonstrated that AZT promoted superoxide anion generation in circulating neutrophils (Fig. 16.1A), and it can be presumed that the NADPH oxidase system also participated in the elevated superoxide anion generation.[41] In addition, part of the oxidative cardiac injury caused by AZT may result from increased inflammatory WBC (macrophages/monocytes/neutrophils) infiltration (Fig. 16.1B), which may activate resident fibroblasts and induce direct oxidative injury to cardiac tissue myocytes.

## Potential Mechanism(s) Underlying the Protective Benefits of Mg-Supplementation

The deleterious effects of AZT were largely overcome by increased dietary Mg that may have potential use in the clinical setting as a cotherapy in patients receiving HAART regimens. Clinical evidence has shown that Mg supplements inhibited tissue calcium accumulation, improved myocardial metabolism, and reduced myocardial cell death in patients with coronary artery disease.[42] Moreover, we[43] previously demonstrated that acute Mg-Sup (threefold higher perfusate Mg concentration vs. control) afforded significant protection against postischemic cardiac dysfunction and oxidative tissue injury in a perfused-working rat heart model. Others showed that dietary Mg-Sup substantially lowered blood pressures in hypertensive patients[44] and spontaneously hypertensive rats.[45] AZT may increase systolic blood pressure in rats,[38] and the observed cardioprotection of Mg-Sup might be partly contributed by its antihypertensive effects. In the previous study,[5] dietary Mg-Sup led to a substantial increase (33%) in circulating total Mg levels in rats. Since it has been suggested that Mg acts as a natural "calcium antagonist",[46] its anticalcium effect may suppress the $Ca^{2+}$-mediated priming and activation of NADPH oxidase in circulating neutrophils from AZT-treated rats. Likewise, calcium is required for COX-1 activation and subsequent $PGE_2$ production,[47] and the attenuation of this process by high Mg (Fig. 16.2C) may also be associated with its anticalcium effect. In addition, the observation that Mg-Sup reduced WBC cardiac infiltration (Fig. 16.1B) also implicates its potential attenuating effect on AZT-induced elevation of adhesion molecules.[5]

One must consider that part of the beneficial effects of high $Mg^{2+}$ may also relate to its ability to competitively displace low-molecular-weight iron (prooxidant) from membrane phospholipid binding sites, and thus, prevent site-specific hydroxyl radical formation.[43] A previous study[31] showed that isolated EC oxidative injury mediated by AZT or select AZT-metabolites was iron-dependent, and that subnormal extracellular Mg (<0.5 mM) augmented, whereas higher Mg (>1 mM) attenuated, this injury. In keeping with this finding, results from the I/R heart model[43] revealed that different Mg supplements were able to significantly reduced membrane LPO damage, measured as iron-catalyzed alkoxyl radical formation, in the postischemic rat heart. Thus, these studies clearly demonstrate the protective effects of high-dietary Mg against AZT-induced oxidative stress, cardiac pathology, and systemic inflammation in rodents.

## In Vivo Oxidative Stresses Caused by Ritonavir and Efavirenz and Protection by Mg

As reviewed in the previous sections, both RTV and EFV were reported to promote elevated superoxide production in vitro in cultured EC, hepatocyte, and isolated tissue models. To determine whether RTV or EFV promoted systemic oxidative stress in vivo and enhanced cardiac inflammation and dysfunction, the Lewis X Brown-Norway F1 hybrid rat model was used since this strain was shown to be sensitive to RTV-induced hyperlipidemia in a manner similar to that of humans.[48] RTV (Abbott) and EFV (Bristol-Myers Squibb) were dissolved in organic vehicle and orally administered by gavage to separate animals at a dose of 75 mg/kg/day. The effects of Mg-Sup were assessed in additional groups of rats receiving each drug plus a sixfold higher dietary Mg content. After 5 weeks, some of the animals from each group were sacrificed, and blood and tissue samples were isolated. Plasma levels of 8-isoprostane were determined as a measure of systemic oxidative stress. As shown in Fig. 16.3A, both RTV and EFV elevated plasma 8-isoprostane levels significantly (by 2.25- and 2.8-fold, respectively), supporting the notion that both agents caused systemic oxidative stress in vivo. Whole-blood neutrophils were freshly isolated from rats of each group and assessed for their free radical generating activities. Fig. 16.3B represents the unstimulated basal activities of neutrophil superoxide anion production from RTV- or EFV-treated rats, with or without high-dietary Mg-Sup. Both RTV and EFV significantly elevated basal superoxide generating activities compared to the Mg-normal control without drug treatment. However, it was surprising to see that the basal activity from the EFV-treated group was substantially higher (6.3-fold above control, $P < .01$) than that elicited by RTV (2.5-fold, $P < .05$). On ex vivo PMA (0.125 ug/mL) stimulation (not shown), neutrophils from the RTV-treated group exhibited a further twofold higher superoxide anion activity compared to the control group. On the other hand, neutrophils from EFV-treated rats displayed no further significant stimulation with PMA. It is likely that the neutrophils from the EFV-treated rats were endogenously activated to near saturation, such that PMA caused no further stimulation. Concomitantly, the RBC GSSG level obtained from the EFV-treated animals was elevated nearly sixfold, whereas that for the RTV-treated group was about threefold above control (Fig. 16.3C). In data not shown, the total RBC glutathione levels from EFV rats decreased 20% ($P < .05$), whereas that from RTV-treated animals were unchanged. In RTV or EFV-treated rats on Mg-supplemented diets, the drug-mediated elevations in both plasma isoprostane and RBC GSSG levels were substantially attenuated (Fig. 16.3). Mg-Sup dramatically lowered the basal superoxide producing activities of neutrophils from both RTV- and EFV-treated rats after 5 weeks. These data suggested that EFV caused a higher level of

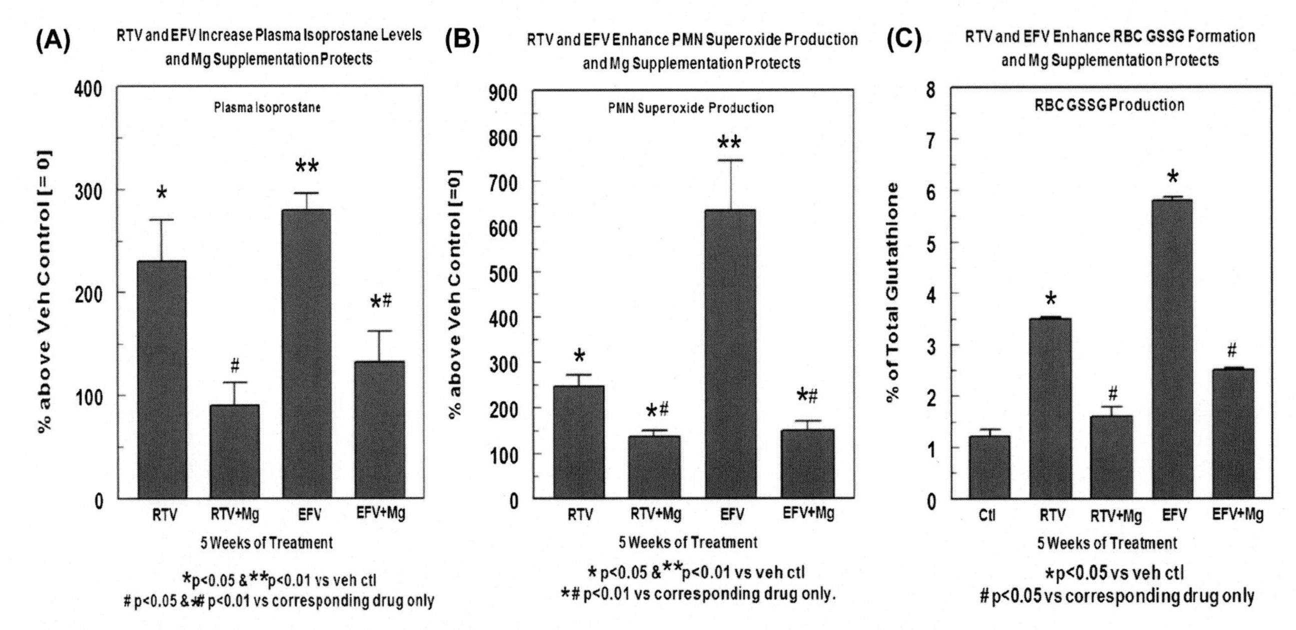


FIGURE 16.3 Effects of ritonavir (RTV) and efavirenz (EFV) with or without dietary Mg-supplementation (sixfold above normal, 0.6% MgO/kg food) on markers of oxidative and inflammatory stress in male Lewis x Brown-Norway hybrid (LBNF1) rats after 5 weeks. Each highly active antiretroviral therapy agent was given daily by gavage (75 mg/kg/day). (A) circulating 8-isoprostane (LPO marker) levels were determined by EIA immunoassay kit (Cayman Chemicals); (B) circulating neutrophil (polymorphonuclear leukocyte(PMN)) basal free radical production (superoxide anion) was determined by superoxide dismutase–inhibitable cytochrome *c* reduction; and (C) RBC GSSG production was assessed by the enzymatic 'recycling' method. Means ± SE of 4–6 rats/group for each endpoint. *Graphic representation of the RTV results modified from Mak IT, Kramer JH, Chen X, Chmielinska JJ, Spurney CF, Weglicki WB. Mg supplementation attenuates ritonavir-induced hyperlipidemia, oxidative stress, and cardiac dysfunction in rats.* Am J Physiol Regul Integr Comp Physiol. *2013;**305**:R1102–11.*

systemic oxidative stress compared to RTV, as indicated by their relative effects on circulating neutrophil activities, isoprostane level, and RBC glutathione status. In cultured ECs, the elevated ROS formation induced by RTV may be related to the partial uncoupling of eNOS functional components[49] and partly due to upregulation of the NADPH oxidase units.[24] For comparison, EFV did not seem to impact eNOS expression but potentially induce endothelial ER (endoplasmic reticulum) stress and autophagy.[50] In addition, EFV was shown to induce significant mitochondrial dysfunction resulting in elevated ROS production in hepatocytes[51] and neuronal cells.[52] Plasma level of nitrite, but not nitrate, has been recognized as a reliable biomarker of active endothelial NO· production.[49] Indeed, using the 2,3-diamino-naphthalene fluorescent assay,[49] it was found that both RTV- and EFV-treatment in rats (5 weeks) caused significant ($P < .05$) reductions (50%, and 42%, respectively) in plasma nitrite levels. While dietary Mg-Sup substantially raised plasma nitrite level of the RTV-treated rats to 92% of control, the same Mg treatment only partially restored the nitrite levels for the EFV-treated rats to 78% of control. Hepatic eNOS mRNA (determined by qRT-PCR) in 5-week animals fell by 20% ($P < .05$) with RTV treatment and <10% (N.S.) with EFV; both decreases were attenuated by Mg-Sup. The data suggested that the lower nitrite level mediated by RTV might be in part related to eNOS downregulation. However, the decreased nitrite in the EFV-treated rats might reflect increased peroxynitrite formation rather than decreased NO synthesis.

It is presumed that the multiple toxicity effects of EFV contribute to the more elevated oxidative indices shown in these studies. Current findings also showed that Mg-Sup was somewhat less effective in preventing EFV-induced oxidative stress compared to RTV-mediated stress (Fig. 16.3). Clinical studies have indicated that therapeutic control of HIV-1 replication by HAART was associated with increased oxidant stress in many patients[53]; an earlier study compared the relative isoprostane levels in HIV patients treated with NRTI, RTV, or EFV, and revealed that EFV treatment elicited the highest plasma F2-isoprostane levels. Moreover, EFV administered to healthy HIV-negative volunteers also led to significant elevation in circulating F2-isoprostane levels[54].

## Mg-Supplementation Attenuates the Effects of Ritonavir and Efavirenz on Cardiac Function

RTV and EFV treatment studies in LBNF1 rats were conducted to assess the progressive development of cardiac dysfunction using noninvasive echocardiography and the potential benefits of dietary Mg-Sup. Unlike the current AZT study (Table 16.1), no significant alterations in functional or anatomical echocardiography parameters were noted at 3weeks of RTV- or EFV treatment. However, at 5weeks, RTV and EFV caused modest LV diastolic dysfunction, as evidenced by 17% ($P < .02$) and 13% ($P < .05$) decreases in mitral valve E/A ratio, respectively (Fig. 16.4 A). In data not shown, mitral valve E/A ratio was similarly depressed in 8-week RTV- and EFV-treated rats. The reduction in the E/A ratio could indicate that the left ventricle was not filling with blood properly during the period between contractions; drug-induced alterations in LV distensibility, filling, and/or relaxation may be responsible for the lower diastolic function. LV diastolic dysfunction is frequently detected in HIV+patients[55] and was independently predicted by HAART exposure, supporting the concept that HAART contributed to myocardial functional decline.[56,57] CO at 5weeks was significantly reduced 12% ($P < .05$) and 14% ($P < .02$) in RTV- and EFV-treated rats, respectively (Fig. 16.4B).

After 5weeks with either drug, slight, yet significant decreases in LV systolic function were observed. LVEF decreased 3.5% ($P < .02$) and 5% ($P < .05$) for RTV and EFV, respectively (Fig. 16.5A and B); and LV % FS fell 6.5% ($P < .05$) and 8.5% ($P < .05$) for RTV and EFV, respectively (Fig. 16.6A and B). When exposure time was extended to 8weeks, the progressive worsening of systolic function was somewhat more prominent in the RTV group (LVEF and LV %FS: 8% and 13.7% lower, $P < .05$ for each) compared with EFV (LVEF and LV %FS: 6% and 10% lower, $P < .05$ for each). These findings are in general agreement with both clinical[55] and experimental[58] findings of LV systolic dysfunction associated with HAART. These data, however, strongly suggest, that PI (RTV)- and NNRTI (EFV)-induced oxidative stress/inflammation events were major contributors to the decline in cardiac function. Others proposed that the PI-mediated contractile depression of Langendorff-perfused hearts from 8-week treated rats was largely a consequence of perturbed calcium handling.[58] These investigators were also unable to detect PI-induced cardiac ultrastructural changes in their model. The present study used two-dimensional, M-mode, and pulsed Doppler images to obtained measurements of LVPW. Echocardiographic findings revealed substantial decreases in LVPW thickness in diastole (LVPWd: 8.5–10%, nonsignificant) and systole (LVPWs: 11–12%, $P < .05$) with both RTV and EFV after 5weeks, and further declines with both agents after 8weeks for LVPWd (9.5–13%, nonsignificant) and LVPWs (Fig. 7A and B: 18% [EFV] and 22% [RTV], $P < .05$). These observations may be indicative of early progression toward a DCM.

Cotreatment with Mg-Sup for 5 or 8weeks substantially attenuated RTV-induced diastolic dysfunction (Fig. 16.4A: E/A ratio by 70%), restored CO to normal vehicle control levels (Fig. 16.4B), completely prevented RTV-mediated declines in LV systolic function (Figs. 16.5A and 16.6A), and lessened thinning of LVPWs by 75% at 8weeks (Fig. 16.7A). Similar diastolic (Fig. 16.4A: E/A improved 51%) and systolic (Figs. 16.5B and 16.6B; 50–100% improved) functional protection by Mg-Sup were exhibited in EFV-treated rats compared with RTV, but the protective effects were not as prominently observed for the other select anatomical (Fig. 16.7B: **LVPWs**) and hemodynamic (Fig. 16.4B: **CO**) parameters. Although the pharmacologic mechanisms of action for RTV and EFV may differ, their deleterious side effects leading to systemic ROS/RNS stress, inflammation, and cardiac functional disturbances/injury appear to be largely corrected by dietary inclusion of Mg.

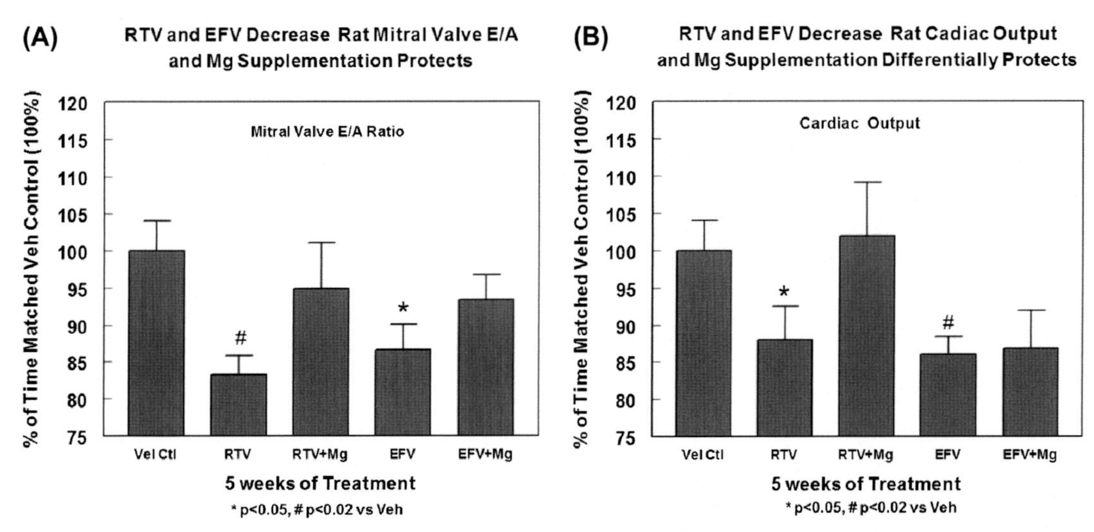


FIGURE 16.4 Changes in (A) mitral valve E/A ratio and (B) CO obtained by echocardiography (GE VingMed System Five with a 10MHz probe, GE Healthcare, Canada) after 5weeks of efavirenz (EFV) or ritonavir (RTV) treatment±dietary Mg supplement (sixfold above normal=0.6% MgO/kg food) in Lewis x Brown Norwegian F1 hybrid rats. Each highly active antiretroviral therapy agent (75mg/kg/day, gavage) caused declines in E/A ratio and CO versus time-matched vehicle control. Mg supplement was partially protective. Means±SE of 4–6 rats. *$P < .05$ and #$P < .02$ versus vehicle control. *Graphic representation of the RTV results modified from Mak IT, Kramer JH, Chen X, Chmielinska JJ, Spurney CF, Weglicki WB. Mg supplementation attenuates ritonavir-induced hyperlipidemia, oxidative stress, and cardiac dysfunction in rats.* Am J Physiol Regul Integr Comp Physiol. *2013;**305**:R1102–11.*

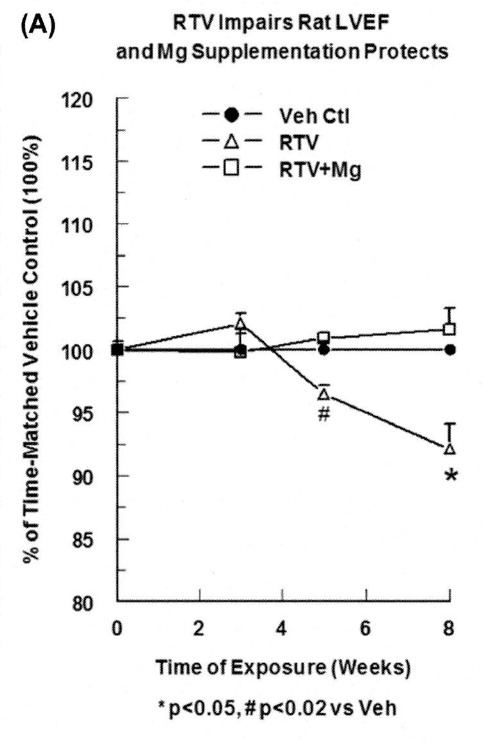


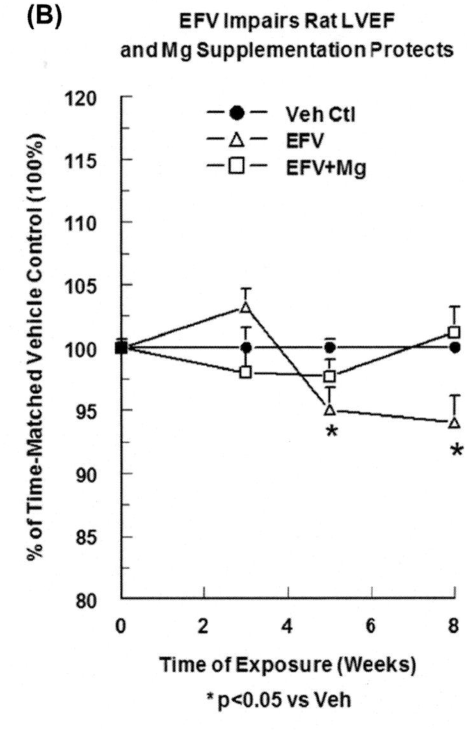


FIGURE 16.5 Changes in echocardiographic systolic functional parameter (left ventricular ejection fraction [LVEF]) during prolonged ritonavir (RTV) and efavirenz (EFV) treatment ± dietary Mg supplement in male LBNF1 rats. Each highly active antiretroviral therapy agent (75 mg/kg/day, gavage) caused progressive declines LVEF versus time-matched vehicle control (up to 8 weeks), and Mg supplement (sixfold above normal) was substantially protective. Means ± SE of 5–6 rats, except at 8 weeks where n = 3–4. Absent SEMs are within symbol size. *$P < .05$ & #$P < .02$ versus vehicle control. Other conditions are described in legends of Figs. 16.3 and 16.4. *The RTV results modified from Mak IT, Kramer JH, Chen X, Chmielinska JJ, Spurney CF, Weglicki WB. Mg supplementation attenuates ritonavir-induced hyperlipidemia, oxidative stress, and cardiac dysfunction in rats.* Am J Physiol Regul Integr Comp Physiol. *2013;**305**:R1102–11.*

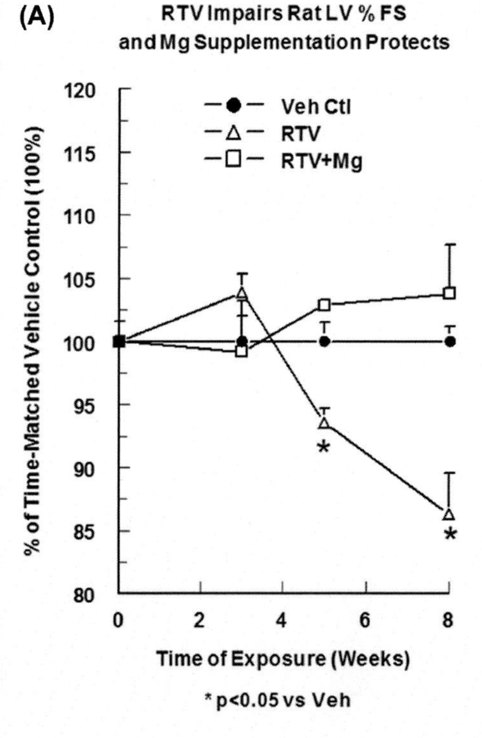


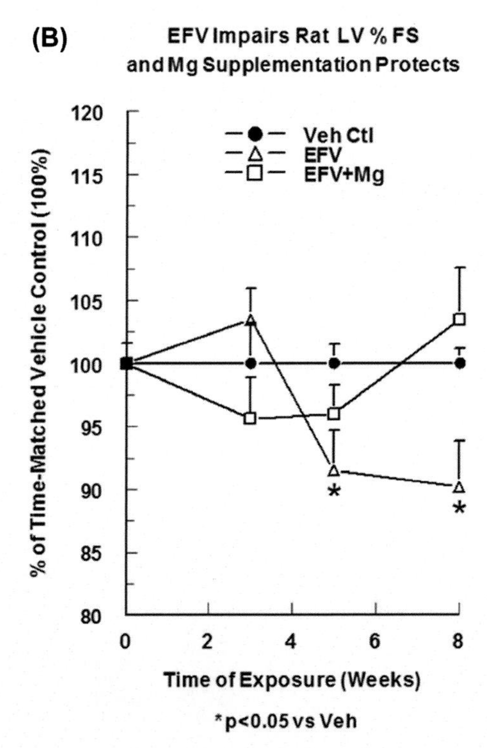


FIGURE 16.6 Changes in echocardiographic systolic functional parameter (% fractional shortening of the left ventricular wall [LV % FS]) during prolonged ritonavir (RTV) and efavirenz (EFV) treatment ± dietary Mg supplement in Lewis x Brown Norwegian F1 hybrid rats. Each highly active antiretroviral therapy agent (75 mg/kg/day, gavage) caused progressive declines in LV % FS versus time-matched vehicle control (up to 8 weeks), and Mg supplement (sixfold above normal) was substantially protective. Means ± SE of 5–6 rats, except at 8 weeks where n = 3–4. Absent SEMs are within symbol size. *$P < .05$ & #$P < .02$ versus vehicle control. Other conditions are described in legends of Figs. 16.3 and 16.4. *The RTV results modified from Mak IT, Kramer JH, Chen X, Chmielinska JJ, Spurney CF, Weglicki WB. Mg supplementation attenuates ritonavir-induced hyperlipidemia, oxidative stress, and cardiac dysfunction in rats.* Am J Physiol Regul Integr Comp Physiol. *2013;**305**:R1102–11.*

## Impact of Mg-Supplementation on Ritonavir- and Efavirenz-Induced Cardiac Fibrotic Development

Cardiac fibroblasts play a critical role in maintaining normal cardiac function, as well as in cardiac remodeling during pathological conditions such as myocardial infarct, hypertension, or oxidative stress. The abundance of fibroblasts (~67%) in the cardiac tissue makes it an easily accessible target for endothelium-derived excessive oxidative stress, leading to uncontrolled fibrosis, which is most likely mediated by activated transforming growth factor-beta

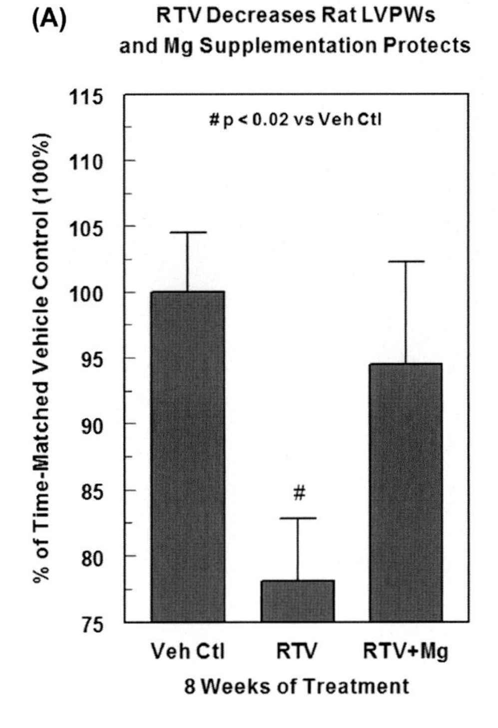


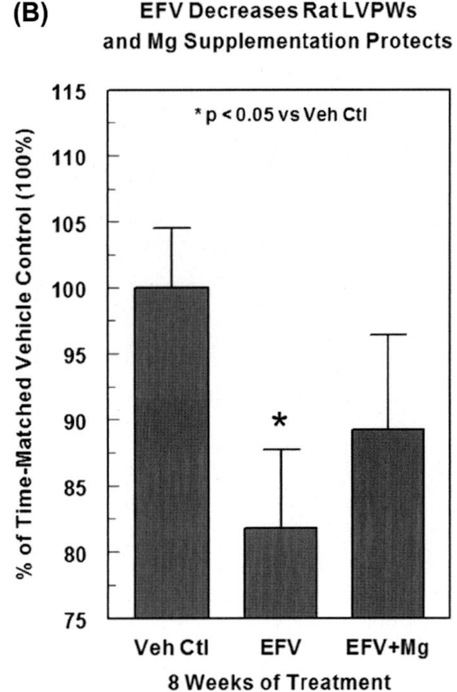


FIGURE 16.7 Changes in echocardiographic anatomical parameters (left ventricular posterior wall dimension in systole [LVPWs]) during prolonged ritonavir (RTV) and efavirenz (EFV) treatment ± dietary Mg supplement in LBNF1 rats. Each highly active antiretroviral therapy agent (75 mg/kg/day, gavage) caused declines in LVPWs versus time-matched vehicle control at 8 weeks, and Mg supplement (0.6% MgO/kg food) was substantially protective. Other conditions are described in legends of Figs. 16.3 and 16.4. Means ± SE of 3–4 rats at 8 weeks. $^*P < .05$, $\#P < .02$ and $^{**}P < .01$ versus vehicle control.

(TGF-β). Experiments using Masson Trichrome staining for collagen in rat heart sections,[48] revealed only mild fibrosis of ventricular tissue in RTV- and EFV- treated rats after 5 weeks. However, by 8 weeks, prominent perivascular fibrosis occurred in both the left and right ventricles of all examined RTV-treated rats (Fig. 16.8, **upper center panel**), and in about 50% of EFV-treated animals (Fig. 16.8 **lower center panel**). The perivascular appearance of fibrosis caused by each drug suggests that reactive interstitial fibrosis was occurring as opposed to replacement fibrosis, which would be apparent throughout the myocardium.[59] The previous study[37] suggested AZT treatment in mice primarily caused replacement fibrosis in the heart after only 3 weeks. It is presumed that HAART-induced infiltration of WBCs in perivascular areas as well as the activated endothelium was the main sources of excessive ROS production, which provoked synthesis and activation of TGF-β; this in turn, may promote the eventual fibrosis during chronic HAART treatment. Indeed, there is clinical[60] and experimental[61] evidence to suggest that the severity of cardiac fibrosis correlates directly with the degree of diastolic dysfunction in several cardiac conditions. Since the structural properties of the heart are influenced by both the myocyte network and interstitial connective tissue, alterations in composition and/or amount of extracellular matrix will likely impact LV diastolic function.[60]

In support of a causative relationship between development of cardiac fibrosis and LV diastolic dysfunction, it was observed that Mg-Sup afforded partial, yet substantial protection against RTV- and EFV-mediated loss of diastolic function (Fig. 16.4A), and this was associated with partial attenuation of collagen formation in both RTV- and EFV-treated rats at 8 weeks (Fig. 16.8); this inhibitory effect on fibrosis may have been preceded by substantial decreases in cardiac WBC infiltration.[5,48]

## SUMMARY POINTS

In summary, studies presented here using the rodent models demonstrated that chronic administration of AZT, RTV, and EFV led to

- Increased systemic oxidative stress, with elevations in plasma F2-isoprostane and RBC oxidized glutathione (GSSG) levels; and inflammation, as suggested by enhanced basal superoxide producing activity of circulating isolated neutrophils and an intermediate stage of cardiac tissue WBC infiltration.
- Progressive cardiac systolic and diastolic dysfunction as well as varying degrees of LV posterior wall thinning, indicative of an early sign of DCM.

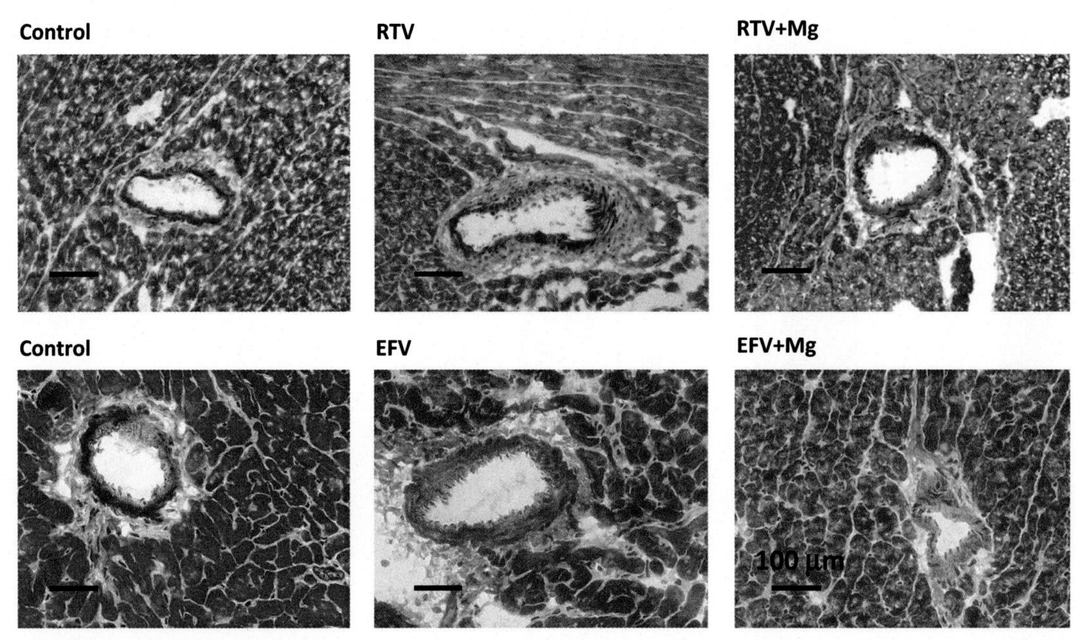


FIGURE 16.8 Light micrographs showing collagen staining (Masson's Trichrome for fibrotic development) in Lewis x Brown Norwegian F1 hybrid rat cardiac ventricles following 8 weeks of ritonavir (RTV) (upper panels) and efavirenz (EFV) (lower panels) treatment with or without dietary Mg-supplementation (sixfold above normal). Representative micrographs were derived from 6 to 8 different fields per heart for three animals per group. Other conditions are described in legends of Figs. 16.3 and 16.5. Magnification 20×.

- Varying degrees of cardiac fibrosis, mainly perivascular in nature, when rats were treated with either RTV or EFV for 5–8 weeks.
- The notion that the prominent oxidative events involved the endothelium and circulating WBCs. At the cellular levels, elevated ROS production mediated by the Highly Active antiretroviral therapy (HAART) agents may involve mitochondrial toxicity, ER stress, and upregulation of NADPH oxidases.
- Despite somewhat diverse modes of actions elicited by the three HAART agents, it was remarkable that dietary Mg-Sup (sixfold higher than normal) was significantly protective against the oxidative/nitrosative/inflammatory stresses induced by each HAART drug.

The protective "mechanisms" provided by Mg-Sup are likely to be complex and multifaceted. Mg may

- Provide direct antiperoxidative chemical action due to its competition with "redox-active" iron at membrane binding sites, thereby limiting potential hydroxyl radical toxicity.
- Act as a natural calcium blocker, and as such, inhibits the calcium influx required not only for NADPH oxidase activation but also for production of several vasoactive metabolites and prostaglandins, which may contribute to HAART-mediated hypertension.
- Attenuate HAART (e.g., AZT and RTV)-mediated downregulation of endothelial nitric oxide (NO$) synthesis and also suppress potential neuronal N-methyl-D-aspartate receptor activation, which would have led to neurogenic inflammation due to rising circulating and tissue substance P levels,[13] and
- Suppress oxidant-induced nuclear factor kappa-beta signaling, which would trigger calcium-dependent production of inflammatory cytokines.

In conclusion, current and previous findings strongly support the therapeutic use of Mg-Sup as an economical, yet effective adjuvant treatment to reduce the risk of systemic oxidative stress and cardiovascular toxicity in HIV-1 patients receiving chronic HAART treatment.

## Funding Support

**NIH-RO1-HL65178, NIH-R21NR012649,** NIH-**R21-HL125038** and The George Washington University McCormick Genomic and Proteomic Center.

## References


1. Lewis W. Pathologic changes in the hearts of patients with AIDS. In: Lipshultz SE, editor. *"Cardiology in AIDS"*. N.Y: Chapman and Hall; 1998. p. 317–29. [Chapter 16].
2. Salhi Y, Costagliola D, Rebulla P, Dessi C, Karagiorga M, Lena-Russo D, de Montalembert M, Girot R. Serum ferritin, desferrioxamine, and evolution of HIV-1 infection in thalassemic patients. *J Acquir Immune Defic Syndr Hum Retrovirol* 1998;**18**:473–8.
3. de Monye C, Karcher DS, Boelaert JR, Gordeuk VR. Bone marrow macrophage iron grade and survival of HIV-seropositive patients. *AIDS* 1999;**13**:375–80.
4. DHHS Panel on Antiretroviral Guidelines for Adults, Adolescents [No authors listed]. *Guildlines for the use of antiretroviral agents in HIV-1 infected adults and adolescents*. 2011. http://aidsinfo.nih.gov.
5. Mak IT, Chmielinska JJ, Kramer JH, Weglicki WB. AZT-induced cardiovascular toxicity—attenuation by Mg-supplementation. *Cardiovasc Toxicol* 2009;**9**:78–85.
6. Dube MP, Lipshultz SE, Fichtenbaum CJ, Greenberg R, Schecter AD, Fisher SD. Effects of HIV infection and antiretroviral therapy on the heart and vasculature. *Circulation* 2008;**118**:e36–40.
7. Lewis W, Copeland WC, Day BJ. DNA depletion, oxidative stress and mutation: mechanisms of dysfunction from NRTIs. Lab. *Invest* 1998;**81**:777–90.
8. Dalakas M, Illa Y, Peseshkpour GH, Laukatis JP, Cohen B, Griffin JL. Mitochondrial myopathy caused by long-term zidovudine therapy. *New Engl J Med* 1990;**322**:1098–105.
9. Szabados E, Fischer G, Toth K, Csete B, Nemeti B, Trombitas K, Habomn T, Ehdrei D, Suunegi B. Role of reactive oxygen species and poly-ADP-ribose polymerase in the development of AZT-induced cardiomyopathy in rats. *Free Rad Biol Med* 1999;**3/4**:309–17.
10. Day BJ, Lewis W. Oxidative stress in NRTI-induced toxicity. *Cardiovasc Toxicol* 2004;**4**:2007–16.
11. Wever K, van Agtmael MA, Carr A. Incomplete reversibility of tenofovir-related renal toxicity in HIV-infected men. *J Acquir Immune Defic Syndr* 2010;**55**(1):78–81.
12. Biagioni Santos MS, Seguro AC, Andrade L. Hypomagnesaemia is a risk factor for nonrecovery of renal function and mortality in AIDS patients. *Braz J Med Biol Res* 2010;**43**:316–23.
13. Kramer JH, Spurney C, Iantorno M, Tziros C, Mak IT, Tejero-Taldo MI, Chmielinska JJ, Komarov A, Weglicki WB. Neurogenic inflammation and cardiac dysfunction due to hypomagnesemia. *Am J Med Sci* 2009;**338**:22–7.
14. Kline ER, Sutliff RL. The roles of HIV-1 proteins and antiretroviral drug therapy in HIV-1-associated endothelial dysfunction. *J Investig Med* 2008;**56**(5):752–69.
15. Hulgan T, Morrow J, D'Aquita RT, Raffanti S, Morgan M, Rebeiro P, Haas DW. Oxidant stress is increased during treatment of human immunodeficiency virus infection. *Clin Infect Dis* 2003;**37**(12):1711–7.
16. Kotler DP. HIV and antiretroviral therapy: lipid abnormalities and associated cardiovascular risk in HIV-infected patients. *JAIDS* 2008;**49**(Suppl. 2):S79–85.
17. Blas-Garcia A, Apostolova N, Ballesteros D, Monleón D, Morales JM, Rocha M, Victor VM, Esplugues JV. Inhibition of mitochondrial function by efavirenz increases lipid content in hepatic cells. *Hepatology* 2010;**52**(1):115–25.
18. Jamaluddin MS, Lin PH, Yao Q, Chen C. NNRT inhibitor efavirenz increases monolayer permeability of human coronary artery endothelial cells. *Atherosclerosis* 2010;**208**:104–11.
19. Knox TA, Oleson L, von Moltke LL, Kaufman RC, Wanke CA, Greenblatt DJ. Ritonavir greatly impairs CYP3A activity in HIV infection with chronic viral hepatitis. *J AIDS* 2008;**49**:358–68.
20. Carr A, Samaras K, Burton S, Law M, Freund J, Chisholm DJ, Cooper DA. A syndrome of peripheral lipodystrophy, hyperlipidemia and insulin resistance in patients receiving HIV protease inhibitors. *AIDS* 1998;**12**:F51–8.
21. Wang X, Chai H, Yao Q, Chen C. Molecular mechanisms of HIV protease inhibitor-induced endothelial dysfunction. *J AIDS* 2007;**44**:493–9.
22. Mondal D, Pradhan L, Ali M, Agrawal KC. HAART drugs induce oxidative stress in human endothelial cells and increase endothelial recruitment of mononuclear cells: exacerbation by inflammatory cytokines and amelioration by antioxidants. *Cardiovasc Toxicol* 2004;**4**:287–302.
23. Conklin BS, Fu W, Lin PH, Lumsden AB, Yao Q, Chen C. HIV protease inhibitor ritonavir decreases endothelium-dependent vasorelaxation and increases superoxide in porcine arteries. *Cardiovasc Res* 2004;**63**:168–75.
24. Chai H, Yang H, Yan S, Li M, Lin PH, Lumsden AB, Yao Q, Chen C. Effects of 5 HIV protease inhibitors on vasomotor function and superoxide anion production in porcine coronary arteries. *J AIDS* 2005;**40**:12–9.
25. Fu W, Chai H, Yao Q, Chen C. Effects of HIV protease inhibitor ritonavir on vasomotor function and endothelial nitric oxide synthase expression. *J AIDS* 2005;**39**:152–8.
26. De la Asuncion JG, del Olmo ML, Sastre J, Millan A, Pellin A, Pallardo FV, Vina J. AZT treatment induces molecular and ultrastructural oxidative damage to muscle mitochondria. Prevention by antioxidant vitamins. *J Clin Invest* 1998;**102**:4–9.
27. Wang Y, Watson RR. Is vitamin E supplementation a useful agent in AIDS therapy? *Prog Food Nutr Sci* 1993;**17**:351–75.
28. Papparella I, Ceolotto G, Berto L, Cavalli M, Bova S, Cargnelli G, Ruga E, Milanesi O, Franco L, Mazzoni M, Petrelli L, Nussdorfer GG, Semplicini A. Vitamin C prevents zidovudine-induced NADPH oxidase activation and hypertension in the rat. *Cardiovasc Res* 2007;**73**:432–8.
29. Food, Drug Administration, U.S. Department of Health and Human Services, Center for Drug Evaluation and Research (CDER). Conversion of animal doses to human equivalent doses based on body surface area. In: *Guidance for industry estimating the maximum safe starting dose in initial clinical trials for therapeutics in adult healthy volunteers*. 2005. p. 1–27. [Table 1] http://www.fda.gov/cder/guidance/index.htm.
30. de Miranda P, Burnette TC, Good SS. Tissue distribution and metabolic disposition of zidovudine in rats. *Drug Metabolism Disposition* 1990;**18**:317–20.
31. Mak IT, Nedelec LF, Weglicki WB. Pro-oxidant properties and cytotoxicity of AZT-monophosphate and AZT. *Cardiovasc Toxicol* 2004;**4**:109–15.
32. Kramer JH, Dadgar S, Hall J, Mak IT, Weglicki WB. Rat tissue iron content is differentially altered by AZT and Mg deficiency (MgD) (abstract). *J Mol Cell Cardiol* 2004;**36**:631–2.
33. Pollack S, Weaver J. Azidothymidine (AZT)-induced siderosis. *Am J Hematol* 1993;**43**:230–3.

34. Scalfaro P, Chesaux JJ, Buchwalder PA, Biollaz J, Micheli JL. Severe transient neonatal lactic acidosis during prophylactic zidovudine treatment. *Intensive Care Med* 1998;**24**:247–50.
35. Rickerts V, Brodt H, Staszewski S, Stille W. Incidence of myocardial infarctions in HIV-infected patients between 1983 and 1998. The Frankfurt HIV-cohort study. *Eur J Med Res* 2000;**5**:329–33.
36. Kramer JH, Lightfoot FG, Weglicki WB. Cardiac tissue Fe: effects on postischemic function and free radical production, and its possible role during preconditioning. *Cell Mol Biol* 2000;**6**(8):1313–27.
37. Mak IT, Goldfarb MG, Weglicki WB, Haudenschild CC. Cardiac pathologic effects of AZT in Mg-deficent mice. *Cardiovasc Toxicol* 2004;**4**:169–78.
38. Papparella I, Ceolotto G, Berto L, Cavalli M, Bova S, Cargnelli G, Ruga E, Milanesi O, Franco L, Mazzoni M, Petrelli L, Nussdorfer GG, Semplicini A. Vitamin C prevents zidovudine-induced NADPH oxidase activation and hypertension in the rat. *Cardiovasc Res* 2007;**73**:432–8.
39. Sutliff RL, Dikalov S, Weiss D, Parker J, Raidel S, Racine AK, Russ R, Haase CP, Taylor WR, Lewis W. Nucleoside reverse transcriptase inhibitors impair endothelium-dependent relaxation by increasing superoxide. *Am J Physiol* 2002;**283**(6):H2363–70.
40. Lund KC, Wallace KB. Adenosine 3′5′-cAMP-dependent phosphoregulation of mitochondrial complex I is inhibited by nucleoside reverse transcriptase inhibitors. *Toxicol Appl Phamacol* 2008;**226**:94–106.
41. Brechard S, Tschirhart EJ. Regulation of superoxide production in neutrophils: role of calcium influx. *J Leukoc Biol* 2008;**84**:1223–37.
42. Shechter M. Does magnesium have a role in the treatment of patients with coronary artery disease. *Am J Cardiovasc Drugs* 2003;**3**:231–9.
43. Murthi SB, Wise RM, Weglicki WB, Komarov AM, Kramer JH. Mg-gluconate provides superior protection against postischemic dysfunction and oxidative injury compared to Mg-sulfate. *Mol Cell. Biochem* 2003;**245**:141–8.
44. Kawano Y, Matsuoka H, Takishita S, Omae T. Effects of Magnesium supplementation in hypertensive patients. *Hypertension* 1998;**32**:260–5.
45. Touyz RM, Milne FJ. Magnesium supplementation attenuates, but not prevents, development of hypertension in spontaneously hypertensive rats. Am. *J Hypertens* 1999;**12**:757–65.
46. Altura BT, Altura BM. Role of magnesium ions in contractility of blood vessels and skeletal muscles. *Magnesium Bull* 1981;**3**:102–6.
47. Choudlary S, Kumar A, Kale RK, Raisz LG, Pilbeam CC. Extracellular calcium induces COX-2 in osteoblasts via a PKA pathway. *Biochem Biophys Res Comm* 2004;**322**:395–402.
48. Mak IT, Kramer JH, Chen X, Chmielinska JJ, Spurney CF, Weglicki WB. Mg supplementation attenuates ritonavir-induced hyperlipidemia, oxidative stress, and cardiac dysfunction in rats. *Am J Physiol Regul Integr Comp Physiol* 2013;**305**:R1102–11.
49. Chen X, Mak IT. Mg supplementation protects against ritonavir-mediated endothelial oxidative stress and hepatic eNOS downregulation. *Free Rad Biol Med* 2014;**69**:77–85.
50. Weiß M, Kost B, Renner-Müller I, Wolf E, Mylonas I, Brüning A. Efavirenz causes oxidative stress, endoplasmic reticulum stress, and autophagy in endothelial cells. *Cardiovasc Toxicol* 2016;**16**(1):90–9.
51. Bertrand L, Toborek M. Dysregulation of endoplasmic reticulum stress and autophagic responses by the antiretroviral drug efavirenz. *Mol Pharmacol* 2015;**88**(2):304–15.
52. Blas-Garcia A, Polo M, Alegre F, Funes HA, Martinez E, Apostolova N, Esplugues JV. Lack of mitochondrial toxicity of darunavir, raltegravir and rilpivirine in neurons and hepatocytes: a comparison with efavirenz. *J Antimicrob Chemother* 2014;**69**:2995–3000.
53. Hulgan T, Morrow J, D'Aquila RT, Raffanti S, Morgan M, Rebeiro P, Haas DW. Oxidant stress is increased during treatment of human immunodeficiency virus infection. *Clin Infect Dis* 2003;**37**(12):1711–7.
54. Gupta SK, Slaven JE, Kamendulis LM, Liu Z. A randomized, controlled trial of the effect of rilpivirine versus efavirenz on cardiovascular risk in healthy volunteers. *J Antimicrob Chemother* 2015;**70**(10):2889–93.
55. Mondy KE, Gottdiener J, Overton ET, Henry K, Bush T, Conley L, Hammer J, Carpenter CC, Kojic E, Patel P, Brooks JT. High prevalence of echocardiographic abnormalities among HIV-infected persons in the Era of highly active antiretroviral therapy. *Clin Infect Dis* 2011;**52**(3):378–86.
56. Nelson MD, LaBounty T, Szczepaniak L, Szczepaniak E, Smith L, St John L, Gottlieb J, Park J, Sannes G, Li D, Dharmakumar R, Yumul R, Hardy D, Conte AH, (abstract). Cardiac steatosis and left ventricular dysfunction is associated with exposure to human immunodeficiency virus highly active antiretroviral therapy: a 3-tesla cardiac magnetic resonance imaging study. *J Am Coll Cardiol* 2014;**63**(12):A1006.
57. Grandi AM, Nicolini E, Giola M, Gianni M, Maresca AM, Marchesi C, Guasti L, Balsamo ML, Venco A, Grossi PA. Left ventricular remodelling in asymptomatic HIV infection on chronic HAART: comparison between hypertensive and normotensive subjects with and without HIV infection. *J Hum Hypertens* 2012;**26**:570–6.
58. Reyskens KM, Fisher TL, Schisler JC, O'Connor WG, Rogers AB, Willis MS, Planesse C, Boyer F, Rondeau P, Bourdon E, Essop MF. Cardiometabolic effects of HIV protease inhibitors (lopinavir/ritonavir). *PLoS One* September 30, 2013;**8**(9):e73347.
59. Weber KT. Cardiac interstitium in health and disease: the fibrillary collagen network. *J Am Coll Cardiol* 1989;**13**:1637–52.
60. Moreo A, Ambrosio G, De Chiara B, Pu M, Tran T, Mauri F, Raman SV. Influence of myocardial fibrosis on left ventricular diastolic function: noninvasive assessment by cardiac magnetic resonance and echo. *Circ Cardiovasc Imaging* 2009;**2**(6):437–43.
61. Ryoke T, Gu Y, Mao L, Hongo M, Clark RG, Peterson KL, Ross Jr J. Progressive cardiac dysfunction and fibrosis in the cardiomyopathic hamster and effects of growth hormone and angiotensin-converting enzyme inhibition. *Circulation* 1999;**100**(16):1734–43.


# Further Reading


1. Entman ML, Youker K, Shoji T, Kukielka G, Shappell SB, Taylor AA, Smith W. Neutrophil induced oxidative injury of cardiac myocytes. A compartmented system requiring CD11b/CD18-ICAM-1 adherence. *J Clin Invest* 1992;**90**:1335–45.

CHAPTER

# 17

# Selenium Supplementation, Antioxidant Effects, and Immune Restorative Effects in Human Immunodeficiency Virus*

*Justin Dong, Anand Muthiah, Parveen Hussain, Miya Yoshida, Vishwanath Venketaraman*

**Western University of Health Sciences, Pomona, CA, United States**

OUTLINE



## Abstract

As the incidence and prevalence of human immunodeficiency virus (HIV) rises, research into antiretroviral therapy seeks to improve morbidity and mortality outcomes in patients infected with the retrovirus. The role of micronutrients in the human body ranges through the entire spectrum of vital biochemical reactions necessary for life. More recently, the exact role that micronutrients play within our body has enabled us to see the role they play in health and disease. The nonmetal micronutrient selenium is an element contained within selenocysteine, a precursor building block of selenoproteins. Selenium is an integral part of peroxidase and reductase enzymes, selenoproteins, and compounds necessary for the maintenance of cell homeostasis. Our focus is on the antioxidant and immune system effects that selenium supplementation plays in HIV-positive individuals. HIV has been shown to decrease levels of selenoproteins and selenium levels. In this chapter we discuss the implications of supplying selenium in HIV-positive patients to improve immune system response, slow the rate of CD4 T cell decline, and finally, exploring the concept of reducing morbidity and mortality rates caused by the virus.

**Keywords:** Antioxidant; HIV; Immune; Restoration; Selenium; Selenoproteins; Supplement.

* All authors contributed equally.



http://dx.doi.org/10.1016/B978-0-12-809853-0.00017-1

## List of Abbreviations

**GPx** Glutathione Peroxidase
**HIV** Human Immunodeficiency Virus
**ROS** Reactive Oxidative Species

# INTRODUCTION

Infection by the Human immunodeficiency virus (HIV) has impacted nearly 36.9 million people worldwide and 1.2 million people have died from AIDS-related illnesses alone in 2014.[1] These statistics highlight the need for research and development of therapeutic treatments to combat the disease. There have been many defensive biological mechanisms proposed throughout the years and a pattern begins to emerge that points toward dietary supplementation.

Selenium, an essential micronutrient in mammals, became established as an important aspect of focus in 1973 following its discovery of being an important constituent of the enzyme glutathione peroxidase (GPx).[2] Several abnormalities and human diseases have been correlated to selenium with the most notable being Keshan disease, a congestive cardiomyopathy found in children that was first described in China and subsequent coxsackie B virus infection.[2] It is believed that with selenium deficiency, oxidative stress caused mutations in the viral genome resulting in a more virulent form.[3] Investigators further carried out studies that identified the importance of selenium supplementation and eventually became fundamental in the eradication of Keshan disease.[2] With these discoveries, awareness was raised for the positive effects of selenium and its possible ability in combating other viral infections such as HIV.

In the human genome, there are 25 genes identified for selenoproteins, with a majority of the genes being expressed by the cells of the immune system.[4] Linked with selenoproteins, many of which are antioxidant selenoenzymes, selenium is known to play a major role in biological functions such as antioxidant defense, redox signaling, and immune response through its incorporation at the 21st amino acid in selenoproteins.[5] Selenium takes on the biological form of selenocysteine [Sec], a cysteine analog that contains selenium in place of the usual sulfur, which undergoes cotranslational incorporation after conversion of O-phosphoseryl-transfer RNA into selenocysteyl-tRNA during protein synthesis.[4] Selenoproteins such as iodothyronine deiodinases, GPxs, and thioredoxin reductases (TrxRs) have all played crucial roles in normal biological functions, immune response, and catalytic activity. Studies have shown that replacement or deactivation of Sec resulted in dramatic decreases in selenoprotein enzymatic activity and thus increase in disease.[6] This chapter seeks to highlight the importance of selenium and its capability in fighting HIV infections.

# ANTIOXIDANT EFFECTS

Of particular interest is the role of selenium in the reduction of cellular oxidative stress. Reactive oxygen species (ROS) are highly reactive compounds that are produced in normal states by the cell, which has cellular mechanisms to eliminate them. In the advent of an increase in ROS or a decrease in protective measures in the cell, ROS can cause significant damage and apoptosis.

The HIV virus induces a potent response from the human immune system, which releases inflammatory cytokines and ROS as a result of oxidative phosphorylation. ROS also increase viral replication in infected cells through the actions of nuclear factor light-chain enhancer of activated B cells (NF-kB) and activator protein-1.[7]

Selenium is required for the action of GPx, an enzyme responsible for catalyzing the reduction of hydrogen peroxide and organic hydroperoxides. The action of GPx allows the cell to protect itself from oxidative stress. Starting with NADPH, electrons are transferred through a series of redox reactions involving glutathione reductase and glutathione (Fig. 17.4). This reaction clears the cell of free radicals, thus helping to reduce the effects of oxidative stress. Furthermore, GPx activity has been shown to increase with selenium supplementation in HIV-infected lymphocytes,[8] protecting against oxidative damage. GPx along with other antioxidant enzymes have also been found to decrease viral activation.[8]

In cells supplemented by selenium, hydrogen peroxide–induced activation of NF-kB is reduced, while the increase in GPx activity decreases tumor necrosis factor-alpha (TNF-α)-mediated HIV activation.[7] Increased TNF-α levels are often correlated with high viral load and low CD4+ counts in chronic viruses. Sustained signaling via TNF-receptor (TNF-R) has been associated with chronic inflammation in advanced stages of HIV.[3,9] It is possible that chronic HIV infection leads to increases in TNF, which may activate NF-kB to thus continue the cycle of viral activation.[10] In studies of acute HIV infection, the addition of selenium prior to that of TNF-α was able to partially inhibit viral replication in chronically infected lymphocytes and monocytes.[11] Without exposure to exogenous TNF-α, there was no noticeable effect in lymphocytes,[11] indicating that the upregulation of antioxidant enzymes by selenium may indirectly influence the immune response to HIV (Fig. 17.2).

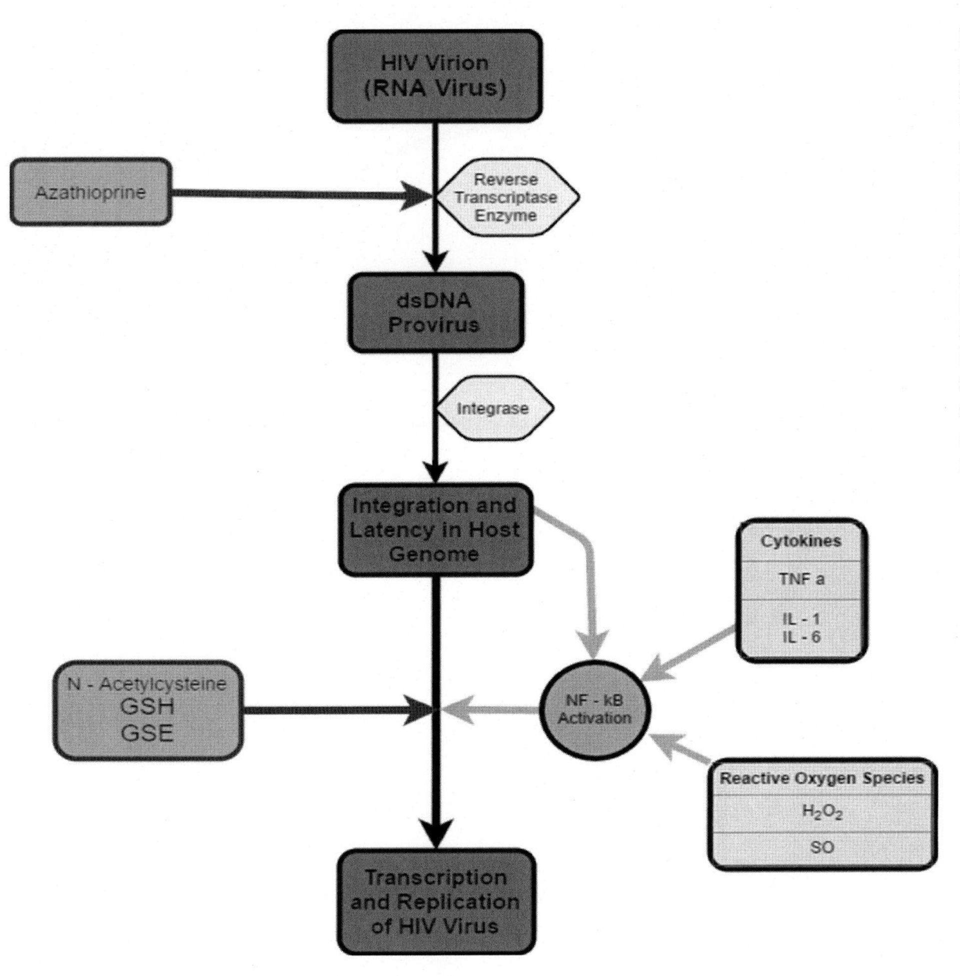


FIGURE 17.1 Flowchart demonstrating activation and inhibition of human immunodeficiency virus (HIV) replication. (1) Azathioprine, a nucleoside analog of HIV reverse transcriptase. (2) Overproduction of proinflammatory cytokines tumor necrosis factor-alpha (TNF-α), IL-1, and IL-6 activate nuclear factor kappa-beta (NF-kB) transcription factor. (3) Reactive oxidative species (ROS) stimulates NF-kB transcription factor and promotes HIV replication. (4) N-acetylcysteine, glutathione (GSH), and glutathione ester (GSE) supplementation improves immune response and decreases HIV transcription and replication.

Selenium is also a well-known constituent of TrxR, a key player in another redox system that guards against oxidative stress. Similar to glutathione reductase, TrxRs incorporate FAD and a disulfide catalytic domain in their structure and are NADPH-dependent.[12] They are capable of reducing oxidized thioredoxins, which undergo further redox reactions (Fig. 17.1). Thioredoxins then provide necessary electrons to ribonucleotide reductase and thioredoxin peroxidase, as well as reducing transcription factors to modify gene transcription.[9] An overlap with the glutathione system is highly likely. High levels of TrxR have been noted in connection with oxidative stress and inflammation in neoplastic cells.[13] It has also been shown to assist in reducing oxidative damage from hydrogen peroxide in epithelial cells.[14] More recently, it has been postulated by Duan et al. that parthenolide, an anticancer lactone, specifically targets the Sec residue on TrxR1, causing inhibition of the enzyme and leading to ROS-mediated apoptosis in cancer cells.[15]

## IMMUNE SYSTEM RESTORATION

Adequate levels can stimulate the functions of neutrophils, enhance NK cell-mediated cytolytic activity, and induce proliferation of T and B lymphocytes leading to the production of cytokines and antibodies, respectively.[16] These findings point toward an integral role of selenium in modulating the immune response. Selenoproteins are highly abundant in immune cells and they are observed to be upregulated during immune cell activation as they are required for Ca2+ flux during activation of T cells, neutrophils, and macrophages.[17,18] In particular T cells and macrophages exhibit the highest levels of selenoproteins within the immune system, which compromises their function when there is a deficiency in selenium.[18] Selenoproteins not only carry antioxidant functions, which are beneficial during disease in which oxidative stress increases, but they are also involved in other cellular functions such as protein folding and cell signaling.[18] During deficiency nonessential selenoproteins are preferentially lost over essential selenoproteins; implying that these are critical for the immune response of the cell.[18] Additionally, studies demonstrated that a lack of selenoprotein in lymphocytes led to a decrease in the number of mature T cells and defects in antibody production.[4]

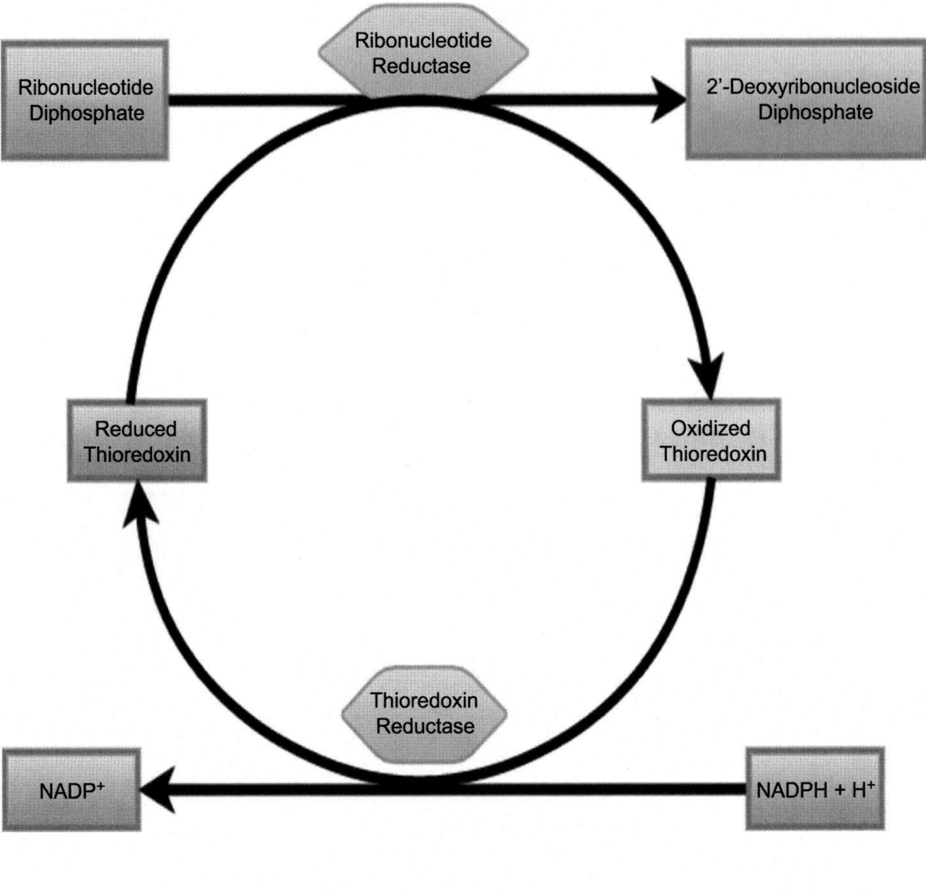


FIGURE 17.2 Flavoprotein thioredoxin redox cycle. Thioredoxin is reduced by thioredoxin reductase and oxidized by ribonucleotide reductase.

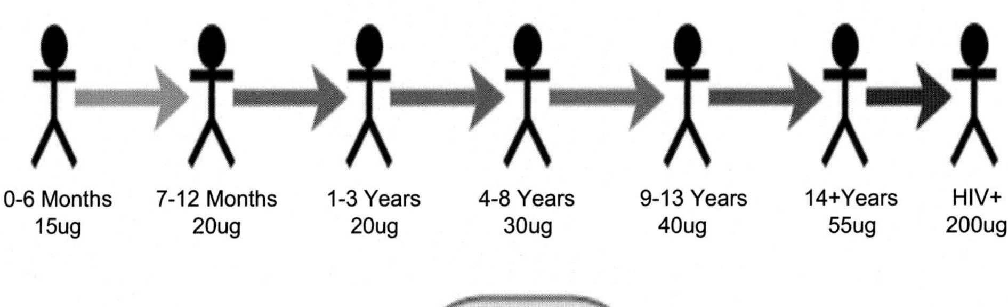

FIGURE 17.3 Selenium Recommended Dietary Allowance for young and older adults. Recommended selenium intake for young children, teenagers, older adults, and an HIV-positive adult.

## DAILY DIETARY REQUIREMENT OF SELENIUM

Selenium has long been known to play a key role in a number of enzymes that participate in reproduction, metabolism, and antioxidant reactions.[19] Animal studies have shown that a deficiency in selenium may lead to a vulnerability to oxidative stress. Large amounts can be toxic; however, trace amounts are necessary for most organisms. Selenium can be found as inorganic compounds as selenite, selenide, and selenate but persists mostly as organic compounds as selenomethionine, Sec, and methylselenocysteine. Inorganic selenium in soil is converted to organic selenium within plants.[19]

Most selenium resides in skeletal muscle, which accounts for nearly 28%–46% of selenium within the body.[19] Plasma or serum selenium is the most frequent method by which selenium intake levels are tested. Plasma selenium levels typically include selenoproteins, selenoprotein P, GPx, and selenomethionine.[20] A deficiency in selenium leads to a deficiency of selenoprotein synthesis. If there is sufficient or high selenium intake, the level of selenoproteins is

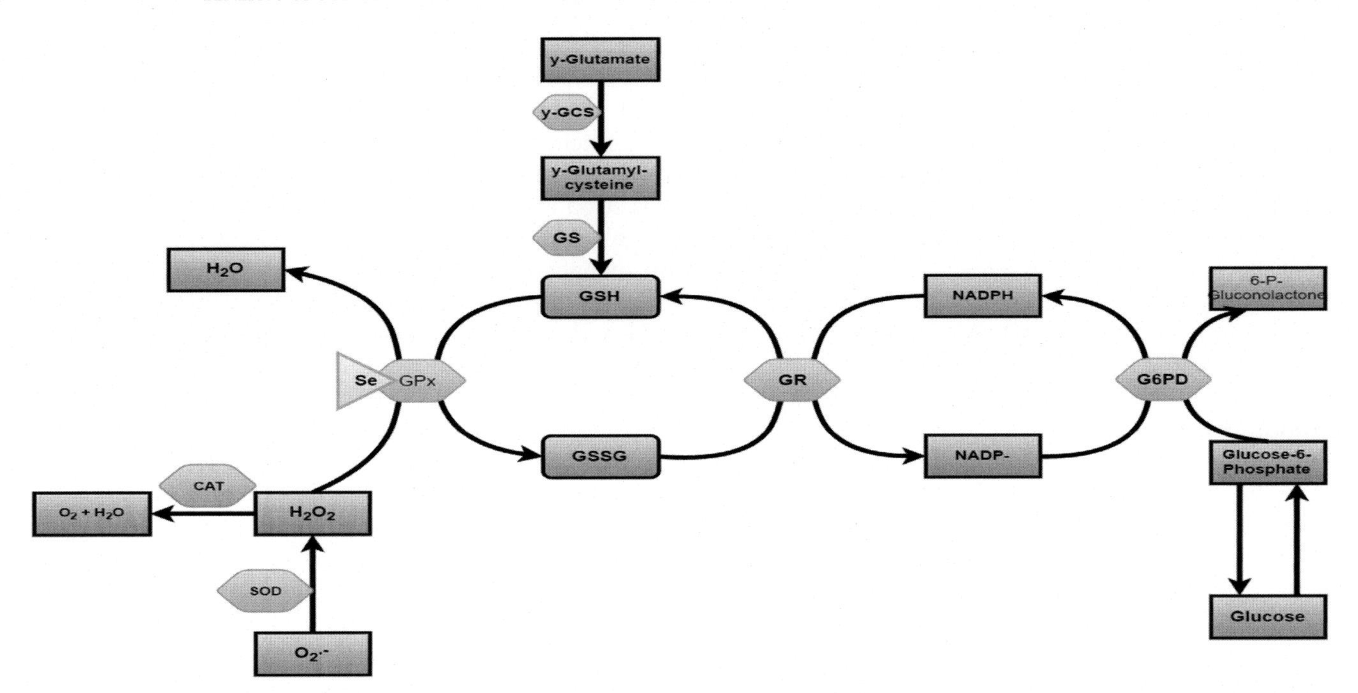


FIGURE 17.4 Glutathione redox cycle. Glutathione (GSH) is tripeptide antioxidant responsible for alleviating reactive oxidative species (ROS). Reduced glutathione (rGSH) is synthesized by glutathione reductase and NADPH cofactor. Selenium is an essential micronutrient in glutathione peroxidase, which oxidizes rGSH, while rGSH reduces ROS. *GPx*, glutathione peroxidase; *GSSG*, glutathione disulfide; *SOD*, superoxide dismutase.

maintained at a minimal level.[20] Selenomethionine makes up the majority of selenium at levels above 8 μg/dL.[20] It can pass through the same metabolism pathway as methionine. This is the source of selenium that is more liable to shrink or grow when selenium intake varies.

Selenium levels have been known to be particularly high in seafood, meat, and cereal.[21] Grains, breads, and dairy products may also contain significant levels of selenium. Because soil content of selenium plays a large role in the intake of animal and plant selenium, levels may vary from country to country. Selenium can also be given as a supplement, usually as selenomethionine, selenium-enriched yeast, sodium selenite, or sodium selenate.[19] Approximately 90% of selenomethionine can be absorbed, yet only 50% can be absorbed from sodium selenite.[19]

Serum selenium concentrations of 8 μg/dL are generally considered to be sufficient for selenoprotein synthesis.[19] The US Food and Nutrition Board determined the Recommended Dietary Allowance, the average daily level considered acceptable for the nutritional requirement of 97%–98% of healthy adults, to be 55 μg (Fig. 17.3).

North Americans tend to consume upward of 80–140 μg of selenium per day. Deficiencies in selenium may occur in countries with low soil concentrations, such as China and New Zealand.[21] Keshan disease, an endemic cardiomyopathy in China, is thought to occur in selenium-deficient children. Chronic dialysis patients may also experience a selenium deficiency due to loss of appetite and removal of selenium from the blood during dialysis. In addition, those infected with HIV have been found to have low selenium levels, which may be due to either decreased consumption or exorbitant loss.[19]

## BENEFITS OF SUPPLEMENTING HUMAN IMMUNODEFICIENCY VIRUS–POSITIVE INDIVIDUALS WITH SELENIUM

There have been several clinical trials to date that have tested the response of selenium supplementation. A dose of 200 μg selenium per day appears to be sufficient to provide protective benefits against prostate, lung, and colorectal cancer and may even assist in increasing CD4+ T cell counts in HIV-infected patients[17,22] and appears to be the most commonly used dosage in clinical trials. Oral selenium supplements are widely available online and in stores, typically as selenomethionine, selenium citrate, or selenium-enriched yeast. Selenium-enriched yeast is mostly composed of selenomethionine.[20] Studies have shown that selenomethionine has a higher bioavailability in comparison to selenite in both replete and deficient individuals,[20] making selenomethionine the supplement of choice.

The results from many different investigations show that patients living with HIV often have low serum selenium levels in comparison to controls.[19,21,23] Low selenium levels have been linked to low CD4+ counts, opportunistic infections, and more advanced stages of HIV.[24,25] Decreased serum selenium has even been associated with increased mortality in HIV patients.[24,26] Different theories regarding the cause of this deficiency involve malnutrition, poor appetite, and diarrhea. The role of GPx and TrxR in regulating HIV-mediated oxidative stress may cause an increased need for selenium, which is reflected in the results from these studies. When transfected into canine kidney cells, the activity of this HIV-encoded GPx actually increased from 21% to 43%.[27,28] It is possible that HIV appropriates the host selenium storage to create its own selenoproteins, decreasing the available selenium. In that regard, it is no surprise that micronutrient studies have shown a large correlation between HIV infection and deficiencies in selenium during advanced stages of the disease.

Since individuals with HIV infection have been shown to have diminished levels of selenium and selenoproteins, which correlated with a decrease in CD4+ T cells counts and increased viral pathogenicity selenium supplementation is believed to have beneficial effects that can help maintain adequate levels of CD4+ T cell counts and regulate host immunity.[20,22]

In one study, increased HIV viral shedding in the genitourinary tract has been seen in patients that are selenium deficient. In these patients, GPx activity was also lower than HIV-infected patients with adequate levels of selenium.[22] Similarly, another study found that selenium supplementation resulted in an increase in the expression of selenoproteins by immune cells, while deficiency reduces levels of selenoprotein mRNA involved in signaling pathways during immune cell activation.[29] Within CD4+ T cells, activation results in increased uptake of selenium, which leads to a significant increase in proliferation via IL-2 and interferon-gamma.[29] It was also observed that selenium supplementation influenced the differentiation of CD4+ T cells toward the Th1 pathway, which can suppress the progression of HIV by boosting host immunity against viral RNA.[20,22,30] In particular selenium reduced the acetylation of the HIV promoter region K12 and K16, which are involved in viral replication.[10] While selenium has not been shown to directly decrease viral load, it has been shown to significantly dampen the rate of CD4+ T cell decline in patients infected with HIV, which can have potential therapeutic applications.[2,19] Through these indirect mechanisms supplementation can suppress the progression of the virus and help maintain immunity in individuals that are already immune compromised.

Selenium supplementation has been shown to positively affect the immune system and improves immune function of CD4+ T cells. In addition, selenium supplementation was also observed to have beneficial effects in the controlling the proliferation of HIV. As a result, selenium supplementation begins to emerge as a potential adjunct therapy for individuals infected with HIV.

In a review of 12 listed trials across three continents, patients with high levels of selenium supplemented with 200 ug exhibited lower HIV viral loads and increased CD4+ T cell levels. A different study of pregnant Tanzanian women infected with HIV showed that maternal viral load and mortality was not affected by selenium supplements, but child survival rate increased.[22] A Thai study of low CD4 T cell HIV-positive patients showed no changes in viral load but did show increased rates of survival. Selenoenzyme TrxR1 has been identified as a negative regulator of the HIV-encoded Tat protein required for viral replication.[22] Through these mechanisms we can see selenoproteins playing a crucial role in immunity. Findings support selenium supplementation as adjunctive therapy to help partially restore immune function of CD4 T cells lost in HIV infection. Micronutrient supplementation of selenium provides an inexpensive adjuvant for slowing the pathogenesis of HIV.[17] Selenium levels have been shown to increase in HIV-positive patients on protease inhibitors, alluding to the potential for benefit in patients with limited resources.[9]

In clinical trials, selenium supplementation has also been demonstrated to reduce hospital admissions as well as the percent of those hospitalizations due to infections in HIV-infected patients.[31] Higher levels of selenium have been linked to decelerated rates of mental decline in HIV-infected individuals.[23]

In 2007, a randomized control trial by Hurwitz et al. showed that a daily supplementation of 200 μg selenium for 9 months appeared to elevate serum selenium levels, which was associated with decreased HIV viral load and an indirect increase in CD4+ T cell counts.[17] These findings remained significant after correction for other disease-related factors such as age, gender, ART regimen, and adherence. Additional studies regarding selenium deficiency and increased levels of oxidative stress have shown strong positive correlation. One study showed that individuals who had both HIV and TB had decreased levels of selenium more so than TB or HIV alone.[32]

With selenium supplementation there are limitations that have been demonstrated in several studies. The antioxidant effects of selenium impact the progression of HIV-infected individuals in advanced stages, however, studies have shown that a single supplementation of selenium has no statistically significant difference from the use of placebo in treating early manifestations of HIV infection and decreased CD4+ count.[33] It was a single supplementation of a combination of multivitamins with B vitamins, C and E vitamins, and selenium that reduced the risk of developing severely low CD4+ counts, HIV symptoms, and death.[33] Though the mechanisms of a combined supplementation have not been fully determined, the effects of a multivitamin and selenium have shown significant evidence-based results as compared to a placebo, in combating HIV disease progression.

As we have seen in Keshan disease, a deficiency of selenium may allow a virus to become unusually virulent. Given that there is also evidence that supplementation of micronutrients and antioxidants may improve immune function and alleviate oxidative stress in HIV and cancer,[29,34] selenium supplementation may benefit patients with such diseases. In addition, selenium supplementation at the recommended doses appears to have little to zero toxicity, making it a safe adjunct to more conventional therapies.[19]

## CONCLUSION

In conclusion, selenium supplementation has potential benefits in improving immune response in combating HIV and reducing oxidative stress. Selenium supplementation is still being investigated as a possible form of therapy to improve immune response and reduce oxidative stress in HIV individuals undergoing antiretroviral therapy.[35] Selenium is a trace mineral and cofactor in GPx and TrxR, both these enzymes reduce oxidized molecules and protect the cell from oxidative damage caused by ROS.[35]

Current research avenues are investigating the mechanistic effects of increasing serum selenium levels and in combating HIV are closer to unveiling the mechanism underlying selenium role in preventing CD4+ T cell depletion.[23] Although the exact mechanism is not as clear some literature suggests that HIV promotes an abnormal production of proinflammatory cytokines, which creates an oxidative imbalance causing oxidative stress.[23] Selenium is an essential cofactor required for GPx directly decreasing hydrogen peroxide and organic hydroperoxides while protecting cells from ROS.[36] Additionally, some studies suggest that HIV produces its own selenoenzymes while depleting selenium, therefore HIV selenoenzymes genes influence the upregulation of proinflammatory cytokines.[23] Some of the benefits of increasing serum selenium concentration detailed include regulation of NF-kB activity and control the transcription of HIV, which is often upregulated causing CD4+ T cell depletion.[23]

Although, studies show that HIV-positive individuals tend to have malnutrition and low levels of serum selenium, however solely providing selenium supplementation is not useful without an individual taking their antiviral drugs.[37] Randomized controlled trials have demonstrated that HIV individuals who take selenium supplementations had decreased rates of hospitalization while increasing CD4+ T cell counts.[37] Studies support the damaging effects of low serum selenium that can further proliferate HIV transcription and progress HIV to AIDS.[23] By providing supplementation of selenium in addition to antiviral drug treatment can stimulate CD4+ T cell proliferation further improving the immune response to combat HIV.[23]

Further clinical trials and research are required to explore the potential benefits of selenium supplementation in HIV patients.[35] The role of selenium supplementation in this chapter remains hopeful for the future and supports the advantages of introducing selenium supplementation in a treatment plan for HIV individuals.[35]

## SUMMARY POINTS

- Selenium is an essential cofactor found in selenoproteins that participates in biological roles such as antioxidant defense, redox signaling, and immune response.
- Some of the major selenoproteins include iodothyronine deiodinases, GPx, and TrxRs, these enzymes play an integral role in maintaining normal immune response, biological function, and catalytic activity.
- Selenium deficiency is associated with severe abnormalities and diseases such as Keshan disease, congestive cardiomyopathy discovered originally in Chinese children.
- Selenium levels are high in foods such as seafood, meat, and cereals. Inorganic and organic compounds also contain selenium levels.
- Majority of selenium concentration is found in skeletal muscles, and the selenium levels required for selenoprotein synthesis are approximately 8 μg/dL.
- A deficiency in selenium levels in a host can disrupt selenoprotein synthesis, which can impair immune response and antioxidant defense.
- Selenium supplementation has protective benefits against lung, prostrate, colorectal cancer, and can improve antioxidant defense in HIV individuals with CD4+ T lymphocyte deficiency.
- Selenium deficiency in chronic HIV individuals results in reduced enzymatic activity of GPx, further increasing ROS that results in an increase in HIV viral transcription and replication.

- Selenium supplementation in HIV individuals has shown to reduce ROS, decrease viral activation, and reduce CD4+ T cell apoptosis.
- The role of selenium supplementation seems promising and has potential benefits for supplementation with treatment of individuals who are immunocompromised.

# References

1. UNAIDS. (n.d.). From: http://www.unaids.org/en/resources/campaigns/HowAIDSchangedeverything/factsheet.
2. Burk RF. Selenium, an antioxidant nutrient. *Nutr Clin Care* 2002;**5**(2):75–9.
3. Aukrust P, et al. Tumor necrosis factor (TNF) system levels in human immunodeficiency virus-infected patients during highly active antiretroviral therapy: persistent TNF activation is associated with virologic and immunologic treatment failure. *J Infect Dis* 1999;**179**(1):74–82.
4. Steinbrenner H, Al-Quraishy S, Dkhil MA, Wunderlich F, Sies H. Dietary selenium in adjuvant therapy of viral and bacterial infections. *Adv Nutr* 2015;**6**(1):73–82.
5. Lu J, Holmgren A. Selenoproteins. *J Biol Chem* 2009;**284**(2):723–7. http://dx.doi.org/10.1074/jbc.R800045200.
6. Moghadaszadeh B, Beggs AH. Selenoproteins and their impact on human health through diverse physiological pathways. *Physiology (Bethesda)* 2006;**21**:307–15.
7. Poli G, Fauci AS. The effect of cytokines and pharmacologic agents on chronic HIV infection. *AIDS Res Hum Retroviruses* 1992;**8**(2):191–7.
8. Sappey C, et al. Stimulation of glutathione peroxidase activity decreases HIV type 1 activation after oxidative stress. *AIDS Res Hum Retroviruses* 1994;**10**(11):1451–61.
9. De Pablo-Bernal RS, et al. TNF-alpha levels in HIV-infected patients after long-term suppressive cART persist as high as in elderly, HIV-uninfected subjects. *J Antimicrob Chemother* 2014;**69**(11):3041–6.
10. Beyer M, et al. Tumor-necrosis factor impairs CD4(+) T cell-mediated immunological control in chronic viral infection. *Nat Immunol 2016* 2016;**17**(5):593–603.
11. Hileman CO, Dirajlal-Fargo S, Lam SK, Kumar J, Lacher C, Combs Jr GF, McComsey GA. Plasma selenium concentrations are sufficient and associated with protease inhibitor use in treated HIV-infected adults. *J Nutr* 2015;**145**(10):2293–9.
12. Couto N, Wood J, Barber J. The role of glutathione reductase and related enzymes on cellular redox homoeostasis network. *Free Radic Biol Med* 2016;**95**:27–42.
13. Mustacich D, Powis G. Thioredoxin reductase. *Biochem J* 2000;**346**(Pt 1):1–8.
14. Shrimali RK, Irons RD, Carlson BA, Sano Y, Gladyshev VN, Park JM, Hatfield DL. Selenoproteins mediate T cell immunity through an antioxidant mechanism. *J Biol Chem* 2008;**283**(29):20181–5.
15. Duan D, Zhang J, Yao J, Liu Y, Fang J. Targeting thioredoxin reductase by parthenolide contributes to inducing apoptosis of HeLa cells. *J Biol Chem* 2016;**291**(19):10021–31.
16. Kiremidjian-Schumacher L, Stotzky G. Selenium and immune responses. *Environ Res* 1987;**42**(2):277–303.
17. Hurwitz BE, et al. Suppression of human immunodeficiency virus type 1 viral load with selenium supplementation: a randomized controlled trial. *Arch Intern Med* 2007;**167**(2):148–54.
18. Jiamton S, Pepin J, Suttent R, Filteau S, Mahakkanukrauh B, Hanshaoworakul W, Jaffar S. A randomized trial of the impact of multiple micronutrient supplementation on mortality among HIV-infected individuals living in Bangkok. *AIDS* 2003;**17**(17):2461–9.
19. NIH. *Selenium: Dietary Supplement Fact Sheet. 2016.* February 11, 2016. Available from: https://ods.od.nih.gov/factsheets/Selenium-Health Professional/.
20. Burk RF, et al. Effects of chemical form of selenium on plasma biomarkers in a high-dose human supplementation trial. *Cancer Epidemiol Biomarkers Prev* 2006;**15**(4):804–10.
21. Beck MA. Selenium and vitamin E status: Impact on viral pathogenicity. *J Nutr* 2007;**137**(5):1338–40.
22. Clark LC, et al. Effects of selenium supplementation for cancer prevention in patients with carcinoma of the skin. A randomized controlled trial. *Nutr Prev Cancer Study Group JAMA* 1996;**276**(24):1957–63.
23. Stone CA, Kawai K, Kupka R, Fawzi WW. Role of selenium in HIV infection. *Nutr Rev* 2010;**68**(11):671–81.
24. Constans J, et al. Membrane fatty acids and blood antioxidants in 77 patients with HIV infection. *Rev Med Interne* 1993;**14**(10):1003.
25. Look MP, et al. Serum selenium, plasma glutathione (GSH) and erythrocyte glutathione peroxidase (GSH-Px)-levels in asymptomatic versus symptomatic human immunodeficiency virus-1 (HIV-1)-infection. *Eur J Clin Nutr* 1997;**51**(4):266–72.
26. Constans J, et al. Serum selenium predicts outcome in HIV infection. *J Acquir Immune Defic Syndr Hum Retrovirol* 1995;**10**(3):392.
27. Kupka R, et al. Selenium status, pregnancy outcomes, and mother-to-child transmission of HIV-1. *J Acquir Immune Defic Syndr* 2005;**39**(2):203–10.
28. Zhao L, et al. Molecular modeling and in vitro activity of an HIV-1-encoded glutathione peroxidase. *Proc Natl Acad Sci USA* 2000;**97**(12):6356–61.
29. Fawzi WW, et al. A randomized trial of multivitamin supplements and HIV disease progression and mortality. *N Engl J Med* 2004;**351**(1):23–32.
30. Harrison's principles of internal medicine choice. *Curr Rev Acad Libraries* 2008;**46**(1):138.
31. Burbano X, et al. Impact of a selenium chemoprevention clinical trial on hospital admissions of HIV-infected participants. *HIV Clin Trials* 2002;**3**(6):483–91.
32. Kassu A, et al. Alterations in serum levels of trace elements in tuberculosis and HIV infections. *Eur J Clin Nutr* 2006;**60**(5):580–6.
33. Baum MK, et al. Effect of micronutrient supplementation on disease progression in asymptomatic, antiretroviral-naive, HIV-infected adults in Botswana: a randomized clinical trial. *JAMA* 2013;**310**(20):2154–63.
34. Grober U, et al. Micronutrients in oncological intervention. *Nutrients* 2016;**8**(3).
35. Constans J, Conri C, Sergeant C. Selenium and HIV infection. *Nutrition* 1999;**15**(9):719–20.
36. Hurwitz BE, Klaus JR, Llabre MM, Gonzalez A, Lawrence PJ, Maher KJ, Schneiderman N. Suppression of human immunodeficiency virus type 1 viral load with selenium supplementation: a randomized controlled trial. *Arch Intern Med* 2007;**167**(2):148–54.
37. Pitney CL, Royal M, Klebert M. Selenium supplementation in HIV-infected patients: is there any potential clinical benefit? *J Assoc Nurses AIDS Care* 2009;**20**(4):326–33.

## Further Reading

1. Carlson BA, Yoo MH, Shrimali RK, Irons R, Gladyshev VN, Hatfield DL, Park JM. Role of selenium-containing proteins in T-cell and macrophage function. *Proc Nutr Soc* 2010;**69**(3):300–10.
2. Hori K, et al. Selenium supplementation suppresses tumor necrosis factor alpha-induced human immunodeficiency virus type 1 replication in vitro. *AIDS Res Hum Retroviruses* 1997;**13**(15):1325–32.
3. Huang Z, Rose AH, Hoffmann PR. The role of selenium in inflammation and immunity: from molecular mechanisms to therapeutic opportunities. *Antioxid Redox Signal* 2012;**16**(7):705–43.
4. Jaspers I, Zhang W, Brighton LE, Carson JL, Styblo M, Beck MA. Selenium deficiency alters epithelial cell morphology and responses to influenza. *Free Radic Biol Med* 2007;**42**(12):1826–37.
5. Morris D, Khurasany M, Nguyen T, Kim J, Guilford F, Mehta R, Venketaraman V. Glutathione and infection. *Biochim Biophys Acta* 2013;**1830**(5):3329–49.
6. Spector A, et al. The effect of $H_2O_2$ upon thioredoxin-enriched lens epithelial cells. *J Biol Chem* 1988;**263**(10):4984–90.
7. Wilson AG. Epigenetic regulation of gene expression in the inflammatory response and relevance to common diseases. *J Periodontol* 2008;**79**(8 Suppl.):1514–9.
8. Zhang W, Zheng X, Wang X. Oxidative stress measured by thioredoxin reductase level as potential biomarker for prostate cancer. *Am J Cancer Res* 2015;**5**(9):2788–98.

C H A P T E R

# 18

# Vitamin D, Oxidative Stress, and the Antiretroviral Tenofovir

*Antonio C. Seguro, Pedro H. França Gois, Daniele Canale*
**University of São Paulo School of Medicine, São Paulo, Brazil**

O U T L I N E



## Abstract

Tenofovir (TFV) is a nucleotide analogue of adenosine monophosphate used as a first-line medication in the treatment of acquired immune deficiency syndrome and chronic hepatitis B infection. Mitochondrial toxicity seems to be the main factor involved in the TFV toxicity. TFV inhibits DNA polymerase $\gamma$, leading to decreased mitochondrial DNA and increased oxidative stress. This has been related to mitochondrial structural changes and depletion. The most predominant kidney manifestations of TFV exposure include elevation in serum creatinine and proximal tubular dysfunction. In animal studies, oxidative stress inhibition reversed TFV nephrotoxicity and might represent a new therapeutic approach. Vitamin D deficiency (VDD) is prevalent among HIV-infected individuals. Superimposed VDD and TFV use aggravates nephrotoxicity due to changes in the redox state and involvement of renin–angiotensin–aldosterone system. Thus, it is important to monitor vitamin D levels, as well as to treat VDD, in HIV-infected patients under TFV therapy.

**Keywords:** Free radicals; Kidney disease; Nephrotoxicity; Oxidative stress; Tenofovir; Vitamin D; Vitamin D deficiency.

## List of Abbreviations

**3TC** Lamivudine
**25[OH]D** Calcidiol
**1,25[OH]$_2$D** Calcitriol
**AIDS** Acquired immune deficiency syndrome
**AKI** Acute kidney injury
**ABC** Abacavir
**ATP** Adenosine triphosphate
**ART** Antiretroviral therapy
**D4T** Stavudine
**DDC** Zalcitabine
**DDI** Didanosine
**eNOS** Endothelial nitric oxide synthase
**FTC** Emtricitabine


http://dx.doi.org/10.1016/B978-0-12-809853-0.00018-3

**FS** Fanconi syndrome
**GFR** Glomerular filtration rate
**GSH** Glutathione
**HAART** Highly active antiretroviral therapy
**HIV** Human immunodeficiency virus
**MetS** Metabolic syndrome
**mtDNA** Mitochondrial DNA
**MRP4** Multidrug resistance-associated protein 4
**NAC** N-acetyl-cysteine
**NaPi-IIa** Sodium-phosphate cotransporter subtype IIa
**NFκB** Nuclear factor kappa B
**NRTIs** Nucleoside reverse transcriptase inhibitors
**OAT1** Organic anion transporter 1
**OAT3** Organic anion transporter 3
**Pol γ** DNA polymerase subunit γ
**PPARs** Peroxisome-proliferator activator receptors
**PT** Proximal tubule
**RAAS** Renin–angiotensin–aldosterone system
**ROS** Reactive oxygen species
**RXR** Retinoid X receptor
**TFV** Tenofovir
**TBARS** Thiobarbituric acid reactive substances
**VDD** Vitamin D deficiency
**VDR** Vitamin D receptor
**ZDV** Zidovudine

## INTRODUCTION

Tenofovir (TFV) is a nucleotide analogue of adenosine monophosphate, which competes with the natural substrate deoxyadenosine 5′-triphosphate for incorporation into DNA during reverse transcription of human immunodeficiency virus (HIV), resulting in viral DNA chain termination.[1,2] TFV is a first-line medication in the treatment of acquired immune deficiency syndrome (AIDS), as part of several combinations in highly active antiretroviral therapy. In 2008, TFV was also approved for treatment of chronic hepatitis B in adults.[3,4]

In the past decade, there has been an increasing interest in oxidative stress as a major role in the pathogenesis and progression of HIV infection. In human T cell lines, oxidative stress can activate the transcription factor NFκ-B, which promotes the expression of target genes involved in the host immune and inflammatory response.[5,6] Moreover, intracellular glutathione (GSH), a major endogenous antioxidant, has been reported to be profoundly reduced in peripheral blood mononuclear cells and lymphocytes of patients with AIDS.[7] Thus, reduced GSH levels may contribute to the immunodeficiency, as well as to the increased inflammatory reactions, in HIV-infected subjects, whereas it profoundly affects early signal transduction events in T lymphocytes.[8] With regard to TFV, data from animal studies showed that TFV exposure has been associated with higher serum thiobarbituric acid reactive substances (TBARS), a byproduct of lipid peroxidation, and lower GSH levels.[9,10]

In this chapter, we will discuss the main features of TFV nephrotoxicity, particularly regarding the involvement of the oxidative stress and vitamin D deficiency (VDD).

## TENOFOVIR, OXIDATIVE STRESS, AND NEPHROTOXICITY

Both HIV infection and nucleoside reverse transcriptase inhibitors (NRTIs) have been associated with mitochondrial toxicity.[11] Drug-induced mitochondrial dysfunction is a major side effect of NRTIs treatment, which can affect almost every organ system in the body. In some cases, the adverse effects can lead to a life-threatening clinical presentation.

The main function of mitochondria is to produce energy in all human cells except erythrocytes. For this purpose, oxidative phosphorylation of fatty acids and pyruvate is crucial to produce energy in the form of adenosine triphosphate.[12] Under normal conditions, mitochondrial enzymes, such as cytochrome oxidases, promote a tight coupling between oxidation and phosphorylation.[12] Mitochondrial dysfunction leads to imbalances of reactive oxygen species (ROS) generation and antioxidant defense response, resulting in peroxidation of DNA, protein, and lipid membranes.[13] DNA polymerase γ (Pol γ) is the only DNA polymerase in the mitochondria and is responsible for structural integrity of the mitochondrial DNA (mtDNA).[14] Therefore, due to its close association with mtDNA, Pol γ is a target of oxidative damage.[15]

Several in vitro and in vivo studies have demonstrated the association between various NRTIs and mitochondrial toxicity.[15–18] NRTIs have been reported to inhibit Pol γ, causing depletion of mtDNA.[11] Inhibition of Pol γ, and consequently reduction of mtDNA, has been advocated as one of the main mechanisms by which NRTIs interfere with mitochondrial functions.[19]

Regarding TFV, the "Pol γ hypothesis" is a matter of debate. Birkus et al. failed to demonstrate reduction in the mtDNA levels in experiments using renal proximal tubular cells treated with TFV .[20] These findings were in accordance with the data obtained by Biesecker et al., who assessed mtDNA in kidney samples of rats, rhesus monkeys, and woodchucks. In contrast, mtDNA depletion has been described in a rat model of TFV toxicity.[21] Furthermore, Kohler et al., using a murine HIV-transgenic model, found decreased mtDNA content associated with mitochondrial structural changes in the proximal tubules (PT).[22] In humans, dysmorphic and swollen mitochondria on electron microscopy are considered the main features of mitochondrial toxicity in the PT.[23] Moreover, kidney mtDNA depletion was described in patients taking TFV in combination with didanosine (DDI), suggesting that DDI exposure might have boosted TFV-induced mitochondrial toxicity.[24] Taken together, it seems that NRTIs have different strengths of inhibiting Pol γ, as follows: zalcitabine >> DDI > stavudine ≥ zidovudine >>> TFV = lamivudine = emtricitabine = abacavir.[12,20] Fig. 18.1 illustrates the potential molecular mechanisms involved in the pathogenesis of TFV nephrotoxicity.

It is well established that TFV nephrotoxicity depends directly on the plasma concentration of the drug, which in turn is influenced by the ingested dose, volume of distribution, and renal excretion.[25,26] High intracellular concentrations of TFV in the renal tubular cells can interfere with cell function and cause harmful side effects, especially with prolonged use.[27,28] TFV is eliminated by the kidneys through a combination of glomerular filtration and tubular secretion, particularly at PT level. As shown in Fig. 18.2, TFV enters the basolateral membrane of the PT cells by organic anion transporters 1 and 3 (OAT1 and OAT3) and is mainly secreted into the tubular lumen by the multidrug resistance-associated protein 4 (MRP4).[26,27,29] As such, the main target of TFV-induced nephrotoxicity is the PT, impairing the ability for solute transport in this nephron segment.

The most predominant kidney manifestations of TFV exposure include elevation in serum creatinine, proximal tubular dysfunction, Fanconi syndrome (FS), and acute kidney injury (AKI).[30,31] In patients on TFV-containing therapy, urinary phosphate loss is the most clinically relevant manifestation, which may lead to hypophosphatemia and, consequently, to osteomalacia, bone pain, and fractures.[32,33] FS is the complete dysfunction of the PT, characterized by loss of urinary solutes such as phosphate, bicarbonate, low-molecular-weight proteins, and amino acids, which would be normally absorbed in this nephron portion.[34] The reported incidence of FS associated with TFV therapy is

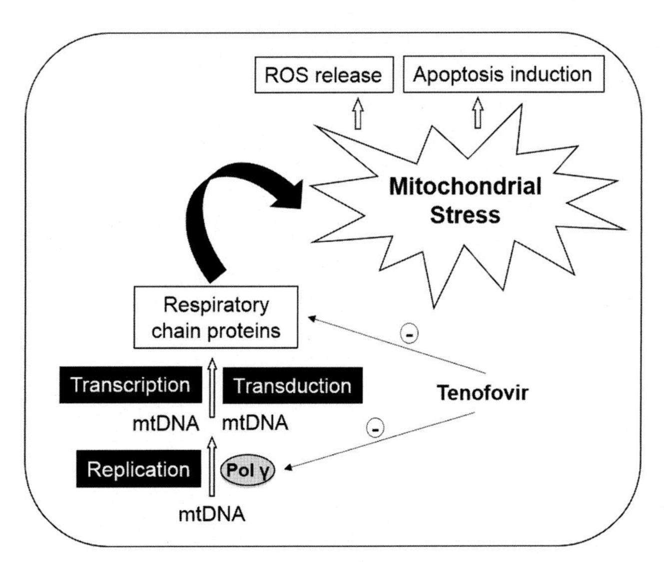


FIGURE 18.1 Proposed model for the signaling mechanism of tenofovir nephrotoxicity. In the mitochondria of the proximal tubules, tenofovir inhibits DNA polymerase γ (Pol γ), leading to decreased mitochondrial DNA (mtDNA). Consequently, reduced transduction of respiratory chain proteins results in release of reactive oxygen species (ROS) and activation of the apoptotic pathway. This has been related to mitochondrial structural changes and depletion. *Adapted from Fernandez-Fernandez B, Montoya-Ferrer A, Sanz AB, Sanchez-Niño MD, Izquierdo MC, Poveda J, Sainz-Prestel V, Ortiz-Martin N, Parra-Rodriguez A, Selgas R, Ruiz-Ortega M, Egido J, Ortiz A. Tenofovir nephrotoxicity: 2011 update.* AIDS Res Treat *2011:354908.*

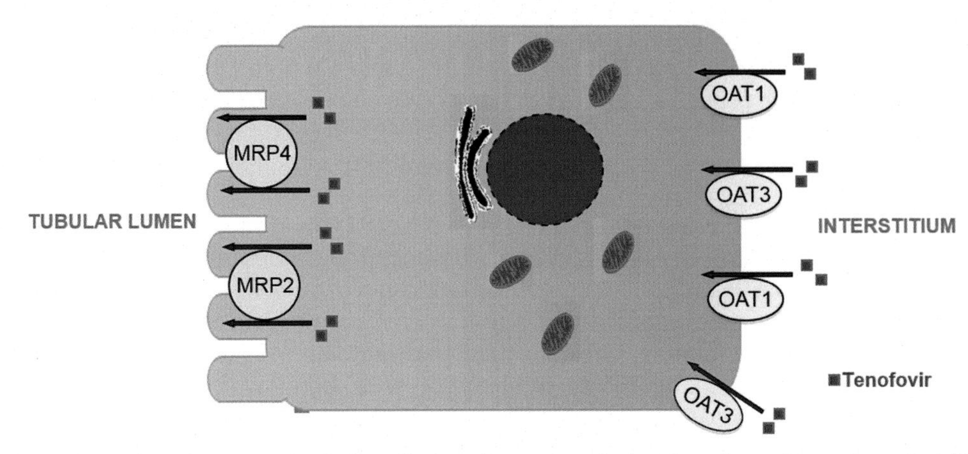


FIGURE 18.2 Tenofovir handling by proximal tubular cells. Tenofovir enters the basolateral membrane through OAT1 and OAT3 and is mainly secreted into the tubular lumen by the MRP4. The multidrug resistance-associated protein 2 (MRP2) may also play a role on tenofovir transport in the apical membrane. 20 to 30% of tenofovir is eliminated unchanged in the urine owing to its secretion in the proximal tubule. *Adapted from Fernandez-Fernandez B, Montoya-Ferrer A, Sanz AB, Sanchez-Niño MD, Izquierdo MC, Poveda J, Sainz-Prestel V, Ortiz-Martin N, Parra-Rodriguez A, Selgas R, Ruiz-Ortega M, Egido J, Ortiz A. Tenofovir nephrotoxicity: 2011 update.* AIDS Res Treat *2011:354908.*

less than 1%.[35–37] However, it might be underestimated due to the lacking of well-defined diagnostic criteria for this syndrome. Data from several observational studies have shown that subclinical dysfunction of the PT is the most common renal disorder associated with TFV treatment, with an estimated incidence of up to 22%.[37,38] Nevertheless, it is still unclear whether mild renal impairment induced by TFV can contribute to the progression of kidney damage and chronic kidney disease.[34]

Liborio et al. reported that rats treated with high doses of TFV exhibited increased blood pressure associated with reduced renal protein expression of endothelial nitric oxide synthase (eNOS), renal vasoconstriction, and AKI. In the same study, smaller doses of TFV also increased renal vascular resistance (RVR), albeit not sufficient to induce AKI.[39]

In summary, TFV-induced PT dysfunction can range from mild to severe. Additionally, it may precede or even occur without any decline of renal function. Mitochondrial toxicity seems to be the main factor involved in the kidney toxicity; however, the precise pathological mechanisms remain to be elucidated.

## ROSIGLITAZONE AND N-ACETYL-CYSTEINE AS POTENTIAL TREATMENTS OF TENOFOVIR NEPHROTOXICITY

Peroxisome-proliferator activator receptors (PPARs) form a family of nuclear hormone-activated receptors and transcription factors that regulates the function and expression of genes related to energy homeostasis and inflammation.[32] To date, there are three known members: PPAR-$\alpha$, PPAR-$\beta/\delta$, and PPAR-$\gamma$.[32] Of interest, the PPAR-$\gamma$ is widely distributed throughout tissues. In the kidneys, PPAR-$\gamma$ was found to be expressed in the medullary collecting ducts, PT, mesangial cells, and renal microvasculature.[32,40]

PPAR-$\gamma$ agonists are drugs traditionally used to treat type 2 diabetes mellitus. However, in the past decade, there have been a number of studies using PPAR-$\gamma$ agonists for the treatment of oxidative stress–mediated diseases.[41] As illustrated in Fig. 18.3, PPAR-$\gamma$ may directly modulate various antioxidant and prooxidant genes in response to oxidative stress.

Liborio et al. found that treatment with rosiglitazone, a PPAR-$\gamma$ agonist, increased the renal protein expression sodium–phosphate cotransporter subtype IIa (NaPi-IIa) and the sodium/hydrogen exchanger 3 in the kidney, reversing the electrolyte and acid–base alterations induced by TFV. Moreover, the authors showed that rosiglitazone attenuated renal vasoconstriction following treatment with high doses of TFV.[39] Although redox balance was not assessed in this study, it is possible to hypothesize that rosiglitazone beneficial effects might rely on the inhibition of the oxidative stress cascade. Therefore, this is an important issue for future research.

N-acetyl-cysteine (NAC) is a pharmaceutical drug and nutritional supplement traditionally used either as a mucolytic agent or in the management of acetaminophen overdose. Its use has been also very popular in the prevention of

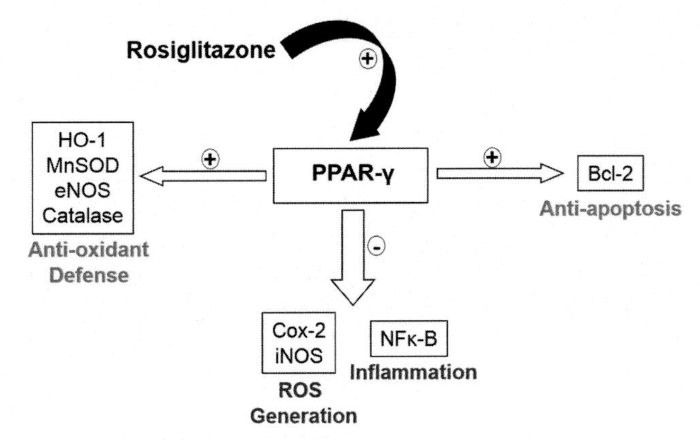


FIGURE 18.3 Peroxisome-proliferator activator receptor γ (PPAR-γ), expressed in the proximal tubules, targets genes, and physiological responses. PPAR-γ directly modulates various antioxidant, prooxidant, and inflammation-related genes in response to oxidative stress. PPAR-γ agonists, e.g., rosiglitazone, have been studied as potential therapy of oxidative stress-mediated diseases. *eNOS*, endothelial nitric oxide synthase; *NFk*-B, nuclear factor kappa-B. *Adapted from Polvani, S., Tarocchi, M., Galli, A. 2012. PPARγ and oxidative stress: Con(β) Catenating NRF2 and FOXO.* PPAR Res *2012:641087.*

TABLE 18.1 Protective Effect of N-Acetyl-Cysteine on Chronic Tenofovir Nephrotoxicity in Rats—Renal Function, Blood Pressure, Oxidative Stress, and Angiotensinogen Assessments

| Evaluated Data | C | TFV | TFV+NAC |
|---|---|---|---|
| GFR (mL/min/100gBW) | 0.63±0.04 | 0.40±0.03[ac] | 0.59±0.04 |
| MAP (mmHg) | 126±5 | 139±3[bc] | 114±5 |
| Plasma TBARS (nmol/mL) | 1.92±0.18 | 5.04±0.34[ac] | 3.51±0.16[a] |
| Urinary TBARS (nmol/24h) | 11.7±1.2 | 137.9±7.5[ac] | 88.2±11.2[a] |
| GSH (μmol/mL) | 2.70±0.15 | 2.18±0.07[ac] | 2.68±0.08 |
| AGT (%) | 100±9 | 175±22[ad] | 95±5 |

*AGT*, angiotensinogen renal protein expression; *BW*, body weight; *GFR*, glomerular filtration rate; *GSH*, serum reduced glutathione; *MAP*, mean arterial pressure; *NAC*, N-acetyl-cysteine; *TBARS*, thiobarbituric acid reactive substances; *TFV*, tenofovir. C (n = 8 Wistar rats): control group; TFV (n = 10 Wistar rats): received TFV (50mg/Kg of diet) administered for 4months; TFV + NAC (n = 10): received TFV (50mg/Kg of diet)+NAC (600mg/L of drinking water) for 4months. Data are mean±SEM.
[a]P<.001.
[b]P<.01 versus C.
[c]P<.001.
[d]P<.01 versus TFV+NAC.

radiocontrast-induced nephropathy.[42] NAC is an antioxidant thiol that can enter to the chain of GSH synthesis and serves as a source of sulfhydryl groups for the cells, acting as a scavenger of free radicals.[43] In experimental studies, we have reported beneficial effects of NAC in numerous diseases, whose pathogenesis includes oxidative stress, such as in chronic renal failure, ischemic, and septic AKI.[44–46]

We have conducted an experimental study, aiming to evaluate the effect of chronic use of TFV on renal function and oxidative stress. An additional group of rats received TFV and NAC. After 4months of treatment, we found that animals treated with TFV exhibited lower glomerular filtration rate, increased blood pressure, oxidative stress and higher renal tissue expression of angiotensin II. Strikingly, NAC attenuated all these alterations, probably due to its potent antioxidant effect (Table 18.1).[11]

Overall, this combination of results provides some support for the conceptual premise that TFV nephrotoxicity might be reversed by oxidative stress inhibition. However, these data must be interpreted with caution. Although NAC is a medication with a high-safety profile, rosiglitazone has been associated with fluid retention and increased risk of heart failure and stroke in humans. Moreover, the above-mentioned studies were performed in rat models and need to be translated to humans. We hope that these data motivate further clinical studies to validate these findings.

## VITAMIN D DEFICIENCY IN TENOFOVIR NEPHROTOXICITY: ROLE OF OXIDATIVE STRESS AND RENIN–ANGIOTENSIN–ALDOSTERONE SYSTEM

In addition to the renal impairment caused by TFV, it has been recently shown that low vitamin D levels are associated with the progression of HIV-related diseases in patients receiving antiretroviral therapy (ART).[9,47,48] Clinical trials also reported that VDD is a major public health care burden worldwide[49,50] and is highly prevalent among HIV-infected individuals,[51,52] emphasizing the necessity of a strong recommendation regarding the monitoring of vitamin D levels in this population. Fig. 18.4 illustrates the prevalence of VDD around the globe.

It is well established that vitamin D is an essential nutrient for mineral homeostasis and is responsible for several physiological activities since the machinery for vitamin D production and its receptor have been described in multiple tissues.[53] As demonstrated in Fig. 18.5, vitamin D can be obtained orally or by the action of sunlight exposure on the skin, where the photolytic conversion of 7-dehydrocholesterol to previtamin $D_3$ occurs. Previtamin $D_3$ is immediately converted to vitamin $D_3$ (cholecalciferol) through a nonenzymatic thermal isomeration. Although the exposure of the skin to UV rays is the major source of cholecalciferol, some foods naturally contain or are enriched with vitamin $D_3$ and $D_2$ (ergocalciferol). Both cholecalciferol and ergocalciferol are metabolized in the liver by 25-hydroxilase and are converted to 25-hydroxivitamin D (25[OH]D) or calcidiol. Rapidly, calcidiol is released from the liver into the blood stream and is subsequently converted by the proximal renal tubule cells to calcitriol or 1,25 dihydroxivitamin D (1,25[OH]$_2$D) through the enzyme 1α-hydroxilase. Calcitriol is the biologically active form of vitamin D and interacts with vitamin D receptor (VDR) to exert its functions.[52,54,55]

Vitamin D status is usually assessed by measuring 25[OH]D-circulating levels and is characterized as follows: VDD (25[OH]D $<$ 10 ng/mL) and vitamin D insufficiency (25[OH]D 10–30 ng/mL). The optimal level of vitamin D is recommended to be over 30 ng/mL.[56,57] Hypovitaminosis D has been associated with several renal and cardiovascular diseases mainly because of its effects on oxidative stress, lipid metabolism, and renin–angiotensin–aldosterone system (RAAS).[54] Tarcin et al. reported that the concentration of TBARS is inversely associated with vitamin D levels, suggesting that vitamin D has a positive effect on reducing oxidative stress and improving endothelial function.[58] VDD also increases advanced oxidation protein products, nitric oxide metabolites, and oxidative stress index and decreases total radical-trapping antioxidant parameter, compromising the redox

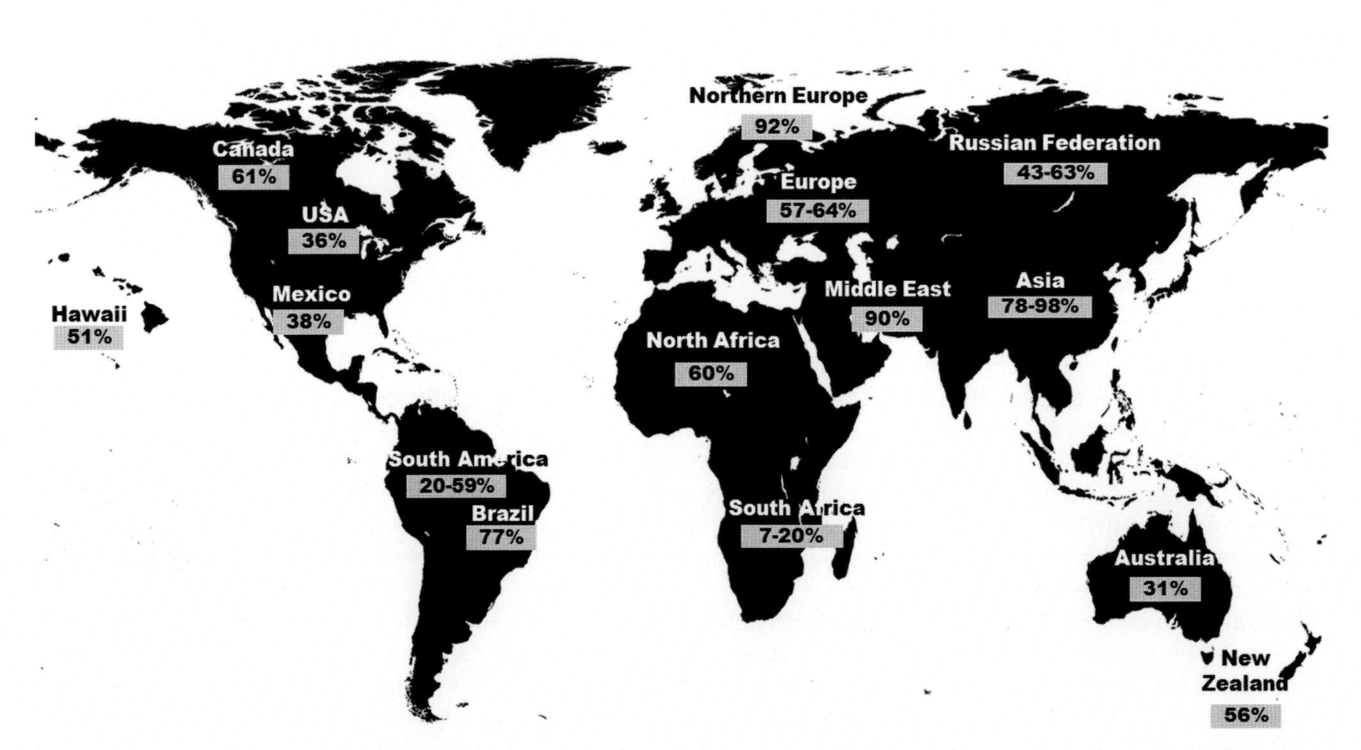


FIGURE 18.4 Prevalence of vitamin D deficiency/insufficiency characterized as 25[OH]D below 20 ng/mL in general population around the world. *Adapted from Hossein-Nezhad A, Holick MF. Vitamin D for Health: a global perspective.* Mayo Clin Proc *2013;**88**:720–55; Palacios C, Gonzalez L. Is vitamin D deficiency a major global public health problem?* J Steroid Biochem Mol Biol *2014;**144**:138–45.*

state in patients.[59] Furthermore, HIV-infected individuals present low concentrations of total GSH, a major intracellular antioxidant agent, leading to an aggravation of the antioxidative immune competence of these subjects.[60]

Several studies have connected poor vitamin D status to the prevalence of metabolic syndrome (MetS) and its risk factors.[61,62] Gagnon et al. demonstrated that low levels of calcidiol were connected to increased waist circumference, fasting plasma glucose, serum triglycerides concentration, and insulin resistance, which are likely the basis of this higher MetS risk.[63] In addition, alterations in lipid metabolism have been described since the arising of HIV-epidemic in both naïve and on treatment patients. It is important to emphasize that different types of antiretroviral regimens may interfere distinctively in lipid metabolism of each individual. TFV may disrupt fatty acid oxidation and energy production by disabling respiratory chain enzymes or mtDNA, resulting in oxidative stress and ensuing anaerobic metabolism, lactic acidosis, and triglycerides accumulation.[64] On the other hand, satisfactory levels of vitamin D or vitamin D supplementation may prevent mitochondrial toxicity induced by ART, ameliorating muscle pain, and lipid metabolism disorders.[65]

Hypertension is also a highly prevalent risk factor in the development of MetS. Vitamin D may regulate blood pressure by interacting with the RAAS. Experimental and clinical trials have shown that both calcidiol and calcitriol negatively regulate plasma renin activity and circulating angiotensin II.[53,66] The combination of RAAS and vitamin D may have implications beyond just hypertension in terms of cardiac abnormalities; renal fibrosis and inflammation are also processes that are managed at least in part by over activity of the RAAS.[53] Gois et al. demonstrated that maternal exposure to TFV during pregnancy resulted not only in

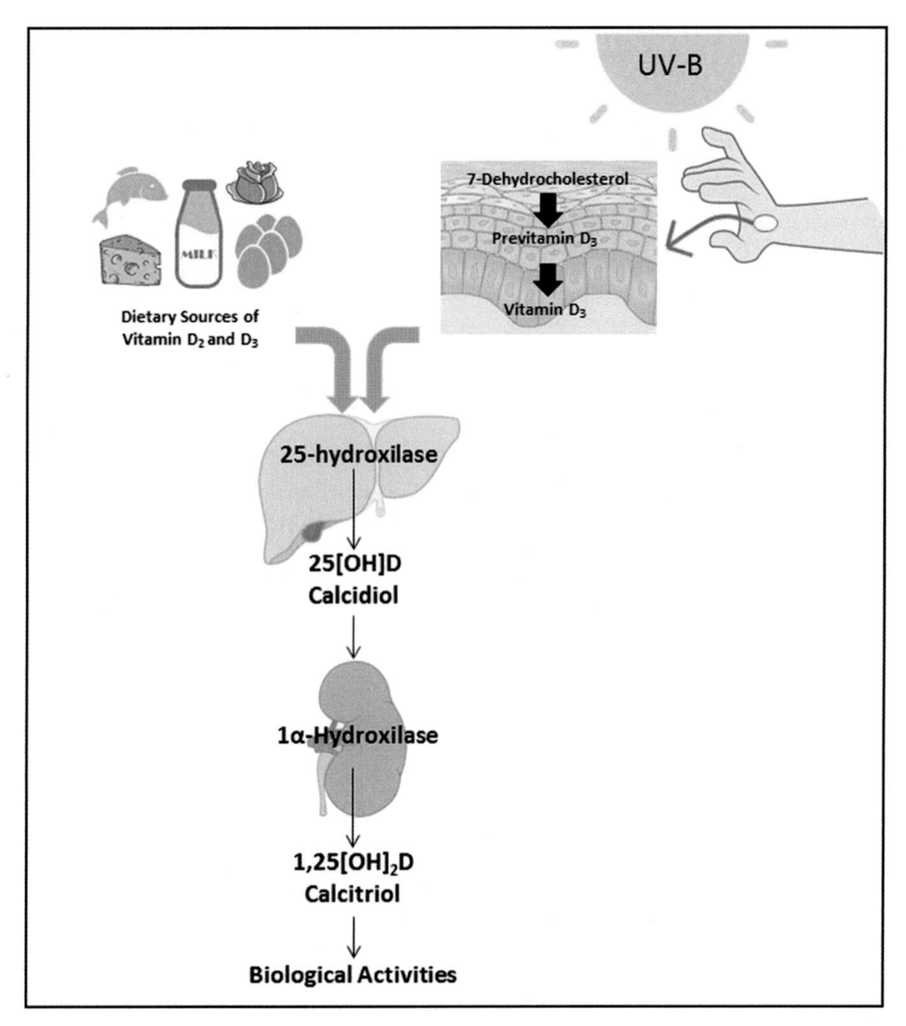


FIGURE 18.5 Vitamin D production and metabolism. *Adapted from Shroff R, Knott C, Rees L. The virtues of vitamin D–but how much is too much?* Pediatr Nephrol *2010;***25***:1607–20.*

(A)

| | Control | VDD | TFV | VDD+TFV |
|---|---|---|---|---|
| TBARS | 1.60±0.07 | 2.03±0.29 | 2.38±0.19[a] | 3.26±0.23[cd] |
| GSH | 2.90±0.22 | 2.05±0.04[b] | 2.26±0.22[a] | 1.45±0.12[cd] |

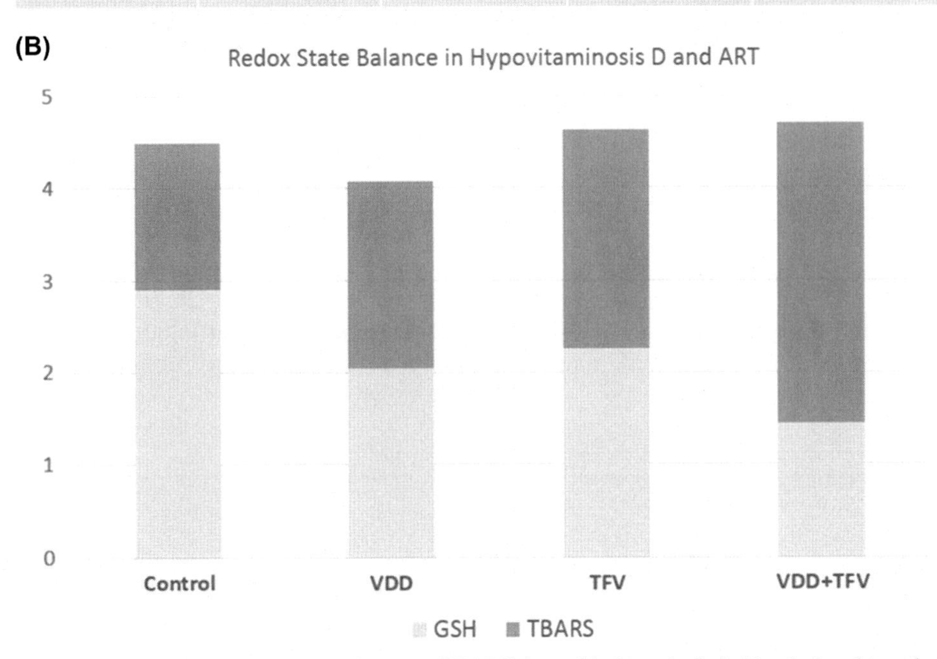


FIGURE 18.6 (A) Serum thiobarbituric acid reactive substances (TBARS) (mmol/mL) and whole-blood glutathione levels (GSH—μmol/mL) of control, vitamin D-deficient animals (VDD), control animals treated with tenofovir (TFV), and vitamin D-deficient animals treated with TFV (VDD + TFV). Values are mean ± SEM. (B) Relationship between oxidative stress marker, TBARS, and antioxidant agent, GSH, in control, VDD, TFV, and VDD + TFV. Note the increased concentration of TBARS and diminished levels of GSH, demonstrating the imbalance in the redox state caused by the combination of VDD and TFV therapy. *ART*, antiretroviral therapy. *Data obtained from Canale D, de Bragança AC, Gonçalves JG, Shimizu MH, Sanches TR, Andrade L, Volpini RA, Seguro A. Vitamin D deficiency aggravates nephrotoxicity, hypertension and dyslipidemia caused by tenofovir: role of oxidative stress and renin-angiotensin system.* PLoS One *2014;**9**:e103055.*

hypertension and dyslipidemia in the offspring but also in deleterious effects on renal tissue.[67] These alterations are probably due to an increase in most of the components of the RAAS and an upregulation of renal sodium transporters.

Supporting the data previously shown, Canale et al. reported that in an animal model of VDD treated with TFV, both systemic and intrarenal RAAS were over activated, and these alterations were accompanied by elevated RVR and decreased renal eNOS expression.[9] These results suggest the possibility of the involvement of nitric oxide cascade in the onset of hypertension since nitric oxide regulates blood pressure by its effects on vascular tone, renal hemodynamics, sodium balance, and extracellular fluid volume. Besides the arising of hypertension, oxidative stress is also associated with the development of renal injury observed by higher lipid peroxidation and lower antioxidant component level. Fig. 18.6 demonstrates the relationship between oxidative stress marker and antioxidant system under the influence of low levels of vitamin D and TFV therapy.

Altogether, it is plausible to assume that vitamin D is crucial for a wide variety of organ systems and its optimal levels may interfere positively in maintaining homeostasis, in terms of oxidative stress and the RAAS (see Fig. 18.7). Since hypovitaminosis D is highly prevalent across different regions of the world in general population, vitamin D monitoring and administration in HIV-infected individuals appear appropriate to reduce or even prevent long-term cardiovascular and renal outcomes. Further studies are required for the clarification of specific mechanisms involved in the progression of HIV-related diseases in vitamin D-deficient patients.

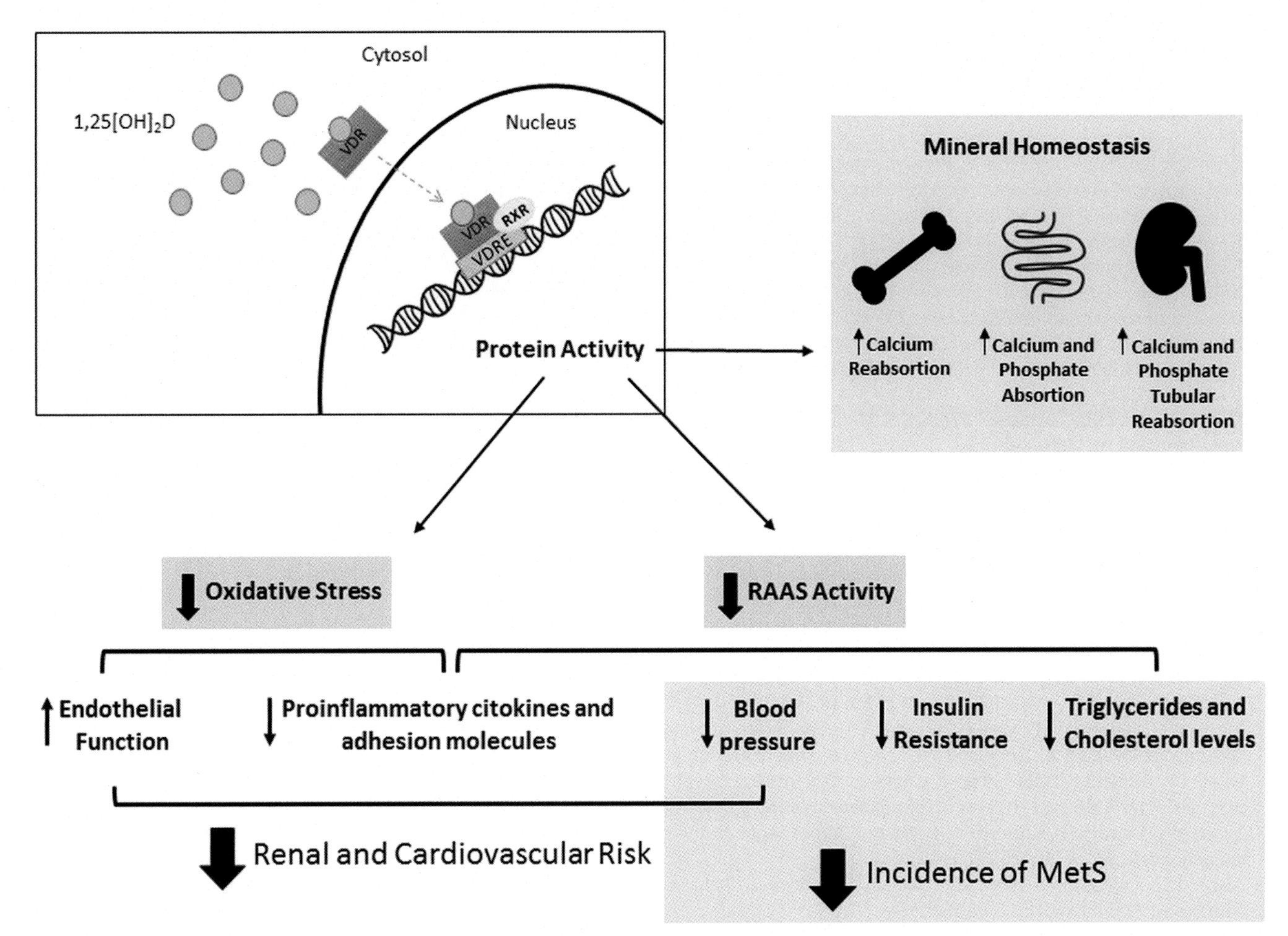


FIGURE 18.7 Vitamin D and its biological activities. Most of the effects of Calcitriol are mediated by vitamin D receptor (VDR). VDR acts as heterodimer with the retinoid X receptor (RXR). This heterodimeric complex interacts with specific DNA regions to control transcriptional activities of target genes, resulting in multiple biological activities of vitamin D. *MetS*, metabolic syndrome. *Adapted from Petchey WG, Johnson DW, Isbel NM. Shining D' light on chronic kidney disease: mechanisms that may underpin the cardiovascular benefit of vitamin D.* Nephrology (Carlton) *2011;**16**:351–67; Shroff R, Knott C, Rees L. The virtues of vitamin D–but how much is too much?* Pediatr Nephrol *2010;**25**:1607–20.*

# SUMMARY POINTS

- TFV therapy has been associated with higher serum TBARS, a by-product of lipid peroxidation and lower GSH levels, an endogenous antioxidant;
- Mitochondrial toxicity seems to be the main factor involved in the TFV toxicity;
- TFV inhibits Pol $\gamma$, leading to decreased mtDNA and increased oxidative stress;
- Dysmorphic and swollen mitochondria on electron microscopy are considered the main features of mitochondrial toxicity in the renal PT;
- Elevation in serum creatinine, proximal tubular dysfunction, FS, and AKI are the most common kidney manifestations of TFV exposure;
- Rosiglitazone and NAC therapy might reverse TFV nephrotoxicity by reducing oxidative stress;
- VDD is a major public health care burden worldwide and is highly prevalent among HIV-infected individuals;
- Superimposed VDD and TFV use aggravates nephrotoxicity due to changes in the redox state and involvement of RAAS;
- It is mandatory to monitor vitamin D levels, as well as to treat VDD, in HIV-infected patients under TFV therapy.

# References


1. Antiretroviral Pregnancy Registry Steering Committee. *Antiretroviral pregnancy registry International Interim report for 1 January 1989 through 31 January 2013*. Wilmington, NC: Registry Coordinating Center; 2013. http://www.apregistry.com/forms/interim_report.pdf.
2. WHO. Antiretroviral drugs for treating pregnant women and preventing HIV infection in infants. Recommendations for a public health approach (2010 version). http://www.who.int/hiv/pub/mtct/antiretroviral2010/en.
3. Department of Health, Human Services. *Panel on antiretroviral guidelines for adults and adolescents guidelines for the use of antiretroviral agents in HIV-1-infected adults and adolescents*. February 2013. www.aidsinfo.nih.gov/ContentFiles/AdultandAdolescentGL.pdf.
4. Sorrell MF, Belongia EA, Costa J, Gareen IF, Grem JL, Inadomi JM, Kern ER, McHugh JA, Petersen GM, Rein MF, Strader DB, Trotter HT. National institutes of health consensus development conference statement: management of hepatitis B. *Ann Intern Med* 2009;**150**:104–10.
5. Schreck R, Rieber P, Baeuerle PA. Reactive oxygen intermediates as apparently widely used messengers in the activation of the NF-kappa B transcription factor and HIV-1. *EMBO J* 1991;**10**:2247–58.
6. Hiscott J, Kwon H, Génin P. Hostile takeovers: viral appropriation of the NF-kappaB pathway. *J Clin Invest* 2001;**107**:143–51.
7. Baruchel S, Wainberg MA. The role of oxidative stress in disease progression in individuals infected by the human immunodeficiency virus. *J Leukoc Biol* 1992;**52**:111–4.
8. Dröge W, Eck HP, Gmünder H, Mihm S. Modulation of lymphocyte functions and immune responses by cysteine and cysteine derivatives. *Am J Med* 1991;**91**:140S–4S.
9. Canale D, de Bragança AC, Gonçalves JG, Shimizu MH, Sanches TR, Andrade L, Volpini RA, Seguro AC. Vitamin D deficiency aggravates nephrotoxicity, hypertension and dyslipidemia caused by tenofovir: role of oxidative stress and renin-angiotensin system. *PLoS One* 2014;**9**:e103055.
10. Shimizu MH, Canale D, de Bragança AC, Volpini RA, Andrade L, Luchi W, Seguro AC. Protective effect of N-acetyl-cysteine on chronic tenofovir nephrotoxicity. In: *Cell-lancet translational medicine: 'what will it take to achieve an AIDS-free World? A translational medicine conference on HIV research*. 2013. p. 68.
11. Moyle G. Mechanisms of HIV and nucleoside reverse transcriptase inhibitor injury to mitochondria. *Antivir Ther* 2005;**10**(Suppl. 2):M47–52.
12. Brinkman K. *Mitochondrial toxicity of HIV nucleoside reverse transcriptase inhibitors*. UpToDate® 2016. http://www.uptodate.com.
13. Nagiah S, Phulukdaree A, Chuturgoon A. Mitochondrial and oxidative stress response in hepG2 cells following acute and prolonged exposure to antiretroviral drugs. *J Cell Biochem* 2015;**116**:1939–46.
14. Hudson G, Chinnery PF. Mitochondrial DNA polymerase-gamma and human disease. *Hum Mol Genet* 2006;**15**:R244–52.
15. Graziewicz MA, Day BJ, Copeland WC. The mitochondrial DNA polymerase as a target of oxidative damage. *Nucleic Acids Res* 2002;**30**:2817–24.
16. Bissuel F, Bruneel F, Habersetzer F, Chassard D, Cotte L, Chevallier M, Bernuau J, Lucet JC, Trepo C. Fulminant hepatitis with severe lactate acidosis in HIV-infected patients on didanosine therapy. *J Intern Med* 1994;**235**:367–71.
17. Lewis W, Gonzalez B, Chomyn A, Papoian T. Zidovudine induces molecular, biochemical, and ultrastructural changes in rat skeletal muscle mitochondria. *J Clin Invest* 1992;**89**:1354–60.
18. Miller KD, Cameron M, Wood LV, Dalakas MC, Kovacs JA. Lactic acidosis and hepatic steatosis associated with use of stavudine: report of four cases. *Ann Intern Med* 2000;**133**:192–6.
19. Kakuda TN. Pharmacology of nucleoside and nucleotide reverse transcriptase inhibitor-induced mitochondrial toxicity. *Clin Ther* 2000;**22**:685–708.
20. Birkus G, Hitchcock MJ, Cihlar T. Assessment of mitochondrial toxicity in human cells treated with tenofovir: comparison with other nucleoside reverse transcriptase inhibitors. *Antimicrob Agents Chemother* 2002;**46**:716–23.
21. Biesecker G, Karimi S, Desjardins J, Meyer D, Abbott B, Bendele R, Richardson F. Evaluation of mitochondrial DNA content and enzyme levels in tenofovir DF-treated rats, rhesus monkeys and woodchucks. *Antivir Res* 2003;**58**:217–25.
22. Kohler JJ, Hosseini SH, Hoying-Brandt A, Green E, Johnson DM, Russ R, Tran D, Raper CM, Santoianni R, Lewis W. Tenofovir renal toxicity targets mitochondria of renal proximal tubules. *Lab Invest* 2009;**89**:513–9.
23. Herlitz LC, Mohan S, Stokes MB, Radhakrishnan J, D'Agati VD, Markowitz GS. Tenofovir nephrotoxicity: acute tubular necrosis with distinctive clinical, pathological, and mitochondrial abnormalities. *Kidney Int* 2010;**78**:1171–7.
24. Côté HC, Magil AB, Harris M, Scarth BJ, Gadawski I, Wang N, Yu E, Yip B, Zalunardo N, Werb R, Hogg R, Harrigan PR, Montaner JS. Exploring mitochondrial nephrotoxicity as a potential mechanism of kidney dysfunction among HIV-infected patients on highly active antiretroviral therapy. *Antivir Ther* 2006;**11**:79–86.
25. Van Rompay KK, Durand-Gasselin L, Brignolo LL, Ray AS, Abel K, Cihlar T, Spinner A, Jerome C, Moore J, Kearney BP, Marthas ML, Reiser H, Bischofberger N. Chronic administration of tenofovir to rhesus macaques from infancy through adulthood and pregnancy: summary of pharmacokinetics and biological and virological effects. *Antimicrob Agents Chemoth* 2008;**52**:3144–60.
26. Rodríguez-Nóvoa S, Labarga P, D'avolio A, Barreiro P, Albalate M, Vispo E, Solera C, Siccardi M, Bonora S, Di Perri G, Soriano V. Impairment in kidney tubular function in patients receiving tenofovir is associated with higher tenofovir plasma concentrations. *AIDS* 2010;**24**:1064–6.
27. Breton G, Alexandre M, Duval X, Prie D, Peytavin G, Leport C, Vildei JL. Tubulopathy consecutive to tenofovir-containing antiretroviral therapy in two patients infected with human immunodeficiency virus-1. *Scand J Infect Dis* 2003;**36**:527–8.
28. Karras A, Lafaurie M, Furco A, Bourgarit A, Droz D, Sereni D, Legendre C, Martinez F, Molina JM. Tenofovir-related nephrotoxicity in human immunodeficiency virus-infected patients: three cases of renal failure, Fanconi syndrome and nephrogenic diabetes insipidus. *Clin Infect Dis* 2003;**36**:1070–3.
29. Rodríguez-Nóvoa S, Labarga P, Soliano V. Pharmacogenetics of tenofovir treatment. *Pharmacogenomics* 2009;**10**:1675–85.
30. de la Prada FJ, Prados AM, Tugores A, Uriol M, Saus C, Morey A. Acute renal failure and proximal renal tubular dysfunction in a patient with acquired immunodeficiency syndrome treated with tenofovir. *Nefrologia* 2006;**26**:626–30.
31. Peyrière H, Reynes J, Rouanet I, Daniel N, de Boever CM, Mauboussin JM, Leray H, Moachon L, Vincent D, Salmon-Céron D. Renal tubular dysfunction associated with tenofovir therapy: report of 7 cases. *J Acquir Immune Defic Syndr* 2004;**35**:269–73.
32. Guan Y. Peroxisome proliferator-activated receptor family and its relationship to renal complications of the metabolic syndrome. *J Am Soc Nephrol* 2004;**15**:2801–15.

33. Fernandez-Fernandez B, Montoya-Ferrer A, Sanz AB, Sanchez-Niño MD, Izquierdo MC, Poveda J, Sainz-Prestel V, Ortiz-Martin N, Parra-Rodriguez A, Selgas R, Ruiz-Ortega M, Egido J, Ortiz A. Tenofovir nephrotoxicity: 2011 update. *AIDS Res Treat* 2011;**2011**:354908.
34. Hall AM. Update on tenofovir toxicity in the kidney. *Pediatr Nephrol* 2013;**28**:1011–23.
35. Nelson MR, Katlama C, Montaner JS, Cooper DA, Gazzard B, Clotet B, Lazzarin A, Schewe K, Lange J, Wyatt C, Curtis S, Chen SS, Smith S, Bischofberger N, Rooney JF. The safety of tenofovir disoproxil fumarate for the treatment of HIV infection in adults: the first 4 years. *AIDS* 2007;**19**:1273–81.
36. Woodward CL, Hall AM, Williams IG, Madge S, Copas A, Nair D, Edwards SG, Johnson MA, Connolly JO. Tenofovir-associated renal and bone toxicity. *HIV Med* 2009;**10**:482–7.
37. Labarga P, Barreiro P, Martin-Carbonero L, Rodriguez-Novoa S, Solera C, Medrano J, Rivas P, Albalater M, Blanco F, Moreno V, Vispo E, Soriano V. Kidney tubular abnormalities in the absence of impaired glomerular function in HIV patients treated with tenofovir. *AIDS* 2009;**23**:689–96.
38. Dauchy FA, Lawson-Ayayi S, de La Faille R, Bonnet F, Rigothier C, Mehsen N, Miremont-Salamé G, Cazanave C, Greib C, Dabis F, Dupon M. Increased risk of abnormal proximal renal tubular function with HIV infection and antiretroviral therapy. *Kidney Int* 2011;**80**:302–9.
39. Libório AB, Andrade L, Pereira LV, Sanches TR, Shimizu MH, Seguro AC. Rosiglitazone reverses tenofovir-induced nephrotoxicity. *Kidney Int* 2008;**74**:910–8.
40. Yang T, Michele DE, Park J, Smart AM, Lin Z, Brosius 3rd FC, Schnermann JB, Briggs JP. Expression of peroxisomal proliferator-activated receptors and retinoid X receptors in the kidney. *Am J Physiol* 1999;**277**:F966–73.
41. Polvani S, Tarocchi M, Galli A. PPARγ and oxidative stress: Con(β) Catenating NRF2 and FOXO. *PPAR Res* 2012;**2012**:641087.
42. Drager LF, Andrade L, Barros de Toledo JF, Laurindo FR, Machado César LA, Seguro AC. Renal effects of N-acetylcysteine in patients at risk for contrast nephropathy: decrease in oxidant stress-mediated renal tubular injury. *Nephrol Dial Transpl* 2004;**19**:1803–7.
43. Zafarullah M, Li WQ, Sylvester J, Ahmad M. Molecular mechanisms of N-acetylcysteine actions. *Cell Mol Life Sci* 2003;**610**:6–20.
44. Shimizu MH, Coimbra TM, de Araujo M, Menezes LF, Seguro AC. N-acetylcysteine attenuates the progression of chronic renal failure. *Kidney Int* 2005;**68**:2208–17.
45. Campos R, Shimizu MH, Volpini RA, de Bragança AC, Andrade L, Lopes FD, Olivo C, Canale D, Seguro AC. N-acetylcysteine prevents pulmonary edema and acute kidney injury in rats with sepsis submitted to mechanical ventilation. *Am J Physiol Lung Cell Mol Physiol* 2012;**302**:L640–50.
46. de Araujo, M., Andrade, L., Coimbra, T.M., Rodrigues, A.C. Jr, Seguro, A.C. Magnesium supplementation combined with N-acetylcysteine protects against postischemic acute renal failure. J Am Soc Nephrol 16, 2005, 3339–3349.
47. Mehta S, Giovannucci E, Mugusi FM, Spiegelman D, Aboud S, Hertzmark E, Msamanga GI, Hunter D, Fawzi WW. Vitamin D status of HIV-infected women and its association with HIV disease progression, anemia, and mortality. *PLoS One* 2010;**5**:e8770.
48. Sudfeld CR, Wang M, Aboud S, Giovannucci EL, Mugusi FM, Fawzi WW. Vitamin D and HIV progression among Tanzanian adults initiating antiretroviral therapy. *PloS One* 2012;**7**:e40036.
49. Hossein-Nezhad A, Holick MF. Vitamin D for Health: a global perspective. *Mayo Clin Proc* 2013;**88**:720–55.
50. Palacios C, Gonzalez L. Is vitamin D deficiency a major global public health problem? *J Steroid Biochem Mol Biol* 2014;**144**:138–45.
51. Childs K, Welz T, Samarawickrama A, Post FA. Effects of vitamin D deficiency and combination antiretroviral therapy on bone in HIV-positive patients. *AIDS* 2012;**26**:253–62.
52. Pinzone MR, Di Rosa M, Malaguarnera M, Madeddu G, Foca E, Ceccarelli G, d'Ettorre G, Vullo V, Fisichella R, Cacopardo B, Nunnari G. Vitamin D deficiency in HIV infection: an underestimated and undertreated epidemic. *Eur Rev Med Pharmacol Sci* 2013;**17**:1218–32.
53. Petchey WG, Johnson DW, Isbel NM. Shining D' light on chronic kidney disease: mechanisms that may underpin the cardiovascular benefit of vitamin D. *Nephrology (Carlton)* 2011;**16**:351–67.
54. Dusso AS, Thadhani R, Slatopolsky E. Vitamin D receptor and analogs. *Semin Nephrol* 2004;**24**:10–6.
55. Shroff R, Knott C, Rees L. The virtues of vitamin D–but how much is too much? *Pediatr Nephrol* 2010;**25**:1607–20.
56. Holick MF. Vitamin D status: measurement, interpretation, and clinical application. *Ann Epidemiol* 2009;**19**:73–8.
57. Ulerich L. Vitamin D in chronic kidney disease–new insights. *Nephrol Nurs J* 2010;**37**:429–31.
58. Tarcin O, Yavuz DG, Ozben B, Telli A, Ogunc AV, Yuksel M, Toprak A, Yazici D, Sancak S, Deyneli O, Akalin S. Effect of vitamin D deficiency and replacement on endothelial function in asymptomatic subjects. *J Clin Endocrinol Metab* 2009;**94**:4023–30.
59. Sales de Almeida JP, Liberatti LS, Nascimento Barros FE, Kallaur AP, Batisti Lozovoy MA, Scavuzzi BM, Panis C, Reiche EM, Dichi I, Colado Simao AN. Profile of oxidative stress markers is dependent on vitamin D levels in patients with chronic hepatitis C. *Nutrition* 2016;**32**:362–7.
60. Borges-Santos MD, Moreto F, Pereira PC, Ming-Yu Y, Burini RC. Plasma glutathione of HIV(+) patients responded positively and differently to dietary supplementation with cysteine or glutamine. *Nutrition* 2012;**28**:753–6.
61. Brenner DR, Arora P, Garcia-Bailo B, Wolever TM, Morrison H, El-Sohemy A, Karmali M, Badawi A. Plasma vitamin D levels and risk of metabolic syndrome in Canadians. *Clin Invest Med* 2011;**34**:E377.
62. Devaraj S, Jialal G, Cook T, Siegel D, Jialal I. Low vitamin D levels in Northern American adults with the metabolic syndrome. *Horm Metab Res* 2011;**43**:72–4.
63. Gagnon C, Lu ZX, Magliano DJ, Dunstan DW, Shaw JE, Zimmet PZ, Sikaris K, Ebeling PR, Daly RM. Low serum 25-hydroxyvitamin D is associated with increased risk of the development of the metabolic syndrome at five years: results from a national, population-based prospective study (The Australian Diabetes, Obesity and Lifestyle Study: AusDiab). *J Clin Endocrinol Metab* 2012;**97**:1953–61.
64. Izzedine H, Launay-Vacher V, Deray G. Antiviral drug-induced nephrotoxicity. *Am J Kidney Dis* 2005;**45**:804–17.
65. Grober U, Kisters K. Influence of drugs on vitamin D and calcium metabolism. *Dermatoendocrinol* 2012;**4**:158–66.
66. Vaidya A, Brown JM, Williams JS. The renin-angiotensin-aldosterone system and calcium-regulatory hormones. *J Hum Hypertens* 2015;**29**:515–21.
67. Gois PH, Canale D, Luchi WM, Volpini RA, Veras MM, Costa Nde S, Shimizu MH, Seguro AC. Tenofovir during pregnancy in rats: a novel pathway for programmed hypertension in the offspring. *J Antimicrob Chemother* 2015;**70**:1094–105.

CHAPTER

# 19

# Vitamin E and Testicular Damage Protection in Highly Active Antiretroviral Therapy

*Onyemaechi O. Azu*[1,2], *Edwin C.S. Naidu*[1]

[1]University of KwaZulu Natal, Durban, South Africa; [2]University of Namibia, Windhoek, Namibia

OUTLINE



## Abstract

Despite the devastating toll of the human immunodeficiency virus (HIV) and acquired immunodeficiency syndrome (AIDS) on morbidity and mortality, the changing epidemiologic landscape has been largely due to the successful introduction and rollout of highly active antiretroviral therapy (HAART). Together with lifestyle changes, HAART has led to improved quality of life and optimism that HIV/AIDS can be managed similarly to other chronic diseases. Despite these positive outcomes, HAART has been associated with an increasing number of metabolic and anthropometric abnormalities including endocrine perturbations potentially contributing to deranged reproductive lifestyle. Evidence from various studies implicates oxidative stress as mediating many HAART-related reproductive derangements, with discordances regarding the use of supplemental antioxidants such as vitamin E in mitigating these effects. This review describes the evolving dynamics regarding HAART impact on testicular morphology and the unresolved issues relating to the roles of vitamin E as a free radical scavenger. Reproductive function is an unmet need in patients who may desire to have a normal reproductive life and therefore warrants attention.

**Keywords:** Antioxidants; HAART; Testicular structure; Vitamin E.



*HIV/AIDS*
http://dx.doi.org/10.1016/B978-0-12-809853-0.00019-5

## List of Abbreviations

**ATP** Adenosine triphosphate
**CAT** Catalase
**GPx** Glutathione peroxidase
**HAART** Highly active antiretroviral therapy
**HIV** Human immunodeficiency virus
**LPO** Lipid peroxidation
**mtDNA** Mitochondrial DNA
**NF-kB** Nuclear transcription factor kappa B
**NNRTIs** Nonnucleoside reverse transcriptase inhibitors
**NRTIs** Nucleoside reverse transcriptase inhibitors
**OS** Oxidative stress
**PIs** Protease inhibitors
**ROS** Reactive oxygen species
**SOD** Superoxide dismutase

# INTRODUCTION

Humankind has never witnessed the magnitude of devastation following the human immunodeficiency virus/ acquired immunodeficiency syndrome (HIV/AIDS) epidemic, which resulted in the death of over 25 million with over 33 million others battling with the virus.[1] In South Africa alone, approximately 6 million people are living with HIV and AIDS with over 2 million on antiretroviral treatment.[2] These devastating loss of lives are precariously exacerbated by worsening economic crises since the 1930s[3] with frightening financial investments confronting the pandemic rolling over $13.8 billion as of 2008.[4] UNAIDS[5] report further detailed that while the percentage of people living with HIV/AIDS (PLWHAs) appears to have stabilized, the greater number living with the virus continues to escalate owing to new infections. In overall, sub-Saharan Africa (SSA) bears the brunt of the HIV/AIDS pandemic with over 67% cases of PLWHAs globally. In addition to sustenance and expansion response to HIV/AIDS treatment is the challenge of universal access, which remains a dream for millions of people and faces serious technical, economic, and political challenges on a number of fronts.[3]

However, the natural history of HIV infection changed dramatically, following the introduction of highly active antiretroviral therapy (HAART) in the areas of the world that are able to afford these therapies.[6] HAART became available in the United States and other developed countries in 1996.[7] With the advent of HAART, overall prognosis has improved dramatically[8] with significant impact on the management of HIV infection, suppression of viral replication, and reduction of the morbidity and mortality associated with AIDS, and many people affected by the virus are now living with a manageable chronic condition.[9,10] HAART has demonstrated remarkable success in reducing the overall health care costs for HIV-positive persons.[11] The success of current regimens has transformed HIV infection into a chronic condition requiring management over the course of years and decades. To successfully manage people with HIV, it is crucial to find a balance between antiretroviral drug potency, tolerability, safety, and convenience while providing durable viral suppression.[12]

Although HAART regimens may not be effective for all infected persons due to emerging issues of drug-resistant strains of HIV and uncomfortable adverse effects,[13,14] many PLWHAs receiving HAART have substantially lowered viral load (VL) through strict adherence to treatment regimens.[14] Consequently, the incidence of AIDS-related deaths has decreased considerably in countries in which HAART has been widely available[15] including South Africa. Despite these positive advances in HIV medicine, there remains a perplexing insufficient data on various consequences of HAART per se relating to resistance, metabolic syndromes predisposing to secondary complications, toxicities, and the general question of whether antioxidant/micronutrient supplementation may be helpful in reducing the burden of HIV infection, improve the quality of life, and contribute to improvement of nutritional status of PLWHAs. This will continue to pulsate the minds of scientists in time to come.

Since the beginning of the HIV/AIDS epidemic, the use of vitamins and other nutritional supplements has been one of the methods adopted by PLWHAs as a complementary therapy to improve their general health status and quality of life as well as possibly reverse or slow down HIV disease progression and increase their survival rate.[16] In addition, reproductive desires have emerged as clinically important in PLWHAs[10] with HIV-positive men desiring to procreate.[17] Therefore, identifying the actual impact of HAART on the testis, its mechanism, and possible attenuating pathways by the use of vitamin E remain worthy and an unmet medical need.

## CLASSIFICATION OF HAART

The identification of HIV as the causative agent for the AIDS[18] paved the way for the breakthrough in treatment of HIV/AIDS with the development of HAART through the pioneering efforts and collaborations between the National Cancer Institute (United States) and Burroughs–Wellcome Company (United Kingdom).[19] Therapeutic options started with nucleoside reverse transcriptase inhibitors (NRTIs), such as lamivudine (3TC) and protease inhibitors (PIs), which enhanced the survival of patients. At present, HAART available for clinical practice consists of drugs from three classes, targeting two different viral proteins: NRTIs, nonnucleoside reverse transcriptase inhibitors (NNRTIs) and PIs. Current treatment regimens against HIV-1 include NRTIs in combination with PIs and/or NNRTIs; fusion and CCR5 coreceptor inhibitors and integrase inhibitors are the most recent additions to the spectrum of available agents.[20] NRTIs include zidovudine (AZT), stavudine, 3TC, and abacavir, among others. NNRTIs include efavirenz and nevirapine, whereas ritonavir, indinavir, saquinavir, and atazanavir are all PIs (Table 19.1).[9] Although NNRTIs, such as nevirapine, have been shown to be superior alternatives to PIs because of the serious metabolic complications caused by PIs in HIV-infected patients and therefore first-line therapies,[21] they have been associated with early adverse effects.[22]

The eradication of HIV from the body remains yet a challenge as clinical trials are reporting that current vaccine strategies are not completely effective[23] owing to several issues of resistant strains, organ toxicities, side effects, and low adherence to therapy.[19] Many authors[24] are in agreement that drug resistance is associated with increased mortality in patients who receive first-line HAART, with NNRTI resistance representing the highest risk. These have all tilted the playing field from single monotherapy to combination of several classes of the drugs (Table 19.1).

After HAART became the standard of care, ways of simplifying treatment regimens began to emerge. In 1997, Combivir a combination pill of two nucleoside antiretrovirals (AZT and 3TC) was the first of a series of fixed-dose combination pills aiming primarily to improve adherence to treatment, a major cause of treatment failure and emergence of drug resistance.[20]

## MALE REPRODUCTION IN THE HAART ERA—WHAT HAVE WE GLEANED FROM THE PAST DECADES?

Human testicular anatomy is uniquely adapted to perform the dual role of supporting reproductive (semen) and excretory (urine) functions, and these roles are highly regulated by the different components of this highly integrated system including the testes and accessory glands (prostate, seminal vesicles, bulbourethral gland) with the urethral pathway. Testicular microanatomy is made up of numerous highly coiled tubules, the seminiferous tubules, which make up 60%–95% of the 30-mL testicular volume representing the site for spermatogenesis of the postpubertal testicles.[25] Enclosed in a three-layered capsule (of tunica),[26] it plays a significant role in the process of reproduction by its uniqueness in structure, location, and function.[27] The entire process of spermatogenesis involves series of morphological and biochemical transformation that will eventually result in the formation of matured sperm cells capable of fertilization. Due to the complex nature of spermatogenic process, involving rhythmic waves and cycles, matured sperm cells are gradually pushed up from the basal compartment (Fig. 19.1) to the luminal section of the seminiferous tubules in a coordinated manner, which falls prey to various physiopathological disturbances predisposing to various forms of subfertility. Oxidative stress (OS) emanating from torsion, cancer, hypogonadism, or disorders of spermatogenesis caused by toxicant exposure, varicocele, or aging has been associated with this process, targeting various cell types of the seminiferous tubule.[27]

At the ultrastructural level, the mechanical support of the architecture of the testis is anchored on the pillars—the Sertoli cells (Fig. 19.1) that initiates and forms the blood-testis barrier (BTB) in conjunction with basement membrane specializations. Some authors[28] believe that cell polarity of the tubular BTB results in the nuclei of Sertoli cells being located near the basement membrane, but studies have demonstrated that the BTB is highly dynamic thereby facilitating the transit of preleptotene spermatocytes across the "anatomical" barriers.[29] The anatomical segregation in the seminiferous epithelium creates an immunologic barrier so that antigens, many of which appear transiently during spermiogenesis, are sequestered from the systemic circulation to avoid the production of antisperm antibodies[30].

The sperm cell plasma membrane is rich in polyunsaturated fatty acids (PUFA), making it especially vulnerable to free radical injuries.[31] Recently cellular redox activity is found to be important for sperm to attain a functionally competent state to fertilize an oocyte.[32] Whether HIV infection directly and/or therapy (from HAART) contributes

TABLE 19.1 Classes of Various Highly Active Antiretroviral Therapy (HAART) Regimen in Use

| HAART Group | | | | | | |
|---|---|---|---|---|---|---|
| Single | Double/ Combination | Class of ART | Targets Cells/ Tissue | Mechanism(s) of Action | Adverse Effect(s) | References |
| Zidovudine (AZT) | | Nucleoside reverse transcriptase inhibitor (NRTI) | T-cell mitochondria skeletal muscle Heart | Inhibits oxidative phosphorylation to promote apoptosis Oxidative stress (OS) | Myopathy, cardiomyopathy, steatosis, lactic acidosis | Matarrese et al.[84] Lewis et al.[88] and Day and Lewis[89] |
| Zalcitabine | | NRTI | Peripheral nerve Liver Skeletal muscle | Inhibition of mitochondrial replication, lipid accumulation in nerves OS | Peripheral neuropathy, lactic acidosis | Day and Lewis[89] |
| Lamivudine (3TC) | | NRTI | Nerve Cultured cells | OS | Neuropathy | Day and Lewis[89] |
| Didanosine | | NRTI | Peripheral nerve | OS | Peripheral neuropathy | Day and Lewis[89] |
| Stavudine | | NRTI | Peripheral nerve, Liver | OS | Peripheral neuropathy, steatosis, lactic acidosis | Lewis et al. [89a] and Day and Lewis[89] |
| Abacavir (ABC) | | NRTI | Liver | Mitochondrial toxicity | Hypersensitivity reaction, rash, depression | Montessori et al.[41] |
| Emtricitabine (FTC) | | NRTI | Liver | Mitochondrial toxicity | neuropsychiatric adverse events | Montessori et al.[41] |
| Apricitabine | | NRTI | | Undergoing phase II or III clinical trials | – | Kumar and Singh[19] |
| Racivir | | NRTI | | Undergoing phase II or III clinical trials | – | Kumar and Singh[19] |
| Amdoxovir | | NRTI | | Undergoing phase II or III clinical trials | – | Kumar and Singh[19] |
| Elvucitabine | | NRTI | | Undergoing phase II or III clinical trials | – | Kumar and Singh[19] |
| Tenofovir diisoproxil fumarate (TDF) | | NRTI | Kidney, mitochondria | OS | Renal failure, gastrointestinal upset | Montessori et al.[41] Cassetti et al.[85] |
| Saquinavir | | Protease inhibitor (PI) | T-cell mitochondria | Inhibits cell cycle progression and apoptosis | Nausea, diarrhea | Matarrese et al. [85a] |
| Darunavir (TMC-114, DRV) | | PI | Poor bioavailability | All FDA approved | – | Kumar and Singh[19] |
| Tipranavir | | PI | Poor bioavailability | All FDA approved | – | Kumar and Singh[19] |
| Ritonavir (RTV) | | PI | Poor bioavailability | All FDA approved; cytochrome P450 intervention on drug metabolism | Nausea, diarrhea, flushing | Kumar and Singh[19] |
| Indinavir | | PI | Poor bioavailability; kidney | All FDA approved | Renal calculi, hemolytic anemia hyperbilirubinemia | Kumar and Singh[19] |

| | | | | | | |
|---|---|---|---|---|---|---|
| Nelfinavir | | PI | Adipocyte, Poor bioavailability | All FDA approved OS | Nausea, diarrhea | Kumar and Singh[19] |
| Amprenavir | | PI | Poor bioavailability | All FDA approved | Hypersensitivity, perioral paresthesia | Kumar and Singh[19] |
| Lopinavir | | PI | Poor bioavailability | All FDA approved | – | Kumar and Singh[19] |
| Fosamprenavir | | PI | Poor bioavailability | All FDA approved | – | Kumar and Singh[19] |
| Atazanavir | | PI | Poor bioavailability | All FDA approved | – | Kumar and Singh[19] |
| Nevirapine | | Nonnucleoside reverse transcriptase inhibitor (NNRTI) | Liver, kidney | OS | CNS abnormalities, rash, increased liver enzymes rare | Andries et al.[86] and Kumar and Singh[19] |
| Efavirenz | | NNRTI | Liver | OS | Lipid and metabolic disorders, CNS toxicities, hangover | Andries et al.[86] Apostolova et al.[40] and Kumar and Singh[19] |
| Delavirdine | | NNRTI | Liver | OS | CNS disorder rare, rash | Andries et al.[86] Kumar and Singh[19] |
| Etravirine | | NNRTI | Liver | OS | Rash, CNS disorder, dizziness | Andries et al.[86] Kumar and Singh[19] |
| Enfuvirtide | | Fusion inhibitors | CD4 cells | Exploits the gp41 conformational transition that follows gp120-CD4 binding and coreceptor binding and precedes pore formation | Drug resistance, peripheral neuropathy | Wild et al.[87] |
| Maraviroc | | Coreceptor inhibitors | Liver | Blocking the CCR5 chemokine receptor | Diarrhea, nausea, headache, and fatigue | Kumar and Singh[19] |
| Raltegravir | | Integrase inhibitors | – | Prevents proviral DNA-strand transfer | dizziness, insomnia, somnolence | Kumar and Singh[19] |
| **Fixed drug combination** | | | | **All FDA approved** | | |
| Combivir | AZT + 3 TC | – | – | All FDA approved | Gastrointestinal tract (GIT) disorders including diarrhea, vomiting | Este and Cihlar[20]; Kumar and Singh[19] |
| Kivexa/Epzicom | ABC + 3 TC | – | – | All FDA approved | | Este and Cihlar[20] and Kumar and Singh[19] |
| Truvada | TDF + FTC | – | – | All FDA approved | | Este and Cihlar[20] and Kumar and Singh[19] |
| **Three drug combination** | | | | | | |
| Trizivir | AZT + 3 TC + ABC | – | – | All FDA approved | GIT disorders including diarrhea, vomiting | Este and Cihlar[20] and Kumar and Singh[19] |
| Atripla | TDF + RTV + Efavirenz | – | – | All FDA approved; inhibition of mitochondrial DNA (mtDNA) polymerase $\gamma$ in axons and Schwann cells | CNS toxicity | Este and Cihlar[20] and Kumar and Singh[19] |

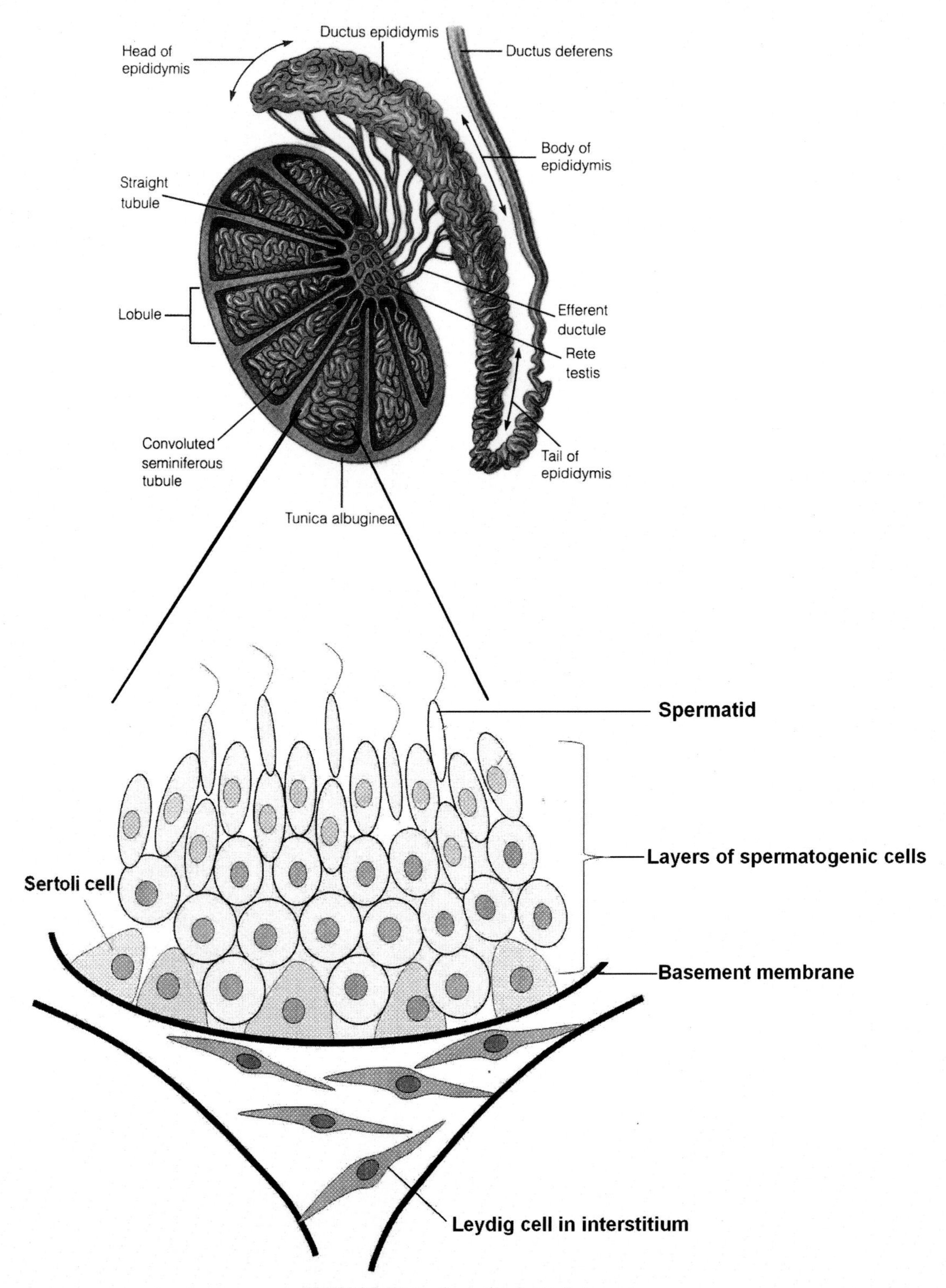


FIGURE 19.1 Testicular (micro)anatomy.

to gonadal failure in both sexes has been poorly elucidated,[33] but the biologic alterations in reproductive physiology have been noted to be responsible for subfertility in women infected with HIV-1.[10]

With the advent of HAART, the spectrum of endocrine disorders relating to male reproductive health has changed markedly in the developed and developing countries of SSA. Up till recently, andrological perturbations have been utilized as the hallmark of sexual dysfunction in HIV medicine with hypogonadism prevalence following HAART ranging from 20% to 30%.[34] However, studies by Dube et al.[35] indicate that effective antiretroviral therapy has been shown to increase testosterone concentrations. Nevertheless, the introduction of HAART has indeed changed the spectrum and modified the prevalence of several comorbidities including androgen deficiency,[36] which seems to occur less frequently than in the past.[37] Therefore, the clinical consequences of hypogonadism whether in HIV-infected or -uninfected populations remain critical areas for further inquiry as we move toward a more positive outlook with regards to improved reproductive health for PLWHAs.

Furthermore, the concept of "sanctuary sites" in the testis as responsible for the low concentrations of antiretroviral drugs in the testes possesses very dire implications regarding the complete eradication of the virus from host tissues/cells. Kis et al.[38] had attributed this to the expression of adenosine triphosphate (ATP)–binding cassette drug efflux transporters at BTBs (just as in the blood-brain barrier), which can actively extrude several drugs, including antiretroviral drugs, from the brain. Unfortunately, this controversy will continue in the near future as issues of drug-resistant viral strains and therapeutic failures continue to be seen in practice tilting the balance in favor of the virus HIV.[39]

## TESTICULAR TOXICITIES OF HAART

Due to the special characteristics of HIV disease itself, the development of antiretroviral drugs in the past was particularly rapid and focused essentially on clinical efficacy (to reduce associated mortality with the scourge). At present, however, the disease has become a controlled and manageable chronic condition[10] hence shifting emphasis to long-term adverse effects caused by the therapy on patients.[40] Both short- and long-term adverse consequences abound with any antiretroviral drug, but the risk varies from drug to drug and from patient to patient. Some of these clinical adverse effects include AIDS-related insulin resistance, lipodystrophy syndrome, and hyperglycemia.[41] The most common and troublesome toxicities of NRTIs and NNRTIs are hepatotoxicity,[13,42] which may be due to damage of mitochondria, especially in patients treated with AZT, stavudine, or didanosine[43]; anemia; and neutropenia.[44] Because HIV is most prevalent among persons of reproductive age and about one-third of those persons would like to have children, reproductive desires have emerged as clinically important in patients with HIV/AIDS.[10] Therefore, the side effects of HAART on fertility, especially in a growing population, become a significant cause for concern.

From Table 19.2, there appear discrepancies in the potential effects of HAART on testicular and seminal parameters based on the various studies. Many authors[45–47] believe that mitochondrial target of HAART laden with NRTIs (3TC and abacavir) putatively compromises the energy reserve of sperm cells with consequent declines in motility (and thus fertility). Given the relevant role played by the mitochondria, exposure to NRTIs alters mitochondrial energy-generating ability in spermatozoa via increased reactive oxygen species (ROS) production resulting in a decreased mitochondrial transmembrane potential (Deltapsim). Reduced Deltapsim leads to the release of some specific apoptotic factors, such as cytochrome C, which initiates programmed cell death.[48] But counter results from some other authors reveal no negative differences in semen parameters that may affect fertility following HAART.[49] While acknowledging these discrepancies, it is noteworthy to point out that much of the observations and reports have been based on different sample populations or experimental cohorts and under diverse etiopathological underpins that might have influenced the outcomes. It is reasonable to believe that this argument may still remain unresolved till such a time when exhaustive tools would have been employed to evaluate the testis with respect to these changes and correlate with other associated parameters of fertility. In the meantime, couples in whom the man is infected by HIV will continue to seek assisted reproductive technology to allow safe procreation, and therefore it is imperative that these counteracting discordances are resolved.[46]

## ANTIOXIDANTS IN HAART ERA—ARE WE THERE YET?

During the last 25 years, a large body of experimental evidence has accumulated from pharmacological intervention studies that suggest an important role for ROS in numerous pathophysiological processes. While a variety of chemical mechanisms of reactive oxygen–induced damage to lipids, proteins, and DNA is fairly well understood, the molecular pathology of oxidant stress-induced tissue injury in vivo remains unclear in many cases.[50] OS results due to an imbalance between the production of ROS and the defense systems, which functions to scavenge or destroy them.[51] Gonadal dysfunction has been long recognized since the early days of the AIDS pandemic,[52] with many studies attributing OS as playing critical role in the pathogenesis of AIDS.[53] With

TABLE 19.2 Highly Active Antiretroviral Therapy (HAART) Toxicity on Testicular and Seminal Parameters

| Investigation | Nature of Study | Outcome(s) | Reference(s) |
|---|---|---|---|
| Ejaculates from 39 human immunodeficiency virus (HIV)-seropositive men and 51 seronegative controls were compared using conventional semen analysis parameters. Centers for Disease Control (Atlanta, GA) classification and peripheral blood CD4 lymphocyte count of seropositive men were recorded at the time of semen analysis. Five patients provided semen before and after zidovudine (AZT) treatment. | Semen analysis in HIV infection | Semen from seropositive men contained significantly fewer motile sperm (48.8% vs. 58.8%, $P=.001$), more round cells (62% vs. 28% subjects with density > 1 × 106/mL, $P=.003$), and was more viscous (1.3 vs. 0.5 cm, $P<.05$) than semen from seronegative controls. Total sperm output, however, was preserved and was not correlated with lymphocyte count or symptomatic disease. Administration of AZT had no deleterious effect on sperm output or other semen variables. HIV infection per se does not significantly reduce sperm production, regardless of clinical stage or mode of acquisition; however, HIV-seropositive men produce more viscous semen containing fewer motile sperm and more round cells. | Crittenden et al.[90] |
| The effects of clinical stage of infection and antiviral therapy on the detection of HIV-1 nucleic acids in semen were investigated by the polymerase chain reaction. | Semiquantitative analysis of six men followed for 8 weeks after the start of nucleoside therapy | A decrease in HIV-1 RNA in seminal plasma was demonstrated in two subjects. | Hamed et al.[95] |
| HIV-1 seropositive men participating in clinical studies at the Fenway Community Health Center (Boston, MA), the University of San Francisco (San Francisco, CA), and Brown University (Providence, RI). | Cross-sectional analysis of semen parameters of 166 HIV-1-seropositive men in various stages of disease progression as defined by peripheral CD4+ cell count. Clinical symptoms and AZT therapy status were obtained from medical records and clinical interviews. | HIV-1-seropositive men that were not on AZT therapy and were in early disease stage (>200 CD4+ cells/mm3) had normal semen parameters as defined by World Health Organization (WHO) criteria. In contrast untreated men in advanced disease stage (<or = 200 CD4+ cells/mm3) had significant reductions in sperm concentration and total sperm count and an increased percentage of abnormal sperm forms. Men receiving AZT antiretroviral therapy, regardless of disease stage, had normal semen parameters similar to those of untreated early disease stage patients. Seminal WBC concentrations were not affected significantly by disease progression but were reduced in patients receiving AZT. | Politch et al. (1994)[95a] |
| Peripheral blood and semen samples from seven men with HIV-1 infection who were receiving HAART and who had no detectable viral RNA | Analysis of samples—HIV-1 in the semen of men receiving HAART | Proviral DNA was detected in seminal cells in four. Replication-competent viruses were recovered from peripheral-blood cells in three men and from the seminal cells in two of these three men. The viruses recovered from the seminal cells had no genotypic mutations suggestive of resistance to antiretroviral drugs and were macrophage-tropic, a feature that is characteristic of HIV-1 strains that are capable of being sexually transmitted. | Zhang et al.[97] |
| Semen characteristics were evaluated in 189 HIV-infected men requesting ART. At the time of semen analysis all men were healthy and 177 were receiving antiretroviral therapy. Comparisons were made with HIV-seronegative men, partners of women requiring in vitro fertilization because of tubal infertility, after matching for age and sexual abstinence delay. | Analysis of samples—Semen alterations in HIV-1-infected men | Reduced percentages of rapidly progressive sperm [median (range), 10% (0%–30%) compared with 15% (5–30%) in the controls, $P<.001$] and increased concentrations of nonspermatic cells [3 × 106/ml (0.2–16 × 106/ml) compared with 1.1 × 106/ml (0.1–14 × 106/ml) in the controls, $P<.001$]. Lower ejaculate volumes [2.8 mL (0.6–9.3 mL) compared with 3.6 mL (1.1–11 mL), $P<.05$].<br>Lower total sperm counts [262.5 × 106 (0–1003 × 106) compared with 310.5 × 106 (48.3–1679 × 106), $P<.05$]. | Dulioust et al.[46] |

| | | | |
|---|---|---|---|
| Study analyzes whether new cellular infections occur in vivo in seminal cells of HIV-1-infected men on suppressive HAART. Peripheral blood mononuclear cells (PBMC) and seminal cells were isolated from a cohort of HIV-1-seropositive men taking suppressive HAART (<50 copies HIV RNA/ml blood plasma). Viral growth assays were performed in vitro and semiquantitative PCR to detect HIV-1 2-LTR circular DNA in PBMC and seminal mononuclear cells. | In vitro analysis—Residual HIV-1 disease in seminal cells of HIV-1-infected men on suppressive HAART: latency without ongoing cellular infections. | Study suggests that in HIV-1-infected men treated with suppressive HAART, new cellular infections occur in PBMC, but that new infections do not take place in seminal cells in vivo. Findings suggest that mainly latent HIV-1 occurs in seminal cells of men on suppressive HAART, which may be a compartment-specific mechanism of residual HIV-1 disease. | Nunnari et al.[93] |
| To investigate the semen parameters in HIV-1-infected patients and to compare their sperm characteristics with those of a control group of fertile, noninfected men. Factors implicated in semen alterations in HIV-1 patients were also analyzed.<br>HIV-infected men (n = 190), of whom 91% were undergoing antiretroviral therapy, and 218 fertile men were studied. Infertility risk factors were recorded and clinical examinations were performed for both groups. Records of history of HIV infection, antiretroviral treatment, and HIV-1 RNA detection in the blood and HIV-1 genome detection in the semen were obtained for the infected patients. | Cohort study—Decreased semen volume and spermatozoa motility in HIV-1-infected patients under antiretroviral treatment | Semen volumes, percentages of progressive motile spermatozoa, total sperm counts, and polymorphonuclear cell counts were decreased, while the pH values and spermatozoa multiple anomaly indices were increased in HIV-infected patients. Even after adjustment for possible sources of bias, the decreases in semen volume and progressive motility and the increase in pH remained significant. The present study demonstrates sperm motility and ejaculate volume alterations in HIV-1-infected patients, most of whom were receiving antiretroviral therapy. In HIV-1 patients, further longitudinal studies are required to analyze the impact of treatment regimen on sperm parameter alterations. | Bujan et al.[91] |
| Retrospective analysis of the sperm-washing database from the treatment of 245 couples with 439 cycles of intrauterine insemination assessed the effects of patient factors (age, maternal FSH, rank of attempt), markers of HIV-disease [time since diagnosis, CD4 count, viral load (VL), use HAART), cycle factors (natural vs. stimulated, number of follicles, fresh vs. frozen sperm) and sperm parameters on clinical and ongoing pregnancy rate. | Retrospective study—A decade of sperm washing: clinical correlates of successful insemination outcome. | 73% of men were on antiretroviral therapy. Semen volume, sperm concentration, total count and progressive motility and postwash concentration, progressive motility, and total motile count inseminated were significantly higher in successful cycles. | Nicopoullos et al.[92] |
| This study was designed to investigate the in vitro effects of didanosine, AZT, saquinavir, and indinavir, commonly used in HAART, on human sperm fertility parameters. Thirty semen samples from healthy men were collected and prepared by gradient density method. Aliquots of 90% fractions with >80% motile spermatozoa were incubated for 1, 3, and 6h with different concentrations of the antiretroviral drugs (20, 40, and 80 μg/ml). Sperm motility was evaluated by computer-assisted sperm analysis system.<br>Sperm mitochondrial potential was evaluated using 3,3′-dihexyloxacarbocyanine iodide (DIOC 6) and the acrosome reaction was examined using pisum sativum agglutinin method. | In vitro assessment of the adverse effects of antiretroviral drugs on the human male gamete | A dose-dependent decrease in sperm motility was observed with saquinavir. Saquinavir also induced a significant time and dose-dependent decrease in mitochondrial potential and an increase in spontaneous acrosome reaction. These findings indicate that, in vitro, higher doses of saquinavir have adverse effects on sperm motility, mitochondrial potential, and acrosome reaction. | Ahmad et al.[98] |

*Continued*

TABLE 19.2 Highly Active Antiretroviral Therapy (HAART) Toxicity on Testicular and Seminal Parameters—cont'd

| Investigation | Nature of Study | Outcome(s) | Reference(s) |
|---|---|---|---|
| To investigate the semen parameters in HIV-1-infected patients with and without HAART and to compare their sperm characteristics with those of healthy men. Serodiscordant couples with an HIV-1-infected man request assisted reproductive technology (ART) to achieve safe conception. Patients were divided into three groups: HIV-infected men taking HAART, HIV-infected patients who did not take HAART until now, and a control group with 93 men consulting our fertility center together with their wives because of tubal sterility. Semen samples were examined (volume, sperm concentration, motility, morphology). | Prospective study of 226 men attending university fertility center of Mannheim (May 1996; July 2003). | The HIV-infected men as a whole group and the subgroup of men with HAART had a lower ejaculate volume, less slow progressive, and more abnormally shaped spermatozoa compared with the control group. The HIV-infected men without an antiretroviral therapy had a significant lower ejaculate volume compared with the control group; the other parameters were not altered significantly. Differences between the subgroups with and without HAART were not significant. | Kehl et al.[99] |
| To determine which antiretroviral drugs are associated with changes in the characteristics of semen and the impact of these ARVs according to their score penetration into the male genital compartment. Data from 144 men infected with HIV-1 enrolled in an assisted reproductive technology program were analyzed retrospectively. | Cross-sectional study—Effect of antiretroviral drugs on the quality of semen | There was no difference on sperm parameters between nucleoside reverse transcriptase inhibitor (NRTI), nonnucleoside reverse transcriptase inhibitor (NNRTI), or protease inhibitor (PI) regimen. In patients receiving NRTIs or PIs no differences were observed between antiretrovirals of these classes. However, in patients receiving NNRTIs, nevirapine (n=22) was associated with a higher percentage of progressively motile spermatozoa ($P<.0001$) versus efavirenz (n=38) as well as vitality ($P=.0004$). No relationship was observed between semen quality and the penetration score. NRTIs and PIs were not associated with any semen changes. Nevirapine was associated with a better quality of semen versus efavirenz. It would be of interest to validate, improve, and test our penetration score in a prospective study. | Lambert-Niclot et al.[45] |
| 116 HIV-positive men (without hepatitis B or C coinfections) were prospectively enrolled from outpatient department for infectious diseases. Patients received a comprehensive andrological work-up. Complete semen analysis was performed according to WHO 2010 recommendations, with each semen variable of the study population being compared with the WHO reference group (n-2000). Correlation analysis was done to investigate the influence of HIV surrogate parameters on semen quality. | Prospective study—Semen quality in HIV patients under stable antiretroviral therapy is impaired compared to WHO 2010 reference values and on sperm proteome level | Each semen variable, about 25% of patients, had values below the fifth percentile of the WHO 2010 reference group. Disease-related parameters (CD4+cell count, VL, CDC stage, duration of disease, duration of ART, number and type of antiretroviral drugs) were not significantly correlated with any sperm parameter. Sperm proteome analysis identified 14 downregulated proteins associated with sperm motility and fertility. It provides evidence of impaired conventional semen parameters and altered sperm protein composition. HIV surrogate parameters are not suitable for predicting semen quality. | Pilatz et al.[94] |
| Study compared semen characteristics across 378 HIV-1-infected patients receiving different antiretroviral regimens or never treated by antiretroviral drugs, in whom an initial semen analysis was done between 2001 and 2007. Methods: The patients were partners from serodiscordant couples requesting medical assistance to procreate safely. Semen parameters were assessed through standard semen analysis. Additional analyses: (measurement of sperm motion parameters using computer-assisted semen analysis, seminal bacteriological analysis, seminal biochemical markers, and testosterone plasmatic levels). All analyses were performed in the Cochin academic hospital. | Case–control study—Impaired sperm motility in HIV-infected men: An unexpected adverse effect of efavirenz. | Sperm motility significantly varied according to treatment status. The median percentage of rapid spermatozoa was 5% in the group of patients receiving a regimen including efavirenz versus 20% in the other groups ($P<.0001$). Sperm velocity was reduced by about 30% in this group ($P<.0001$). As efavirenz is widely used in current HAART, the possibility of common cellular impacts underlying adverse effects of efavirenz in sperm cells and neurons deserved investigation. | Frapsauce et al.[96] |

the stabilization of global prevalence, however, many clinical researchers are pointing to the use of antioxidant treatment as a means of neutralizing or suppressing the progression of AIDS. Nonetheless, drug-induced OS is implicated as a mechanism of toxicity in numerous tissues and organ systems with HAART also well characterized to mediate its effects via OS mechanisms.[54]

Pathologically, toxicant injury affecting male reproductive competencies have directly or indirectly affected many of the mechanisms or processes that regulate spermatozoal motility. Therefore, potent drugs utilized in HAART have targeted the mitochondria,[47] thereby leading to various spermatogenic abnormalities, teratozoospermia, and asthenozoospermia. The fact that HAART-generated free radicals can be mopped up by endogenous antioxidants (which in the case of HIV seropositive patients are already depleted) or supplementation with exogenous antioxidant modalities paves the way for the discuss on the possible beneficial effects of augmentation of HIV patients undergoing HAART with nutritional supplements and vitamins. Preventative antioxidants, such as metal chelators and metal-binding proteins, block the formation of new oxygen/nitrogen species, whereas scavenger antioxidants, such as vitamins E and C, beta-carotene, and other antioxidant dietary supplements, glutathione and enzymes, remove ROS already generated by cellular oxidation. For the purposes of this review, we shall focus primarily on vitamin E and HAART-associated pathophysiology related to testicular anatomy.

## VITAMIN E

Since the discovery of this vitamin in rats in 1922[55] and its rediscovery in the 1950s as factor 2 by Klaus Schwarz,[56] there has been tremendous interests driven by its cellular antioxidant potentials. The term vitamin E summarizes different compounds, basically tocopherols and tocotrienols. The reactivity of vitamin E with organic peroxyl radicals is associated with the redox properties of the chromane ring and accounts for vitamin E's antioxidant activity, which is believed to be its major biochemical function.[57] Vitamin E encompasses a group of potent, lipid-soluble, chain-breaking antioxidants richly abundant in plasma[58] and red blood cells[59] against peroxidative damage. While structural analyses have revealed variations in activities of vitamin E in man and rats, α-tocopherol is the most abundant form in nature.[60] It has the highest biological activity and reverses vitamin E deficiency symptoms in humans.[61] While the molecular functions fulfilled specifically by α-tocopherol is yet to be fully described, it is unlikely they are limited to general antioxidant functions.

## SOURCES OF VITAMIN E

The human body does not synthesize most of the vitamins (essential nutrients that are required by the body for its various biochemical and physiological processes) and hence they must be supplied by the diet in the required amount.[62] Table 19.3 shows a list of possible sources of dietary supplementation richly laden with vitamin E.

## ANTIOXIDANT POTENTIALS OF VITAMIN E

Vitamin E antioxidant activity is linked to its ability to prevent chronic diseases (e.g., cardiovascular diseases and cancer), which are OS driven. Epidemiological studies show that high vitamin E intakes are correlated with a reduced risk of cardiovascular diseases, whereas intakes of other dietary antioxidants (such as vitamin C and β-carotene) are not, suggesting that vitamin E plays specific roles beyond that of its antioxidant function.[63] For example, the critical role played by vitamin E in testicular protection was highlighted by Mason[64] where testicular degeneration was observed in vitamin E-deficient male rats owing to inadequate protection against free radicals generated unavoidably during cytochrome P450-catalyzed synthesis of testosterone in Leydig cells. We have shown in our laboratory that vitamin E offers better cytoprotection on testicular and hepatic tissues when compared with others such as vitamin C or plant-based antioxidant extracts in an antiretroviral model.[65,66]

Recycling of vitamin E occurs due to the reduction of its phenoxyl radicals and contributes to its potency. Ascorbic acid (vitamin C) has been shown to reduce vitamin E phenoxyl radical back to the phenolic form, and during this interaction, ascorbic acid is oxidized to semidehydroascorbyl radical, which can be effectively reduced by dihydrolipoic acid (but not glutathione), thereby contributing in the recycling of vitamin E.[67] However, little is known about the metabolism of α-tocopherol in humans with studies indicating that oxidative cleavage of the phytyl side chain is a major metabolic pathway in humans,[68] but further research is warranted to clarify this.

TABLE 19.3 Dietary Sources of Vitamin E

| Food/Plant Source | Vitamin E Content (Quantity) (mg) |
|---|---|
| Sunflower seeds dry roasted | ¼ Cup (8.4) |
| Almonds | 1 ounce (24 nuts) (7.4) |
| Spinach, cooked | 1 cup (6.7) |
| Spinach, raw | 1 cup (6.7) |
| Safflower oil | 1 tablespoon (4.6) |
| Mango, raw | 1 whole (1.9) |
| Peanut butter | 1 tablespoon (1.4) |
| Wheat germ oil | 20.3 |
| Corn oil | 1.9 |
| Tomato raw | 0.7 |
| Hazelnuts dry roasted | 4.3 |
| Broccoli chopped, boiled | 1.2 |
| Red peppers raw | 2.4 |
| Asparagus cooked | 2.2 |
| Avocado raw | 2.0 |
| Collard greens cooked | 2.1 |
| Sword fish cooked | 2.1 |

*USDA National Nutrient Database for Standard Reference, Vitamin E Content of Selected Foods. http://www.hsph.harvard.edu/nutritionsource/food-sources-of-vitamin-e Accessed 22nd February 2016.*

Vitamin E is an efficient antioxidant not only due to its high reactivity toward oxidizing radicals but also because of the low reactivity of its phenoxyl radicals toward critical biomolecules.[67] The association of vitamin E deficiencies, OS, and HIV infection/HAART has been reported by many authors,[66,69] with results suggesting that combination of vitamin E and AZT may be beneficial in reducing VL as well as suppressing bone marrow toxicity. Other reports indicate that by doubling the vitamin E intake of patients, the risk of progression to AIDS declined. Therefore, in addition to its free-radical scavenging activity, vitamin E impact on the regulation of various cytokines in the immune process is believed to play a major role in inhibiting HIV-1 replication. By enhancing immune functions through its antioxidant properties, it prevents peroxidation of rapidly proliferating cells of the immune system.[70] This is mediated through the nuclear factor KB pathway,[71] which is activated by oxidative stressors such as hydrogen peroxide and lipid hydroperoxides.

While the foregoing discuss has focused on the ability of vitamin E to scavenge free radicals (FRs) emanating from pathologies related to HIV infection or therapy per se, biological systems, however, provide the favorable conditions for controlled oxidative reactions because of the existence of unsaturated fats in cell membranes (CM) and the abundance of oxidative reactions in normal metabolism. The susceptibility of a given tissue/cell to OS would therefore depend on a number of factors including nutrient antioxidants, enzymatic and nonenzymatic scavengers, and factors related to inactivation of oxidation by-products.[72] While vitamin E has an important role as an antioxidant, it also possesses functions that are independent of antioxidant activity. Of the four tocopherols and four tocotrienols known, α-tocopherol has both antioxidant and nonantioxidant effects.

## OXIDATIVE STRESS, HAART, AND VITAMIN E—CULPRITS AND REMEDIES

With over 85% of cellular oxygen uptake by mitochondria needed for energy (ATP) generation, one consequence of mitochondrial dysfunction is an increased generation of ROS and oxidative damage.[73] As the mitochondria continues to be a major target for drug-induced cytotoxicity occurring via a wide range of mechanisms such as inhibition or uncoupling of oxidative phosphorylation, OS, or opening of mitochondrial permeability transition pore,[74] many of the HAART drugs known to be culprits of OS-mediated mechanisms are continually been researched with the view to finding mitigating options. A growing body of evidence suggests that in animals and in humans, NRTI-related mitochondrial dysfunction could be improved by antioxidant supplementation.[75] In the light of this evidence, it seems plausible that the major culprits for oxidative damage are the antiretroviral drugs with the mitochondria as the "recipient" of the noxious effects (see Tables 19.1, 19.2 and 19.4).

With the transition to more universal access to HAART, a better understanding of the role of micronutrient supplementation in HIV-positive persons receiving HAART has become a priority. The provision of simple, inexpensive micronutrient supplements as an adjunct to HAART may have several cellular and clinical benefits, such as a reduction in mitochondrial toxicity and OS and an improvement in immune reconstitution.[76] It has become clear that in addition to contributing to the pathogenesis of many human diseases, OS is a common denominator to many cellular and metabolic perturbations in antiretroviral therapy. OS occurs when there is an overload of free radicals and oxidants leading to their accumulation in the body. In biologic system, there is equilibrium in free radical production that is counterbalanced by several mechanisms (both enzymatic and nonenzymatic antioxidants), which oftentimes is breached with resultant OS status. The inability of the human biological system to detoxify and reduce oxidants or to repair detrimental damage disrupts physiological homeostasis. Table 19.4 highlights key research done involving HAART and vitamin E.

Some studies have indicated that micronutrient supplementation can reduce cellular and metabolic complications of HAART.[75,76] Gavrila et al.[77] studied 120 HIV-positive adults receiving HAART and found that a greater intake of vitamin E was associated with fewer outcomes of HAART-associated metabolic complications owing to changes in the ratio of plasma reduced to oxidized glutathione and oxygen free radicals. The priority attached to supplementation of micronutrients in patients under HAART has been driven by the successes recorded from various studies. But because the pathophysiology of HIV/AIDS is so complex that therapeutic approaches with antioxidants in HIV/AIDS treatment present a challenge to researchers, it becomes critical that the understanding of the mechanistic information regarding the interactions of the various modalities of biological and exogenous antioxidants in the course of therapy would significantly broaden the scope of treatment with consequent reduction of the morbidity and mortality associated with HIV/AIDS.

## PUTATIVE MECHANISM OF INTERACTION BETWEEN VITAMIN E AND HAART

It is important to note that the potential interactions between vitamin E in HAART protocols may be beneficial or detrimental to sperm cells or other components of the testicular architecture depending on several intricate characteristics that the male reproductive tract exhibits. Seminal plasma is rich with many antioxidant defense capacity for preventing sperm oxidative attack following ejaculation. However, during spermatogenesis and epididymal storage, the sperm are not in contact with seminal plasma antioxidants and must rely on epididymal/testicular antioxidants and their own intrinsic antioxidant capacity for protection. Sperm cells are therefore vulnerable to oxidative damage during epididymal transit corroborated by studies supporting declines in sperm motility following HAART in patients.[47,78] Therefore, while seminal plasma antioxidants may help minimize ejaculated sperm OS, they have no capacity to prevent oxidative damage–initiated "upstream" at the level of the testis and epididymis and vitamin E supplementation may become relevant. It is interesting to note that in normal subjects not under OS vitamin E supplementation has no effect on markers of oxidative damage .[79] This is possible because the level of antioxidant protection by the interaction of various antioxidants is so efficient that under normal circumstances, the levels of oxidized lipids are kept in check.

In understanding the potential mechanistic pathways whereby vitamin E could dissipate its actions (antioxidant/ or prooxidant) in biological systems and under normal or pathological states, it is necessary to contextualize these complex chemical processes within the ambit of the redox state cycle. By virtue of its lipid–lipid interactions with unsaturated fatty acids in CM, vitamin E stabilizes biological membranes with high levels of PUFAs. Despite its low concentration in CM, vitamin E is the main lipid-soluble antioxidant in the body.[80] Hence its effects differ significantly depending on the milieu (on CM or cytosol). However, studies on molecular model building further alludes that due to its specific physicochemical interactions between the phytyl side chain and fatty acyl chains of PUFA (mostly from arachidonic acid), vitamin E is able to perform its antioxidant role and may itself also form a thin separating membrane between two aqueous phases.[81]

Fig. 19.2 schematically illustrates the pathways, synergism, and potential areas for HAART-targeted perturbations in the mechanism of vitamin E action. The initiation of a FR chain reaction occurs when PUFA on CM looses a hydrogen atom by reacting with an FR. The propagation of this reaction sees the formation of a peroxyl FR (PUFA-OO*), which may take a hydrogen to other PUFA-initiating formation of the peroxyl radical and hence CM disruptions occur. Vitamin E ($VE^h$) reacts with PUFA-OO* forming PUFA-OOH (not an FR) and VE*

TABLE 19.4 Studies Showing Effects of Vitamin E and HAART

| Investigation | Type of Study | Outcome | Reference(s) |
|---|---|---|---|
| Female C57BL/6 mice were infected with LP-BM5 retrovirus causing murine AIDS, which is functionally similar to human AIDS. Vitamin E effects on immune functions, cytokine production, and nutritional concentrations in retrovirus-infected mice were determined. | Experimental animal study | A 15-fold increase in dietary vitamin E largely restored concentrations of some micronutrients (vitamins A and E, zinc, and copper) in the liver, intestine, serum, and thymus. Partially restored production of interleukin (IL)-2 and IFN-γ by splenocytes prevented retrovirus-induced suppression of splenocyte proliferation and natural killer cell activity. | Wang et al.[107] |
| Murine retrovirus infection induces loss of vitamin E and immune dysfunction with loss of cytokine production by T-helper (Th) cells. Therefore interferon-γ (IFN-γ) was given during dietary vitamin E supplementation to effectively prevent murine retrovirus-induced immunosuppression, cytokine dysregulation, and development of murine AIDS. | In vitro studies | Administration of IFN-γ during vitamin E supplementation significantly prevented development of retrovirus-induced suppression of splenic natural killer cell activity and T cell proliferation; slowed retrovirus-induced elevation of Th-2 cytokine [IL-4, IL-5, and IL-10] production and monokine (IL-6 and tumor necrosis factor-α) secretion by splenocytes prevented loss of Th1 cytokine (IL-2 and IFN-γ) secretion by splenocytes from retrovirus-infected mice alleviating splenomegaly and hypergammaglobulinemia. | Wang et al.[106] |
| This study examines whether zidovudine (AZT) causes oxidative damage to DNA in patients and to skeletal muscle mitochondria in mice and whether this damage may be prevented by supranutritional doses of antioxidant vitamins. | Experimental mice model | Dietary supplements with vitamins C and E at supranutritional doses protect against oxidative damage to skeletal muscle mitochondria caused by AZT. | De la Asunción et al.[75] |
| We propose that testicular damage caused by HAART can be attenuated by antioxidant treatment by investigating the testicular histomorphologic and stereological effects of antiretroviral drugs and its interaction with antioxidants using an experimental animal model. Sprague–Dawley rats were treated orally with 0.9% normal saline as placebo, a HAART cocktail of stavudine, lamivudine, and nevirapine using the adjusted human therapeutic doses of 200, 600, and 350–400 mg/day, respectively, and antioxidants ascorbic acid (vitamin C) and I.M a-tocopherol (vitamin E). | Experimental animal protocol | Testis of animals treated with placebo, ascorbic acid alone, and α-tocopherol alone as well as vitamin E + HAART displayed normal testicular microanatomy. Groups treated with HAART alone, HAART + vitamin C + vitamin E, and vitamins C + HAART showed extensive seminiferous tubular atrophy, necrosis, and hypocellularity in the histoarchitectural patterns. The results show that vitamin E could be useful in protecting testicular tissue from toxicities of HAART regimes as these results mirror stereological data for the groups. | Azu et al.[65] |
| This study was conducted to investigate the possible protective effect of Hypoxis hemerocallidea (African potato (AP)) against HAART-induced hepatotoxicity. A total of 63 pathogen-free adult male Sprague–Dawley rats were treated according to protocols. | Experimental animal protocol | Markers of liver injury assayed showed significant increase ($P < .003$, 0.001) in aspartate amino transferase in AP alone as well as HAART + vitamins C and E groups, respectively. Adjuvant HAART and AP and vitamins C and E also caused significant declines in ALT and ALP levels. Serum gamma glutamyl transferase was not markedly altered. Disturbances in histopathology ranged from severe hepatocellular distortions, necrosis, and massive fibrosis following cotreatment of HAART with vitamins C and E as well as HAART alone. | Azu et al.[66] |
| Vitamin E: total (9.24 3.4 mg/L); 10 of 29 (34%) deficient (6 mg/L); concentrations not presented for HAART and non-HAART groups. 30 HIV-positive adults, mostly injection-drug users (23 receiving HAART for 3 years, 7 not receiving HAART). | Cross-sectional study in France. | Mean plasma concentrations of vitamins A and E were not significantly different between those with a CD4 count and 250 cells/L, between those with viral load (VL) and 5000 copies/mL, and between those receiving and not receiving HAART. | Rousseau et al.[109] |

| | | | |
|---|---|---|---|
| 175 HIV-positive injection drug users (30 receiving HAART, 65 receiving dual- or monotherapy, 80 not receiving any HIV medications)—Tocopherol 2: HAART (1076468 g/dL); no HIV medications (778209 g/dL) | Cross-sectional study in the United States. | Adjusted mean serum concentrations of α-tocopherol ($P=.0008$) was significantly higher in those receiving HAART than in those not taking any HIV medications. | Tang et al.[105] |
| The relationship among habitual exercise, diet, and the presence of metabolic abnormalities (body fat redistribution, dyslipidemia, and insulin resistance) was studied in 120 HIV–infected subjects with use of bivariate and multivariate regression-analysis models. | Cross-sectional study | Diastolic blood pressure was significantly and inversely associated with supplemental or total but not habitual dietary intake of vitamin E.<br>Exercise and vitamin E intake were independently and negatively associated with several phenotypic manifestations of HIV-associated metabolic syndrome, whereas other macro- or micronutrients did not have comparable significance. | Gavrila et al.[77] |
| To determine if asymptomatic stable chronic hyperlactatemia in HIV-infected patients under HAART, including nucleoside reverse transcriptase inhibitor (NRTI), could be improved by antioxidant supplementation.<br>To match two groups of patients taking NRTI for at least 24 months: 15 without and 15 with antioxidant supplementation (vitamin E, β-carotene, N-acetylcysteine, selenium, Gingko biloba extracts, and nutritional supplements). It was made up of vitamin E (1 g/d). | Mixed (experimental and matching) | Blood oxidative stress (OS) markers, i.e., vitamin E, vitamin A, and β-carotene tended to be higher in the supplemented group. Antioxidant supplementation improves the asymptomatic stable chronic hyperlactatemia observed in HIV-infected patients taking HAART including NRTI for a long time. | Lopez et al.[103] |
| 30 HIV-positive adults receiving HAART. Daily vitamin A (5000 IU), vitamin C (50 mg), and vitamin E (100 IU) for 6 months. | Randomized, placebo-controlled trial in Poland. | Six months of an antioxidant multivitamin supplement significantly increased antioxidant defenses, reduced OS, and possibly improved immunologic status. | Jaruga et al.[102] |
| 10 HIV-positive adults receiving HAART for 12 months; 9 had lipoatrophy, and 1 had sustained hyperlactemia at enrollment. Daily vitamin C (1000 mg) and vitamin E (800 IU) and twice daily N-acetyl cysteine (600 mg) for 24 weeks. | Nonrandomized Trials, open-label pilot study without placebo control in the United States. | 24 weeks of a supplement worsened fasting glucose and insulin resistance and did not significantly improve peripheral fat, lipoatrophy, immunologic status, or plasma VL. | McComsey et al.[108] |
| 29 HIV-positive adults with CD4 count 500 cells/L. 26 initiated HAART and 3 initiated dual combination therapy at study enrollment. Daily vitamin E (800 mg α-tocopherol) for 6 months. | Randomized, placebo-controlled trial in Brazil. | Six months of vitamin E supplementation improved lymphocyte viability but did not affect immune cell count or plasma VL. | Spada et al.[104] and De Souza et al.[100] |
| The effect of multivitamin (MV) supplementations among patients on HAART. Eligible subjects were randomized to receive placebo or MV supplementation including vitamins B-complex, C, and E. Participants were followed for up to 18 months. Primary endpoints were change in CD4 cell count, weight, and quality of life. Secondary endpoints were (i) development of a new or recurrent HIV disease progression event, including all-cause mortality; (ii) switching from first- to second-line antiretroviral therapy; and (iii) occurrence of an adverse event. | RCT, double-blind, placebo-controlled trial. | No significant differences were observed in primary endpoints or in occurrence of adverse events between the trial arms.<br>Conclusions: one recommended daily allowance of MV supplementation was safe but did not have an effect on indicators of disease progression among HIV-infected adults initiating HAART. | Guwatudde et al.[101] |

(a radical), which may also combine with PUFA-OOH to produce PUFA-OO*. This latter action has been considered as one of the prooxidant actions of vitamin E,[82] of which the administration of vitamin C ($VC_{reduced}$) in the presence of glutathione (reduced) is regenerated back to $VE^h$ with vitamin C becoming oxidized in the process (Fig. 19.2). Phospholipase $A_2$ catalyzes the supply of PUFA-OOH release into the cytosolic milieu whereupon the latter is acted on by the enzymatic antioxidant components (catalase, superoxide dismutase, and glutathione peroxidase) into PUFA-OH.[80]

In examining the cellular and metabolic disturbances sequel to commencement of HAART in patients, many of the components, especially the NRTIs, are known to impair mitochondrial DNA replication. This exacerbates FR release with resultant OS-mediated damage. Supporting this is evidence from a longitudinal study of HIV-positive children who were switched from dual-NRTI regimen to HAART with resultant significant increase in antioxidant capacity 2 months after switch.[83] Therefore, several steps in the interactions between vitamin E and drugs used in HAART could adversely alter FR formation, lipid peroxidation, mitochondrial dysfunction, altered calcium handling, DNA damage, or activation of proapoptotic signaling cascades/inhibition of survival signaling. At the molecular level, FR damages amino acids, lipid, and DNA. Mitochondrial dysfunction and associated alterations in energetics together with effects on survival/apoptotic signaling cascades may lead to a proapoptotic response, and these common mechanisms may be key to many toxicities seen in HAART drugs with AZT-dependent skeletal myopathy in mind.

The understanding of how antioxidants are related to AIDS is still in its early stages and more research is needed to decipher the optimal conditions by which micronutrients can help modulate or immune-potentiate the body's defenses and, in addition, how the host nutritional status influences the degree of virulence of some pathogens. In addition, the putative interactions of specific antioxidant with drugs used in HAART cocktails require a careful consideration as there are patient-to-patient and country-to-country differences in uptake and use of antiretrovirals with the tendency toward various combinations that may not necessarily confer the desired outcomes.

## FUTURE PERSPECTIVES

While certain foods are good sources of exogenous antioxidants and therefore necessary to be included as adjuvants in the treatment of illnesses accompanied by OS-mediated endpoints, there is the need to continuously interrogate specific aspects of the diverse roles that may be played by vitamin E in biological system. In which case, with the discovery of the role of ROS in the development of AIDS and its therapy with HAART, there is broadening of search for mitigating agents toward amelioration of the FR-mediated perturbations. These approaches with antioxidants in HIV/AIDS treatment will continue to present a challenge to researchers because HIV/AIDS itself is a vexing disease with many implications. Therefore, understanding the interactions of the various modalities of biological and exogenous antioxidants in the course of therapy would significantly broaden the scope of treatment with consequent reduction of the morbidity and mortality associated with HIV/AIDS.

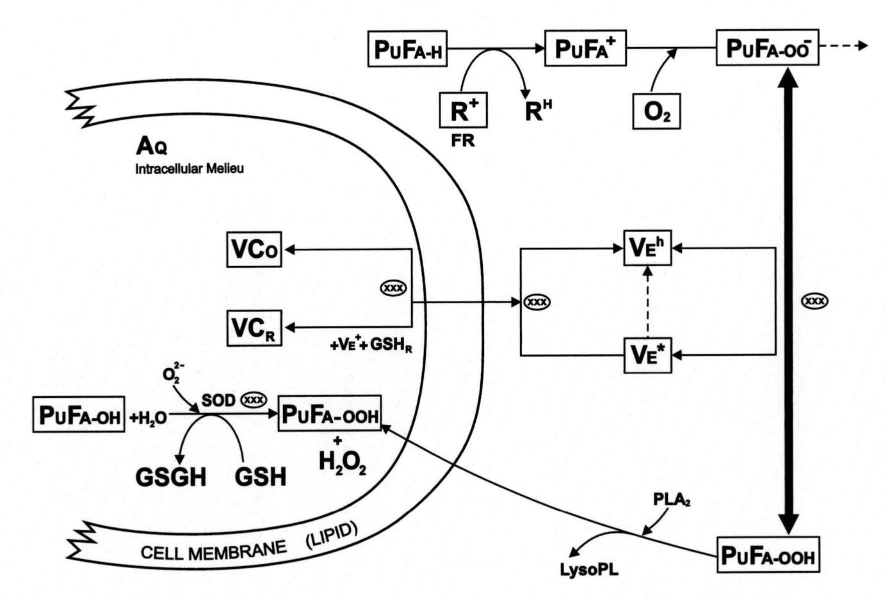


FIGURE 19.2 Putative mechanism of action of vitamin E antioxidative actions.

## SUMMARY POINTS

- Global success in HIV/AIDS therapy has been successful due to introduction of HAART.
- ROS and FRs play a major role in the pathophysiological actions of HAART warranting interventions using antioxidants.
- While the male genital tract and testis have inherent protective mechanisms against FR ravages due to HAART, vitamin E plays significant antioxidant role protecting the testicular architecture from HAART-induced toxicities.
- Testicular microanatomy offers potential targets to components of HAART with consequent depletion of mitochondrial energy level and impaired reproductive outcomes.
- While research efforts try to unravel the complexities regarding intervention/supplementation with vitamin E in HAART protocols, there will be need to explore every angle that offers clue to better therapeutic outcomes of patients.

## Acknowledgments

The support of Eberechi OO Azu is gratefully acknowledged. Mr. Justin Naidu is acknowledged for his technical assistance with the illustrations.

## References


1. WHO. *Towards Universal Access: Scaling up priority HIV/AIDS interventions in the health sector*. 2008.
2. Shisana O, Rehle T, Simbayi LC, Zuma K, Jooste S, Zungu N, Labadarios D, Onoya D. *South African National HIv prevalence, incidence and Behaviour Survey, 2012*. Cape Town: HRSC Press; 2014. p. 32.
3. Granich R, Crowley S, Vitoria M, Smyth C, Kahn JG, Bennett R, Lo Y-R, Souteyrand Y, Williams B. Highly active antiretroviral treatment as prevention of HIV transmission: review of scientific evidence and update. *Curr Opin HIV AIDS* 2010;**5**(4):298–304.
4. UNAIDS. Financial resources required to achieve universal access to HIV prevention, treatment, care and support. Geneva, Switzerland ;**2007**. 2009. http://data.unaids.org/pub/Report/2007/20070925_advocacy_grne2_en.pdf.
5. UNAIDS. Geneva: AIDS epidemic update; 2008.
6. Kaplan JE, Hanson D, Dworkin MS, Frederick T, Bertolli J, Lindegren ML, Holmberg S, Jones JL. Epidemiology of human immunodeficiency virus-associated opportunistic infections in the United States in the era of highly active antiretroviral therapy. *Clin Infect Dis* 2000;**30**(Suppl. 1):S5–14.
7. Crepaz N, Hart TA, Marks G. Highly active antiretroviral therapy and sexual risk behavior: a meta-analytic review. *JAMA* 2004;**292**(2):224–36.
8. Auvert B, Males S, Puren A, Taljaard D, Carael M, Williams B. Can highly active antiretroviral therapy reduce the spread of HIV? a study in a township of South Africa. *J Acquir Immune Defic Syndr* 2004;**36**:613–21.
9. Kayode AAA, Kayode OT, Aroyeun OA, Stephen MC. Hematological and hepatic enzyme alterations associated with acute administration of antiretroviral drugs. *J Pharmacol Toxicol* 2011;**6**(3):293–302.
10. Kushnir VA, Lewis W. Human immunodeficiency virus/acquired immunodeficiency syndrome and infertility: emerging problems in the era of highly active antiretrovirals. *Fertil Steril* 2011;**96**:546–53.
11. Mocroft A, Ledergerber B, Katlama C, Kirk O, Reiss P, d'Arminio Monforte A, Knysz B, Dietrich M, Phillips AN, Lundgren JD. EuroSIDA study group. Decline in the AIDS and death rates in the EuroSIDA study: an observational study. *Lancet* 2003;**362**:22–9.
12. Kress KD. HIV update: emerging clinical evidence and a review of recommendations for the use of highly active antiretroviral therapy. *Am J Health Syst Pharm* 2004;**61**(3):3–14.
13. Soriano V, Puoti M, Garcia-Garsco P, Rockstroh JK, Benhamou Y, Barreiro P, McGovern B. Antiretroviral drugs and liver injury. *AIDS* 2008;**22**:1–13.
14. Deeks SG, Smith M, Holodniy M, Kahn JO. HIV-1 protease inhibitors. A review for clinicians. *JAMA* 1997;**277**(2):145–53.
15. Mocroft A, Vella S, Benfield TL, Chiesi A, Miller V, Gargalianos P, d'Arminio Monforte A, Yust I, Bruun JN, Phillips AN, Lundgren JD, For the EuroSIDA Study Group. Changing patterns of mortality across Europe in patients infected with HIV-1. *Lancet* 1998;**352**:1725–30.
16. Oguntibeju OO, Esterhuyse AJ, Truter EJ. Truter Possible benefits of micronutrient supplementation in the treatment and management of HIV infection and AIDS. *Afr J Pharm Pharmacol* 2009;**3**(9):404–12.
17. Terashima K, Takawa Y, Niwa M. Powerful antioxidative agents based on Garcinoic acid from Garcinia kola. *Bioorg Med Chem* 2002;**10**:1619–25.
18. Popovic M, Sarngadharan MG, Read E, Gallo RC. Detection, isolation and continuous production of cytopathic retroviruses (HTLV-III) from patients with AIDS and pre-AIDS. *Science* 1984;**224**:497–500.
19. Kumari G, Singh RK. Highly active antiretroviral therapy for treatment of HIV/AIDS patients: current status and future prospects and the Indian scenario. *HIV AIDS Rev* 2012;**11**:5–14.
20. Este´ JA, Cihlar T. Current status and challenges of antiretroviral research and therapy. *Antivir Res* 2010;**85**:25–33.
21. Saag MS, Powderly WG, Schambelan M, Schambelan M, Benson CA, Carr A, Currier JS, Dube´ MP, Gerber JG, Grinspoon SK, Grunfeld C, Kotler DP, Mulligan K. Switching antiretroviral drugs for treatment of metabolic complications in HIV-1 infection: summary of selected trials. *Top HIV Med* 2002;**10**:47–51.
22. Wit FW, Weverling GJ, Weel J, Jurriaans S, Lange JM. Incidence and risk factors for severe hepatotoxicity associated with antiretroviral therapy. *J Infect Dis* 2002;**186**:23–31.
23. Saag MS, Kilby JM. HIV-1 and HAART: a time to cure, a time to kill. *Nat Med* 1999;**5**:609–11.

24. Hogg RS, Bangsberg DR, Lima VD, Alexander C, Bonner S, Yip B, Wood E, Dong WW, Montaner JS, Harrigan PR. Emergence of drug resistance is associated with an increased risk of death among patients first starting HAART. *PLoS Med* 2006;**3**(9):e356. http://dx.doi.org/10.1371/journal.pmed.0030356.
25. Sharpe RM. Paracrine control of the testis. *Clin Endocrinol Metab* 1986;**15**:185–207.
26. Brooks JD. Anatomy of the lower urinary tract and male genitalia. In: Kavoussi LR, Novick AC, Partin AW, Peters CA, Wein AJ, editors. *Campbell's Urology*. 9th ed. Saunders Elsevier; 2007. p. 38–80.
27. Wong CH, Cheng CY. The blood-testis barrier: its biology, regulation and physiological role in spermatogenesis. *Curr Top Dev Biol* 2005;**71**:263–96.
28. Wong EW, Cheng CY. Polarity proteins and cell-cell interactions in the testis. *Int Rev Cell Mol Biol* 2009;**278**:309–53.
29. Cheng CY, Mruk DD. A local autocrine axis in the testes that regulates spermatogenesis. *Nat Rev Endocrinol* 2010;**6**:380–95.
30. Su L, Mruk DD, Lee WM, Cheng CY. Drug transporters and blood-testis barrier function. *J Endocrinol* 2011;**209**:337–51.
31. O WS CH, Chow PH. Male genital tract antioxidant enzymes–their ability to preserve sperm DNA integrity. *Mol Cell Endocrinol* 2006;**250**:80–3.
32. de Lamirande E, Gagnon C. Capacitation-associated production of superoxide anion by human spermatozoa. *Free Radic Biol Med* 1995;**18**:487–95.
33. Mhawech P, Onorato M, Uchida T, Borucki MJ. Testicular atrophy in 80 HIV-positive patients: a multivariate statistical analysis. *Int J STD AIDS* 2001;**12**:221–4.
34. Kalyani RR, Gavini S, Dobs AS. Male hypogonadism in systemic disease. *Endocrinol Metab Clin North Am* 2007;**36**:333–48.
35. Dube MP, Parker RA, Mulligan K, et al. Effects of potent antiretroviral therapy on free testosterone levels and fat-free mass in men in a prospective, randomized trial: a5005s, a sub-study of AIDS Clinical Trials Group Study 384. *Clin Infect Dis* 2007;**45**(1):120–6.
36. Crum NF, Furtek KJ, Olson PE, Amling CL, Wallace MR. A review of hypogonadism and erectile dysfunction among HIV-infected men during the pre- and post-HAART eras: diagnosis, pathogenesis, and management. *AIDS Patient Care STDS* 2005;**19**:655–71.
37. Dobs AS, Few WL, Blackman MR, Harman SM, Hoover DR, et al. Serum hormones in men with human immunodeficiency virus-associated wasting. *J Clin Endocrinol Metab* 1996;**81**:4108–12.
38. Kis O, Robillard K, Chan GNY, Bendayan R. The complexities of antiretroviral drug-drug interactions: role of ABC and SLC transporters. *Trends Pharmacol Sci* 2010;**31**:22–35.
39. Saksena NK, Wang B, Zhou L, Soedjono M, Ho YS, Conceicao V. HIV reservoirs in vivo and new strategies for possible eradication of HIV from the reservoir sites. *HIV AIDS (Auckl)* 2010;**2**:103–22.
40. Apostolova N, Gomez-Sucerquia LJ, Moran A, Alvarez A, Blas-Garcia A, et al. Enhanced oxidative stress and increased mitochondrial mass during efavirenz-induced apoptosis in human hepatic cells. *Br J Pharmacol* 2010;**160**:2069–84.
41. Montessori V, Press N, Harris M, Akagi L, Montaner JS. Adverse effects of antiretrocviral therapy for HIV infection. *Can Med Assoc J* 2004;**170**:229–38.
42. Abrescia N, D'Abbraccio M, Figoni M, Busto A, Maddaloni A, De Marco M. Hepatotoxicity of antiretroviral drugs. *Curr Pharm Des* 2005;**11**:3697–710.
43. Verucchi GL, Calza L, Biagetti C, Attard L, Costagliola R, Manfredi R, Pasquinelli G, Chiodo F. Ultrastructural liver mitochondrial abnormalities in HIV/HCV-coinfected patients receiving antiretroviral therapy. *J Acquir Immune Defic Syndr* 2004;**35**:326–8.
44. Fistche MA, Parker KP, Pettineli C. A randomized controlled trial of reduced daily dose of zidovudine in patients with AIDS. *N Eng J Med* 1990;**323**:200–2.
45. Lambert-Niclot S, Poirot C, Tubiana R, Houssaini A, Dominguez S, Prades M, Bonmarchand M, Calvez V, Flandre P, Peytavin G, Marcelin A. Effect of antiretrovirals on semen quality. *J Med Virol* 2011;**83**(8):1391–4.
46. Dulioust E, Le Du A, Costagliola D, Guibert J, Kunstmann JM, Heard I, Juillard JC, Salmon D, Leruez-Ville M, Mandelbrot L, Rouzioux C, Sicard D, Zorn JR, Jouannet P, De Almeida M. Semen alterations in HIV-1 infected men. *Hum Repro* 2002;**17**(8):2112–8.
47. van Leeuwen E, Wit FW, Repping S, Eeftinck Schattenkerk JK, Reiss P, et al. Effects of antiretroviral therapy on semen quality. *AIDS* 2008;**22**:637–42.
48. Sergerie M, Martinet S, Kiffer N, et al. Impact of reverse transcriptase inhibitors on sperm mitochondrial genomic DNA in assisted reproduction techniques. *Gynecol Obstet Fertil* 2004;**32**:841–9.
49. Krieger JN, Coombs RW, Collier AC, Koehler JK, Ross SO, et al. Fertility parameters in men infected with human immunodeficiency virus. *J Infect Dis* 1991;**164**:464–9.
50. Jaeschke H. Mechanisms of oxidant stress-induced acute tissue injury. *Proc Soc Exp Biol Med* 1995;**209**:104–11.
51. Favier A, Sappey C, Leclerc P, Faure P, Micoud M. Antioxidant status and lipid peroxidation in patients infected with HIV. *Chem Biol Interact* 1994;**91**:165–80.
52. Brown TT, Cole SR, Li X, Kingsley LA, Palella FJ, Riddler SA, Visscher BR, Margolick JB, Dobs AS. Antiretroviral therapy and the prevalence and incidence of diabetes mellitus in the multicentre AIDS cohort study. *Arch Intern Med* 2005;**165**:1179–84.
53. Papadopulos-Eleopulos E. Reappraisal of AIDS – is the oxidation induced by the risk factors the primary cause? *Med Hypotheses* 1988;**25**(3):151–62.
54. Deavall DG, Elizabeth AM, Judith M, Horner RR. Drug-induced oxidative stress and toxicity. *J Toxicol* 2012. http://dx.doi.org/10.1155/2012/645460.
55. Evans HM, Bishop KS. On the existence of a hitherto unrecognized dietary factor essential for reproduction. *Science* 1922;**56**:650–1.
56. Schwarz K. Role of vitamin E, selenium, and related factors in experimental nutritional liver disease. *Fed Proc* 1965;**24**:58–67.
57. Sies H, Stahl W. Vitamins E and C, b-carotene and other carotenoids as antioxidants. *Am J Clin Nutr* 1995;**62**(Suppl.):1315S–21S.
58. Burton GW, Joyce A, Ingold KU. First proof that vitamin E is the major lipid-soluble, chain breaking antioxidant in human blood plasma. *Lancet* 1982;**2**:327–8.
59. Burton GW, Joyce A, Ingold KU. Is vitamin E the only lipid-soluble, chainbreaking antioxidant in human blood plasma and erythrocyte membranes? *Arch Biochem Biophys* 1983;**22**:28–90.
60. Sheppard AJ, Pennington JAT, Weihrauch JL. Analysis and distribution of vitamin E in vegetable oils and foods. In: Packer L, Fuchs J, editors. *Vitamin E in health and disease*. New York: Marcel Dekker Inc.; 1993. p. 9–31.

61. Brin MF, Pedley TA, Emerson RG, Lovelace RE, Gouras P, MacKay C, Kayden HJ, Levy J, Baker H. Electrophysiological features of abetalipoproteinemia: functional consequences of vitamin E deficiency. *Neurology* 1986;**36**:669–73.
62. Iqbal K, Khan A, Khattak MMAK. Biological significance of ascorbic acid (vitamin C) in human health-a review. *Pakistan J Nutr* 2004;**3**:5–13.
63. Stampfer M, Hennekens C, Manson J, Colditz G, Rosner B, Willett W. Vitamin E consumption and the risk of coronary disease in women. *N Engl J Med* 1993;**328**:1444–9.
64. Mason KE. Differences in testis injury and repair after vitamin A deficiency, vitamin E deficiency and inanition. *Am J Anat* 1933;**52**:153–239.
65. Azu OO, Naidu ECS, Naidu JS, Chuturgoon A, Masia T, Nzemande N, Singh SD. Testicular histomorphologic and stereological alterations following short-term treatment with highly active antiretroviral drugs in an experimental animal model. *J Androl* 2014;**2**(5):772–9.
66. Azu OO, Jegede AI, Offor U, Onanuga IO, Kharwa S, Naidu ECS. Hepatic histomorphological and biochemical changes following highly active antiretroviral therapy in an experimental animal model: does Hypoxis hemerocallidea exacerbate hepatic injury? *Toxicol Rep* 2016;**3**:114–22.
67. Halliwell B. Reactive oxygen species in living systems: source, biochemistry and role in human disease. *Am J Med* 1991;**91**:14–22.
68. Schultz M, Leist M, Petrzika M, Gassmann B, Brigelius-Floh R. Novel urinary metabolite of a-tocopherol, 2,5,7,8-tetramethyl-2(2′-carboxyethyl)-6-hydroxychorman, as an indicator of an adequate vitamin E supply. *Am J Clin Nutr* 1995;**62**(15):275–345.
69. Pacht ER, Diaz P, Clanton T, Hart J, Gadek JE. Serum vitamin E decreases in HIV-seropositive subjects over time. *J Lab Clin Med* 1997;**130**:293–6.
70. Tengerdy RP. The role of vitamin E in immune response and disease resistance. *Ann N Y Acad Sci* 1990;**587**:24–33.
71. Suzuki YI, Packer L. Inhibition of NF-kappa B activation by vitamin E derivatives. *Biochem Biophys Res Commun* 1993;**193**:277–83.
72. Jones DP, Kagan VE, Aust SD, Reed DJ, Omaye ST. Impact of nutrients on cellular lipid peroxidation and antioxidant defense system. *Fundam Appl Toxicol* 1995;**26**:1–7.
73. Beal MF. Energetics in the pathogenesis of neurodegenerative diseases. *Trends Neurosci* 2000;**23**:296–304.
74. Labbe G, Pessayre D, Fromenty B. Drug-induced liver injury through mitochondrial dysfunction: mechanisms and detection during preclinical safety studies. *Fundam Clin Pharmacol* 2008;**22**:335–53.
75. De la Asunción JG, del Olmo ML, Sastre J, Millán A, Pellin A, Pallardó FV, et al. AZT treatment induces molecular and ultrastructural oxidative damage to muscle mitochondria. Prevention by antioxidant vitamins. *J Clin Invest* 1998;**102**:4–9.
76. Drain PK, Kupka R, Mugusi F, Fawzi WW. Micronutrients in HIV-positive persons receiving highly active antiretroviral therapy. *Am J Clin Nutri* 2007;**85**(2):333–45.
77. Gavrila A, Sotirios T, Doweiko J, et al. Exercise and vitamin E intake are independently associated with metabolic abnormalities in human immunodeficiency virus–positive subjects: a cross-sectional study. *Clin Infect Dis* 2003;**36**:1593–601.
78. Tremellen K. Oxidative stress and male infertility-a clinical perspective. *Hum Reprod Update* 2008;**14**:243–58.
79. Meagher EA, Barry OP, Lawson JA, Rokach J, Fitz Gerald GA. Effects of vitamin E on lipid peroxidation in healthy persons. *JAMA* 2001;**285**:1178–82.
80. Herrera E, Babrbas C. Vitamin E: action, metabolism and perspectives. *J Physiol Biochem* 2001;**57**(1):43–56.
81. Seufert WD, Beauchesne G, Belanger M. *Biochem Biophys Acta* 1970;**211**:356.
82. Nagaoka S, Inuoue M, Nishioka C, Nishioku Y, Tsunoda S, Ohguchi C, Ohara K, Mukai K, Nagashima U. *J Phys Chem B* 2000;**104**:856–62.
83. De Martino M, Chiarelli F, Moriondo M, et al. Restored antioxidant capacity parallels the immunologic and virologic improvement in children with perinatal human immunodeficiency virus infection receiving highly active antiretroviral therapy. *Clin Immunol* 2001;**100**:82–6.
84. Matarrese P, Tinari A, Gambardella L, Mormone E, Narilli P, Pierdominici M, Cauda R, Malorni W. HIV protease inhibitors prevent mitochondrial hyperpolarization and redox imbalance and decrease endogenous uncoupler protein-2 expression in gp 120-activated human T lymphocytes. *Antivir Ther* 2005;**10**(Suppl. 2):M29–45.
85. Cassetti I, Madruga JVR, Suleiman JMAH, Etzel A, Zhong L, Cheng AK, Enejosa J. The safety and efficacy of tenofovir DF in combination with lamivudine and efavirenz through 6 years in antiretroviral-naïve HIV-1-infected patients. *HIV Clin Trials* 2007;**8**(3):164–72.

85a. Matarrese P, Gambardella L, Cassone A, Vella S, Cauda R, Malorni W. Mitochondrial membrane hyperpolarization hijacks activated T lymphocytes toward the apoptotic-prone phenotype: homeostatic mechanisms of HIV protease inhibitors. *J Immunol* June 15, 2003;**170**(12):6006–15.

86. Andries K, Bethune, Kukla MJ, Azijn H, Lewi PJ, Janssen PAJ, Pauwels R. R165335–TMC125, a novel nonnucleoside reverse transcriptase inhibitor (NNRTI) with nanomolar activity against NNRTI resistant HIV strains. In: *40th Interscience conference on antimicrobial agents and chemotherapy. cited 7 times. Toronto, Canada, September 17-20 abstract no. 1840*. 2000.
87. Wild C, Greenwell T, Matthews T. A synthetic peptide from HIV-1 gp41 is a potent inhibitor of virus-mediated cell-cell fusion. *AIDS Research Human Retroviruses* 1993;**9**(11):1051–3.
88. Lewis W, Day BJ, Copeland WC. Mitochondrial toxicity of NRTI antiviral drugs: an integrated cellular perspective. *Nat Rev Drug Discov* 2003;**2**:812–22.
89. Day BJ, Lewis W. Oxidative stress in NRTI-induced toxicity: evidence from clinical experience and experiments in vitro and in vivo. *Cardiovasc Toxicol* 2004;**4**:207–16.

89a. Lewis W, Copeland WC, Day BJ. Mitochondrial DNA depletion, oxidative stress, and mutation: mechanisms of dysfunction from nucleoside reverse transcriptase inhibitors. *Lab Invest* 2001;**81**(6):777–90.

90. Crittenden JA, Handelsman DJ, Stewart GJ. Semen analysis in human immunodeficiency virus infection. *Fertil Sterility* 1992;**57**(6):1294–9.
91. Bujan L, Sergerie M, Moinard N, Martinet S, Porte L, Massip P, Pasquier C, Daudin M. Decreased semen volume and spermatozoa motility in HIV-1-infected patients under antiretroviral treatment. *J Androl* 2007;**28**(3):444–52.
92. Nicopoullos JD, Almeida P, Vourliotis M, Goulding R, Gilling-Smith C. A decade of sperm washing: clinical correlates of successful insemination outcome. *Hum Reprod* 2010;**25**(8):1869–76.
93. Nunnari G, Otero M, Dornadula G, Vanella M, Zhang H, Frank I, Pomerantz RJ. Residual HIV-1 disease in seminal cells of HIV-1-infected men on suppressive HAART: latency without on-going cellular infections. *AIDS* 2002;**16**(1):39–45.
94. Pilatz A, Discher T, Lochnit G, Wolf J, Schuppe HC, Schüttler CG, Hossain H, Weidner W, Lohmeyer J, Diemer T. Semen quality in HIV patients under stable antiretroviral therapy is impaired compared to WHO 2010 reference values and on sperm proteome level. *AIDS* 2014;**28**(6):875–80.

95. Hamed KA, Winters MA, Holodniy M, Katzenstein DA, Merigan TC. Detection of human immunodeficiency virus type 1 in semen: effects of disease stage and nucleoside therapy. *J Infect Dis* 1993;**167**(4):798–802.
95a. Politch JA, Mayer KH, Abbott AF, Anderson DJ. The effects of disease progression and zidovudine therapy on semen quality in human immunodeficiency virus type 1 seropositive men. *Fertil Steril* 1994;**61**:922–8.
96. Frapsauce C, Grabar S, Leruez-Ville M, Launay O, Sogni P, Gayet V, Viard JP, De Almeida M, Jouannet P, Dulioust E. Impaired sperm motility in HIV-infected men: an unexpected adverse effect of efavirenz? *Hum Reprod* 2015;**30**(8):1797–806.
97. Zhang H, Dornadula G, Beumont M, Livornese L, Uitert BV, Henning K, Pomerantz RJ. Human immunodeficiency virus type 1 in the semen of men receiving highly active antiretroviral therapy. *N Engl J Med* 1998;**339**:1803–9.
98. Ahmad G, Moinard N, Jouanolou V, Daudin M, Gandia P, Bujan L. In vitro assessment of the adverse effects of antiretroviral drugs on the human male gamete. *Toxicol Vitro* 2011;**25**(2):485–91.
99. Kehl S, Weigel M, Müller D, Gentili M, Hornemann A, Sütterlin M. HIV-infection and modern antiretroviral therapy impair sperm quality. *Arch Gynecol Obstetrics* 2011;**284**(1):229–33.
100. De Souza JO, Treitinger A, Baggio GL, et al. Alpha-tocopherol as an antiretroviral therapy supplement for HIV-1-infected patients for increased lymphocyte viability. *Clin Chem Lab Med* 2005;**43**:376–82.
101. Guwatudde D, Wang M, Ezeamama AE, Bagenda D, Kyeyune R, Wamani H, Manabe YC, Fawzi WW. The effect of standard dose multivitamin supplementation on disease progression in HIV-infected adults initiating HAART: a randomized double blind placebo-controlled trial in Uganda. *BMC Infect Dis* 2015;**15**(1):348.
102. Jaruga P, Jaruga B, Gackowski D, et al. Supplementation with antioxidant vitamins prevents oxidative modification of DNA in lymphocytes of HIV-infected patients. *Free Radic Biol Med* 2002;**32**:414–20.
103. Lopez O, Bonnefont-Rousselot D, Edeas M, Emerit J, Bricaire F. Could antioxidant supplementation reduce antiretroviral therapy-induced chronic stable hyperlactatemia? *Biomed Pharmacother* 2003;**57**:113–6.
104. Spada C, Treitinger A, Reis M, et al. An evaluation of antiretroviral therapy associated with alpha-tocopherol supplementation in HIV infected patients. *Clin Chem Lab Med* 2002;**40**:456–9.
105. Tang AM, Smit E, Semba RD, et al. Improved antioxidant status among HIV-infected injecting drug users on potent antiretroviral therapy. *J Acquir Immune Defic Syndr* 2000;**23**:321–6.
106. Wang JY, Liang B, Watson RR. Vitamin E supplementation with interferon-γ administration retards immune dysfunction during murine retrovirus infection. *J Leukoc Biol* 1995;**58**:698–703.
107. Wang Y, Huang DS, Liang B, Watson RR. Nutritional status and immune responses in mice with murine AIDS are normalized by vitamin E supplementation. *J Nutr* 1994;**124**(10):2024–32.
108. McComsey G, Southwell H, Gripshover B, Salata R, Valdez H. Effects of antioxidants on glucose metabolism and plasma lipids in HIVinfected subjects with lipoatrophy. *J Acquir Immune Defic Syndr* 2003;**33**:605–7.
109. Rousseau MC, Molines C, Moreau J, et al. Influence of highly active antiretroviral therapy on micronutrient profiles in HIV-infected patients. *Ann Nutr Metab* 2000;**44**:212–6.

CHAPTER

# 20

# Assessment of Antioxidant Potential of Dietary Components


*Shashank Kumar[1], Rapalli Krishna Chaitanya[1], Victor R. Preedy[2]*

[1]Central University of Punjab, Bathinda, India; [2]Kings College London, London, United Kingdom


OUTLINE





*HIV/AIDS*
http://dx.doi.org/10.1016/B978-0-12-809853-0.00020-1

## Abstract

Antioxidants neutralize or mitigate the harmful effects of free radicals. Such antioxidants may be classed as either enzymatic or nonenzymatic. Some components in the diet act as important antioxidants as they may have direct antioxidant activity or are a component of antioxidant systems. Free radical-mediated stress arises when body fails to ameliorate the excess generation of free radicals. In such circumstances the need for supplementary or other dietary antioxidants arises. As a consequence, it is necessary to assay antioxidants status in subjects or antioxidant potential of novel dietary components. Several techniques have been developed to measure the antioxidant potential of dietary components. We describe antioxidants in general, then various platforms using spectroscopic, chromatographic, electrochemical, and photochemical methods. The following in assays and protocols are reviewed: hydrogen atom transfer, oxygen radical absorbance capacity, diphenyl-1-picrylhydrazyl radical scavenging, trolox equivalent antioxidant capacity, ferric-reducing antioxidant power, total radical-trapping antioxidant parameter, metal-chelating capacity, hydroxyl radical antioxidant capacity, diene conjugates, thiobarbituric acid reactive substances, hexanal, 2,2′-azinobis-(3-ethylbenzothiazoline-6-sulfonic acid, phycoerythrin, bleomycin–iron, copper-1,10-phenanthroline complex, peroxynitrite, lipid-soluble antioxidants, beta-carotene bleaching, hydroxyl radical scavenging, superoxide anion radical scavenging capacity, ferrous oxidation-xylenol orange, ferric thiocyanate, nonenzymatic in vivo and enzymatic in vivo assays.

**Keywords:** Antioxidants; Assays; Biosensors; Chromatography; Dietary compounds; Electrochemical; Free radical; Spectroscopy.

## List of Abbreviations

**CV** Cyclic voltammetry
**DMPO** 5,5-dimethyl-1-pyrroline N-oxide
**EPR** Electron paramagnetic resonance
**FRAP** Ferric reducing antioxidant power
**GC** Gas chromatography
**GGT** Glutamyl transpeptidase activity
**GPx** Glutathione peroxidase
**GR** Glutathione reductase
**GSH** Reduced glutathione
**GSSG** Glutathione disulfide
**GSt** Glutathione-S-transferase
**HAT** Hydrogen atom transfer
**HORAC** Hydroxyl radical antioxidant capacity
**HPLC** High performance liquid chromatography
**LPO** Lipid peroxidation
**NBT** Nitro blue tetrazolium
**ORAC** Oxygen radical absorbance capacity
**PCL** Photochemiluminescence
**RT** Retention time
**SET** Single electron transfer
**SOD** Superoxide dismutase
**TMM** tetramethylmurexide
**TPTZ** 2,4,6-tripyridyl-s-triazine
**USDA** US Department of Agriculture
**XO** Xylenol orange

# INTRODUCTION

Excess of free radicals causes oxidative stress and tissue damage in the body. Antioxidant compounds possess the ability to delay or inhibit these free radical-mediated processes. Thus, antioxidants decrease the risk of diseases by virtue of their reduction potential, metal-ion chelation, and radical scavenging activity (Fig. 20.1). Some, but not all, dietary components are known to act as antioxidants in biological systems. These enhance the endogenous enzyme systems, which have antioxidant activities such as glutathione peroxidase (GPx), superoxide dismutase (SOD), and catalase (CAT). Nonenzymatic compounds such as uric acid, albumin, and metallothioneins are also altered. Techniques based on colorimetry, spectrophotometry, chromatography, fluorimetry, and photometry are employed to assess the antioxidant potential of dietary components (Table 20.1).

Current research is focused on exploration of dietary antioxidant properties, their abundance, and role in oxidative stress-associated diseases such as cancer, cardiovascular disease, and neurodegeneration. Disease states such as HIV/AIDs are also associated with excessive generation of free radicals, either due to the disease process, therapeutic regimes or poor diets (Fig. 20.1). Additional antioxidant supply is necessitated when the endogenous antioxidative systems fails. In this regard, nutritional supplements/pharmaceutical products that are composed of antioxidant compounds play a vital role in maintaining optimal health and are recommended by some organizations.

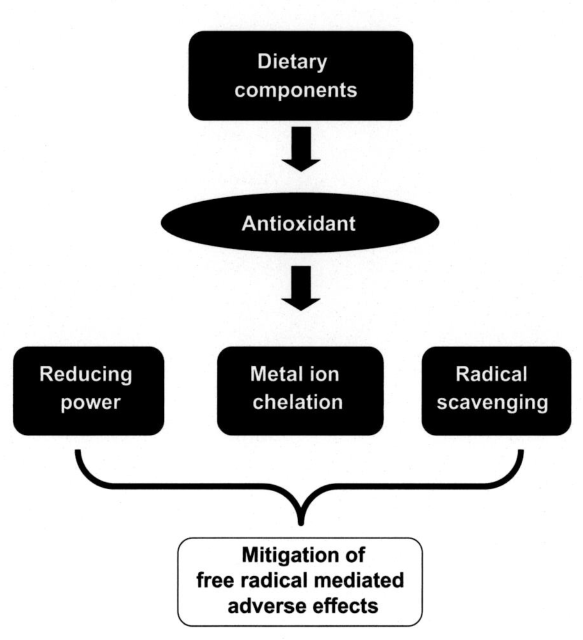


FIGURE 20.1 **Mechanism of antioxidant activity of dietary components.** Dietary components having antioxidant potential may exert their antioxidant activity by virtue of reducing power, metal ion chelation, and radical scavenging activity. By neutralizing the radical and ion-mediated damage dietary antioxidants may prevent the initiation and progression of various diseases.

TABLE 20.1 Techniques Used in the Assessment of Antioxidant Potential

| Technique | Name of Assay or Platform | Principal Involved |
|---|---|---|
| Colorimetry | **DPPH** (2,2-diphenyl-1-picrylhydrazyl))<br>**ABTS** (2,2′-Azinobis-(3-ethylbenzothiazoline-6-sulfonic acid)<br>**FRAP** (Ferric reducing antioxidant power) | Antioxidant reacts with cationic, anionic radical, or metal complex producing colour in the solution that can be measured at particular wavelength. |
| Fluorescence based | **ORAC** (Oxygen I radical absorbance capacity)<br>**HORAC** (Hydroxyl radical antioxidant capacity) | Antioxidant reacts with different radicals produced by some radical-inducing agents. The reaction decreases the fluorescence intensity of a fluorescence compound used in the study |
| Chromatography | **GC** (Gas chromatography)<br>**HPLC** (High-performance liquid chromatography) | Separation of the compounds in a mixture is based on the repartition between a liquid stationary phase and a gas mobile phase |
| Electrochemical techniques | **Cyclic voltammetry**<br>**Amperometry**<br>**Biamperometry** | Intensity of current measured involving oxidation–reduction reactions |

Very often assays and physical platforms are described in terms of their acronyms such as DPPH, ABTS, etc. The full terms are described in the tables.

Dietary antioxidants include vitamins (C and E) and plant-based natural compounds (phenolics, carotenoids (which also have provitamin A activity), proanthocyanidins, benzoic acid derivatives, flavonoids, coumarins, stilbenes, lignans, lignins, and many others).[1] These components are absorbed with differential efficacy in the body and further subjected to various modes of xenobiotic and metabolism. Also, the dietary compounds may differ in their chemical interactions between with macromolecules within the diet. Digestive enzymes and bacterial microflora release potential bioavailable dietary components from the food matrix.[2] There is no comprehensive database available so far that lists out all the antioxidants present in food due to the enormous diversity of these compounds so Table 20.2 just lists some within the various categories. Further, the total antioxidant potential of food is not dependent on any one molecule's antioxidant potential, but it is the result of redox and synergic interaction between different moieties present in the diet.[3]

TABLE 20.2 List of Dietary Antioxidants

| Category | Examples |
|---|---|
| Vitamins | Vitamin C |
| | Vitamin E |
| Pigments | Beta-carotene |
| | Lycopene |
| | Lutein |
| Phytochemicals | Phenolics |
| | Flavonoids |
| | Anthocyanins |
| Metals | Zinc |
| | Selenium |

The list of dietary antioxidants is not exhaustive but illustrative only. Caution is required in interpreting the table: for example, selenium is a component of the enzyme glutathione peroxidase, which has the ability to detoxify lipid hydroperoxides. A deficiency of dietary selenium will reduce activities of blood glutathione peroxidase thus increasing oxidative stress. Vitamin E (comprising four tocotrienols and four tocopherols) on the other hand is a chain-breaking antioxidant that also requires vitamin C in the cycle of vitamin E regeneration. Many dietary components classified as antioxidants have complex biological roles, which also impact on molecular events in the cell.

Various phenomenon (heat, light, ionizing radiation, metal ions, and metalloproteins) are known to initiate oxidation process continuously in the presence of target substrate (oxygen, phospholipids, cholesterol, polyunsaturated fatty acids, and DNA) via a free radical-mediated chain reaction consisting of multiple steps i.e., initiation (Eq. 20.1), propagation (Eq. 20.2), branching (Eq. 20.3), and termination (Eqs. 20.4–20.6) steps.[4] The simple illustration of the reaction may be depicted as follows:

## Initiation

$$LH + R\cdot \rightarrow L\cdot + RH \tag{20.1}$$

## Propagation

$$\begin{gathered} L\cdot + O_2 \rightarrow LOO\cdot \\ LOO\cdot + LH \rightarrow L\cdot + LOOH \end{gathered} \tag{20.2}$$

## Branching

$$\begin{gathered} LOOH \rightarrow LO\cdot + HO\cdot \\ 2\ LOOH \rightarrow LOO\cdot + LO\cdot + H_2O \end{gathered} \tag{20.3}$$

## Termination

$$LO\cdot + LO\cdot \rightarrow \text{Nonradical species} \tag{20.4}$$

$$LOO\cdot + LOO\cdot \rightarrow \text{Nonradical species} \tag{20.5}$$

$$LO\cdot + LOO\cdot \rightarrow \text{Nonradical species} \tag{20.6}$$

(LH = substrate molecule; R· = initiating oxidizing radical; L· = allyl radical; LO· = alkoxyl radical; LOO· = lipid peroxyl radical; LOOH = lipid hydroperoxides; HO· = hydroxyl radical). The point suffix indicates the free radical or unpaired electron.

## TECHNIQUES INVOLVED IN ANTIOXIDANT ASSESSMENT OF DIETARY COMPONENTS

Dietary components have various degree of solubility in different solvents. In terms of solubility these compounds may be polar, nonpolar, or midpolar. Prior to antioxidant assessment, a particular compound is extracted in an appropriate solvent or solvent system by using Soxhlet apparatus or other methods. The commonly used solvents are hexane, benzene, chloroform, ethyl acetate, acetone, methanol, or water. The extraction procedure concentrates the desired compound in a given sample. Extracts can further be concentrated by evaporation of the solvent. The extract is then subjected to various techniques to assess the antioxidant potential of dietary components (Fig. 20.2).

## ANALYTICAL PLATFORMS

There are various analytical platforms and each have their own advantages (such as ease of use, levels of detection, ease of automation, etc.) and disadvantages (sophisticated machinery needs specialist training, may be expensive, prone to tube-blockage, etc.). Fig. 20.3 shows there is a rank order for the usage of different analytical platforms: chromatographic techniques are most popular. This is followed by colorimetric, fluorescence, electrochemical, biosensors, and photochemiluminescence (PCL), respectively (Fig. 20.3). However, the usages of different analytical platforms are also dependent on advances in physics, chemistry, and computing. In the ensuing text we describe some selective methods.

### Electrochemical Techniques

Two dynamic electrochemical measurement techniques such as cyclic voltammetry (CV) and biamperometry are broadly applied to assess the antioxidant content in the given sample. CV experiments involve measuring the function of the electrode potential against time. Electrode potential is scanned linearly from the

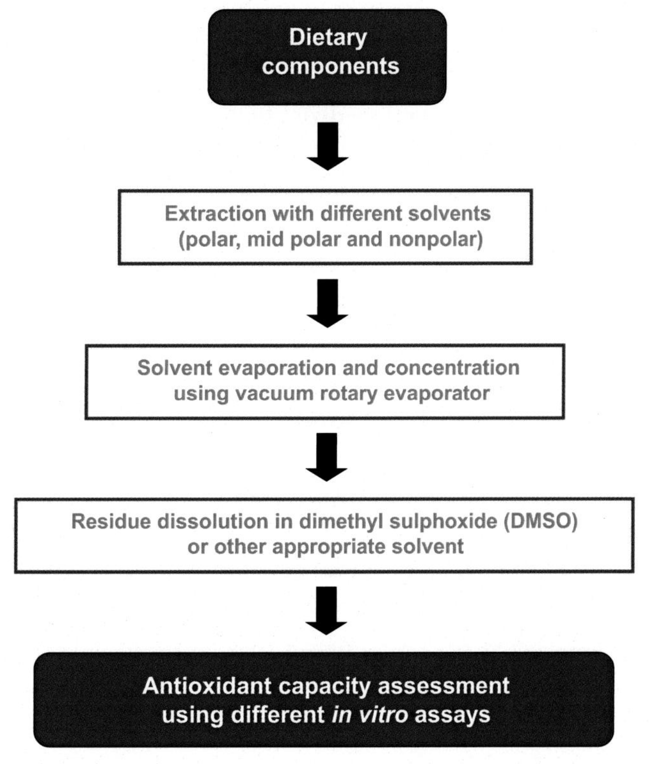


FIGURE 20.2 **Flow chart for the in vitro antioxidant activity assessment of dietary components.** Figure showing the steps of sample preparation and extraction using different polarity of solvents. The solvent is determined according to the chemical nature of the dietary component.

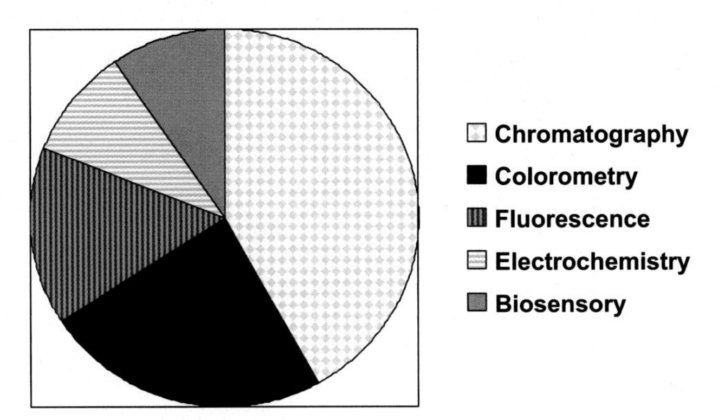


FIGURE 20.3 **Pie chart showing popularity of different techniques used for antioxidant assessments.** Relative proportion of various techniques *viz.*, chromatography (n = 2419); colorimetry (n = 1378); fluorescence (n = 871); electrochemistry (n = 547) and biosensory (n = 547)-based antioxidant assay. The number of chemiluminescence-based antioxidant assay (n = 6), used for assay of antioxidant assessment of compounds was too small to be shown in the pie chart. Data were generated by PubMed though it is expected that other data bases will produce the same relative proportions.

initial to final stage and once again to the initial value while respective current intensities are recorded. CV has been used to assess antioxidant potential of dietary components in vitro. This technique can also be applied to tissue homogenates, plasma, and plant extracts. The sensitivity of this method has been reported in terms of vitamin C.[5] The results obtained by the CV technique are comparable to that of spectrophotometric methods viz., ABTS (2,2′-azinobis-(3-ethylbenzothiazoline-6-sulfonic acid)) and DPPH (2,2-diphenyl-1-picrylhydrazyl) assays.

The biamperometric method is based on the basic principles of electrochemistry. In this technique, the current flowing between identical working electrodes with small potential difference, immersed in a reversible redox couple solution, is measured. The measurement is based on the redox couple indicated by the reaction of the analytes used. The sensitivity of the experiment is based on the specific reactions of particular redox pairs and analytes. Commonly used redox couples in biamperometric measurements are $I_2/I^-$, $Fe^{3+}/Fe^{2+}$, and $Fe(CN)_6^{3-}/Fe(CN)_6^{4-}$. Besides these, the DPPH•/DPPH redox pair is also used for the antioxidant measurement of samples. The reduced form of DPPH is produced when the antioxidant (analyte) reacts with the DPPH radical. The intensity of the current measured by the method is directly proportional to the residual concentration of the DPPH• in the test solution.[4] In this method, two identical working electrodes are used where the reduction and oxidation of DPPH• radical and DPPH occurs, respectively. This process maintains the reduced and oxidized form of the DPPH in equilibrium. Addition of antioxidant moieties into the solution decreases the concentration of DPPH radical, which generates the cathode current, which is measured.

## Biosensor-Based Methods

The electron transfer property of oxidoreductases during catalytic reactions has often been exploited in biosensor applications. These enzymes are stable and do not require coenzymes or cofactors. Several studies reported the determination of antioxidant capacity using biosensor methods.[6–8] Superoxide radical, nitric oxide, glutathione, uric acid, ascorbic acid, and phenolic compounds can be monitored by using biosensors. DNA-based biosensors evaluate total antioxidant potential of the samples electrocatalytically.[9] In this method, a partially damaged DNA layer is adsorbed onto the electrode surface by OH radicals generated by the Fenton reaction. The subsequent electrochemical oxidation of the intact adenine bases generates an oxidation product that catalyzes the oxidation of nicotinamide adenine dinucleotide (NADH). Addition of antioxidant compounds scavenges hydroxyl radicals leaving adenine bases unoxidized and thus, increase the electrocatalytic current. Enzymes viz., tyrosinases, laccases, or peroxidases are used as biosensors for the detection of antioxidant moieties. Dietary polyphenols are detected by a specialized biosensor using an immobilized polyphenol oxidase.[4] Polyphenols exert their antioxidant activity through their hydroxyl groups. Tyrosinase-based biosensors detect the total amount of OH groups present in the polyphenols. This method provides an indirect evaluation of the antioxidant potential of sample that possesses hydroxyl groups. The results may be reported as trolox or gallic acid equivalents in mg/L.[4]

## Chromatographic Methods

Multiple chromatographic techniques are being used in this field of antioxidant assessments. These methods include liquid, gas, affinity, and exchange chromatography. Antioxidant ingredients of dietary components have been analyzed by gas ion-chromatography with flame ionization detector and gas chromatography (GC) coupled with mass spectrometry. GC is used for volatile substances that do not decompose during the analytical process. Separation occurs between a stationary (liquid) and mobile (gas) phase. Microscopic layers of liquid and inert gas serve as mobile and stationary phases. The retention time (RT) of the analyte is quantitatively measured. Ionization and thermal conductivity detectors are the most commonly used GC detectors. High-performance liquid chromatography (HPLC) is another useful technique used for the estimation of antioxidant moieties in dietary components. This technique utilizes stationary phases, a pump, column, and a detector. The pump is used to move the mobile phase and analytes across the column with high pressure to provide a characteristic RT for the analyte. Usually, a diode array detector is used to obtain the additional information regarding the characterization of the analyte in the form of spectroscopic data. In one study, the 2,2′-azino-bis-3-ethylbenzothiazoline-6-sulphonic acid–based radical scavenging activity of coffee was measured by using an HPLC system with postcolumn online antioxidant detection.[11] The sample was separated, and different elutes were subjected to a photodiode array detector and mixed with ABTS cation radical solution (which has a deep blue color). The absorbance of the solution was read at 720nm by the detector. Quenching of radicals by antioxidant moieties resulted in the disappearance of the blue color, which was detected by a negative peak on the HPLC trace. If there were different antioxidant moieties in the given sample then each moiety would depict individual peak shifts. So the total HPLC-derived antioxidant potential of the given sample can be achieved by adding all the contributory peaks.[11] Fluorescence detection combined with HPLC was used to determine the antioxidant potential of propyl gallate, nordihydroguaiaretic acid, butylated hydroxyanisole, tert-butylhydroquinone, and octyl gallate in edible oils and foods.[4] Antioxidant potential of turmeric oil was also determined by various chromatographic methods.[10]

## Fluorescence Spectroscopy

When a substance absorbs light or other electromagnetic radiation it emits the energy in the form of fluorescence. The emitted light has lower energy than the absorbed one. In other words, the light that has been emitted has a longer wavelength. When an excited electron relaxes to its ground state it emits photon energy in the form of fluorescence. This principle has been exploited to determine the antioxidant content of components.[4] Two different methods have been employed, based on the antioxidant component that needs to be analyzed. One includes recording of fluorescence and excitation spectra at different wavelengths, and the other method requires a strict pH control as the fluorescence intensity depends strongly on the pH value. Fluorescence methods are also used to explain the lateral organization of sterol in the biological membrane, which in turn affects the potentiality of antioxidant components.[12] This sheds light as to why some lipid-soluble antioxidants reflect adverse effects. As the cell membrane is an essential requirement for the proper functioning of cells, any disruption to it hampers its homeostatic metabolism and physiological processes. Few antioxidant compounds are soluble in both water and lipids. Through fluorometry, it is now known that these compounds disrupt the sterol organization by insertion into membrane bilayers leading to detrimental effect on the cells.

## Photochemiluminescence Methods

PCL assays involve the detection of free radicals by chemiluminescence detection generated by photochemical cleavage. Luminol is a photo inducer and auto oxidizer. It works as both the radical detection reagent and photosensitizer. In PCL assays, the auto oxidation is inhibited by a single or groups of antioxidant compounds at the nanomolar range. The antioxidant potential of the sample can be measured by studying the lag phase at different concentrations. The results are expressed as mmol equivalents of antioxidant activity of a reference compound (i.e., trolox) by using appropriate calibration curves. Some PCL methods have been developed with a combination of two different protocols. For example, measurement of antioxidant potential of water soluble (flavonoids, ascorbic acid, aminoacids) and lipid soluble (tocopherols, tocotrienols, carotenoids) components.[13] The PCL assay has few advantages over other methods:

- Easy and rapid to perform
- Do not require high temperatures for radical generation
- High sensitivity

# SPECIFIC METHODS FOR ANTIOXIDANT ASSESSMENTS

## Hydrogen Atom Transfer Assay

The transfer of a hydrogen atom is an important step to prevent the radical chain reactions. Hydrogen atom transfer (HAT)-based assays quantitate the ability of an antioxidant component to quench free radicals by virtue of hydrogen donation leading to the formation of a stable moiety. To understand the chemistry of the HAT assay, the following equation can be considered (Eq. 20.7):

$$AH + ROO\cdot \rightarrow ROOH + AO\cdot \quad (20.7)$$

Here, AH is an antioxidant component, ROO•, AO• are free radicals, and ROOH is a stable component. The equation reveals that AH donates a hydrogen atom to ROO• and transforms into a relatively stable free radical species i.e., AO•. Thus, there is a lesser possibility that the antioxidant free radical species (AO•) may be involved in the propagation of further radical reactions with initiation substrates.[14] Both phenolic and nonphenolic antioxidant compounds can undergo this mechanism when measuring their antioxidant potential. The aromatic ring in phenolic compounds shares the delocalized electron, which makes the radical stable.[15] If the hydrogen atom is weakly held to the antioxidant compound then the probability of being detached from its parent component increases, hence will react faster with the free radical. Therefore, in HAT-based assays, the bond dissociation enthalpy of the hydrogen-donating group of the antioxidant determines the relative antioxidant potential of the particular component.[16]

## Single Electron Transfer Assays

Single electron transfer (SET) assays are used to detect the SET ability of a potential antioxidant to participate in the reduction of free radicals.[16] The assay is based on the color produced during the assay. The addition of an antioxidant compound to the experimental solution will decrease/increase the color intensity based on the type of assay. The transfer of single electrons from an antioxidant to active oxygen species results in a radical-cationic antioxidant complex. The complex is deprotonated through the interaction with aqueous medium. SET reactions involve the same reaction setup as discussed for HAT assays (described above). However, in terms of radical scavenging potential, SET reactions can be further be subjected to radical-propagation reactions. The half-life of the radical-cationic antioxidant complex is extended in SET type of assays.[14] The relative reactivity of the samples' antioxidant potential in SET assays is based on the deprotonation and ionization potential of the reactive functional group, which makes the reaction pH dependent. SET reactions occur at alkaline conditions. As pH increases then ionization potential decreases, which reflects the increased electron donating ability with deprotonation.[15]

## Oxygen Radical Absorbance Capacity Assay

Oxygen radical absorbance capacity (ORAC) is one of the most common antioxidant activities used in research and has been applied to both clinical and food-based studies. For example, besides dietary supplements, this method has been used to measure the antioxidative capacity of fruits and vegetables, wines, juices, and nutraceuticals. In addition, it is used to determine total antioxidant ability of plasma or serum samples. A report published by US Department of Agriculture provided a list of ORAC values for about 60 different foods in the American diet (http://www.orac-info-portal.de/download/ORAC_R2.pdf). In principle, when a protein is subjected to oxidation by a radical it loses its conformation. Due to the loss of conformation, the fluorescence of the protein will decrease, which is quantitatively measured.[17] β-phycoerythrin was the originally used protein, which reacts with peroxyl radicals leading to the formation of a nonfluorescent product in the ORAC assay. Later, the method was modified with the use of synthetic nonprotein fluorescein (FL) (3′,6′-dihydroxyspiro[isobenzofuran-1[3H], 9′[9H]-xanthen]-3-one) as the fluorescence probe[18]. In a microplate well, an antioxidant and a free radical producing moiety are added along with a fluorescent molecule (fluorescein) followed by heating. Thermal degradation produces the free radical and reacts with antioxidant compounds and thereby decreases the hydrogen atom donating potential per antioxidant compound. This results in the loss of fluorescence as there is a decrease in radical concentration. Fluorescence intensity curves against time function are recorded at different excitation/emission wavelengths. A standard water-soluble antioxidant compound (e.g., trolox) is generated and compared with the area under the curve obtained by the addition of antioxidant. Thus, the results are expressed as standard antioxidant compound equivalents i.e., μM standard antioxidant compound equivalents (TE) per gram of sample.

## Diphenyl-1-Picrylhydrazyl Radical Scavenging Assay

The DPPH assay is one of the oldest indirect methods used for the determination of antioxidant potential of various samples. This assay involves the HAT mechanism[19]. The assay relies on the principal that the purple-colored radical (DPPH·) chromogen in a solution of ethanol converts into a pale yellow hydrazine (DPPH-H) solution in the presence of a radical scavenging compound. This can be monitored at 515–528 nm[15].

## Trolox Equivalent Antioxidant Capacity Assay

Miller et al.[20] demonstrated the use of the trolox equivalent antioxidant capacity (TEAC) assay for assessment of infant plasma antioxidant capacity. In principle, the ABTS anion is oxidized by peroxyl radicals leading to the formation of the ABTS cation radical with an intense color change, which is measured at an absorption maxima of 415, 645, 734, and 815 nm.[15] The presence of antioxidant compounds in the test solution will interfere with the interaction of ABTS anions and peroxyl radicals thereby decreasing the ABTS cation formation. This in turn reflects the antioxidant potential of the sample. Interference in the final result due to sample turbidity and/or other absorbing materials at given wavelengths is minimum at 734 nm. This assay is widely employed to assess the antioxidant potential of many food components.[15]

## Ferric Reducing Antioxidant Power Assay

This assay is based on the oxidation and reduction of an iron complex. In this assay, the yellow-colored ferric 2,4,6- tripyridyl-s-triazine $[Fe^{3+}–(TPTZ)_2]^{3+}$ complex is reduced to the blue-colored ferrous complex $[Fe^{2+}–(TPTZ)_2]^{2+}$ in the presence of an antioxidant compound by electron donation. This is measured at 593 nm.[21] The assay is set up in acidic conditions and relates linearly with total reducing capacity of electron-donating moieties i.e., antioxidants. Originally, the method was used to assess the antioxidant capacity of human plasma, but later it was adopted to assess the antioxidant potential of dietary components with few modifications. The disadvantage of the FRAP assay is that the reaction is based on SET and therefore cannot detect antioxidants that involves HAT to reflect their antioxidant potential. A combination of the FRAP assay with any other method[15] addresses this issue.

## Total Radical-Trapping Antioxidant Parameter Assay

The total radical-trapping antioxidant parameter (TRAP) assay has been used to assess human plasma antioxidant capacity. In this assay, azo compounds such as 2,2′-azobis(2-methylpropionamidine) dihydrochloride (AAPH) and α,α′-azodiisobutyramidine dihydrochloride are added to the plasma/sample, which in turn produces peroxyl radicals. The oxidation is monitored by the oxygen consumed in the reaction. The antioxidative potential of the sample is measured by quantifying the adsorption of oxygen using oxygen electrodes. The red algal protein, R-phycoerythrin, is used as a fluorescence probe to detect the decrease in oxygen concentration in the experimental setup.[22] The assay comprises of lag time (induction period) compared with an internal standard i.e., trolox. The basic difference between ORAC and TRAP is that the former assay measures the area under the kinetic curve and the later measures lag time.[23]

## Metal-Chelating Capacity Assay

Metal ions (e.g., iron and copper) are known to generate reactive oxygen species in biological systems. Iron is an essential mineral. The homeostatic mechanism in cellular metabolism maintains the required iron pool in the body. Slight imbalance in the free iron content in the biological fluids can be disastrous to the system. Iron is an active prooxidant known to promote peroxidation of lipids, which interrupts the integrity of cellular membranes and metabolism. The Fenton reaction is an iron-mediated free radical generation process, which occurs both in vitro and in vivo. It induces decomposition of hydroperoxides and the production of hydroxyl radicals.[15] In this assay, a transition metal ion is used to generate free radicals, which would complex with an antioxidant compound via coordination bonding. The formation of the complex between metal ions and antioxidant compounds is known as chelation. Therefore, the greater the chelating ability of antioxidant compounds, the higher is its antioxidant potential. When metal ion reacts with a substrate (tetramethylmurexide or ferrozine) it produces an intense color, which is measured at 485 and 562 nm. In the presence of an antioxidant compound, there is an inhibition of the metal ion–substrate chelation interaction thereby decreasing the color intensity.

## Hydroxyl Radical Antioxidant Capacity Assay

This assay also measures metal ion chelating ability of antioxidant compounds. However, Co(II) complex is the metal ion source.[24] There is a good correlation between data generated from hydroxyl radical antioxidant capacity (HORAC), ORAC, and TRAP.[24]

## Diene Conjugates

Lipid peroxidation (LPO) causes oxidative stress in the pathophysiology of various diseases. Quantification of conjugated dienes is a useful technique to study LPO, which evaluates the antioxidant potential of an inhibitory moiety. In this method, the oxidation of polyunsaturated fatty acids is initiated by a metal (copper, iron); chemical components in the assay medium (AAPH, DPPH) and heat application lead to the formation of diene conjugates.[25] The mechanism involves the abstraction of hydrogen from the $CH_2$ group, and stabilization of the product by molecular rearrangement leading to the formation of conjugated diene.

## Thiobarbituric Acid Reactive Substances Assay

Thiobarbituric acid reactive substance (TBARS) assay is another method to detect lipid oxidation. This assay measures malondialdehyde (MDA), which is a split product of an endoperoxide of unsaturated fatty acids resulting from oxidation of lipid substrates. The MDA reacts with thiobarbituric acid (TBA) forming a pink chromogen (TBARS), which is measured at 532–535 nm. In the course of time, few modifications have been incorporated into this method. However, one modification is in debate, i.e., the addition of ethanol in the test solution, as there is evidence that ethanol itself may act as an antioxidant.[26] In this assay, the substrate becomes oxidized with the addition of a metal ion (copper, iron), a free radical generating compound (AAPH) followed by addition of TBA. The extent of oxidation can be measured spectrophotometrically. The addition of any antioxidant moiety to the test solution inhibits the oxidation process, and the reduced chromogen formation indicates the antioxidant capacity. The result is quantified with a calibration curve using MDA or in term of percentage inhibition.[27]

## Measurement of Hexanal

Hexanal (an unsaturated aldehyde) is one of the oxidative products formed by the lipid oxidation process. The decomposition of primary oxidation products gives rise to secondary products including hexanal. Antioxidant activity can be calculated as the percentage inhibition of one or more secondary oxidation products relative to controls. Both sensory and physicochemical methods have been used for the hexanal determination.[28] This method has an advantage of analyzing a single, well-defined, end product of the LPO process, which is lacking in other peroxidation detection methods.[15]

## 2,2′-Azinobis-(3-Ethylbenzothiazoline-6-Sulfonic Acid Assay

ABTS is a highly water soluble and chemically stable compound. It acts as a peroxidase substrate and produces a metastable cation when subjected to oxidation by $H_2O_2$ or ferrylmyoglobin. The ABTS and its stable cation show absorption maxima at 342 and 419 nm, respectively.[29] The presence of an antioxidant compound inhibits the formation of $ABTS^{4+}$, which is measured spectrophotometrically. The lesser the absorption of the test solution at 734 nm, the greater is the potential of the antioxidant compound. The measurement of the hydrogen donating potential of the antioxidant sample at 734 nm minimizes the interference due to sample turbidity and other absorbing materials. Electron or hydrogen donating potential of antioxidants, to scavenge the ABTS radical cation in comparison to that of trolox is denoted as TEAC. It is equal to millimolar concentration of trolox solution with the antioxidant capacity equivalent to a 1 mM solution of the test substance.[29]

## Phycoerythrin Assay

B-phycoerythrin and R-phycoerythrin are derived from natural sources (some bacteria and algae) and have been used as the target for free radical-mediated damage in in vitro assays. The fluorescence of the phycoerythrin protein is quenched by the peroxyl radical generated by thermal decomposition of AAPH. The introduction of an antioxidant compound to the test solution will react and neutralize the peroxyl radical thereby increasing the fluorescence of the protein toward normal. The method has been significantly modified since studies on this reaction began. In one modification, phycoerythrin is used to assess the antioxidant potential of the sample against hydroxyl radical-mediated oxidative damage. OH radicals are generated from an ascorbate–$Cu^{2+}$ system at copper-binding sites on macromolecules. This results in site-specific macromolecule damage. Differences in the areas under the phycoerythrin decay curves, between a sample and the blank, are expressed in trolox equivalents in the final results.[15]

## Bleomycin–Iron Dependent Assay

Bleomycin binds to DNA by its bithiazole and terminal amine residues. It also forms a complex with metals ions using the amino-alanine pyrimidine-hydroxy histidine portion of the molecule. This assay was first used to measure nontransferrin bound iron in biological samples. Later, it was used to assess the prooxidant potential of food additives and/or nutritive components.[30] In the presence of $O_2$ and a reducing agent, the DNA can be degraded by the bleomycin–iron complex via ferric bleomycin peroxide. Hydroxyl radicals (produced sometimes due to decomposition) and bleomycin–iron (III) complex by themselves are not capable of inducing DNA damage. Therefore, this assay requires a reducing agent/hydrogen peroxide and oxygen for DNA damage to occur. The damage results in the release of free bases and base propenals. At low pH, these bases rapidly decompose on heating and react with TBA to form a TBA–MDA chromogen. The addition of dietary components to the reaction decreases the chromogen formation and so reflects its antioxidant potential.[31]

## Copper-1,10-Phenanthroline Complex Mediated Assay

The copper-phenanthroline assay was designed to assess copper ions in biological samples. Later, it was applied to the assessment of the prooxidant action of food additives and/or nutrient components.[30] In this assay, $H_2O_2$ is produced by a copper-phenanthroline system, which damages the DNA. Hydroxyl radicals are involved in the damage of DNA caused by the copper-phenanthroline system. This damage is confined mainly to the DNA bases, unlike bleomycin–iron mediated DNA damage. To increase the DNA damage in this assay, reducing agents such as ascorbate and mercaptoethanol have been used. This assay is preferred where DNA solubility has been rendered by organic solvents.[31]

## Peroxynitrite Involving Reaction-Based Assay

Peroxynitrite ($ONOO^-$) is an oxidant produced by the reaction between nitric oxide (NO) and superoxide radicals ($O_2{\cdot}^-$). It is known to produce oxidative stress by virtue of LPO, methionine, and sulfhydryl group oxidation in proteins, antioxidant depletion, and DNA damage. It is involved in nitration of tyrosine residues, which is considered as a marker for peroxynitrite-dependent damage in biological systems.[32,33] Enhanced levels of 3-nitrotyrosine are associated with various human diseases. Peroxynitrite radical scavenging by antioxidants based on tyrosine nitration serves as a useful tool to assess the antioxidant potential of dietary components.[34]

## Lipid-Soluble Antioxidant Assay

This assay is similar to the FRAP method and is used to monitor the lipid-soluble antioxidants in dietary components. In this method, the organic residue is redissolved in propanol:acetone (2:1 v/v) containing 1% (v/v) Triton X-100. The rest of the procedure is as described for the FRAP assay.[35]

## Beta-Carotene Bleaching Assay

This assay is used to assess the antioxidant potential of both volatile and nonvolatile compounds.[36] Linoleic acid and Tween-40 are dissolved in chloroform by boiling followed by the addition of beta-carotene. The mixture is evaporated till dryness followed by addition of oxygenated water, which forms an emulsion. Dietary components and standard antioxidant compounds are dissolved in ethanol to prepare another emulsified solution. Both the solutions are mixed, and the absorbance is recorded at 15 min interval, at 470 nm wavelength. The result is represented as percentage inhibition.[37]

## Hydroxyl Radical Scavenging Assay

In a biological system, the hydroxyl radical is one of the most reactive free radicals. It can be generated by the Fenton reaction between ferrous iron and $H_2O_2$.[38] The reaction between dimethyl sulfoxide and $H_2O_2$ is also used to generate hydroxyl radicals.[39] Hydroxyl radical-mediated damage is assessed by different types of probes viz., deoxyribose, benzoate, and salicylate in colorimetric or fluorometric techniques. The antioxidant compounds, which possess the ability to scavenge the hydroxyl radical also, inhibit the radical-mediated damage in the assay. The results of the hydroxyl radical scavenging (HRSA) are generally expressed as a percentage of HRSA activity of the test sample. Electron paramagnetic resonance (EPR) techniques have also been exploited to assess HRSA activity with the help of spin-trapping agents, in addition to using probes. In this technique, a nitrone/nitroso compound reacts with free radicals to form a relatively stable adduct, which is measured with EPR spectroscopy, which produces a distinguishable adduct-specific spectrum. DMPO (5,5-dimethyl-1-pyrroline N-oxide) is commonly used to trap hydroxyl radicals leading to the formation of relatively stable DMPO-OH adducts, which are detected and quantified by EPR.[38] HRSA is a powerful tool for assessing the antioxidant potential of dietary components.

## Superoxide Anion Radical Scavenging Capacity Assay

In this assay, the $O_2{\cdot}^-$ is generated through enzymatic/nonenzymatic superoxide anion reaction systems. Superoxide radical and uric acid are produced by a reaction catalyzed by xanthine oxidase acting on hypoxanthine or xanthine using $O_2$ as a cofactor.[40] In this reaction, an electron is transferred from NADH to $O_2$ present in the test solution. Occasionally, NADH is oxidized by phenazine methosulfate to produce $O_2{\cdot}^-$. Nitro blue tetrazolium (NBT) is used as a probe for the quantification of superoxide radical concentrations, by virtue of NBT reduction into a purple-colored formazan.[39] The antioxidant sample is incubated with phenazine methosulfate-NADH-NBT to assay its superoxide radical scavenging potential. The absorbance of the mixture is recorded at 562 nm against a blank. Similar to the HRSA, the superoxide radical anion assay also requires fluorometric probes. DMPO is used in both the assays to detect and trap the radicals because it cannot distinguish between superoxide and hydroxyl radicals. The use of EPR spectroscopy in combination with an appropriate spin trap agent may increase the reliability of the result.

## Ferrous Oxidation-Xylenol Orange Assay

This assay is based on the oxidation of ferrous ions (e.g., hydroperoxides) into the ferric form followed by a reaction with a reagent containing xylenol orange (XO). This leads to the formation of a ferric-XO blue–purple color complex, which shows absorbance maxima at 550 nm. The method has been used to detect hydroperoxides in various samples. The presence of an antioxidant component in the given sample inhibits hydroperoxide formation by its electron donating ability to the ferric ion.[41]

## Ferric Thiocyanate Assay

This assay is similar to ferrous oxidation-xylenol orange (FOX) except that the formed ferrous ion is monitored as a thiocyanate complex at 500 nm.[42] This assay also requires linoleic acid as a hydroperoxide source. The inhibition of the ferric thiocyanate (FTC) complex formation by compounds in samples is proportional to the antioxidant potential of the sample. The results of the assay are highly reproducible, but compounds having absorption maxima at 500 nm may produce false positive results in the assay. The assay in, combination with other assays viz., TBA, DPPH, FRAP, and ABTS, has been used to assess the antioxidant potential of various natural compounds.

Besides the above-discussed assays there are other assays that can be utilized to assess the antioxidant potential of dietary antioxidants viz., aldehyde/carboxylic acid assay; ascorbate, and ascorbate oxidase assays.[35,43] Recently, researchers have reported the assessment method for the dietary intake of antioxidant mineral such as selenium and zinc.[44] The antioxidant potential of these metals could be measured using various biological parameters (viz., oxidation of biological membranes, hemoglobin acetylation, DNA damage, formation of protein carbonyl content, etc.). These markers of oxidative damage are then correlated with the mineral intake.

## METHODS FOR ANTIOXIDANT ASSESSMENTS: IN VIVO METHODS

In vivo antioxidant potential of the dietary components can be tested by administering them into laboratory animals viz., rat, mice, etc. at definite standardized dose for a specific time. The collected tissue and/or blood samples are assayed either enzymatically or nonenzymatically (Fig. 20.4) as discussed below:

### Nonenzymatic In Vivo Assays

The ferric reducing ability of plasma is a simple, rapid, and useful assay to measure antioxidant potential in vivo. The principle is similar to the in vitro FRAP assay. In this method, the blood sample is collected (from the retro-orbital venous plexus or via other routes such as cardiac puncture under anesthesia) of experimental models into a heparinized tube. Samples containing antioxidants enhance the color intensity of the FRAP reagent by virtue of the formation of ferrous ion formation, which is measured at 593nm.[45]

Reduced glutathione (GSH) is involved in cataract formation and renal amino acid reabsorption. Glutathione plays a role via oscillating between its two states i.e., reduced and oxidized forms. Its ratio is proposed to be a marker of oxidative balance. Dietary antioxidants maintain the balance between the two states. The quantification of GSH in tissue homogenates has been developed to assess antioxidant potential in vivo.[46] Ellman's reagent (5,5′-dithiobis-2-nitrobenzoic acid) dissolved in phosphate buffer is used in this assay.

LPO assays are also employed as in vivo makers of oxidative stress. MDA reacts with TBA to produce a pink color chromogen, which is measured spectrophotometrically at 532nm wavelength.[47]

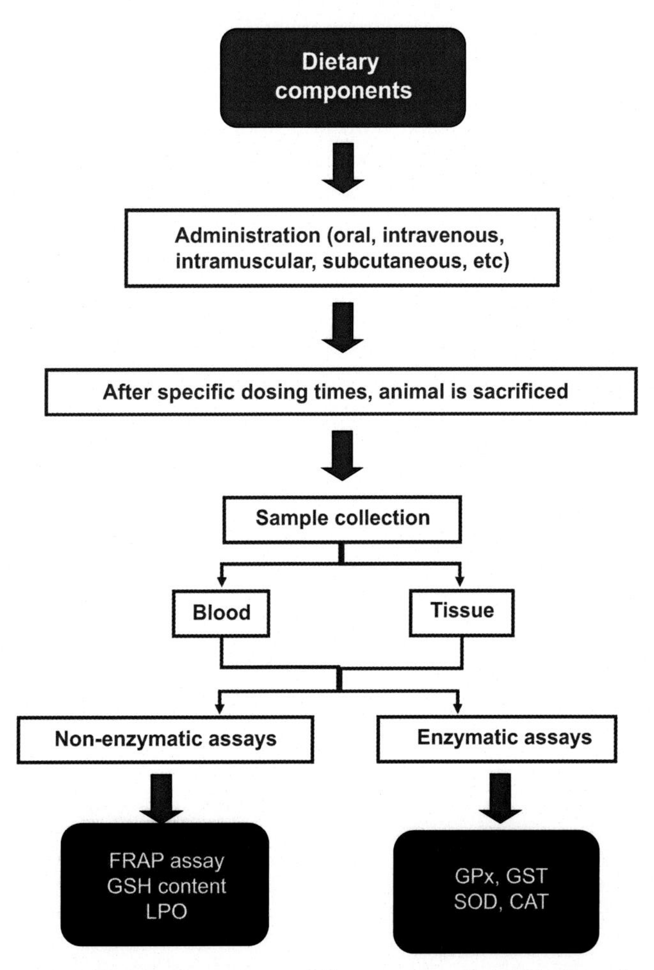


FIGURE 20.4 **Flow chart for the in vivo antioxidant assessment of dietary components.** This figure shows the steps involved in dietary component administration in experimental animals, sample collection, and different enzymatic and nonenzymatic assays for the in vivo antioxidant assessment of dietary components. *CAT*, catalase; *FRAP*, Ferric reducing antioxidant power; *GPx*, glutathione peroxidase; *GSH*, reduced glutathione; *LPO*, lipid peroxidation; *SOD*, superoxide dismutase.

## Enzymatic In Vivo Assays

GPx is a seleno-enzyme that catalyzes the reaction of hydroperoxides with GSH to form glutathione disulfide and hydroperoxide. On the basis of its occurrence i.e., cellular, extracellular, biomembrane, and gastro-intestine, there are four isoenzymes of GPx. It has been reported that the decrease in GPx activity is associated with imbalance between oxidative stress and antioxidants.[48] Reduced dietary selenium also reduces plasma GPx activity.

Glutathione-S-transferase is known to initiate detoxification of alkylating agents by catalyzing the reaction with the sulfhydryl group of glutathione, leading to the neutralization of their electrophilic sites and water-soluble byproducts.

SOD and catalase enzymes are involved in superoxide anion and hydrogen peroxide–neutralizing phenomenon. In general, red blood cell lysates are used as samples for the assay, details of which are described elsewhere.[49,50] Similarly, two other enzymes viz., glutamyl transpeptidase activity and glutathione reductase are also being used to assess the antioxidant potential of administered components in rodents.

In all these assays consideration must be given to the fact that storage conditions may affect the assay results so pilot assays need to be conducted.

# SUMMARY POINTS

- A healthy diet-containing dietary antioxidants or components that contribute to an enhanced antioxidant status are necessary to eliminate the risk of various lifestyle and genetic diseases.
- Excess of free radicals produces an oxidative environment in the body resulting in initiation and progression of various diseases.
- Antioxidants mitigate free radical-mediated damage in the body.
- Various platforms are employed to assess antioxidant potential of dietary components including electrochemical, biosensory, chromatographic, fluorescence, and PCL methods.
- The following assays and protocols can be used to asses dietary components or the effects of treatments: HAT, ORAC, DPPH radical scavenging, TEAC, FRAP, total radical-trapping antioxidant parameter (TRAP), metal-chelating capacity, HORAC, diene conjugates, TBARS, hexanal, ABTS, phycoerythrin, bleomycin–iron, copper-1,10-phenanthroline complex, peroxynitrite, lipid-soluble antioxidants, beta-carotene bleaching, HRSA, superoxide anion radical scavenging capacity, FOX, FTC, nonenzymatic in vivo, and enzymatic in vivo assays.
- Assessment of antioxidant potential in dietary components may help lessen the effects of an imbalanced diet. Alternatively such information may be used to mitigate the effects of excessive free radicals due to disease, toxic agents, or drugs themselves, which cause free radical-mediated damage as unwanted side effects.

## References

1. Lindsay DG, Astley SB. European research on the functional effects of dietary antioxidants-EUROFEDA. *Mol Aspects Med* 2002;**23**:1–38.
2. Parada J, Aguilera JM. Food microstructure affects the bioavailability of several nutrients. *J Food Sci* 2007;**72**:21–32.
3. Ramadan MF, Moersel JT. Impact of enzymatic treatment on chemical composition, physicochemical properties and radical scavenging activity of goldenberry (*Physalis peruviana* L.) juice. *J Sci Food Agric* 2007;**87**:452–60.
4. Pisoschi AM, Negulescu GP. Methods for total antioxidant activity determination: a Review. *Biochem Anal Biochem* 2011;**1**:1–10.
5. Chevion S, Roberts MA, Chevion M. The use of cyclic voltammetry for the evaluation of antioxidant capacity. *Free Radic Biol Med* 2000;**28**:860–70.
6. Giardi MT, Rea G, Berra B. *Bio farms for nutraceuticals: functional food and safety control by biosensors*. Landes Bioscience and Springer Science; 2010.
7. Barroso MF, De-los-Santos-Álvarez N, Delerue-Matos C, Oliveira MBPP. Towards a reliable technology for antioxidant capacity and oxidative damage evaluation: electrochemical (bio) sensors. *Biosens Bioelectron* 2011;**30**:1–12.
8. Cortina-Puig M, Noguer T, Marty JL, Calas-Blanchard CC. Electrochemical biosensors as a tool for the determination of phenolic compounds and antioxidant capacity in foods and beverages. In: *Biosensor in food processing, safety and quality control*. CRC Press; 2010.
9. Barroso MF, de-los-Santos-Álvareza N, Lobo-Castanón MJ, Miranda- Ordieres AJ, Delerue-Matos C, Oliveira MB, Tunon-Blanco P. DNA-based biosensor for the electrocatalytic determination of antioxidant capacity in beverages. *Biosens Bioelectron* 2011;**26**:2396–401.
10. Jayaprakasha GK, Jena BS, Negi PS, Sakariah KK. Evaluation of antioxidant activities and antimutagenicity of turmeric oil: a byproduct from curcumin production. *Z Naturforsch* 2002;**57**:828–35.
11. Stalmach A, Mullen W, Nagai C, Crozier A. On-line HPLC analysis of the antioxidant activity of phenolic compounds in brewed, paper-filtered coffee. *Braz J Plant Physiol* 2006;**18**:253–62.
12. Olsher M, Chong PLG. Sterol superlattice affects antioxidant potency and can be used to assess adverse effects of antioxidants. *Anal Biochem* 2008;**382**:1–8.

13. Besco E, Braccioli E, Vertuani S, Ziosi P, Brazzo F, Bruni R, Sacchetti G, Manfredini S. The use of photochemiluminescence for the measurement of the integral antioxidant capacity of baobab products. *Food Chem* 2007;**102**:1352–6.
14. Wright JS, Johnson ER, Di Labio GA. Predicting the activity of phenolic antioxidants: Theoretical method, analysis of substituent effects, and application to major families of antioxidants. *J Am Chem Soc* 2001;**123**:1173–83.
15. Sveinsdottir H, Hamaguchi PY, Bakken HE, Kristinsson HG. Methods for assessing the antioxidative activity of aquatic food compounds. In: Kristinsson HG, editor. *Antioxidants and functional components in Aquatic foods*. John Wiley & Sons, Ltd; 2014. p. 151–74.
16. Gulcin I. Antioxidant activity of food constituents: an overview. *Arch Toxicol* 2012;**86**:345–91.
17. Zulueta A, Esteve MJ, Frígola A. ORAC and TEAC assays comparison to measure the antioxidant capacity of food products. *Food Chem* 2009;**114**:310–6.
18. Ou B, Hampsch-Woodill M, Prior RL. Development and validation of an improved oxygen radical absorbance capacity assay using fluorescein as the fluorescent probe. *J Agric Food Chem* 2001;**49**:4619–26.
19. Braude EA, Brook AG, Linstead RP. Hydrogen transfer. Part V. Dehydrogenation reactions with diphenylpicrylhydrazyl. *J Chem Soc (Resumed)* 1954:3574–8.
20. Miller NJ, Rice-Evans C, Davies MJ, Gopinathan V, Milner A. A novel method for measuring antioxidant capacity and its application to monitoring the antioxidant status in premature neonates. *Clin Sci* 1993;**84**:407–12.
21. Benzie IFF, Strain JJ. The ferric reducing ability of plasma (FRAP) as a measure of "antioxidant power": the FRAP assay. *Anal Biochem* 1996;**239**:70–6.
22. Wayner DDM, Burton GW, Ingold KU, Locke S. Quantitative measurement of the total peroxyl radical trapping antioxidant capability of human blood plasma by controlled peroxidation: the important contribution made by plasma proteins. *FEBS Lett* 1985;**187**:33–7.
23. Magalhaes LM, Segundo MA, Reis S, Lima JL. Methodological aspects about in vitro evaluation of antioxidant properties. *Analytica Chim Acta* 2008;**613**:1–19.
24. Ciz M, Cizova H, Denev P, Kratchanova M, Slavov A, Lojek A. Different methods for control and comparison of the antioxidant properties of vegetables. *Food Control* 2010;**21**:518–23.
25. Heinonen IM, Lehtonen PJ, Hopia AI. Antioxidant activity of berry and fruit wines and liquors. *J Agric Food Chem* 1998;**46**:25–31.
26. Criqui MH. Ethanol or antioxidants? *BioFactors* 1997;**6**:421–2.
27. Du Z, Bramlage J. Modified thiobarbituric acid assay for measuring lipid oxidation in sugar-rich plant tissue extracts. *J Agric Food Chem* 1992;**40**:1566–70.
28. Antolovich M, Prenzler PD, Patsalides E, McDonald S, Robards K. Methods for testing antioxidant activity. *Analyst* 2002;**127**:183–98.
29. Rice-Evans CA, Miller NJ. Total antioxidant status in plasma and body fluids. *Methods Enzymol* 1994;**234**:279–93.
30. Aruoma OI. Nutrition and health aspects of free radicals and antioxidants. *Food Chem Toxicol* 1994;**32**:671–83.
31. Aruoma OI, Spencer JPE, Rossi P, Aeschbach R, Khan A, Mahmood N, Munoz A, Murcia A, Butler J, Halliwell B. An evaluation of the antioxidant and antiviral action of extracts of rosemary and Provençal herbs. *Food Chem Toxicol* 1996;**34**:449–56.
32. Beckman JS. Oxidative damage and tyrosine nitration from peroxynitrite. *Chem Res Toxicol* 1996;**9**:836–44.
33. Lancaster J. *The biological chemistry of nitric oxide*. New York: Academic Press; 1995.
34. Aruoma OI. Methodological considerations for characterizing potential antioxidant actions of bioactive components in plant foods. *Mutat Res* 2003;**523**:9–20.
35. Hunter KJ, Fletcher JM. The antioxidant activity and composition of fresh, frozen, jarred and canned vegetables. *Innovative Food Sci Emerging Tech* 2002;**3**:399–406.
36. Pratt DE. Natural antioxidants of soybean and other oil-seeds. In: Simic MG, Karel M, editors. *Autoxidation in food and biological systems.*, New York: Plenum Press; 1980. p. 283–92.
37. Mallet JF, Cerati C, Ucciani E, Gamisana J, Gruber M. Antioxidant activity of fresh pepper (Capsicum annuum) cultivares. *Food Chem* 1994;**49**:61–5.
38. Lloyd RV, Hanna PM, Mason RP. The origin of the hydroxyl radical oxygen in the Fenton reaction. *Free Radic Biol Med* 1997;**22**:885–8.
39. Liang N, Kitts DD. Antioxidant property of coffee components: assessment of methods that define mechanisms of action. *Molecules* 2014;**19**:19180–208.
40. Chung HY, Baek BS, Song SH, Kim MS, Huh JI, Shim KH, Kim KW, Lee KH. Xanthine dehydrogenase/xanthine oxidase and oxidative stress. *Age* 1997;**20**:127–40.
41. Girao H, Mota C, Pereira P. Cholesterol may act as an antioxidant in lens membranes. *Curr Eye Res* 1999;**18**:448–54.
42. Kikuzaki H, Nakatani N. Antioxidant effects of some ginger constituents. *J Food Sci* 1993;**58**:1407–10.
43. Moon JK, Shibamoto T. Antioxidant assays for plant and food components. *J Agric Food Chem* 2009;**57**:1655–66.
44. Serra-Majem L, Pfrimer K, Doreste-Alonso J, Ribas-Barba L, Sa´nchez-Villegas A, Ortiz-Andrellucchi A, Henrı´quez-Sa´nchez P. Dietary assessment methods for intakes of iron, calcium, selenium, zinc and iodine. *Br J Nut* 2009;**102**:S38–55.
45. Benzie IFF, Strain JJ. The ferric reducing ability of plasma (FRAP) as a measure of 'antioxidant power': the FRAP assay. *Anal Biochem* 1996;**239**:70–6.
46. Ellman GL. Tissue sulfhydryl groups. *Arch Biochem Biophys* 1959;**82**:70–7.
47. Ohkawa H, Onishi N, Yagi K. Assay for lipid peroxidation in animal tissue by thiobarbituric acid reaction. *Anal Biochem* 1979;**95**:351–8.
48. Paglia DE, Valentin WN. Studies on the quantitative and qualitative characterization of erythrocyte glutathione peroxidase. *J Lab Clin Med* 1967;**70**:158–69.
49. McCord J, Fridovich I. Superoxide dismutase, an enzymic function for erythrocuprin. *J Biol Chem* 1969;**244**:6049–55.
50. Aebi H. Catalase. *Methods Enzymol* 1984;**105**:121–6.

CHAPTER

# 21

# Recommended Resources on the Role of Oxidative Stress and Dietary Antioxidants in HIV Infection and AIDS

*Rajkumar Rajendram*[1,2], *Vinood B. Patel*[3], *Victor R. Preedy*[1]

[1]King's College London, London, United Kingdom; [2]King Abdulaziz Medical City, Ministry of National Guard Health Affairs, Riyadh, Saudi Arabia; [3]University of Westminster, London, United Kingdom

OUTLINE



## Abstract

Infection with the human immunodeficiency virus (HIV) is currently a major public health threat worldwide. This epidemic was first identified in 1981 when unusual clusters of Pneumocystis pneumonia were reported in Los Angeles. Since then the pace of the developments in this field has been phenomenal. The diagnosis and treatment of HIV and acquired immune deficiency syndrome (AIDS) have become more and more complicated as the understanding of the pathophysiology of the disease has grown. It is now nearly impossible even for experienced scientists and clinicians to remain up-to-date. For those new to the field it is difficult to know which of the myriad of available sources are reliable. To assist colleagues who are interested in understanding more about the role of oxidative stress and dietary antioxidants in the treatment of HIV infection we have therefore produced tables containing up-to-date resources. We list information on regulatory bodies, organizations and professional societies, websites, journals, books, protocols, and other miscellaneous guidelines and recommendations that are most relevant to the evidence-based study of HIV, AIDS, nutrition, and oxidative stress. The experts who assisted with the compilation of these tables of resources are acknowledged.

**Keywords:** Antioxidant; Books; Evidence; HIV; Journals; Oxidative stress; Professional societies; Regulatory bodies; Resources.

## INTRODUCTION

The AIDS is caused by a HIV. Infection with HIV is currently a major global public health threat. This epidemic was identified on June 5, 1981, when the United States Center for Disease Control and Prevention reported unusual clusters of *Pneumocystis* pneumonia in Los Angeles.[1] Over the next 2 years clusters of this and other diseases, which usually only occur in immunocompromised patients were reported throughout the United States.[2]

By August 1982, the disease was referred to as AIDS,[3] and in May 1983, a French group lead by Luc Montagnier isolated the retrovirus that caused it.[4] In 1986, the International Committee on Taxonomy of Viruses decided that this virus should be named HIV,[5] and in 2008 Luc Montagnier was awarded the Nobel Prize in Physiology or Medicine.

Antioxidants are depleted in many of the tissues of patients with HIV, and oxidative stress may contribute to different stages of the viral life cycle.[6] Oxidative stress may increase viral replication, exacerbate the inflammatory response, and reduce immune cell proliferation.[6] Free radical production in patients treated with antiretrovirals is



*HIV/AIDS*
http://dx.doi.org/10.1016/B978-0-12-809853-0.00021-3

higher than in untreated patients with HIV and in healthy uninfected patients.[6] So HIV infection alone or in combination with antiretroviral treatment may induce oxidative stress. However, antioxidants could be used to scavenge free radicals and may enhance antiretroviral therapy.[6] As with any field of biomedical study, though, caution should be taken regarding the transfer of knowledge obtained from modeling systems (in vitro or in vivo) to the clinical situation, which involves HIV/AIDS patients.

Since the start of the epidemic in 1981 the diagnosis and treatment of HIV and AIDS have become more and more complicated as the understanding of the pathophysiology of the disease has grown. It is now nearly impossible even for experienced scientists and clinicians to remain up-to-date. For those new to the field it is difficult to know which of the myriad of available sources are reliable. To assist colleagues who are interested in understanding more about the role of oxidative stress and dietary antioxidants in the treatment of HIV infection, we have therefore produced tables containing reliable, up-to-date resources in this chapter. The experts who assisted with the compilation of these tables of resources are acknowledged below.

Tables 21.1–21.5 list the most up-to-date information on the regulatory bodies, organizations and professional societies (Table 21.1), websites (Table 21.2), journals (Table 21.3), books (Table 21.4), techniques, protocols, and other miscellaneous recommendations (Table 21.5) that are most relevant to an evidence-based study of the effect of oxidative stress and dietary antioxidants on HIV infection and AIDS.

TABLE 21.1 Regulatory Bodies, Organizations and Professional Societies

| | |
|---|---|
| ***HIV AND AIDS*** | |
| AIDS Action Europe | www.aidsactioneurope.org |
| Bill & Melinda Gates Foundation | www.gatesfoundation.org |
| British HIV Association | www.bhiva.org/ |
| Centers for Disease Control and Prevention | www.cdc.gov/hiv |
| Centre for the AIDS Programme of Research in South Africa (CAPRISSA) | www.caprisa.org/Default |
| Cuban Society of Pharmacology | www.scf.sld.cu |
| Elton John AIDS Foundation | www.ejaf.org |
| ICONA | www.fondazioneicona.org |
| International AIDS society | www.iasociety.org |
| National Heart, Lung, and Blood Institute (NHLBI) AIDS Program | www.nhlbi.nih.gov/research/funding/aids/about/hiv-research-dcvs.htm |
| National Institute for Health and Care Excellence | www.nice.org.uk |
| National Institutes of Health | www.nih.gov |
| Terrence Higgins Trust | www.tht.org.uk |
| United Nations Joint Programme on HIV/AIDS (UNAIDS) | www.unaids.org |
| World Health Organization | www.who.int |
| ***NUTRITION, OXIDATIVE STRESS, AND ANTIOXIDANTS*** | |
| Academy of Nutrition and Dietetics | www.eatright.org |
| American Society for Nutrition | www.nutrition.org |
| European Society for Clinical Nutrition and Metabolism (ESPEN) | www.espen.org |
| Free Radical Program & Aging Program | www.researchgate.net/institution/Oklahoma_Medical_Research_Foundation/department/Free_Radical_Biology_and_Aging_Program |
| Global Child Nutrition Foundation | www.gcnf.org |
| HIV AIDS—Academy of Nutrition and Dietetics | www.eatright.org/resources/health/diseases-and-conditions/hiv-aids |
| International Society of Antioxidant in Nutrition and Health | www.isanh.com |
| Medicines and Healthcare Products Regulatory Agency (MHRA) | www.mhra.gov.uk |
| Society for Redox Biology and Medicine | www.sfrbm.org |

This table lists the regulatory bodies, organizations and professional societies involved with various aspects of HIV infection, AIDS, nutrition, oxidative stress, and dietary antioxidants. The location of some sites may alter during the course of time and some organizations may undergo name changes.

TABLE 21.2 Relevant Internet Resources

| | |
|---|---|
| AIDS.GOV | www.aids.gov/hiv-aids-basics/staying-healthy-with-hiv-aids/taking-care-of-yourself/nutrition-and-food-safety |
| Boots HIV and AIDS Guide | www.webmd.boots.com/hiv-aids/guide/hiv-finding-health-index |
| Mayo Clinic | www.mayoclinic.org |
| National Library of Medicine | www.nlm.nih.gov |
| Medscape | www.medscape.com |
| National Institutes of Health: Office of Dietary Supplements | ods.od.nih.gov |
| Pubmed | www.ncbi.nlm.nih.gov/pubmed/ |
| WebMD | www.webmd.com/hiv-aids/guide/nutrition-hiv-aids-enhancing-quality-life#1 |
| United States Department of Agriculture National Agricultural Library | https://www.nal.usda.gov/fnic/hiv-and-food-safety |
| Unicef | www.unicef.org/nutrition/index_HIV.html |

This table lists some internet resources relevant to various aspects of HIV infection, AIDS, nutrition, oxidative stress, and dietary antioxidants. Other resources can be found with the sites listed in Table 21.1.

TABLE 21.3 Journals Covering HIV, AIDS, and Nutrition or Oxidative Stress

| ***(A) HIV INFECTION AND AIDS*** |
|---|
| PLoS One |
| Journal of Acquired Immune Deficiency Syndromes |
| AIDS |
| AIDS and Behavior |
| AIDS Care Psychological and Socio Medical Aspects of AIDS HIV |
| AIDS Research and Human Retroviruses |
| Journal of Virology |
| Clinical Infectious Diseases |
| International Journal of STD and AIDS |
| BMC Infectious Diseases |
| BMC Public Health |
| Journal of Infectious Diseases |
| Journal of the International AIDS Society |
| HIV Medicine |
| Retrovirology |
| Sexually Transmitted Infections |
| AIDS Patient Care and STDs |
| Current Opinion in HIV and AIDS |
| Journal of the Association of Nurses in AIDS Care |
| Journal of Antimicrobial Chemotherapy |
| ***(B) HIV, AIDS, AND NUTRITION OR OXIDATIVE STRESS*** |
| PLoS One |
| Free Radical Biology and Medicine |
| Journal of Biological Chemistry |

*Continued*

TABLE 21.3 Journals Covering HIV, AIDS, and Nutrition or Oxidative Stress—cont'd

| |
|---|
| Journal of Neurovirology |
| Current HIV Research |
| Journal of Acquired Immune Deficiency Syndromes |
| Journal of Neuroimmune Pharmacology |
| Current Medicinal Chemistry |
| AIDS |
| Antioxidants and Redox Signaling |
| Journal of Neurochemistry |
| Medical Hypotheses |
| Journal of Virology |
| AIDS Research and Human Retroviruses |
| American Journal of Clinical Nutrition |
| Journal of Infectious Diseases |
| Chemico Biological Interactions |
| Current Pharmaceutical Design |
| FEBS Letters |
| Free Radical Research |
| ***(C) HIV, AIDS IN NUTRITION JOURNALS*** |
| Nutrition |
| Journal of Health Population and Nutrition |
| Maternal and Child Nutrition |
| Journal of Pediatric Gastroenterology and Nutrition |
| Journal of Nutrition |
| American Journal of Clinical Nutrition |
| British Journal of Nutrition |
| International Journal of Sport Nutrition and Exercise Metabolism |
| South African Journal of Clinical Nutrition |
| European Journal of Clinical Nutrition |
| Food and Nutrition Bulletin |
| Clinical Nutrition |
| Nutrition Journal |
| Applied Physiology Nutrition and Metabolism |
| Critical Reviews in Food Science and Nutrition |
| Nutrition in Clinical Practice |
| World Review of Nutrition and Dietetics |
| Annals of Nutrition and Metabolism |
| Clinical Nutrition ESPEN |
| Journal of The International Society of Sports Nutrition |

This table lists the top 20 journals publishing original research on (A) HIV infection and AIDS, (B) with reference to aspects of nutrition and dietary antioxidants or (C) specifically with reference to nutrition-related journals. The list was generated from SCOPUS (www.scopus.com) using general descriptors of HIV, AIDS, treatment regimens (e.g., antiretrovirals), or nutrition (antioxidants). The journals are listed in descending order of the total number of articles published in the past 5 years. Of course, different indexing terms or different databases will produce different lists so this is a general guide only.

TABLE 21.4 Relevant Books and Other Publications

| *HIV AND AIDS* |
|---|
| Kavoussi L.R., Novick A.C., Partin A.W., Peters C.A., Wein A.J., Editors. Campbell's Urology: Anatomy of the lower urinary tract and male genitalia. Saunders Elsevier, USA, 2007. |
| Volberding P. Sande's HIV/AIDS Medicine: Medical Management of AIDS 2013, Saunders, USA, 2012. |
| Levy J.A. HIV and Pathogenesis of AIDS. ASM Press, USA, 2007. |
| ***OXIDATIVE STRESS AND ANTIOXIDANTS*** |
| Andreescu S., Hepel M. Oxidative stress: diagnostics, prevention and therapy volume 1 American Chemical Society, Oxford University Press, USA, 2011. |
| Andreescu S., Hepel M. Oxidative stress: diagnostics, prevention and therapy volume 2 American Chemical Society, Oxford University Press, USA, 2015. |
| Dichi I., Breganó J.W., Simão A.N.C., Cecchini R. Role of Oxidative Stress in Chronic Diseases. CRC Press, USA, 2014. |
| Montagnier L., Olivier R., Pasquier C. Oxidative Stress in Cancer, AIDS, and Neurodegenerative Diseases. CRC Press, USA, 1997. |
| Sheppard A.J., Pennington J.A.T., Weihrauch J.L. Analysis and distribution of vitamin E in vegetable oils and foods. In Sheppard AJ, Pennington JAT, Weihrauch JL (authors) Vitamin E in Health and Disease. Marcel Dekker Inc, New York, USA, 1993. |

This table lists books on HIV infection, AIDS, oxidative stress, and antioxidants.

TABLE 21.5 Guidelines, Protocols, and Other Miscellaneous Recommendations

| | |
|---|---|
| Antiretroviral Pregnancy Registry Steering Committee | www.apregistry.com/forms/interim_report.pdf |
| British HIV Association on Standards of care for people living with HIV | http://www.bhiva.org/documents/Standards-of-care/BHIVAStandardsA4.pdf |
| Department of Health and Human Services. Panel on Antiretroviral Guidelines for Adults and Adolescents Guidelines for the use of antiretroviral agents in HIV-1-infected adults and adolescents | www.aidsinfo.nih.gov/ContentFiles/AdultandAdolescentGL.pdf |
| ESPEN Guidelines on Enteral Nutrition: Wasting in HIV and other chronic infectious diseases | http://espen.info/documents/ENHIV.pdf |
| Medical Foundation and Sexual Health on recommended standards for NHS HIV services | www.medfash.org.uk/uploads/files/p17abl6hvc4p71ovpkr81ugsh60v.pdf |
| Royal College of Nursing and Wales National Health Service. Good practice in infection prevention and control | www.wales.nhs.uk/sites3/Documents/739/RCN%20infection%20control.doc.pdf |
| World Health Organization and Guidelines on commencing antiretroviral therapy and on preexposure | apps.who.int/iris/bitstream/10665/186275/1/9789241509565_eng.pdf |

This table lists techniques, protocols, and other miscellaneous recommendations relevant to HIV infection, AIDS, oxidative stress, and antioxidants.

## SUMMARY POINTS

- Infection with human immunodeficiency virus (HIV) is currently a major global public health threat.
- The HIV epidemic began in 1981, and the virus was identified in 1983.
- HIV infection alone or in combination with antiretroviral treatment may induce oxidative stress.
- The role of oxidative stress in HIV infection and acquired immune deficiency syndrome (AIDS) is an emerging field.
- There are a variety of regulatory bodies, journals, books, and websites that are relevant to an evidence-based approach to oxidative stress in HIV infection and AIDS.

## Acknowledgments

We would like to thank the following authors for contributing to the development of this resource. Azu OO, Gil del Valle L, Gois P, Kelesidis T, Louboutin J-P, Mak IT, Motta I, Muhammad F, Ng T, Seguro AC.

## References

1. Centers for Disease Control. Pneumocystis pneumonia – Los Angeles. *Morb Mortal Wkly Rep Centers Dis Control* 1981;**30**:1–3.
2. Centers for Disease Control. Kaposi's sarcoma and pneumocycstis pneumonia among homosexual men – New York City and California. *Morb Mortal Wkly Rep Centers Dis Control* 1981;**30**:305–8.
3. Centers for Disease Control. Current trends update on acquired immune deficiency syndrome (AIDS) –United States. *Morb Mortal Wkly Rep Centers Dis Control* 1982;**31**(507–508):513–4.
4. Barré-Sinoussi F, Chermann JC, Rey F, Nugeyre MT, Chamaret S, Gruest J, Dauguet C, Axler-Blin C, Vézinet-Brun F, Rouzioux C, Rozenbaum W, Montagnier L. Isolation of a T-lymphotropic retrovirus from a patient at risk for acquired immune deficiency syndrome (AIDS). *Science* 1983;**220**:868–71.
5. Case K. Nomenclature: human immunodeficiency virus. *Ann Intern Med* 1986;**105**:133.
6. Sharma B. Oxidative stress in HIV patients receiving antiretroviral therapy. *Curr HIV Res* 2014;**12**:13–21.

# Index

"*Note:* Page numbers followed by "f" indicate figures and "t" indicate tables."

## I



## K



## L



## M

Printed in the United States
By Bookmasters